THE ENCYCLOPEDIA OF RHODODENDRON SPECIES

This book is dedicated to all those who subscribed to
the pre-publication offer and all who lent photographs.
Without both, nothing would have been achieved.

R. caucasicum in the central Caucasus (Photo M.J.B. Almond).

R. wardii var. *wardii* and *R. calostrotum* ssp. *riparium* on the The Dokar La, S.E. Tibet.

Endpapers, a forest of *Rhododendron selense* var. *jucundum* on the Cangshan, N.W. Yunnan, China.

THE ENCYCLOPEDIA OF RHODODENDRON SPECIES

Peter A. Cox and Kenneth N.E. Cox

Peter A. Cox

GLENDOICK PUBLISHING

ACKNOWLEDGEMENTS

We would like to acknowledge the following without whom this book could not have been completed:

– Hubert Andrew who introduced us to his book-publisher son Hugh who in turn introduced us to Neville Moir. Neville arranged the design and printing of this book; without his expertise and negotiating skills, it is doubful that we would have gone ahead. – Mark Blackadder for the design and layout who at times had to curb his artistic instincts so we could fit it all in! – Fred Hunt for his exacting and thorough proof-reading. – Our hard-working secretary Pauline Brown who handled the subscriptions for the book and the accounts while the book was in production. – Arthur George of Hydon Nurseries for allowing us to use the excellent line drawings from his catalogue. – Paul Valerio at Toppan for setting up the printing of the book and Sarah Adam for handling it during production. – Makiko Sakai for support during the long period of publishing meetings and deadlines.

– Dick Cavender for the immense amount of effort he made in drumming up support in the U.S.A. and Canada, for handling subscriptions there and for handling our distribution in the Americas. – The American Rhododendron Society for allowing us to use their journal to advertise in. – Sonja Nelson, editor of the A.R.S. Journal, for supporting this project all the way along and for giving our book space in the journal.

– The Scottish Rhododenron Society, the R.H.S. Rhododendron Group, The Dutch, German, Danish, Canadian, American, French, Norwegian, Swedish, Japanese, New Zealand and Australian Rhododendron societies and their members who publicised the book and encouraged subscribers. Dick Cavender, Tijs Huisman, Christian Madvig, Dr A. Kerr Grant, Torben Stein, Christian Madvig, Kerstin Andren, Börge Malmgren, Teruo Takeuchi, John Weagle and others who collected together groups of subscribers.

All photographs were taken by the authors except where acknowledged. We greatly appreciate the assistance of the following people who have kindly taken, lent or given us photographs for inclusion in this book:

Michael Almond, Florence Auckland, J.E. Barrett, John Basford, Mary Beasley, Warren Berg, Bergen Botanical Institute, Peter Bland, John Bond, Simon Bowes Lyon, Jerry Broadus and Clarice Clark, Janet Cubey, Yasuyki Doi, Art Dome, Laurie Fickes, Hamish Gunn, Lilian Hodgson, I. Holmåson, Fred Hunt, I. Ashida, D. Jolley, Guan Kaiyun, Kunming Botanical Institute, Roy Lancaster, Ron McBeath, Sandra MacDonald, G. McLellan, G. Miller, Peter Nagy, Jan Oldham, Nigel Price, A. Reid, Rhododendron Species Foundation, Royal Botanic Garden, Edinburgh, Royal Botanic Gardens, Kew, Keith Rushforth, Tony Schilling, R. Shunn, Ian Sinclair, George Smith, Graham Smith, Mike and Polly Stone, Hideo Suzuki, Teruo Takeuchi, Per Wendelbo, Keith White, Shinichi Yoshida, Yang Zhenhong.

– Dr. David Chamberlain for all his help on the taxonomic side of this book, both in discussing the classification and on botanical details in the descriptions.

– The Royal Botanic Garden, Edinburgh photographic department, especially Debbie White for taking the excellent close-up photographs. – Members of staff of the Royal Botanic Garden, Edinburgh who collected the live material from the garden – The many other members of the R.B.G.E. staff who have helped us along the way.

– Last but certainly not least the late Major George Sherriff for being the first to suggest the concept of such a book and also for his encouragement to go out and see rhododendrons in the wild for ourselves.

First published in Great Britain in 1997 by Glendoick Publishing,
Glendoick Gardens Ltd, Glencarse, Perth, PH2 7NS, Scotland.

Copyright © Peter A. Cox and Kenneth N.E. Cox.

All rights reserved. No part of this publication may be reproduced in any form or
by any means without permission of the authors.

ISBN 0 9530533 0X

Type layout and page make up by Mark Blackadder

Printed in China by Toppan Printing Co., Ltd

Contents

Acknowledgements	iv
Introduction	ix
Subgenus Azaleastrum	1
Subgenus Candidastrum	8
Subgenus Hymenanthes	9
Subgenus Mumazalea	203
Subgenus Pentanthera	204
Subgenus Rhododendron	224
Subgenus Therorhodion	362
Subgenus Tsutsusi	364
Glossary	385
Bibliography	391
Index	392

SUPPLEMENT TO SECOND EDITION

R. platypodum
Diels 1900. H4-5. (Subsection Fortunea)

Height 2-8m, shrub or small tree, usually rounded in habit. **Leaves** *thick, broadly elliptic*, 7-13 x 5-7.5cm, apex rounded, mucronate, base rounded, lower surface with minute punctulate hair bases when mature, otherwise glabrous; petiole *broad, up to 1cm, winged*, glabrous. **Inflorescence** c. 12-flowered: rachis up to 4.5cm, pedicel 2-2.5cm, glabrous. **Corolla** 7-lobed, open-campanulate, entirely glabrous, *pinkish-red*, lacking markings, 3.5-4.5cm. Calyx c. 1mm, with minute glabrous rounded lobes. Stamens 14, filaments puberulent below. Ovary and entire style stipitate-glandular. **Distribution:** S.E. Sichuan, Guangxi in thickets and on rocks and cliff edges, 1,800-2,100m (6,000-7,000ft).

This is a very distinctive species, differing from its nearest relatives in subsection Fortunea in its broadly elliptic thicker leaves, the winged petiole and the pink flowers. It has some resemblance to *R. cyanocarpum* in subsection Thomsonia. From the photographs in *Sichuan Rhododendrons of China*, this appears to be an extremely desirable species with spectacular flowers and considerable hardiness. Its habitat on Jinfo Shan in Sichuan is very limited and it may be threatened by tourism. Introduced in 1999 by Peter Cox & Peter Hutchison. Probably April-May.

R. serotinum
Hutch. 1920. H3-4? (Subsection Fortunea)

Height to 5m or more. **Leaves** broadly oblong, 8-16 x 4-7cm, base rounded, obtuse or cordulate, upper surface wax-coated, becoming glossy when waxed or heated, lower surface glabrous. **Inflorescence** 7-8-flowered, rachis 2-3cm, pedicel 3.2-4.5cm, *glandular*. **Corolla** widely funnel-campanulate, 4.8-6cm long, fragrant, white, sometimes flushed rose, (said to be) blotched red; calyx minute, stamens 14-16, style and ovary glandular.

Said to differ from *R. decorum* in the blotched flower and later flowering period (not usually a consistent character), and from *R. hemsleyanum* in its less spreading habit and its narrower, less wavy leaves. Described on p. 56 as *R. hemsleyanum* aff. from S. Yunnan, this taxon is considered to better match *R. serotinum*. *R. serotinum* was described from cultivated material as a very late flowering species, even later than *R. auriculatum* but whether this is the case with C&H 7189 remains to be seen. Introduced 1889, reintroduced 1995. July-September?

Grandia Species Nova
(*R. protistum* aff.) (Subsection Grandia)

This new introduction from N. Vietnam differs from *R. protistum* in its late growth and creamy-yellow flowers and may be given specific status.

R. gongshanense
T.L Ming 1984 H2-3? (Subsection Irrorata)

Height to 10m, young shoots glandular-setose. **Leaves** to *21 x 4.2cm*, narrowly lanceolate or oblanceolate, *rugose*, under surface with tufted hairs, margins recurved. **Inflorescence** *18-21*-flowered, rachis and pedicel sparsely glandular. **Corolla** tubular-campanulate, 3-3.5cm long, red in bud, opening light red, with nectar pouches; ovary sparsely puberulous, style glabrous. **Distribution** N.W. Yunnan, 2,100-2,500m, in broad-leaved warm temperate forest.

Closely related to *R. tanastylum*, *R. ramsdenianum* and *R. kendrickii*. It differs from its nearest relatives in its rugose, larger leaves, and greater number of flowers to the inflorescence. This appears to be a fine introduction for mild areas with handsome foliage and full trusses of attractive flowers, as shown in *Rhododendrons of China Vol. 3*, page 46. Introduced 1997 and 2000. Probably February-March.

R. luciferum
Cowan 1953 (Subsection Lanata) Described on p. 97.

Widespread in the Tsari region of S.E. Tibet, this is a very fine foliage plant. There are a few old cultivated plants which match this species, usually grown under the name *R. lanatum*. (Re-)introduced 1998. February-April.

1. *R. platypodum* on Jinfo Shan, S. E. Sichuan, China. Photo P.C. Hutchison.
2. Grandia SS., Species Nova photographed in Vietnam. Photo T. Hudson.
3. Foliage of *R. gongshanense* on east flank of Dulong-Salween divide.
4. *R. luciferum* on the Sur La, S. Tibet.
5. *R. miniatum* in the Tsari valley, S. Tibet.
6. *R. phaeochrysum* yellow on the Bimbi La, S. Tibet.

Picture 1

Picture 3

Picture 5

Picture 2

Picture 4

Picture 6

R. oligocarpum
Fang ex X.Y. Zhang 1983 (Subsection Maculifera)

Description as *R. maculiferum* (p. 100) with the following differences: **Leaves** oval to obovate. **Corolla** *pale pink to purple*. **Distribution:** Guizhou & Guangxi 1800-2500m, in broad-leaved forest.

Recently described as a species but would probably be more appropriately described as a subspecies of the closely related *R. maculiferum*. Picture on p.101 (*R. maculiferum* Guiz 148).

R. miniatum
Cowan 1937. H4? (Subsection Neriiflora or Fulgensia)

Height 1.5-2m, a compact shrub. Young shoots covered in an evanescent tomentum, eglandular. **Leaves** elliptic, 4-5 x 2-2.6cm, apex rounded, base rounded. Upper surface glabrous, lower surface with a *dense fulvous, lanate indumentum*. Petiole 0.5-0.8cm, tomentose. **Inflorescence** c. 5-flowered; rhachis 2-3mm: pedicels c. 6mm, glabrescent. **Corolla** funnel-campanulate, crimson with darker nectar pouches, 3-3.5cm. Calyx 8-15mm, fleshy cupular, glabrous, lobes crimson, rounded. Ovary glabrous. **Distribution:** apparently confined to Tsari Valley, S.E. Tibet in mixed scrub and on river banks, c. 3,700m.

Studies in the wild suggest that this taxon is best placed in subsection Neriiflora rather than Subsection Fulgensia as originally described. The Ludlow and Sherriff herbarium material is very poor. The most closely related species appear to be *R. neriiflorum* and *R. sperabile* from which it differs in its thick, dark indumentum and smaller flowers. First discovered by Ludlow and Sherriff in 1936, but never introduced, it was re-located in quantity and introduced by Kenneth Cox et al. from Tsari in 1999. It appears to an interesting and distinctive species. Probably early flowering. Introduced (CER 9927) 1999. March-April?

R. phaeochrysum
yellow form Taxon Nova? (Subsection Taliensia)

Habit and foliage as in var. *agglutinatum*. **Corolla** light yellow, usually with red spotting. **Distribution:** Tsari, S.E. Tibet and surrounding valleys, 3,200-4,000m (10,500-13,000ft) usually on open moorland. April-May?

Recorded but never introduced by Ludlow and Sherriff, recent expeditions have revealed a large uniform population of yellow *R. phaeochrysum* which closely matches var. *agglutinatum* in foliage. So far, it is proving very slow-growing. This taxon should probably be given varietal status.

7. *R. leptocladon* KR 2932 at Glendoick, flowering indoors.
8. *R. valentinianum* var. *oblongilobatum* at Rhododendron Species Botanical Garden. Photo S. Hootman.
9. *R. pseudociliipes* at Dan Zhu, Irrawaddy-Salween divide, W. Yunnan, China.

R. fragrans
(Adams) Maximowicz 1870. (*Azalea fragrans, Osmothamnus fragrans, O. pallidus, R. adamsii*) H5? (Section Pogonanthum.)

Height to 0.5m, a small shrub. *Leaf bud scales deciduous*. **Leaves** oblong-elliptic to oblong-ovate, 1-2 x 0.5-1cm, somewhat revolute, upper surface dark glossy green, lower surface pale yellowish with dense, *uniform* overlapping +/- plastered scales. **Inflorescence** 7-10-flowered, pedicels lepidote, calyx lobes 1-3mm, ciliate. **Corolla** pink, lobes usually with darker veins, tube glabrous and elepidote outside, villous-pilose within, the lobes *pilose within for some distance from the throat of the tube*. Stamens 5-(6), ovary scaly, capsule lepidote. **Distribution:** *Siberia & Mongolia*, forming thickets in the alpine zone.

Essentially this is a northerly population of the widespread and very variable *R. primuliflorum*. The deciduous bud scales, uniform scales and pilose corolla lobes are the main distinguishing features. Likely to prove tricky to cultivate successfully. Introduced by Josef Halda in 1997-8 from the Mongal Altai Mtns., Mongolia. He reports that the flowers are scented and that the leaves are used to make tea. The only member of Sect. Pogonanthum from outside the Sino-Himalayan region.

R. dekatanum
Cowan 1937 H4 (Section Rhododendron, Subsection Boothia)

Height 0.60-1.20m, forming a rounded bush. Bark light reddish-brown. Young growth scaly, not hairy. **Leaves** to 5 x 3.3cm, broadly ovate-oblong, *apex often emarginate*, lower surface scales very unequal, contiguous to their own diameter apart. **Inflorescence** 2-3-flowered. **Corolla** *broadly campanulate*, 2-2.6cm long, bright lemon yellow. Calyx *6-8mm*. **Distribution** S. Tibet, (Ludlow & Sherriff 1360) 3,500m (11,500ft), in rhododendron and bamboo forest.

This species is closely related to *R. sulfureum* but differs in its broadly ovate-oblong leaves, the greater density of scales on the leaf lower surface, the larger calyx and the larger, more open corolla. We have grown this for many years, acquired from Muncaster Castle, N.W. England. The cultivated plant matches the type specimen of *R. dekatanum*. It has proved hardy at Glendoick but flowers very early, so is easily frosted. It should be a valuable addition to the range of low yellows for mild climates. Introduced 1936. February-March.

R. leptocladon
Dop 1930 H2-3? (Subsection Maddenia)

Height to 2m?, +/- compact, young shoots with reddish-brown scales. **Leaves** to 10 x 4cm, lower surface with reddish-brown scales, variable in size, ½-1½ times their own diameter apart. **Inflorescence** c. 3-flowered. **Corolla** broadly funnel-shaped, about 8cm across, pale greenish-yellow with a deeper throat. **Distribution** Lao Cai Province, Vietnam, 2,000m on banks and cliffs.

This species had been placed in synonymy with *R. lyi* but it is obviously distinct. Yellow flowers clearly separate it from its relatives. A very promising new introduction with fine flowers, freely produced from a young age. Easily cultivated. Introduced 1992-4.

Picture 7 *Picture 8* *Picture 9*

R. valentinianum var. *oblongilobatum*
R.C. Fang 1982 H3? (Subsection Madennia)

See page 301 under *R. valentinianum* aff. **Corolla** deep yellow, funnel-campanulate. This is amongst the deepest yellow of all non-Vireya species. April-May.

R. pseudociliipes
Cullen 1978 H2? (Subsection Maddenia)

Height to 2m, epiphytic. **Leaves** *4-8 x 1.6-2.7cm*, narrowly elliptic to narrowly obovate, lower surface brownish but with rather lax scales. **Inflorescence** *1(-2)* -flowered. **Corolla** openly funnel-campanulate, 5-7cm long, white or faintly flushed pink. **Distribution** N.W. and W. Yunnan, N.E. Burma, 2,400-3,050m (8,000-10,000ft), rocks, cliffs, thickets and on cliffs.

This recently described species is easily identified by its small leaves and usually 1-flowered inflorescences. It was not apparently introduced to cultivation before 2000.

R. petrocharis
Guiz 120 Diels. 1921 H4 (Subsection Moupinensia)

Height 60cm? compact in the open, *branches bristly for several years*. **Leaves** evergreen, to 3.5 x 2.5cm, orbicular, *margins bristly*, upper surface scales partly deciduous, lower surface glaucous, equal scales about their own diameter apart, reddish brown. **Inflorescence** 1-2-flowered. **Corolla** *more funnel-shaped* than in *R. moupinense*, pale pink to white. **Distribution** N.E. Guizhou.

Described under *R. dendrocharis* aff. on p. 310. March-April.

Subsection Monantha

Epiphytic or free-growing shrubs. Scales large, flat, often unequal, broadly rimmed. Inflorescence terminal, 1-3 flowered. Corolla lepidote, *yellow or purple*. Stamens 10. Stamens and style excerted beyound the corolla lobes. Style impressed. Seeds *winged and finned*.

This subsection is related to subsections Boothia and Uniflora. The distinctive corolla shape and exserted stamens and style and the winged and finned seeds (characteristic of subsection Vireya) are identifying features. This little known group of species may form a link between section Vireya and section Rhododendron.

R. monanthum Balfour. f & W.W. Smith 1916 H3?

Height 0.3-1m, usually epiphytic. Young growth densely scaly. **Leaves** ovate-elliptic, 3-4.5cm long, upper surface dark green, scaly, lower surface finely papillose, brownish or silvery, densely covered with unequal, broadly rimmed flat scales. **Inflorescence** 1(-2 or more) - flowered, pedicels 2-5mm, densely lepidote. **Corolla** *tubular-funnel-shaped to tubular-campanulate*, lobes scarcely spreading, 1.4-2cm long, greenish to bright yellow. Calyx minute, densely scaly, style impressed, glabrous, elepidote, *exceeding the stamens*. **Distribution:** N.E. Burma, Yunnan and S.E. Tibet, 2,450-3,650m (8,000-12,000ft), usually epiphytic.

The most important feature of this species is autumn flowering, September-October in the wild. Two other species are almost certainly better considered forms of *R. monanthum* (*R. flavantherum* and *R. kasoense*). The fourth member of subsection Monantha is the purple-flowered *R. concinnoides*. *R. monanthum* was first introduced in 1997 by Chamberlain, Cox, Hootman, and Hutchison 8208, 8133. October to December in cultivation.

R. sichotense
Pojarka 1913. H5 (Subsection Rhodorastra)

Differs from *R. dauricum* in the leaves, densely glandular on both sides, which are partly retained in winter, at least until flowering time, and the larger corolla (2.1-2.7cm).

The clone of *R. dauricum* known as 'Midwinter', widely cultivated for its very early flowering, may be better considered a form of *R. sichotense*. Introduced several times during the 1990's. January-March?

Species Nova Peter Wharton 67 H3? (Subsection Triflora)

Height 4.5-6m, habit loose. Branchlets with a few bristles and scales at first. **Leaves** evergreen, to 10 x 2.7cm, lanceolate; margins wavy and recurved, *margins and upper surface bristly*, lower surface with rimmed scales with clear centres, 1-3 x their own diameter apart. **Inflorescence** terminal and axillary, 1-3-flowered. **Corolla** widely funnel-shaped, c. 2.3cm long, pink with paler pink lobes. **Distribution** N.W. Guizhou, 1,310m, in dense scrub.

This new introduction is probably most closely allied to *R. yunnanense*, differing in its more bristly leaves, the more densely scaly leaf underside and the typically two-tone flower colour. This has not been fully tested for hardiness but it has survived outdoors at Glendoick for several mild years. Introduced 1994. March-April.

Species Nova Peter Wharton 86 H4? (Subsection Triflora)

Height 2.1m or more, width 1.2m, upright, somewhat sprawling, vigorous shrub. Branchlets almost glabrous. **Leaves** evergreen, to 12 x 4cm, lanceolate, *bronzy-red when young*, upper surface with scales at first 2-4 x their own diameter apart, *later almost glabrous*. Lower surface scales yellow, turning golden, 3-6 x their own diameter apart, later partly deciduous. **Inflorescence** terminal, occasionally terminal and axillary, 5-9-flowered. **Corolla** widely funnel-shaped, c. 2cm long, white to pale pink. **Distribution** C. Guizhou, 1,380m (4,700ft), dense scrub on hill crests.

This is a very distinct member of the Triflora subsection with the shiny leaves almost devoid of scales at maturity. It appears to be hardier than P.W. 67. This makes a handsome plant but the flowers are rather small compared with the leaf size. Introduced 1994. April-May.

---

10. *R. monanthum* growing epiphytically on east flank of Dulong-Salween divide.
11. Species Nova Triflora PW 67 at Glendoick.
12. Species Nova Triflora PW 86 at Glendoick.

Picture 10   *Picture 11*   *Picture 12*

Introduction

WHY THIS BOOK WAS WRITTEN

During the early 1960s, Major George Sherriff, one half of the well-known Ludlow and Sherriff plant hunting team, suggested that Euan Cox and his son Peter Cox should write a monograph on rhododendron species. While considering this an excellent idea, it was simply not possible to put such a volume together at that time. Colour photographic reproduction was still in its infancy and the alternative of a painted botanical illustration of every species in cultivation would have been prohibitively expensive.

In recent years, the authors have become more and more aware of the need for a fully comprehensive, well-illustrated volume on rhododendron species in cultivation which could act as a tool for identification of each species. Keeping the original concept of George Sherriff's in mind, we started to bring the idea to fruition. We had already spent many hours studying the live collection in the Royal Botanic Garden, Edinburgh, and its satellite gardens as well as the dried specimens in the herbarium. This work was done in preparation for our previous publications (see Bibliography) and also for the many plant hunting trips we have both been on to the areas of the world where most of the wild (non-Vireya) rhododendron species are to be found. We have co-operated with Dr David Chamberlain in endeavouring to work out acceptable compromises where there are obvious shortcomings in the existing taxonomy. The classification we use here should be very close to that of the new edition of *The Rhododendron Handbook, Rhododendron Species in Cultivation* (Royal Horticultural Society 1997).

THE TAXONOMY OF RHODODENDRON

In 1753, Linnaeus, the deviser of the binomial system of plant taxonomy, created two genera, *Rhododendron* and *Azalea*, into which the 9 hitherto recognised species were placed. Over the next 240 years, many botanists have furthered the taxonomy of rhododendron. G. Don, C. J. Maximovicz, J.D. Hooker and T. Nakai were some of the most important names during the 19th and early 20th century. From 1916 onwards, the classification of the genus has been largely carried out by botanists working in the herbarium of the Royal Botanic Garden, Edinburgh. From 1916 to 1922 Professor Sir Isaac Bayley Balfour developed the system of 'series' for grouping related or seemingly related rhododendron species. This system was devised to cope with the wealth of material then flooding in from the expeditions of Forrest, Kingdon Ward, Farrer and others. The 'series' classification was designed as a temporary measure which he intended to revise when time was available but unfortunately Balfour died in 1922 before this could be accomplished. Work in Edinburgh, Kew and the

Arnold Arboretum continued, through Tagg, Hutchinson & Rehder, leading to the publication of *The Species of Rhododendron*, edited by John B. Stevenson, first published in 1930 with a second edition in 1947. For many years this book was considered to be the 'bible' for serious students of rhododendron species. The comprehensive revision foreseen by Balfour was shelved for the next 40 years, his successors only making minor alterations to the 'series' system. Dr John M. Cowan and H.H. Davidian published revisions in the *Rhododendron Yearbook* of the Royal Horticultural Society and several new species were described in this and in the *Journal of the American Rhododendron Society*. Davidian, after his official retirement, published *The Rhododendron Species Volumes 1-4*, 1982-1996. These works retain the 'series' and accept the basic tenets of the Balfourian system. Davidian ignores virtually all taxonomic advances both here and in China, Japan and the U.S.A. and has carried out no fieldwork since the reopening of China in 1980. The completed publication of his life's work almost certainly brings the curtain down on an out-of-date and somewhat discredited system of taxonomy which now is only accepted by a handful of gardeners.

While the Balfourian system was widely accepted by gardeners, particularly in Britain, elsewhere botanists such as Herman Sleumer were convinced that its scientific basis was fundamentally flawed. William and Melva Philipson, in *The Rhododendron Story* (1996), go as far as to state that '*The Species of Rhododendron* is not a classification, but a careful avoidance of classification' (p. 32). Cullen in *Notes from the Royal Botanic Garden, Edinburgh, Vol. 39 No 1.* summarises the inadequacies of the Balfourian system in several areas:

1. The characters used to classify rhododendron species tend to be visible – i.e. horticulturally important (such as flower colour) – while often ignoring more fundamental morphological characteristics. Likewise geographical distribution is often ignored.

2. Although herbarium specimens were used as the basis of material for species naming, many species were named from garden material of unknown origin and from atypical plants or 'rogues' (often natural hybrids) which appeared in seed pans. Some examples of such 'species' are *R. peregrinum, R. chlorops, R. cuffeanum, R. microleucum, R. paradoxum, R. inopinum, R. planetum* and *R. lochmium*. In the introduction to the Edinburgh revision, Cullen (op. cit.) states: 'Species described and known only from cultivated material have not been accepted unless they are very distinct (even then, the possibility that they are of accidental hybrid origin must be borne in mind).'

3. The classification is not sufficiently hierarchical to show the true relationship between groups of species.

Sleumer and the Edinburgh revision.	Balfour (and Davidian)
genus	genus
subgenus	
section	
subsection	series
(alliance – not 'official')	subseries
species	species

As will be noticed, the Balfourian system had no equivalent to subgenus and section, showing all 'series' at equal rank with one another. This is extremely misleading as it suggests the same degree of relationship between two taxonomically isolated series/sections such as section Pogonanthum and subsection Maddenia (formerly the Anthopogon and Maddenii series) as between two closely related ones such as subsections Grandia and Falconera (Grande and Falconeri series). A more hierarchical structure had actually been proposed as long ago as 1834 by George Don and in 1870 by C. Maximovicz but these structures were ignored by the dominant 'horticultural' school of botanists in Edinburgh. A structural revision of the genus was proposed by Cowan (*Rhododendron Yearbook 1949* p. 55) but neither he nor Davidian followed it up. The Philipsons (1996) point out that Davidian's insistence in using the lepidote-elepidote distinction as the basis for his taxonomy is scientifically seriously flawed. The Edinburgh revision (published from 1975 onwards) used H. Sleumer's 1949 hierarchical structure as a basis for the taxonomy of rhododendrons and it is easy to see how this is a much more satisfactory approach. Current taxonomy is concerned with the evolution of a genus and therefore the relationship between different subgenera, sections and subsections is extremely important.

4. In the Balfourian classification, the 'species' concept is too narrow. Rhododendron species can be extremely variable and are often very widely distributed. Far too many species were named during the first half of the century; every time a slight variation was found, a new species name was given. The extent of this can be seen in the synonyms now included under such variable species as *R. phaeochrysum, R. oreotrephes* and *R. nivale* ssp. *boreale*.

Before going any further we should define what is meant by a 'species'. As it applies to rhododendrons, an ordinary dictionary is of rather limited use. *The Concise Oxford Dictionary* defines a species as 'a category in the system of classification of living organisms consisting of similar individuals capable of exchanging genes or interbreeding'. The problem here is that this definition could equally well be applied to whole sections or subgenera of the genus rhododendron whose species often resemble one another and can and do interbreed. Dr Cullen in *Notes from the Royal Botanic Garden, Edinburgh, Vol. No. 1, Revision of subgenus Rhododendron*, defines the concepts of species, subspecies and variety as applied to rhododendrons in the Edinburgh revision. This is derived from the work of Du Reitz (1930) and has been applied to much modern botanical revision:

'Species should differ from each other in at least two independent but correlatedly varying characters, and have geographical or ecological distributions different from those of their closest allies; if two or more taxa appear to intergrade, then the resulting treatment depends on the proportion of intermediate specimens. If these are few in number, two (or more) species are recognised, which are considered to hybridise to a small extent. If the proportion of intermediate specimens is larger (up to c.25% of the total) but the units are geographically discriminable with morphological intermediates in a geographically intermediate area, then one species is recognised with two (or more) subspecies within it. Alternatively, if the various units are geographically indiscriminate, then one species is recognised, either undivided, or, if morphological variation is appropriate, divided into two (or more) varieties.'

Cullen's definition of a species gives a very clear picture of the assumptions being made in the Edinburgh revision. Possible misgivings about this approach are that it assumes that the available selection of herbarium specimens are representative of the general population of a taxon. This may not be the case as some species are not sufficiently well represented in herbaria to carry out such statistical tasks, and, in addition, some collectors tend to seek out exceptional and extreme forms which can make the available specimens less than typical.

'Lumpers' v. 'Splitters'

The world of taxonomy can be roughly divided into two broad schools, 'lumpers' and 'splitters'. This distinction is fundamental to the underlying philosophy of taxonomy. Perhaps the best way to understand this is to use the example of *Homo sapiens*. Until the present century, there was a strong scientific impetus to divide ('split') the human race into several species or subspecies: Caucasian, Negro, Asian etc. This was based on obvious visual characteristics and ignored the fact that the there was a gradation between one racial type and another, that all humans could breed with one another and that, however hard they tried, no one could find any other fundamental differences apart from skin colour and facial appearance to divide one racial type from another. The 'lumpers' simply advocated that there was only one variable species *Homo sapiens*, a position which is now of course accepted and taken for granted. A similar debate has taken place in plant taxonomy, none more so than in rhododendrons.

Few would doubt that far too many species of Rhododendron were named by taxonomists earlier this century and the process of 'lumping' began almost as soon as species began to be named. Often, widely-distributed species collected from different areas would end up with several different names, and only later was it discovered that they were the same species. Two examples are *R. edgeworthii* and *R. bullatum* (now both classified under *R. edgeworthii*), and *R. wardii*, *R. croceum*, *R. astrocalyx* and *R. litiense* (now all classified under *R. wardii*). One reason why so many species were named appears to be that collectors such as Forrest were often paid a bonus by their rich patrons for each 'new' species found and it was therefore in the collector's interest to name or have named as many new species as possible. Perhaps the botanists themselves shared in this bonus scheme. It is inconceivable that such sharp-eyed plantsmen as Forrest and Kingdon-Ward did not have grave doubts as to the validity of many of the so-called species whose Latin descriptions poured off the presses in Edinburgh in the earlier part of the century.

Dr Herman Sleumer/ the Edinburgh Revision

As previously indicated, Balfour had always intended his system of 'series' should be completely revised at some point in the future. A more comprehensive and more taxonomically correct classification was developed by Dr Herman Sleumer (published 1949, 1966). This introduced a more realistic hierarchy in the classification of the genus, incorporating the work of Don and Maximovicz, by the inclusion of the ranks of subgenus and section and by introducing subsections in place of series. It was decided in Edinburgh that a group of botanists should finally undertake the mammoth task of a comprehensive revision of the genus Rhododendron beginning in 1972. These monographs followed the framework proposed by Sleumer, with some modifications, and have made use of new morphological, cytological, anatomical and chemical research.

The Sleumer/Edinburgh Revision:

H. Sleumer: section Vireya in *Flora Malesiana* 1968.
W.R. Philipson and M.N. Philipson: subsection Lapponica, 1975.
J. Cullen: subgenus Rhododendron excluding section Vireya, 1980.
D.F. Chamberlain: subgenus Hymenanthes, 1982.
W.R. Philipson and M.N. Philipson: subgenera Azaleastrum, Candidastrum, Mumeazalea and Therorhodion, 1986.
D. F. Chamberlain and S.J. Rae: subgenus Tsutsusi, 1990.
K.A. Kron: subgenus Pentanthera, (part) 1993 .
W.S. Judd and K.A. Kron subgenus Pentanthera (part) 1995.

Cullen describes the classification of rhododendrons as 'a

history of tension between horticulturally-based and herbarium-based classifications' (Notes R.B.G. Edinburgh 39 (1) p.2). The Edinburgh Revision caused a great deal of controversy as it appeared to take a more sweeping 'lumping' approach to the classification of rhododendrons than had previously been advocated. Gardeners around the world strongly objected to losing species names and to having to change names of well-loved plants. Although such name-changes are always annoying, the botanist often has very little option due to the strict codes of nomenclature. There are several reasons why a taxon might have its name changed: sometimes a previously accepted name is found to be illegal (*R. smithii* for instance), in other cases it is discovered that a species was in fact previously described under a less familiar name, here the older name takes precedence (*R. principis* over *R. vellereum* for example). Most commonly, a plant will have its botanical status reduced from specific to subspecific, or varietal rank or be 'sunk' to synonymy because the botanist believes that the taxon is not deserving of the taxonomic status previously accorded it. Making these decisions has become easier through important new research in both herbarium and laboratory. Most importantly, botanists have been able to study whole populations of species in the wild in the Himalaya and China where things can be observed which are not apparent in herbarium specimens alone.

In order to diffuse some of the anger of those who wished to keep a 'horticulturally-based' classification and who, above all, objected to the changing and losing of names, the Royal Horticultural Society in *The Rhododendron Species Handbook 1980* proposed 'a new method of accommodating "sunk" taxa – the Group system – . . . so that the names (epithets) of horticulturally important entities may be retained.' (p. v.) This allows the retention of taxa such as *R. repens*, *R. concatenans* and *R. radicans* (formerly considered species or subspecies) at 'Group' level, even if botanical standing is not justified. The use of groups in the *The Rhododendron Species Handbook 1980* was inconsistent and ignored whole areas of the genus. As the Group system was recognised in The International Code for the *Nomenclature of Cultivated Plants*, Article 26 (1980), and the *Cultivated Plant Code* (1995), we have increased the number of recognised groups in a more systematic way in this book. There is a complication here, as it appears that there is no existing mechanism for describing new taxa at group level. At present only taxa correctly described (in Latin) at varietal status or above and then sunk, are eligible for group status. There are several cases where we think it would be useful to describe new groups and would like a formal mechanism to do this.

The future of rhododendron taxonomy

We believe that in the future, there will have to be a further considerable reduction of 'species' within the genus Rhododendron. This will come about partly as a result of more 'lumping' and partly as a recognition of the many natural hybrids currently masquerading as species. As gardeners, we can and should continue to use all names for horticulturally distinct entities be they at specific, subspecific, varietal or group status or simply by applying clonal names. A garden-worthy plant is a garden-worthy plant and its taxonomy should not affect this at all.

The days of the taxonomy of plants being done almost entirely by the study of dried herbarium material is over. Nowadays botanists use many more strategies and tools in their work. Fieldwork in China and elsewhere has been extremely important in allowing us to understand rhododendron species and their evolution. The cladistic and phenetic analysis of herbarium specimens and wild populations as practised by K.A. Kron in the revision of section Pentanthera demonstrates other more recently developed techniques in action. In the longer term, D.N.A. analysis will probably be the most decisive tool. Work under way at the University of Seattle, Washington, U.S.A., and at the Royal Botanic Garden, Edinburgh, has used D.N.A. to plot evolution pathways to show the interrelationship between species or subsections. The Seattle work has concentrated on the the family Ericaceae showing the relationship between different genera and how they relate to different sections of the genus Rhododendron. At Edinburgh the research has looked at the relationship between subsections. Such research is in its infancy but the potential is enormous. At the end of the day, botanists may well throw up their hands and announce that the 'species concept' within the genus Rhododendron is scientifically untenable or redundant and that it is a concept which will be maintained largely for the benefit of gardeners.

It is unlikely that it will ever be possible to produce a definitive and 'final' classification of rhododendrons. We are constantly learning more about them, both from observations of wild populations and from scientific work in the herbarium and laboratory. At the end of the day, some decisions are educated guesswork and some are unsatisfactory compromise. In the end, we have to accept that we cannot always classify plants into easily-defined pigeon holes; nature just does not behave in this way. As Dr Frederick Coe said in the American Rhododendron Society's Quarterly Bulletin, April 1960: 'The species idea is one of convenience and means little to the plants'.

We are convinced that the Edinburgh Revision, though not by any means perfect, is by far the most comprehensive and satisfactory attempt to produce a classification of the genus *Rhododendron*. Although there are always going to be missgivings about decisions concerning particular species, the broad approach towards a scientifically sound and consistent taxonomy used here can only be applauded and supported. There are still people who refuse to accept any of the changes which have been made, preferring to keep faith with the Balfourian classification. There are several reasons for this: people don't like change and don't want to have to relearn their plant names. People don't like losing names (they don't have to under the group system). Most would have their eyes opened in a very short time on a Chinese hillside. It has always seemed a great pity that China was closed during the days of H.H. Davidian's working years at the Royal Botanic Garden, Edinburgh. Had he had the access that his successors have had to wild material in Asia, he could not possibly remain convinced of the validity of the structure and content of the Balfourian system and his own survey of the genus Rhododendron would have been very different.

Recent taxonomic developments in Asia

From the history previously outlined it will be clear that virtually all taxonomic decisions made on the subject of rhododendrons were made in the west by westerners with little or no consultation with the people of the countries where the plants grow wild. The Chinese, of course, had their own names for many of their plant species, often related to their use in medicine. Such names were very rarely taken into account by western collectors and botanists, who saw it as their right to use western names to apply to the plants they collected and described. The 'cultural/scientific imperialism' implicit in this activity is now recognised and thankfully the tendency is for most western botanists to co-operate with their Asian counterparts and to use local names (e.g. *R. kesangiae* recently named after the Queen Mother of Bhutan). Of course, China, India and other Asian countries have their own (often much older) scientific institutions where their own botanists are describing the rich flora of their own countries. Many new plants are being discovered and named. Within the genus *Rhododendron*, many of the newly described taxa are from southern China and adjacent countries where the European collectors did not venture. Several interesting new taxa have been described in section Tsutsusi, subsection Pseudovireya of section Vireya and in subgenus Azaleastrum.

Ironically some of the current botanical work in China appears to parallel that in Edinburgh earlier this century. Undoubtedly some of the new taxa being described are genuinely new but the torrent of new descriptions pouring out in Chinese botanical publications at the present time suggests that their botanists are being paid at a piece-rate which requires them to keep producing Latin descriptions to earn a living. These 'super-splitters' make it very difficult to evaluate the validity of new taxa. We have included some of these in this book but many will have to be further evaluated.

Natural hybrids

One of the most startling revelations since the reopening of China, has been the quantity of natural hybrids which occur within the genus *Rhododendron*. It is well known that in cultivated collections of species seed collected from individual bushes is usually hybridised if there are other compatible species growing close by. One or two taxonomically isolated species such as *R. schlippenbachii* and *R. camtschaticum* do not appear to cross with other species, but this is the exception. Almost all rhododendron species readily hybridise with other species growing close by whether in the garden or in the wild. In his book *Rhododendrons of the World* 1955, David Leach wrote:'I am convinced that there are literally scores of natural hybrids presently enjoying distinction as species. Yet the plant explorers stoutly maintain that they have never seen a wild hybrid in Asia.' This was a very perceptive comment and something that few collectors or botanists at the time seemed prepared to acknowledge. Indeed, despite so much evidence gained in the last fifteen years in China and the Himalaya, some botanists still seem happier to ignore the facts: H.H. Davidian appears very reluctant to accept the existence of natural hybrids; his four-volume survey of rhododendron species makes barely a mention of them. We firmly believe that, without taking natural hybrids into consideration, one cannot possibly expect to come to a good understanding of the complexity of the taxonomy of the genus Rhododendron. The *Encyclopedia of Rhododendron Species* is the first rhododendron publication to give this phenomenon the prominence that it deserves.

Natural hybrids can be considered in two ways, those that have been found in the wild as individual plants or whole populations and those which, although not located in the wild, appear from wild-collected seed grown in cultivation.

Natural hybrids in the wild

Wherever we have studied rhododendrons in the wild, we have found quantities of natural hybrids. The azaleas on Gregory Bald mountain in E. U.S.A., the subsection Pontica species in E. Turkey, in Bhutan among the large-leaved species, in Nepal between *R. thomsonii* and other species and in Sichuan between *R. pachytrichum* and other species are only some examples. Undoubtedly, disturbance by man greatly increases the incidence of natural hybridity and this was nowhere better illustrated than at Napa Hai, Zhongdian, N.W. Yunnan, where a hillside of rhododendrons had had all its forest cover cut down. Here we found *R. wardii*, *R. vernicosum*, *R. selense*, *R. phaeochrysum* and *R. roxieanum* and a mass of hybrids of almost every combination between the five species. Disturbance by man is not an essential prerequisite however. On the Salween-Mekong divide, where there are no roads and very little disturbance, *R. sanguineum* and its relatives hybridise with many other species, including *R. aganniphum*, *R. forrestii*, *R. eclecteum*, *R. selense* and several others. Sometimes natural hybrids are attacked by diseases such as rust, preventing them from establishing, but sometimes the vigour of the natural hybrids allows them to establish well, sometimes to the detriment of the parent species, forming large populations. This can take place gradually over several generations and may lead to the process of speciation, giving rise to such taxa as *R. semnoides* which is derived from hybrids of *R. arizelum* x *R. praestans*.

Natural hybrids in cultivation

Gradually, over the last 90 or so years, it has become apparent that many of the so-called species in our gardens are natural hybrids. In his books *Rhododendrons and The Various Hybrids* early in this century, J.G. Millais recognised that the plant cultivated as *R. wightii* was a hybrid. For the next few decades this information was conveniently ignored by botanist and grower alike until the real *R. wightii* was introduced from Nepal in the 1970's and was found to be completely different to the cultivated plant. Many, many so-called species are now being revealed as natural hybrids. *R. wongii* (*R. ambiguum* x *R. flavidum*), *R. bathyphyllum* (*R. proteoides* x *R. aganniphum*) and *R. erythrocalyx* (*R. selense* x *R. wardii*) are a few of the many examples. A variable percentage of the seedlings from many seed numbers from the the SBEC 1981 expedition to the Cangshan turned out to be natural hybrids but surprisingly most of these hybrids were not apparent on the mountains where the seed was collected. Hybrids included: *R. neriiflorum* x *R. rex* ssp. *fictolacteum*, *R. taliense* x *R. lacteum*, and *R. haematodes* x *R. rex* ssp. *fictolacteum*. In fact, almost every species crossed with some or all of its near neighbours in flower at the same time, especially among the species in Section Hymenanthes. The fact that the steep upper slopes of the Cangshan are relatively little disturbed by man probably accounts for the scarcity of wild-occurring natural hybrids on this mountain range.

Recognition of the relative commonness of natural hybrids and the gradual evolution or speciation process that can take place is absolutely fundamental to our understanding of and therefore accurate identification of rhododendron species. Perhaps up to a quarter of all currently accepted taxa are in fact hybrids or of hybrid origin. Amongst section Vireya, the number may be even higher. The Balfourian taxonomy does not deal with natural hybrids and indeed appears to deny their existence in allowing seedlings which appear in wild seed lots to be described as new species: *R. paradoxum*, *R. peregrinum*, *R. inopinum*, and *R. planetum* are examples this.

IDENTIFYING RHODODENDRON SPECIES

Identifying rhododendron species is not necessarily an easy task. Some species are very distinctive and can rarely be confused with any others; examples are *R. genesterianum* and *R. camtschaticum*. For the majority of species, however, identification is more difficult, demanding a good eye and a source of reliable information in the form of a botanical description. Such descriptions, in greater or lesser detail, are found in many published works on rhododendrons and form the backbone of this book. Botanists have made detailed descriptions of all species, outlining the characteristics which identify them.

Key characters in rhododendron identification

Height: measurements given are for mature plants. Heights vary from prostrate to almost 30m. It is often hard to give accurate height measurements as species grow at very different rates depending on soil, rainfall and other climatic considerations. Within one species, a great variation in height can be found.

Habit: low and creeping through compact, loose to straggly and/or upright and/or tree-like. Some species may vary considerably in habit and much depends on age as well as whether a plant is grown in the open or in shade. Whether a plant is stoloniferous or not may be taxonomically significant.

Branchlets: colour and thickness of branchlets; the presence or absence of hairs and glands, and the persistence and colour of bud-scales.

Bark: descriptions refer mainly to bark of mature plants. Most species have a roughish grey-brown bark. The majority of species with a coloured (other than grey-brown), peeling or flaking bark occur in subsections Barbata, Falconera, Thomsonia, Glauca and Maddenia.

Buds: shape is sometimes important as is the presence of elongated bud-scales and presence or absence of hairs (section Pentanthera). Section Tsutsusi is characterised by the emergence of flowers and foliage from within the same bud. Bud colour is seldom significant as many species have a mixture of bud colours even within one population. Blackish, reddish, or greenish buds may sometimes help to predict the flower colour.

Leaves: these are usually the most important part of a plant for identification purposes, especially as plants spend most of the year without flowers and may take years to reach flowering size. Size, shape (including whether flat or with margins recurved or wavy), thickness, texture and colour are important; the most important characters are the presence or absence of indumentum, hairs, glands and scales. Indumentum can be of many colours, in one or two layers (unistrate or bistrate), woolly to plastered, continuous or split into sections, thick to thin and can be made up of a variety of different hair types. Leaf indumentum may take many years to fully develop, making young plants particularly hard to identify. Leaves on young seedlings often have a reddish underside. This is simply juvenility and the plant will grow out of it. Hairs and bristles on stems, petioles and leaf surfaces and margins vary from tiny and pubescent to long and bristly. Glands vary from minute to quite large and are found on the tips of hairs. Within subgenus Rhododendron, the small scales, usually most prominent on the leaf underside, are very significant. These can be be uniform or of different sizes or/and colours, and they vary in distribution, ranging from widely scattered to overlapping or even in tiers (as in section Pogonanthum). Foliage is often aromatic in subgenus Rhododendron, e.g. *R. heliolepis*, *R. kongboense* and sometimes scented (rather elusive) in subsection Taliensia.

Petioles: length, thickness, the presence or absence of hairs, glands or scales. Flattened petioles merging into the leaf bases are important in the identification of large-leaved species.

Inflorescence: length of rachis and pedicels and the presence or absence of hairs, glands and scales. Usually the inflorescence is terminal but in subsections such as Scabrifolia and Virgata the inflorescence is partly or wholly produced from axillary buds. The number of flowers in an inflorescence and its shape are often important.

Corolla: size, shape, number of lobes (usually 5), whether the corolla is symmetrical or not and the shape and length of the tube and lobes and whether the latter are divided from each other or not. Colour is often significant but it is important to remember how variable in flower colour certain species can be. The presence or absence of a blotch, spots, and of pubescence on the corolla are often significant. Some species have distinct nectar pouches. Scent may be present, especially in subsections Maddenia and Fortunea.

Calyx: size, shape (including the length and shape of lobes), presence, absence or extent of hairs, glands and scales. The calyx is usually green but can be be similar in colour to the corolla.

Stamens: these usually number 10, and where this number varies it is usually an important diagnostic character. They often vary in length within one flower and may be inserted (remaining within the corolla) or exserted (protruding). Stamens can be of various colours.

Style: the presence of hairs, glands or scales, the length relative to the stamens and whether it protrudes (is exserted) beyond the corolla.

Ovary: presence or absence of hairs, glands and scales and the number of chambers. In subsection Maddenia, it is often important whether the style tapers gradually into the ovary or is impressed (there is an obvious and abrupt change from style to ovary).

Capsule: occasionally of significance as to shape, size and length, number of chambers and whether the calyx is persistent or or not.

Seed: size and shape vary considerably. More important can be the presence of wings (as in subsection Maddenia) or tails (as in section Vireya) on the seeds.

MAKING AN IDENTIFICATION

Armed with a botanical description, approach a plant from a distance if possible; some characters are best seen from afar. The first thing to look for is the height, habit, leaf shape and leaf poise, bark; then the bud shape and size and texture of the foliage, the thickness of the branches, the leaf retention, the presence or absence of indumentum and scales. With flowers, consider the size and shape of the inflorescence, whether terminal or axillary, the number of flowers per truss, the size, colour and shape of the corolla. Lastly closely examine (with a hand-lens if needed) the type of indumentum, hairs or scales on leaves, calyx, stamens etc.

It must be emphasised that morphological characters (as outlined above) can vary enormously within one species. Where possible, we have given the full possible range in such characters as leaf size and shape and flower colour. We often hear remarks such as 'it can't be such-and-such a species as the leaf is too long/wide etc.' There is a certain type of rhododendron collector whose considerable knowledge is often hindered by a pedantic and inflexible approach to new discoveries and changes to taxonomy. Such botanical and horticultural flat-earthism can be extremely counter-productive for good species identification, particularly in the wild. Difficulties arise in that a plant or specimen may fit into the broad concept of a species but one or more particularly important characters may not hold good. This happens time and time again in the wild and occasionally in gardens: what do we call *R. arizleum* with a flattened petiole (S.E. Tibet), or *R. kesangiae* (which normally has distinctive rounded buds) with pointed buds (Bhutan) or *R. longesquamatum* at Dawyck Botanic Garden, S. Scotland, where one plant does not have the characteristic persistent bud scales? There are two ways to go here: either broaden the parameters of the species or describe a new species/subspecies. We would generally tend towards the former. If we bear in mind that the original (or type) description of a species usually represents a small population of a perhaps widely-distributed and very variable taxon, we should not be surprised if subsequent collecting reveals characteristics which do not quite fit the original concept of that species. The descriptions in this book often deviate from previously published work to reflect new discoveries from the wild. In addition, we have noticed that many species produce larger leaves and flowers in mild, moist, western British gardens than they do in the wild and we have re-measured many cultivated specimens to try to cover the full potential range of the dimensions of a species.

The shortcomings of botanical descriptions

One of the problems with botanical descriptions is that the enormous number of technical terms can be impenetrable to the layman. We considered trying to avoid them in the writing of this book but realised that in many cases it is impossible to find another suitable word. We have therefore tried to make the glossary clearer, more thorough and less technical than those found in most other publications. We have also tried to keep the species descriptions relatively short, concentrating on the essential characteristics. This is a deliberate reaction against the impenetrable nature of some recent botanical work. This is sometimes due to the complexity of the science involved (D.N.A. analysis for example), but often it is an issue of presentation. Botanical accounts must follow accepted codes of nomenclature and must, of course, be scientifically accurate. What is often overlooked is much attempt to make the descriptions concise or clearly presented or to be consistent in terminology or degree of detail. Most importantly, the 'essential' features of a species are seldom made to stand out from the mass of data presented. An example of this is the work of K.A. Kron in the revision of Section Pentanthera, published in the *Edinburgh Journal of Botany Vol. 50 No. 3. 1993*. The descriptions contained in this revision are mind-numbingly detailed and full of botanical terms (such as 'adaxial' and 'abaxial' to describe upper and lower leaf surfaces) which are inconsistent with terms used in other parts of the Edinburgh revision. Quite apart from the detail, the way the text is laid out in a single paragraph without any emboldening of headings makes it very hard to follow for anyone trying to extract key information. We are not questioning the scientific accuracy of the work in question, it is just that we feel that in this and in other cases science is let down by poor presentation and communication skills: it is not user-friendly enough. In this book we have tried to give botanical descriptions in as concise and accessible a way as possible and have highlighted with italics the key characteristics of each species which separate it from its near relatives.

The *gestalt* approach

There are probably only a small number of people who are really proficient at identifying rhododendrons. Of course, a close knowledge of the key botanical features of each species is invaluable, but we have found that there is a more useful and less quantifiable skill gained from years of handling rhododendrons in cultivation and in the wild. We have found that working with plants on the nursery gives us a *gestalt* approach to identification. The essence of a species is more than a sum of its botanical description but more a subconscious recognition of a host of visual characteristics which allow an identification to be made. The best way to explain this is to consider the way we recognise a particular person from their face. Most people can probably recognise hundreds of people by their faces alone with great precision, and we do this unconsciously using our memory. We could not explain exactly how we do this nor could we write a facial description of someone we know which would allow someone else to instantly recognise that person. We use a similar recognition process to identify rhododendron species. Rhododendrons give much more in the way of visual clues than can be given in a botanist's description, however detailed, and we use the published description to confirm rather than make our identifications. This is what we mean by a *gestalt* approach and this method is especially important when considering a plant which may be a species or a hybrid. It may not key out properly to any botanical description, but instead may state its origins through a number of clues visible to the more experienced observer. The parents of a natural hybrid are often quite obvious to us but it is often hard to pinpoint exactly what leads us to our conclusions.

Identifying species in cultivation

In addition to considering the use of botanical descriptions to identify rhododendron species, there are other points to consider which will narrow down the possible alternatives and make the task of identification easier. This is equally true whether faced with a few or a great number of plants. It is perhaps most important when faced with the task of identifying unlabelled species in an old collection. In such cases there are certain key questions to ask:

a) When was the specimen or collection planted?

The main part of most rhododendron collections was planted over a limited number of years. If this date can be even roughly pin-pointed, it greatly helps to narrow down the number of species which are likely to be have been planted at that time. If it is known that most plants came from a certain source, for example from seed collected in the Himalaya from 1840 to 1860, this probably eliminates all species from China and all species introduced after those dates. In the text of this book, the dates of introduction and reintroduction of species are given, which should help in the identification of plants in older collections.

b) Where did the plant or plants come from?

1. Wild seed?
Did the previous owner of the garden receive seed or plants grown from seed of plant-hunting expeditions? Nearly all the good collections of species in Britain and elsewhere contain quantities of plants grown from wild origin seed, frequently with

their original collectors' numbers. Often gardens are particularly associated with one collector or even one expedition. Examples of this include the Wilson-collected plants at Dawyck, S. Scotland, the Ludlow and Sherriff plants at the Hobbie Park, Germany, and the last Rock expedition seed which Cecil Smith, Carl Phetteplace and others grew in W. U.S.A. A little detective work, browsing through correspondence can reveal a lot. In New Zealand recently, I (Kenneth Cox) was shown extensive correspondence between F. Kingdon-Ward and the New Zealand Rhododendron Society detailing seed numbers sent to them. Such information is invaluable in aiding identification. If information about collectors and expeditions is available, *The Rhododendron Handbook* 1980 can be referred to. This gives lists of species with their collectors, numbers from each expedition pre 1980. The new edition to be published in 1997 includes later expeditions. In addition, F. Kingdon-Ward and to a lesser extent other collectors (but not G. Forrest unfortunately) have written extensively about their trips and these books and articles can give information about which species were collected on which expeditions.

2. Garden Origin Seed
Rhododendrons are notoriously promiscuous and if they can possibly cross with each other they will. Regrettably and despite criticism, there was a period in which the Royal Horticultural Society, the National Trust and botanic gardens in the U.K. distributed open-pollinated seed, especially abroad, under specific names with no indication to the unsuspecting gardener that the majority of the seed would be hybridised. As a result of this, there have been many hybrids masquerading as species, the discovery of which has dismayed their owners years later. Open-pollinated species seed from gardens is virtually worthless as it is almost always hybridised, unless the species is one which does not cross with others or it is completely isolated from other genetically compatible species flowering at the same time. On the other hand, hand-pollinated (controlled-pollination) garden seed (such as that supplied in seed exchanges) is an excellent source of species material, especially if superior forms of the species are used as parents. Most cultivated *R. pachysanthum* and most *R. souliei* (especially the deep pink selected strain) have originated at Glendoick as hand-pollinated seedlings of plants in our garden, and now seed of these seedlings is being distributed in seed exchanges.

3. Self-sown seedlings
It may astonish those who garden in harsh climates to see the extent that naturally germinated rhododendron seedlings thrive in milder and wetter gardens in Britain and elsewhere. *R. ponticum* and *R. luteum* have long been naturalised in the U.K. but many western seaboard gardens can easily be taken over by forests of self-sown, often hybridised seedlings of many different species and which, if not removed, can create an impenetrable jungle. Stonefield Castle, Argyll, West Scotland is a good example of this and the extraordinary proliferation of seedlings there has long been exploited by neighbours and other visitors who have dug them up and planted them in their own gardens. There are one or two small cottages nearby with 30ft ex-Stonefield *R. arboreum* seedlings towering over them! The problem with self-sown seedlings is that, although they are often hybrids, they can tend to resemble their parents quite closely in foliage and it can be hard to tell them apart. This is particularly true of hybrids between large-leaved species. In gardens, where the seedlings have not been systematically removed, it is often virtually impossible to separate the good from the bad when the inevitable overcrowding has led to a forest of bare trunks with the leaves and flowers far overhead and out of reach.

4. Nurseries
Did the owner of the garden patronise several or even a single nursery? There are garden owners who bought everything from Hilliers for example. Are there any old catalogues or bills that can be found? Often a nursery will distribute the same clones of species for many years and such clones become irrevocably linked to certain nurseries, helping by association in the identification of other plants. Unfortunately there is no guarantee that even good nurseries will always have sold correctly labelled stock. In addition, many wild-collected seed packets are not what they claim to be and it is often many years before such mistakes are discovered.

Of course, many of the plants in a collection may be hybrids which are often even harder to identify than species, especially when not in flower. There are even fewer hybrids experts than species experts and in the case of older hybrids there are no adequate descriptions available. There is no easy way to tell a species from a hybrid, be it man-made or not. We are frequently sent fresh (or not so fresh!) flowering specimens in the post to name. Sometimes they are easily-identified species or hybrids, other times they may be old (once named) hybrids or just unnamed hybrid seedlings where it may only be possible to give some idea of parentage. Gaining a good knowledge of species is really the only way of being able to tell or guess whether a plant is a species or a hybrid. Many common hybrids are quite easily recognised and even when this is not the case it is often possible to guess at least one of the parents, especially of first generation hybrids (with species rather than hybrids as parents).

Identifying species in the wild

A. Preparation

Rhododendrons are often very hard to identify in the wild. They may look dramatically different to their cultivated cousins, they are often drawn-up and straggly and they vary so much it is often hard to see where one species starts and another ends. And just to complicate matters, there are often lots of natural hybrids which further blur the boundaries.

1. Books and articles

Most books which cover rhododendron species include information on the wild distribution of each species. The Edinburgh revision volumes give useful maps with dots on them to indicate locations. These are by no means complete and are sometimes inaccurate but they do give a good quick idea of what is in any given area. *The Rhododendron Species* volumes by H.H. Davidian go into considerable detail on the areas species were collected from and often mention actual mountains or mountain ranges. F. Kingdon-Ward wrote many fine and detailed books about his expeditions and reading relevant parts can be very helpful. There are books about G. Forrest, E.H. Wilson and Ludlow and Sherriff (see bibliography) which also contain valuable information. There are also many very useful, more modern accounts of expeditions to China and the Himalaya, such as books by Roy Lancaster and reports by Warren Berg, Keith Rushforth, Peter Cox and many others in rhododendron journals. From all this literature, it is quite easy to draw up a list of possible species you may encounter in a certain area.

2. Herbaria, botanic gardens and gardens with good labelling

Armed with a list of likely to be encountered species, you can then refer to botanical descriptions or better still check herbarium specimens or live plants so that you will recognise what you find in the wild. Hopefully you will find species which you do not recognise, but you will find that previous collectors, who often spent many weeks, months or even years in an area, have generally done a good job in recording the more significant plants, rhododendrons being one of the most obvious! Time and time again, we receive packets of seed with names of species on them which cannot possibly occur in the location they were collected from. Such mistakes are easily avoided with a minimum amount of homework. It is undoubtedly true that the greatest plantsmen are those who do the most thorough preparative homework. The best herbarium in the U.K. for rhododendrons is the Royal Botanic Garden, Edinburgh. The Royal Botanic Gardens, Kew, and the Royal Natural History Museum, London, also have extensive collections. It is necessary to get permission to study in these herbaria.

B. In the field

With good preparation, you will manage to identify most of what you find. "Be ready to be baffled" is possibly the best advice. Rhododendron species make no attempt to help in their identification. Many so-called species merge with other species and it can be quite disconcerting to find this happening time and time again. Particularly in subsections Lapponica, Maddenia, Triflora, Taliensia, Neriiflora, and in sections Pentanthera and Brachycalyx many species form geographical or altitudinal 'clines' from one species to another. In this volume, wherever possible we draw attention to species which merge with their relatives and hybridise with their neighbours in the wild.

Every expedition confounds our expectations and teaches us more. We usually take a small notebook with essential taxonomic information where we expect to meet with difficulties. This is especially important with hard-to-identify species such as those in subsections Lapponica and Taliensia. It might even be worth drawing up a small multi-feature key to help sort out the more troublesome species. Natural hybrids may even have gone some way in the process of speciation and occur as populations rather than just individuals. Some hybrids are extremely poor and often disease-ridden while others can be magnificent. Two good ones we have found recently are *R. roxieanum* var. *oreonastes* x *R. vernicosum* and *R. decorum* x *R. glischrum*. If you cannot identify something, it may be possible to press a few leaves or a flower and to bring it back for identification. A good written description might be enough to help arrive at a positive identification at a later date.

Hardiness and climate

We have used our own garden and nursery as a standard to measure hardiness. Glendoick is situated on the northern edge of the fertile Carse of Gowrie to the north of the Tay estuary between Perth and Dundee in east Perthshire. The main parts of the garden and nursery lie on the lower south-facing slopes of the Sidlaws with other parts on the adjacent flat carse. Elevation is from 15-60m (50-200ft). The low elevation south-facing slopes encourage early growth and rapid drying out of the soil. Spring frosts can cause considerable damage to the flowers and young growth and bark, especially after a mild winter when the sap flow is often well-advanced as early as the end of March. Early autumn frosts can cause occasional damage to late growth and unripened buds. The minimum temperature recorded is -18°C (0°F) and the maximum is rarely over 27°C (80°F). Many winters are relatively mild with no frosts below -7°C (20°F) but hard winters do occur occasionally. Yearly rainfall averages just under 76cm (30in), falling all the year round, but dry spells can last for a month or more, necessitating watering most years. Heavy snowfalls are rare and frost may penetrate the bare soil to 15cm (6in) or more. Shelter is fairly good from the west round through north to east. Damaging winds occur every few years. The soil on the slopes is a medium to medium-heavy loam with a more silty-sandy loam on the valley floor. Most of the soil is short in organic matter, and natural mulching. The pH is moderately to fairly acid. Drainage is generally good.

PROVINCES OF CHINA

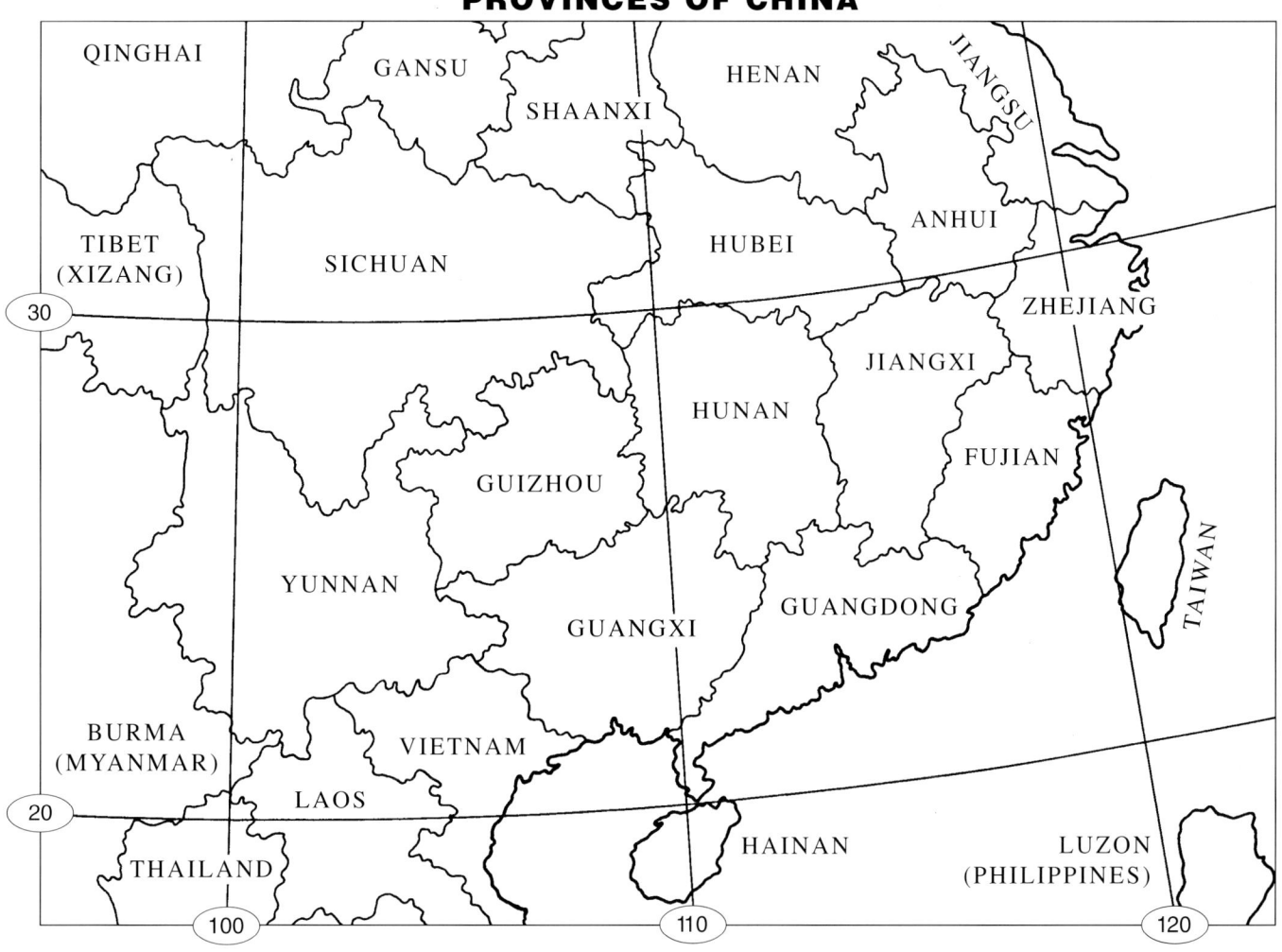

THE SINO-HIMALAYAN REGION

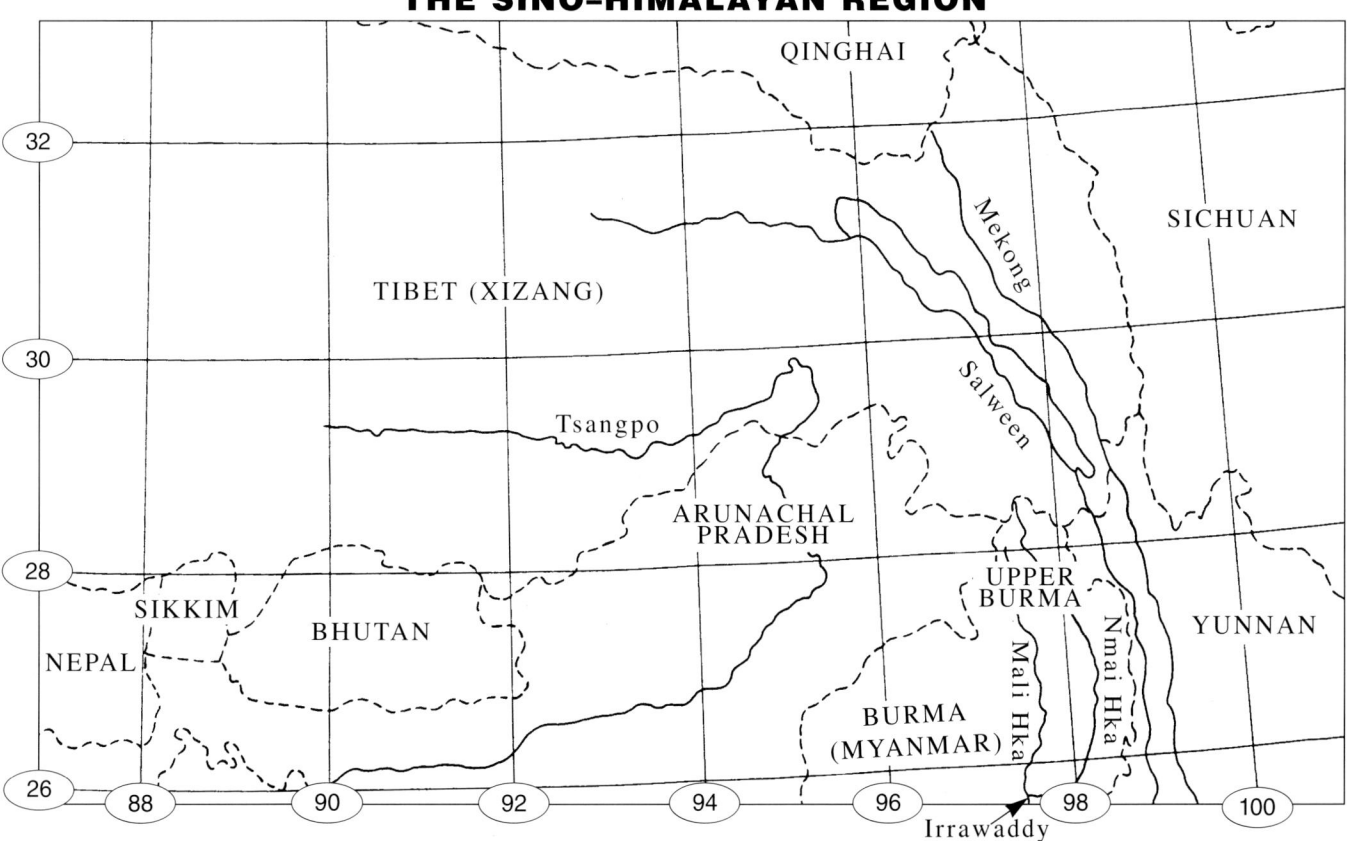

Explanation of descriptions

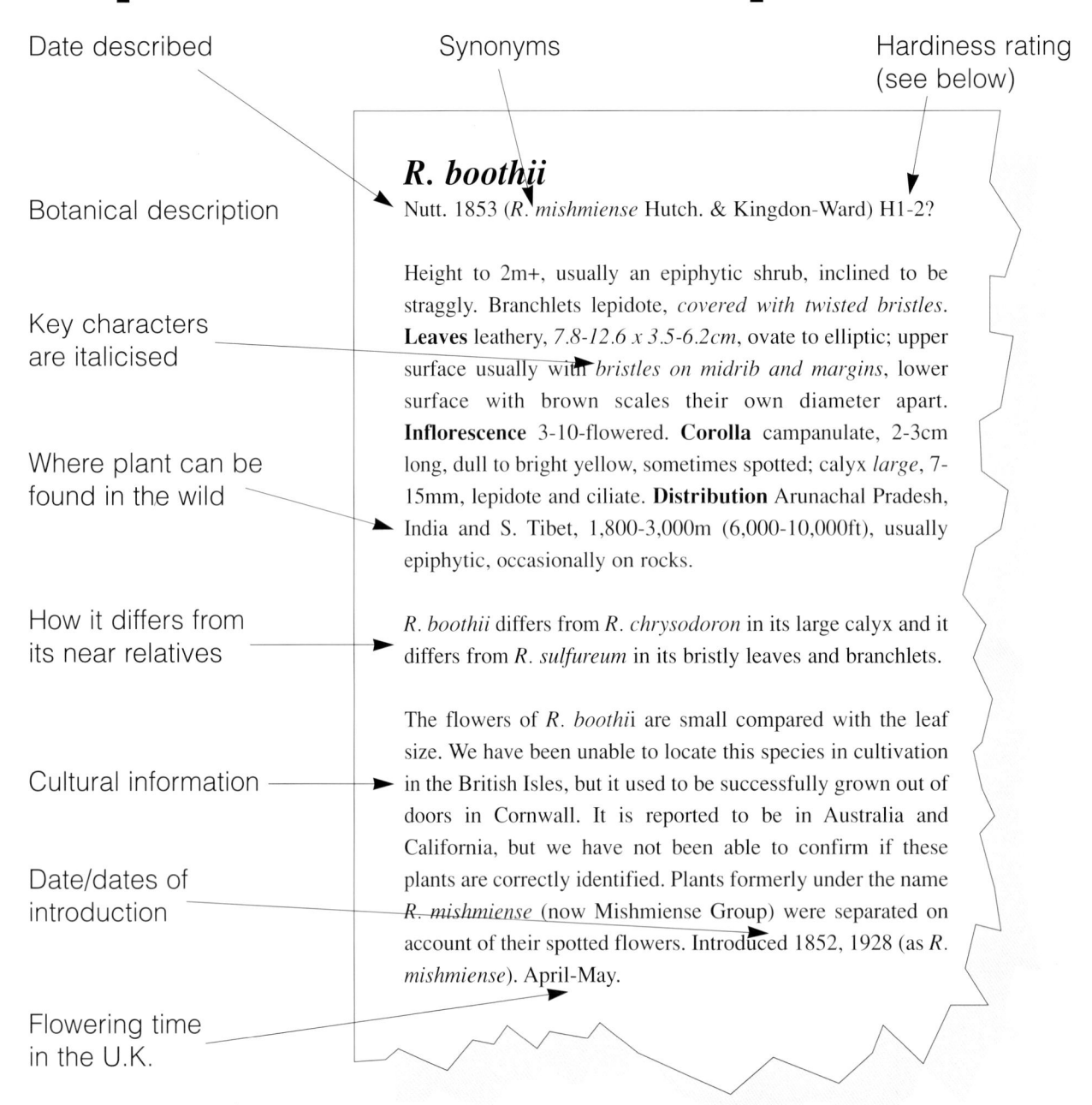

Date described — Synonyms — Hardiness rating (see below)

Botanical description →

R. boothii
Nutt. 1853 (*R. mishmiense* Hutch. & Kingdon-Ward) H1-2?

Key characters are italicised →

Height to 2m+, usually an epiphytic shrub, inclined to be straggly. Branchlets lepidote, *covered with twisted bristles*. **Leaves** leathery, *7.8-12.6 x 3.5-6.2cm*, ovate to elliptic; upper surface usually with *bristles on midrib and margins*, lower surface with brown scales their own diameter apart. **Inflorescence** 3-10-flowered. **Corolla** campanulate, 2-3cm long, dull to bright yellow, sometimes spotted; calyx *large*, 7-15mm, lepidote and ciliate. **Distribution** Arunachal Pradesh, India and S. Tibet, 1,800-3,000m (6,000-10,000ft), usually epiphytic, occasionally on rocks.

Where plant can be found in the wild →

How it differs from its near relatives →

R. boothii differs from *R. chrysodoron* in its large calyx and it differs from *R. sulfureum* in its bristly leaves and branchlets.

Cultural information →

The flowers of *R. boothii* are small compared with the leaf size. We have been unable to locate this species in cultivation in the British Isles, but it used to be successfully grown out of doors in Cornwall. It is reported to be in Australia and California, but we have not been able to confirm if these plants are correctly identified. Plants formerly under the name *R. mishmiense* (now Mishmiense Group) were separated on account of their spotted flowers. Introduced 1852, 1928 (as *R. mishmiense*). April-May.

Date/dates of introduction →

Flowering time in the U.K. →

Hardiness ratings

These ratings cover mid-winter weather only, when plants should be at their hardiest. Sudden cold spells after mild weather or unseasonable late spring or early autumn frosts can damage almost any species in a state of active growth, or which has not hardened off its young leaves. In addition, many species will not acquire their optimum hardiness until they have produced a substantial woody trunk. This may take five or more years. Certain species, notably members of subsection Pentanthera and section Tsutsusi from southern distributions require warm summers to ripen their growth and only achieve full hardiness where wood is properly ripened.

H1 Slightly heated greenhouse (in cold areas) or not below -7° C (20°F).

H2 Most favourable coastal areas of British Isles or equivalent, minimum -12°C (10°F).

H3 Sheltered gardens near to coasts of British Isles, minimum -15°C (5°F).

H4 Hardy in all but coldest inland and high areas of British Isles, minimum -18°C (0°F).

H5 Hardy anywhere in British Isles, in much of eastern Europe and the warmer coastal areas of eastern North America, minimum -21 to -24°C (-5 to -10°F) or colder.

SUBGENUS AZALEASTRUM

Evergreen shrubs or trees. Inflorescence lateral below terminal or pseudoterminal vegetative buds with one to several flowers. Corolla rotate to tubular-campanulate; calyx lobes obsolete or large; stamens 5 or 10.

Section Azaleastrum

Height 0.6-8m, generally small shrubs in cultivation though they may grow large in favourable climates; considerably larger, erect to spreading, sometimes straggly shrubs in the wild. Branchlets slender, *usually minutely puberulous*. **Leaves** evergreen, glabrous except on midrib. Young leaves often highly coloured. **Inflorescence** axillary, *single-flowered*. **Corolla** rotate to tubular-campanulate with spreading lobes, white, pink, rose, pale to magenta-purple, crimson, with or without spots; calyx lobes *large, to 8mm long*; stamens *5*; style usually glabrous; ovary bristly; capsule *very short* with persistent calyx. **Distribution** S, Central and W China, N.E. Upper Burma, and Tibet

The species in this section differ from those in section Choniastrum in their generally smaller stature in cultivation, the more slender, usually minutely puberulous branchlets, the single-flowered inflorescence, the large calyx, the 5 rather than 10 stamens, and the shorter capsule.

The species of section Azaleastrum have rather dainty foliage and pretty and sometimes showy flowers, but can be rather shy-flowering in cultivation. All the species are best suited to warmer climates than the U.K. None are widely cultivated.

R. hongkongense

Hutch. 1930 (*Azalea myrtifolia*) Champ.) H1-2?

Height to 5m in wild, probably only 2-3m in cultivation, a fairly compact to broadly-upright shrub. Branchlets minutely puberulous. **Leaves** 3-6.5 x 1.3-3.3cm, *elliptic to narrowly elliptic*, young leaves often purplish. **Inflorescence** single-flowered, axillary in upper leaf axils. **Corolla** rotate with short tube and spreading lobes, c.5cm across, *white spotted violet*, occasionally with a yellow blotch, (sometimes?) fragrant; calyx c.3mm long, lobes rounded, with stalked glands at base; style glabrous or with a few hairs at base. **Distribution** Hong Kong, Guangdong, 300-1,200m (1,000-4,000ft), on rocky slopes.

R. hongkongense is closely related to *R. ovatum*, differing in the narrower leaves and the always white, spotted flowers which are usually pink-purple in *R. ovatum* and red in *R. vialii*. The presence of fragrance is also a distinguishing feature, although it is not clear if this species is always fragrant.

This species is cultivated in the Royal Botanic Garden, Edinburgh under glass where it is fairly compact and flowers very freely, making a pretty plant. The young growth of this clone is a striking purplish colour. Introduced 1971. March-April.

Picture 1

Picture 2

1. *R. hongkongense* collected by J. Patrick showing the compact habit and freedom of flowering. Royal Botanic Garden, Edinburgh.
2. *R. hongkongense* with its white spotted violet rotate corolla. Royal Botanic Garden, Edinburgh.

AZALEASTRUM

R. leptothrium
Balf.f. & Forrest. 1919 (*R. australe* Balf.f. & Forrest) H2-3

Height 0.6-8m, less in cultivation, an often upright and straggly shrub but sometimes dense. Branchlets minutely puberulous, reddish at first, brown when mature. **Leaves** 2-12 x 1-3.6cm, *narrowly elliptic to lanceolate,* young foliage shiny, rich bronze-green or reddish. **Inflorescence** single-flowered, axillary in upper leaf axils. **Corolla** rotate with a short tube and spreading lobes, c.4.5cm across, *pale rose to rich purplish-rose,* with or without spots, scentless; calyx 3.5-8cm long, *usually oblong;* style glabrous. **Distribution** N.E. Upper Burma, N.W. Yunnan and adjacent Sichuan and Tibet, 1,500-3,400m (5,000-11,000ft), in the open, as part of hedgerows or in conifer or mixed forest.

R. leptothrium differs from *R. ovatum* in its more westerly distribution, its narrower leaves and the deeper-coloured, usually earlier flowers. *R. vialii* differs in the shape and colour of the flowers.

This uncommonly cultivated species is attractive in both flower and young growth which blend well together. It is a little slow to bloom in cultivation but is free-flowering with age. Some forms are relatively hardy but the early growth makes its very vulnerable to late frosts and it is really only suitable for mild, sheltered gardens. It can be spectacular in the wild where it often grows in open, dry and exposed sites, indicating that it may be a useful species for hot, dry climates. Introduced 1914->, 1988->. April-May.

Picture 2

Picture 3

Picture 4

Picture 1

1. *R. leptothrium* Weixi, N.W. Yunnan, China showing the early growth coinciding with the flowers.
2. *R. leptothrium* from Blackhills. This form was formerly wrongly labelled and distributed as *R. vialii*.
3. A pale-flowered *R. leptothrium* at Guavis, North Island, New Zealand. This plant is grown in New Zealand under the name *R. bachii* (now synonymous with *R. ovatum*) but this form is in fact closer to *R. leptothrium* with its narrowly elliptic leaves.
4. *R. leptothrium* showing the narrow, bronzy young leaves. This form, grown under cover at Glendoick, is from seed collected near Weixi, N.W. Yunnan, China.

R. ovatum
Maxim. 1870 (*R. bachii* Lév., *R. lamprophyllum* Hayata)
H3-4

Height 1-4m, an erect to compact shrub. Branchlets minutely puberulous, sometimes glandular. **Leaves** 2.5-6 x 1-2.6cm, *ovate to ovate-elliptic*, glabrous except on the petiole and midrib. Young leaves often bright red or pink. **Inflorescence** 1-flowered, axillary in upper leaf axils. **Corolla** rotate with short tube and spreading lobes when fully opened, 2.5-5cm across, *white to pink to pale purple*, spotted and sometimes with a yellow blotch; calyx 4-7mm long, lobes *broadly rounded*; stamens 5, partly hairy; ovary bristly and glandular; style glabrous. **Distribution** Widespread in Central and S.E. China and Taiwan, open slopes to dense forest to 2,000m (6,500ft).

R. ovatum is related to *R. leptothrium,* differing in its broader leaves, paler flowers, more rounded calyx lobes and its more easterly distribution. *R. vialii* differs in the shape and colour of the flowers.

This species is not really satisfactory in northern climates, where it tends to be shy-flowering, lacking in vigour and prone to foliage chlorosis. It should be much better in warm and hot areas and some forms have proven to be heat resistant. It is neat and attractive when well grown but is comparatively rare in cultivation. Plants are cultivated in New Zealand under the now synonymous *R. bachii* (now Bachii Group). This taxon comes from the western end of the the range of *R. ovatum* and seem to be intermediate between *R. ovatum* and *R. leptothrium* and is characterised by the calyx lobes fringed with glandular hairs. *R. ovatum* introduced 1843 but plants in cultivation probably date from 1900->. May-June.

Picture 1

Picture 2

Picture 3

1. *R. ovatum* in J. Keeley's garden, New Zealand, showing the coloured young foliage with shorter leaves than those of *R. leptothrium*.
2. A nicely spotted, pale-flowered *R. ovatum*, W. U.S.A. (Photo G. Miller).
3. *R. ovatum* with pale pink spotted flowers, Rhododendron Species Botanical Garden, W. U.S.A. (Photo Rhododendron Species Foundation).

R. vialii
Delavay & Franch. 1895. H2?

Height 1-3m, an evergreen shrub, branchlets pubescent. **Leaves** 3-10 x 1.5-3.5cm, obovate or obovate-lanceolate, thin, upper surface midrib slightly pubescent, lower surface glabrous; petiole 1.5-2cm long, flat. **Inflorescence** single-flowered (occasionally 2-flowered?) in uppermost leaf axils; pedicel *densely glandular*. **Corolla** *tubular-campanulate*, 3-3.5cm long, deep red to crimson (or pink), calyx dark red, deeply 5-lobed; stamens 5, *glabrous or slightly pubescent;* style glabrous, reddish. **Distribution** C. & S Yunnan, 1,300-2,000m, in thickets on mountain slopes.

The characteristic corolla shape and colour, the glabrous or near glabrous stamens and the densely glandular pedicel distinguish *R. vialii* from its relatives.

This species is reported to be in cultivation although we have not been able to verify this. Plants with purple flowers, distributed under this name turned out to be *R. leptothrium*. The flower colour is very striking but the early flowering and low altitude habitat make it a plant for mildest climates only. January-March.

AZALEASTRUM

Picture 1

1. *R. vialii* in Yunnan, China, showing its distinctive red, tubular flowers. (Photo Yang Zhenhong)

Section Choniastrum

Height to 15m, bushy to erect shrubs to small trees. Branchlets glabrous or bristly; bark often *smooth or peeling, red to brown*. **Leaves** evergreen, usually *glossy*, to 17cm long, glabrous or bristly, young leaves often *tinged red*. **Inflorescence** 1 or more-flowered from axillary buds. **Corolla** white, rose to purple, narrowly funnel-shaped with narrow tube, medium to large, 1.8-8.8cm across, glabrous, often *fragrant;* calyx *very small;* stamens *10;* ovary glabrous or rarely hairy; style glabrous; capsule elongate. **Distribution** Japan, Taiwan, W, S. & S.E. China, Burma, Tibet, Arunachal Pradesh, Laos, Vietnam to W. Malaya.

The species in Section Choniastrum differ from those of Section Azaleastrum in their smaller calyx, in their greater stature, their larger leaves and flowers and the 10 rather than 5 stamens.

This is a much neglected group of species with great potential for warm climates such as S.E. Australia and the milder parts of New Zealand and California. Photographs in *Rhododendrons of China Vol. 2* give some idea of the beauty and variation which exists but unfortunately several fine species have never been introduced. The 1986 revision of this section was somewhat premature and inadequate as none of the necessary fieldwork in China was done, and there was insufficient access to herbarium material.

R. championae
Hook.f. 1851 H1-2

Height 2-8m, a spreading shrub. Branchlets stiff, *covered with glandular bristles;* bark not peeling. **Leaves**, often tinged reddish-purple, 7-15 x 2.5-5cm, elliptic to obovate; upper surface *roughish with short bristles, particularly near margins,* lower surface loosely bristly; petiole *densely bristly*. **Inflorescence** c.5-flowered. **Corolla** broadly funnel-shaped, 10-11.3cm across, tubes 1.2-1.5cm long, pale rose-pink buds open to white with yellow spots, slightly fragrant; calyx with variable lobes, ciliate; ovary *densely bristly*. **Distribution** S. and S.E. China, rare in Hong Kong, shady ravines and stream sides.

This is a distinctive species, being the only member of the section to have bristles on the stems, adult leaves, petiole and ovary.

R. championae has attractive foliage and large flowers. It is just hardy enough to have survived for periods in some Cornish gardens but has always been rare in cultivation. Introduced 1881, reintroduced recently. April-May.

Picture 1

Picture 2

1. *R. championae* in flower under glass at Glendoick.
2. Foliage of *R. championae* showing the densely bristly leaf margins and petioles.

Picture 3

3. Leaf portions of *R. championae* showing the bristly leaf upper surface and margins. (Photo R.B.G., E.)

Picture 2

2. *R. hancockii* foliage at Kunming Botanical Institute Garden, Yunnan, China.

R. hancockii Hemsl. 1895 H2?

Height 1-4.5m, a shrub or tree; branchlets glabrous. **Leaves** evergreen, 8.5-13.5 x 2.8-5cm, obovate, apex acuminate, upper and lower surfaces glabrous. **Inflorescence** 1-(2)-flowered from axil or axils at branchlet tip. **Corolla** widely funnel-shaped, *deeply 5-lobed*, 5-7 cm long, white, with a yellow flare on one lobe, *fragrant;* calyx small, deeply 5-lobed; stamens 10, unequal, often exserted, pubescent on lower half; style glabrous; ovary *densely tomentose;* capsule hairy. **Distribution** Widespread in S. Yunnan, 1-2000m, in thickets on sunny slopes, also on cliffs and steep banks.

The scented, deeply 5-lobed flowers are quite distinctive. Other important features are the densely tomentose ovary and the relatively large leaves. Doubtfully distinct enough from *R. moulmainense* to retain specific status.

This looks a fine species as we have seen it in the wild in Yunnan and it may well prove an excellent scented species for mild and hot climates. It may not yet be in cultivation. November?-June.

Picture 1

1. *R. hancockii* between Yipinglang and Li Feng, Yunnan, China. The showy, white-flowered bushes could be seen from a considerable distance away.

R. latoucheae

Franch. 1899 (*R. amamiense* Ohwi, *R. wilsonae* Hemsl. & E.H. Wilson, *R. latoucheae* var. *ionanthum* G.Z. Li) H1-3

Height to 7m, an upright to spreading shrub (*R. wilsonae* Wakehurst form is low and spreading). Branchlets glabrous. Bud scales persistent. **Leaves** 5-10 x 1.8-8cm, broadly elliptic to elliptic-lanceolate, glabrous, stiff and shiny. **Inflorescence** *1- (2) -flowered, clustered near the ends of leafy shoots.* **Corolla** widely funnel-shaped, with broad, spreading lobes, c 2.7cm long, tube *short c.1cm,* pink to pale mauve, spotted, scented; ovary glabrous. **Distribution** Hubei S. to Guangxi to Zhejiang and Ryukyu Is., Japan. 500-1,800m (1,600-6,000ft), open forests.

This species differs from *R. moulmainense* and *R. stamineum* in its l(-2)-flowered truss and in its short corolla tube.

This species is pretty but in a modest way compared with *R. moulmainense*. It is cultivated out of doors under *R. wilsonae* (var. *ionanthum*) at Wakehurst Place, Sussex, England, and under *R. latoucheae* and *R. amamiense* in N.W. North America indicating that at least some forms of the species are relatively hardy. Introduced 1900. Late March.

Picture 1

1. *R. latoucheae* Wilsonae Group, at Van Veen Nursery, Oregon, U.S.A.

Picture 2

Picture 3

Picture 4

2. *R. latoucheae* Amamiense Group growing in Japan showing the short tube and 2 flowers per bud. (Photo T. Takeuchi)
3. *R. latoucheae* Amamiense Group in Japan. (Photo T. Takeuchi)
4. Young foliage of *R. latoucheae* Amamiense Group at Warren Berg's garden, Hood Canal, W. U.S.A.

R. moulmainense

Hook.f. 1856 (*R. ellipticum* Maxim., *R. klossii* Ridl., *R. laoticum* Dop, *R. leiopodum* Hayata, *R. leptosanthum* Hayata, *R. leucobotrys* Ridl., *R. mackenzianum* Forrest, *R. nematocalyx* Balf.f. & W.W. Sm., *R. oxyphyllum* Franch., *R. pectinatum* Hutch., *R. siamensis* Diels, *R. stenaulum* Balf.f. & W.W. Sm., *R. tanakai* Hayata, *R. westlandii* Hemsl.) Hl-2

Picture 1

Picture 2

Picture 3

Picture 4

1. *R. moulmainense* Ellipticum Group Page 10230 from Taiwan, Royal Botanic Garden, Edinburgh.
2. *R. moulmainense* Leiopodum Group, in a pot at Glendoick.
3. *R. moulmainense* Stenaulum Group at Caerhays, S.W. England. (Photo R.B.G., E.)
4. *R. moulmainense* Westlandii Group, New Territories, Hong Kong, showing the coloured young growth common in this section.

Height to 15m, very variable, often forming a small tree in the wild, a thin to compact shrub to small tree in cultivation. Buds long and pointed. Branchlets glabrous. Bark red to tawny-purple, smooth or peeling. **Leaves** 6-17 x 2-5cm, elliptic to narrowly elliptic, +/- *glabrous*, juvenile leaves sometimes bristly on margins, adult leaves occasionally with a few bristles on margins. **Inflorescence** (1-)3-8-flowered per axillary bud with up to 22 flowers per shoot. **Corolla** funnel-shaped with narrow tube, *3.5-6cm long, broad spreading lobes 2.1-4cm long*, white, pink to violet with or without yellow or green blotch, scented; pedicel glabrous; calyx usually glabrous; stamens and style *not exserted from corolla;* ovary glabrous. **Distribution** from Arunachal Pradesh, S.E. Tibet, Burma through S. China to Taiwan, Ryukyu Is., Japan and S. Indo China to W. Malaysia, 50-3,700m (165-12,000ft), in evergreen, conifer and mixed forest.

Many former species have been made synonymous under *R. moulmainense* by Sleumer and the Philipsons but future field work and access to Chinese herbaria may well lead to some being re-instated at sub-specific or varietal status. *R. moulmainense* differs from *R. stamineum* in its larger flowers and in the non-exserted style and the stamens. It differs from *R. latoucheae* in its longer corolla tube and multi-flowered inflorescence.

We have seen plants of *R. moulmainense* in cultivation under the following names: *R. ellipticum, R. leiopodum, R. moulmainense, R. stenaulum,* and *R. westlandii*. At its best in a suitable location this is an excellent plant with good foliage, a splendid bark and large scented flowers. We have seen impressive displays, both in Arunachal Pradesh and Yunnan, showing this plant's potential for mild areas. Some introductions just survive in Cornwall and other mild areas. Plants introduced under *R. ellipticum* and *R. leiopodum* from Taiwan are probably the hardiest. It grows rather too large for general cultivation indoors. March-May.

R. stamineum

Franch. 1886 (*R. aucubaefolium* Hemsl., *R. cavaleriei* var. *chaffanjonii* Lév., *R. pittosporifolium* Hemsl.) H2-**3**

Height 3-13m, a shrub or small tree, a usually smallish shrub in cultivation, sometimes straggly. Branchlets glabrous; bark smooth. **Leaves** 6-14 x 2-4.5cm, elliptic to oblanceolate, glabrous and glossy, rather pendulous. Inflorescence 3-5(-8) flowers per bud, totalling 20-22 per shoot. **Corolla** *narrowly tubular-funnel-shaped, 2.5-3.7cm long*, tube *long and narrow, 1-1.5cm, with narrowly oblong lobes,* white to pale pink sometimes with yellow or orange blotch, often scented; pedicel glabrous; calyx glabrous; stamens and style *exserted from corolla;* ovary glabrous or slightly pilose. **Distribution** N.E. Upper Burma through Yunnan and Sichuan to Anhui, 350 -2,700m (1,200-9,000ft), usually in mixed forest.

R. stamineum differs from both *R. latoucheae* and *R. moulmainense* in its narrow corolla lobes and protruding stamens and style and from *R. latoucheae* alone, in its longer corolla tube.

The form of *R. stamineum* we grow at Glendoick is hardy but has rather insignificant flowers. Photographs in *Rhododendrons of China vol. 2.* show how spectacular this scented species can be and it would be well worth introducing superior forms for areas with warm climates. Var. *lasiocarpum* R.C. Fang and C.H. Yang is reported to differ in its densely hairy ovary. This may be the same taxon described as *R. cavaleriei* Lév. Introduced 1900. May-July.

Picture 1

Picture 2

1. *R. stamineum* at the Savill Gardens, Windsor Great Park, S. England, showing the characteristic long corolla tube and the protruding stamens and style. (Photo J. Bond)
2. *R. stamineum* from Wakehurst Place, growing in the open at Glendoick.

SUBGENUS CANDIDASTRUM

A monotypic subgenus

R. albiflorum

Hook.f. 1834 (*Azalea albiflora* Kuntze, *Azaleastrum albiflorum* Rydb., *R. warrenii* (A. Nels.) MacBride) H5

Height 0.9-2.10m, an erect shrub with ascending branches. Branchlets *with loosely adpressed brown hairs*. Bark light grey to greyish-brown after second year. **Leaves** *deciduous, thin*, scattered or clustered at ends of branchlets, 2.5-7 x 1-2.5cm, elliptic-oblong to oblong, somewhat undulate; upper surface bright green, upper surface and lower surface midrib with adpressed, brown hairs. **Inflorescence** *1-2-flowered with axillary buds scattered along the branchlets*; flowers appearing after the leaves. **Corolla** rotate campanulate, hairy inside and out, c. 2cm across, white, rarely spotted; calyx 8-13mm long *with long, brown, adpressed hairs*; ovary bristly and glandular; style straight, impressed, pilose below; capsule enclosed in the persistent calyx. **Distribution** W. North America from British Columbia to Colorado, 1,200- 2,200m (4,000-7,200ft), near tree-line, often around stunted conifers.

This subgenus contains a single, taxonomically isolated species which was previously placed in its own genus. The chief characters are the deciduous leaves, the long, whippy branches, the adpressed hairs on the branchlets, leaves and outside of the calyx, and the axillary, 1-2-flowered inflorescences, scattered along the branchlets. It is difficult to identify as a rhododendron out of flower, particularly when bereft of leaves. Some authorities recognise var. *warrenii* M.A. Lane, distinguished by its glandular-ciliate leaves.

R. albiflorum is quite pretty in flower in the wild but in cultivation it has proven to be being difficult to grow and shy-flowering. It seems to prefer a fairly sunny well-drained site in poor, stony soil. Introduced to Europe 1828. July-August in wild, June-July in cultivation.

Picture 1

Picture 2

1. A well-flowered shoot of *R. albiflorum* showing the axillary flowers, 1-2 per bud, with leaves already well-developed. We have not seen it flowering as freely as this in cultivation. Mt. Baker, Washington State, W. U.S.A.
2. *R. albiflorum* showing the long, erect branches. Mt. Baker, Washington State, W. U.S.A.

SUBGENUS HYMENANTHES

Section Ponticum

Shrubs or trees. **Leaves** evergreen, scales absent. Plant glabrous or with indumentum on young shoots, leaves etc.

Subsection Arborea

Height to 30m, large rounded bushes to tall columnar trees. Branchlets tomentose, sometimes glandular; bark rough, partly flaking. **Leaves** *thick and leathery,* elliptic to oblanceolate, retained 1-3 years; upper surface with indumentum at first, usually dark green and glabrous on maturity, lower surface *with a dense spongy to plastered indumentum.* **Inflorescence** *10-35-flowered, dense.* **Corolla** campanulate to tubular-campanulate, with nectar pouches, red, pink, white, mauve; calyx *minute;* stamens 10; ovary densely tomentose, sometimes glandular; style glabrous. **Distribution** Himalaya, S.W. China, Laos, Vietnam, Thailand, India, Sri Lanka.

The chief characteristics of the closely related species of subsection Arborea are the thick leathery leaves, the indumentum on the lower leaf surface, the densely tomentose young growth and the packed, many-flowered inflorescence.

Subsection Arborea species can make splendid garden plants, often living for over 100 years. They appreciate a moist, mild climate with temperatures rarely below -18°C. They are very showy in flower and are quite common in western British gardens.

R. arboreum, especially as subspecies *cinnamomeum* and R. niveum are common in cultivation. Less common are R. arboreum ssp. *delavayi* and R. lanigerum. R. arboreum ssp. *nilagiricum* and ssp. *zeylanicum* are rare.

R. arboreum
Sm. 1805 H2-4

Height to 30m in the wild, to 18m+ in cultivation, an upright, tall single or multi-trunked tree. Branchlets densely tomentose; bark rough, brown, tan, sometimes purplish. **Leaves** 6.5-20 x 2-5.6cm, *thick and hard,* lanceolate, oblanceolate to narrowly elliptic, retained 2-3 years; upper surface dark to olive-green, glossy or matt when mature, lower surface with a shiny and plastered to spongy and woolly, white or silvery to fawn or red brown indumentum. **Inflorescence** 15-20-flowered, compact. **Corolla** tubular-campanulate, 3-5cm long, deep crimson to scarlet, pink, white, white tinged pink or two-toned, usually spotted, sometimes on all lobes; calyx 1-3mm, hairy, glandular or glabrous; ovary usually tomentose and sometimes glandular.

With a disjunct distribution over large areas of the Indian sub-continent and the Sino-Himalayan region, *R. arboreum* is an an extremely variable species.

Picture 1

Picture 2

1. *R. arboreum* ssp. *arboreum* with Baudha Peak beyond, 3,000m, C. Nepal. (Photo A.D. Schilling)
2. *R. arboreum* ssp. *arboreum*, Gurkha Himalaya, C. Nepal (Photo A.D. Schilling)

R. arboreum has the ability to re-grow from both the base and trunk after a hard winter, fire or when part of the overhead canopy is removed. Lowland forms are usually heat resistant.

R. arboreum divides into several subspecies and varieties:

Ssp. *arboreum* (R. *puniceum* Nutt., R. *windsori* Roxb.)

Leaves 10-19 x 3-5cm; lower surface with compacted indumentum, usually silvery or white. **Corolla** bright red to carmine, rarely pink or white. **Distribution** Himalayan foothills, Kashmir to Bhutan, 1,200-2,600m (4,000-8,500ft), open or mixed forest.

This subspecies generally includes the low elevation forms from the Himalaya, often with paler leaves and blood-red flowers. This subspecies and subspecies *cinnamomeum* are characterised by the poise of the leaves which varies greatly between summer when they are upturned and winter when they droop.

The low elevation forest where ssp. *arboreum* occurs is rapidly being destroyed. While a fine plant, the type of *R. arboreum* ssp. *arboreum* is too tender for most of Britain and is better suited to warmer climates as in coastal California, New Zealand and S.E. Australia. Introduced c. 1809 or 1814, reintroduced many times 1960s->. March-May

Picture 3

Picture 4

3. A hillside in the Gurkha Himalaya, C. Nepal, showing a mixture of both red and pink-flowered forms of *R. arboreum*. (Photo A.D. Schilling).
4. A 25m high *R. arboreum* ssp. *arboreum*, Milke Danda, E. Nepal.

Picture 5

Picture 6

5. A red-flowered *R. arboreum* ssp. *arboreum* T. Spring-Smythe 26 at Glendoick.
6. *R. arboreum* ssp. *arboreum* with two-toned, spotted and blotched flowers at Glendoick.

Ssp. *cinnamomeum*
(Lindley) Tagg in Stevenson 1930.

Leaves a little smaller on average than ssp. *arboreum*, lower surface *with brown to dark rusty brown, looser* indumentum which sometimes wears off mature leaves. **Corolla** usually pink to carmine but also white or red.

This subspecies is a splendid plant in foliage and often in flower; the deepest-coloured indumentum can be very striking. In some forms, the flowers open a harsh bluish-red and fade to pink. Plants formerly named *R. campbelliae* tend to be from relatively low elevations and are correspondingly less hardy. Seed of this subspecies often produces variable seedlings and these may include var. *roseum*. Ssp. *cinnamomeum* is not to be confused with 'Sir Charles Lemon' (probably *R. arboreum* ssp. *cinnamomeum* x *R. campanulatum*) which has broader, larger leaves and a looser truss of white flowers.

Var. *cinnamomeum*
(Lindley) Tagg 1930 (*R. campbelliae*) H3-4

Leaves with a *bistrate* indumentum on the lower surface, the upper layer loose and floccose, the lower whitish to fawn and compacted. **Distribution** E. Nepal, Bengal, Sikkim 2,750-3,650m (9,000-11,500ft), open forests and rocky slopes.

Var. *roseum* Lindley 1829.

Leaves with a fawn, more compacted *unistrate* indumentum, usually paler than in var. *cinnamomeum*. **Distribution** E. Nepal, N.E. India, Bhutan, S. Tibet, from relatively high elevations, 2,400-4,000m (8,000-13,000ft), often forming pure forests.

Var. *roseum* intergrades with ssp. *arboreum* and ssp. *delavayi*. Many natural hybrids of this variety have been found, crossed with *R. campanulatum, R. wallichii, R. barbatum* etc. Confusingly, var. *roseum* now includes white forms, formerly known as var. *album*, now Album Group or forma *album*. These usually have a bistrate indumentum and a white corolla with purple spotting in the throat.

Picture 7

Picture 8

Picture 9

7. A rather 'hot-coloured' pink-flowered *R. arboreum* ssp. *cinnamomeum* Schilling 2297, grown from seed collected in Khumbu, E. Nepal.
8. The lower leaf surface of *R. arboreum* ssp. *cinnamomeum* showing the characteristic rufous indumentum.
9. The leaf lower surfaces of *R. arboreum* from the rusty-brown indumentum of ssp. *cinnamomeum* to the silvery indumentum of ssp. *arboreum*. Photographed on snow in Nepal. (Photo A.D. Schilling).

Picture 10

Picture 11

10. Roy Lancaster standing beneath a huge plant of a white-flowered *R. arboreum* ssp. *cinnamomeum* at Rossie Priory, E. Scotland.
11. *R. arboreum* ssp. *cinnamomeum* var. *roseum* 'Tony Schilling' F.C.C.

Ssp. *delavayi* (Franch.) D.F. Chamb. 1979

Height to 9m. **Leaves** 5-16 x 1.8-3cm; upper surface usually reticulate, dark and glossy, lower surface *with a white to fawn thin spongy* indumentum. **Corolla** usually deep crimson to carmine.

Ssp. *delavayi* differs from ssp. *arboreum* chiefly in its more spongy indumentum and from ssp. *nilagiricum* it its acute rather than rounded leaf apex. There is some disagreement as to how far west this subspecies occurs, particularly as to whether many Bhutanese plants are referable to ssp. *arboreum/cinnamomeum* or to ssp. *delavayi*; they appear to meet and overlap there.

Ssp. *delavayi* is generally more tender than ssp. *cinnamomeum* but hardier than ssp. *arboreum*. At its best it is a splendid plant with rich-coloured flowers and contrasting dark leaves. Introduced 1889, reintroduced 1981->.

ARBOREA

Var. *delavayi*. H3

Leaves *relatively broad*, 5-16 x 2-2.4cm, ratio 1:2.8-4, lower surface indumentum usually brown or fawn, rarely white. **Distribution** N.E. India, Burma, Thailand, Yunnan, Guizhou, 1,500-3,200m (4,950-10,500ft), in conifer and mixed forest, often quite dry, now frequently remnants in areas cleared of forest.

Var. *peramoenum*
(Balf.f & Forrest) D.F. Chamb. 1979 H2-3.

Leaves relatively *narrow*, 7.5-15 x 1.8-3cm, ratio *1:4.5-6.5*, lower surface indumentum fawn or brown. **Distribution** Arunachal Pradesh, W. Yunnan, 2,440-3,355m (8,000-11,000ft).

This variety chiefly differs from var. *delavayi* in its narrower leaves. It sometimes occurs as large stable populations and some forms have very poor, small flowers. March-May

Var. *albotomentosum* Davidian 1989

Leaves c. 8-9 x 3cm; lower surface with *white*, spongy indumentum. **Corolla** scarlet, unspotted. **Distribution** Mt. Victoria, S. Burma, 2,400-3,000m (8,000-10,000ft). Introduced 1956 under K.W. 21976.

The geographically isolated var. *albotomentosum* (recognised as a subspecies by some authorities) is a neat plant with small, compact, glowing scarlet, unspotted trusses. Quite variable in habit, from erect to spreading.

Ssp. *delavayi* aff.

Upper surface of young leaves, stems and petioles at first covered with evanescent *rich red-brown* indumentum, polished, bullate beneath, lower surface with a sparse, somewhat spongy indumentum. Flower buds *very large*, as large as those of *R. niveum* and *R. lanigerum*. **Distribution** N. Vietnam, 2,000-2,500m (6,500-8,250ft).

This is a new introduction in 1992 under K.R. 1990. Although likely to be tender, it could be a valuable introduction for heat tolerance. This may be described as a new subspecies at some time in the future.

Picture 12

Picture 13

Picture 14

Picture 15

12. *R. arboreum* ssp. *delavayi* var. *delavayi* photographed on Shillong Peak, N.E. India, 1965, showing the red flowers usual in this subspecies.
13. Girl holding *R. arboreum* ssp. *delavayi* var. *delavayi*, Cangshan, Yunnan, China, 1981 SBEC expedition.
14. *R. arboreum* ssp. *delavayi* var. *delavayi* at the Royal Botanic Garden, Edinburgh.
15. The rare *R. arboreum* ssp. *delavayi* var. *albotomentosum* at Arduaine, W. Scotland.

Ssp. *nilagiricum* Tagg 1930 H2-3

Leaves *bullate, convex*, apex +/- rounded. Lower surface with a spongy yellowish-brown indumentum. **Corolla** deep crimson, crimson-rose, occasionally pink. **Distribution** Nilgiris and other S. Indian hills, 1,800-2,250m (6,000-7,250ft).

Studies of this subspecies in the wild indicate that it is intermediate between ssp. *delavayi* and ssp. *zeylanicum*, differing from ssp. *zeylanicum* in its narrower, less rounded leaves, its earlier flowers and in its isolated distribution. According to Davidian, it also differs in its eglandular branchlets, lower leaf surface midrib, petiole, pedicel and calyx.

Very rare in cultivation. This species is remarkably resilient in the wild, growing well even when other native flora has been removed and replaced by *Acacia* and *Eucalyptus*. Introduced 1840, reintroduced 1991,95,96. It should be an excellent plant for hot and dry climates. May-June.

Picture 16

Picture 17

16. *R. arboreum* ssp. *nilagiricum* in the Palini hills, near Kodaikanal, Tamil Nadu, S. India.
17. *R. arboreum* ssp. *nilagiricum* on a cliff top in the Palini Hills, near Kodaikanal, Tamil Nadu, S. India.

Ssp. *zeylanicum* (Booth) Tagg 1930

Bark *deeply fissured*. **Leaves** *strongly bullate, convex, elliptic to ovate*, margins *recurved*. **Corolla** crimson to pink. **Distribution** Sri Lanka, 900-2,400m (3,000-8,000ft) to the tops of the mountains.

This differs from the closely related ssp. *nilagiricum* in its more rounded and more bullate leaves and its later flowers. Confined to Sri Lanka, its distribution is completely isolated from all other rhododendron species. According to Davidian, it is glandular while ssp. *nilagiricum* is eglandular.

In very mild gardens, ssp. *zeylanicum* can grow into a long-lived majestic specimen with fine foliage. Its relatively late flowers are an added bonus. Introduced 1830s, reintroduced post 1945. June-July.

Picture 18

Picture 19

18. *R. arboreum* ssp. *zeylanicum* from Sri Lanka at Arduaine, W. Scotland, clearly showing the bullate leaves.
19. Another clone of *R. arboreum* ssp. *zeylanicum* from the famous plants at Arduaine, W. Scotland.

R. lanigerum Tagg 1930 (*R. silvaticum* Cowan) H3-4

Height 2.7-6m, a rounded large bush to small tree. Branchlets tomentose at first; bark partly flaking, greenish, greyish or brownish. Flower buds *large, round*. **Leaves** 10-22 x 5-7.5cm, elliptic to oblanceolate, retained 1-2 years, upper surface creamy-white when young, often dark and shiny when mature, frequently with some traces of indumentum, lower surface with a *whitish to fawn dense fine woolly bistrate* indumentum. **Inflorescence** *20-35-flowered, large and rounded.* **Corolla** campanulate, with nectar pouches, 4-5cm long, pink, through reddish-purple, cherry-red to scarlet crimson; calyx minute, 1-2mm, fleshy; ovary densely tomentose; style glabrous. **Distribution** S. Tibet (Tsangpo Gorge, Pemako) and E. Arunachal Pradesh, 2,600-3,400m (8,500-11,000ft), amongst fir and bamboo.

This species differs from the closely related *R. niveum* in the larger flower buds, the usually longer and narrower leaves and the earlier flowers which are pink to red rather than lilac to mauve. *R. lanigerum* is a fine, often free-flowering species where it can be grown to perfection but its early swelling buds make it unsuitable for colder areas. Plants under the name Silvaticum Group from Pemako in S.E. Tibet have crimson flowers. Introduced 1924, 1928. Reintroduced 1995 (Silvaticum group) February-April

Picture 1

Picture 2

1. *R. lanigerum* on the south side of the Doshong La, S.E. Tibet
2. *R. lanigerum* Kingdon-Ward 8251, a pink-flowered form at the Royal Botanic Garden, Edinburgh, showing the rounded many-flowered truss.

Picture 3

Picture 4

3. *R. lanigerum,* a red-flowered form (formerly *R. silvaticum*), at the Younger Botanic Garden, W. Scotland.
4. *R. lanigerum,* a red-flowered form, at the Younger Botanic Garden, W. Scotland.

R. niveum Hook.f. 1851 H3-4

Height 2.5-6m, a rounded and generally compact bush or small tree, habit looser with age. Branchlets densely white tomentose; bark pale brown to greyish. **Leaves** 10-16.5 x 4-6.3cm, oblanceolate to elliptic, retained 1-3 years; upper surface dark green and largely glabrous on maturity, lower surface with a dense bistrate woolly indumentum, *white* at first, turning to *grey to fawn* on maturity. **Inflorescence** 15-20-flowered, dense and compact. **Corolla** tubular-campanulate, with nectar pouches, 4-5cm long, *deep magenta to deep lilac* with darker nectar pouches; calyx minute 1-2mm; ovary densely tomentose, style glabrous. **Distribution** Rare in the wild, Sikkim and Bhutan, 3,000-3,700m (10,000-12,000ft), bamboo, hemlock, fir and mixed forest.

The flower colour of *R. niveum* a one of the most distinctive of all rhododendron species. It differs from *R. lanigerum* in its smaller flower buds, shorter, slightly broader leaf and usually later flowers which are mauve to lilac rather than pink to red.

This is a fine species with attractive foliage and flowers of an unusual colour which is not to everyone's taste. Easily and quite widely grown, particularly in west-coast British gardens. Introduced 1859, reintroduced 1970->. April-May.

Picture 1

Picture 2

Picture 3

Picture 4

1. *R. niveum* at Glendoick, showing the typically full truss.
2. *R. niveum* a showing the nectar pouches in the corolla throat.
3. *R. niveum* with the characteristic white indumentum on the lower surface of young leaves and the fawn indumentum on a mature leaf.
4. A fine form of *R. niveum* at the R.B.G., Edinburgh.

Subsection Argyrophylla

Height to 12m, usually upright but often compact shrubs to small trees. Branchlets scurfy to tomentose; bark fairly smooth to rough and shaggy. **Leaves** narrowly elliptic to oblanceolate, retained 1-3-(5) years; upper surface glabrous, *lower surface with a plastered to thin, fine-textured, white grey or fawn* indumentum. **Inflorescence** 4-20 (-30)-flowered, usually *loose*. **Corolla** open to funnel-campanulate, nectar pouches absent (except *R. ririei*), white through pink to purple; calyx usually minute, stamens (8-) 10-15 (18-20 in *R. haofui*); ovary densely tomentose, occasionally glabrous; style usually glabrous. **Distribution** China, Tibet, Taiwan.

This is a fairly well-defined subsection, exclusively occurring in China, Tibet and Taiwan, distinguished from subsection Arborea (with which it was formerly combined) in distribution and usually in the more plastered or more finely-textured indumentum. Other characteristics of subsection Argyrophylla are the usually loose truss with slender pedicels, the absence of nectar pouches (except *R. ririei*) and the usually campanulate corolla.

Subsection Argyrophylla contains several distinctive and hardy species which make fine garden plants. Some species have a very localised distribution in the wild. The flowers are usually beautifully shaped and of delicate shades but plants may be slow to bloom and prone to leaf chlorosis. Several recently introduced species from this subsection show considerable promise as garden plants.

Only *R. argyrophyllum*, and *R. insigne* are commonly cultivated. Less common are *R. adenopodum, R. floribundum, R. formosanum, R. hunnewellianum, R. ririei, R. thayerianum*. The remainder are very rare, recently introduced or not yet in cultivation.

R. adenopodum

Franch. 1895 (*R. youngae* Fang). H5

Height 1.5-3m, a usually rounded or spreading bush. Branchlets densely tomentose. **Leaves** 9-18 x 2-5cm, oblanceolate; upper surface glabrous when mature, lower surface with a grey to fawn, bistrate, felted indumentum. **Inflorescence** 6-8-flowered, loose. **Corolla** funnel-campanulate, *shallowly-lobed*, 4-5cm long, pale rose, usually spotted crimson; calyx usually 3-6mm, glandular-ciliate; ovary *densely fulvous-glandular*, style glabrous. **Distribution** E. Sichuan and Hubei, 1,500-2,200m (5,000-7,250ft), in thin woodland.

This species looks rather like a narrow-leaved form of *R. degronianum* but differs in its glandular ovary, pedicel and capsule and in its more shallowly-lobed corolla. It is closely related to the much less hardy *R. simiarum*.

Chamberlain rightly transferred *R. adenopodum* back to subsection Argyrophylla from the old Ponticum series. Although it seems to lie taxonomically between the two subsections, it is better placed here both morphologically and geographically. A useful species for severer climates such as E. North America and Scandinavia, it is hardy, free-flowering but rather slow-growing and inclined to have chlorotic foliage. Introduced 1900-1. April-May.

Picture 1

Picture 2

1. *R. adenopodum* showing the characteristic spreading habit, narrow leaves and loose trusses.
2. *R. adenopodum*, at the Royal Botanic Garden, Edinburgh.

Picture 3

3. *R. adenopodum* in young growth showing the grey felted indumentum.

R. argyrophyllum Franch. 1886 H4-5

Height 2-6(12)m, a rounded to erect shrub or small tree. Branchlets tomentose at first, bark grey to greenish-brown, roughish. **Leaves** 6-16 x 1.5-6cm, elliptic to oblanceolate, retained 2-3 years; upper surface glabrous, lower surface with a *thin fine-textured white, silvery to fawn, compacted* indumentum. **Inflorescence** 4-12-flowered. **Corolla** usually funnel-campanulate, 2-4cm long, white to deep rose, with or without spots; calyx 1-2mm floccose; stamens usually 11-15; ovary tomentose, glandular or eglandular, style glabrous.

The white to silvery (fawn in ssp. *omeiense*) indumentum is the chief characteristic which separates this from its allies. *R. coryanum* looks very similar in leaf but has 20-30 flowers to the truss and a +/- glabrous ovary. *R. formosanum* from Taiwan is very similar to ssp. *omeiense*.

Ssp. *argyrophyllum* (*R. chionophyllum* Diels, *R. argyrophyllum* var. *cupulare* Rehder & E.H. Wilson)

Leaves 6-9cm long. **Corolla** *funnel-campanulate*, 3-3.5cm long; ovary and pedicels eglandular. **Distribution** Sichuan, N.E. Yunnan and S. Shaanxi, 2,000-3,600m (6,500-12,750ft), forests and open slopes.

This variable subspecies is an attractive plant when growing well, but is liable to be chlorotic in cultivation and the pretty flowers may take many years to appear. Introduced 1904-1938, reintroduced 1980->. May.

Ssp. *hypoglaucum* (Hemsl.) D.F. Chamb. 1979

Leaves often more intensely glaucous below. Pedicel and ovary *glandular*; style sometimes glandular at the base. **Distribution** E. Sichuan, W. Hubei, 1,500-2,700m (5,000-9,000ft), thickets and woods.

Introduced 1900-07. May.

Ssp. *omeiense*
(Rehder & E.H. Wilson) D.F. Chamb. 1979

Leaves generally *smaller and narrower* than those of the other subspecies, with *fawn* indumentum. Ovary *eglandular*. **Distribution** C. Sichuan, Emei Shan & elsewhere, 1,800m (6,000ft) in forest along ravines.

Ssp. *omeiense* is not as showy as the best ssp. *argyrophyllum*. It tends to have chlorotic leaves. This subspecies differs little from ssp. *argyrophyllum* and would probably be better treated as a variety of it. Introduced 1980.

Ssp. *nankingense* (Cowan) D.F. Chamb. 1979
(*R. argyrophyllum* var. *leiandrum* Hutch.)

Leaves *11-16cm long*; upper surface slightly *rugulose*, dark green. **Corolla** larger than type, more than 5cm across, rich clear pink to lilac purple. **Distribution** Guizhou, Fan-jin-shan, 2,250m (7,500ft), rocky slopes.

This is often a fine, hardy plant with excellent flowers, easier to grow than some forms of ssp. *argyrophyllum* and less prone to chlorosis. 'Chinese Silver' A.M. with clear pink flowers is the best-known clone. Introduced 1932, reintroduced 1985.

Picture 1

Picture 2

Picture 3

Picture 4

Picture 5

Picture 6

1. *R. argyrophyllum*, an impressive pink-flowered form.
2. A fine white-flowered form of *R. argyrophyllum* at Van Dusen Botanic Garden, Vancouver, Canada.
3. *R. argyrophyllum* ssp. *hypoglaucum* 'Heane Wood' A.M. at Glendoick.
4. The rare *R. argyrophyllum* ssp. *omeiense* Lancaster 530 from Emei Shan, C. Sichuan, at Glendoick, showing the characteristic small, narrow leaves.
5. *R. argyrophyllum* ssp. *nankingense* 'Chinese Silver' A.M. at Glendoick.
6. *R. argyrophyllum* ssp. *nankingense* 'Chinese Silver' A.M. showing the slightly rugulose, dark green leaves.

ARGYROPHYLLA

R. coryanum Tagg & Forrest 1926 H4

Height 3-6m, a usually broadly upright shrub or small tree. Branchlets thin, grey, tomentose and glandular; bark rough. **Leaves** 7-17 x 1.8-4cm, narrowly elliptic to oblanceolate, retained 2-3 years; upper surface dark green and glabrous, lower surface with a thin, compacted, buff indumentum becoming ash-grey, sparsely or moderately glandular. **Inflorescence** *15-30-flowered*. **Corolla** funnel-campanulate, 2.5-3cm long, whitish with red spots; calyx 1-2mm, sometimes glandular-ciliate; ovary *glabrous* or with a few hairs, style glabrous. **Distribution** N.W. Yunnan-S.E. Tibet border, (2,300-) 3,700-4,300m ((7,500-) 12,000-14,000ft), thickets and forest.

Picture 1

1. *R. coryanum* at the Royal Botanic Garden, Edinburgh, showing the many-flowered truss. (Photo H. Gunn).

The many-flowered truss and the usually glabrous ovary makes this an easy species to identify when in flower; out of flower it is hard to separate from *R. argyrophyllum*.

R. coryanum appears to be very rare in cultivation; most plants labelled *R. coryanum* in gardens turn out to be *R. argyrophyllum* or *R. uvarifolium*. Introduced 1921-1923. April-May.

R. denudatum Lév. 1914 H4?

Height 2-3m, an erect shrub or small tree. Branchlets densely floccose; bark slightly rough. **Leaves**, 12.5-20 x 4-7cm, convex, lanceolate to elliptic (?), upper surface olive-green to deep green, *bullate*, lower surface with a bistrate indumentum, upper layer *yellowish, through light brown to cinnamon*, lower layer whitish. **Inflorescence** 8-10-flowered. **Corolla** light pink through rose to wine-red, spotted or blotched, campanulate, c. 4cm long; calyx c.1mm, tomentose; ovary densely whitish-tomentose; style glabrous. **Distribution** Central & S. Sichuan, N.E. Yunnan, N.W. Guizhou, 2,200-3,350m (7,250-11,000ft).

This species is related to *R. floribundum* but differs in its shiny rather than matt upper leaf surface, its deeper coloured indumentum and its often larger, wider leaves. *R. denudatum* is obviously also related to *R. coloneuron*, which is presently but perhaps wrongly classified as a member of subsection Taliensia. *R. denudatum* can be considered intermediate between subsections Argyrophylla and Taliensia.

As seen in the wild (in autumn), *R. denudatum* is a fine foliage plant and the photographs in *Rhododendrons of China Vol. 2* show it as being very floriferous so this should be a valuable new introduction. Introduced 1995.

Picture 1

1. *R. denudatum* photographed in China, showing the typical yellowish-brown indumentum. (Photo Yang Zhenhong)

R. floribundum Franch. 1886 H4

Height 2-7.6m, an erect shrub or small tree. Branchlets *densely white* floccose, bark slightly rough. **Leaves** 7-18 x 2-5.5cm, convex, oblanceolate to elliptic, retained 1-3 years; *upper surface olive-green, bullate*, lower surface with a *white to greyish, woolly* indumentum. **Inflorescence** 5-12-flowered, lax. **Corolla** widely campanulate, c.4cm long, pink to purplish-lavender, *with a large blotch and spots*; calyx 1mm tomentose; ovary densely white tomentose; style usually glabrous. **Distribution** Central & S. Sichuan, 1,200-2,600m (4,000-8,500ft), in mixed forest, now often partially destroyed.

This species is related to *R. denudatum*, differing in its narrower leaves, matt upper leaf surface and paler indumentum. It is distinct from other species in subsection Argyrophylla in its convex, bullate, olive-green leaves with whitish indumentum on the lower surface and in its blotched and spotted, usually purplish flowers.

While *R. floribundum* is attractive when well grown with showy flowers of an unusual colour, it often has chlorotic foliage in cultivation. This condition may be improved by treating with dolomitic lime or gypsum. Probably introduced 1910, reintroduced 1988->. April.

ARGYROPHYLLA

Picture 1

Picture 2

1. *R. floribundum* 'Swinhoe' A.M. a clone from Exbury photographed at the Valley Gardens, Windsor, S. England.
2. A mature bush of *R. floribundum* at the Valley Gardens, Windsor, S. England.

Picture 2

2. A pure white-flowered form of *R. formosanum*.

This species very closely resembles *R. argyrophyllum* ssp. *omeiense* and should perhaps be treated as a subspecies of *R. argyrophyllum*. Davidian states that *R. formosanum* generally has more flowers to the truss.

R. formosanum is free-flowering from a young age but in cultivation, it is not living up to its Taiwanese reputation of being their most beautiful native species and some introductions have proven to be tender. It remains rare in cultivation. Introduced 1969. April-May.

R. formosanum Hemsl. 1895 H3-4

Height 2-5.5m, an erect and rather sparse shrub or small tree. Branchlets sparsely grey tomentose, **Leaves** 7-18 x 1.5-5cm, *narrowly oblanceolate to lanceolate, recurved*; lower surface with a pale buff, compacted indumentum. **Inflorescence** 7-20-flowered. **Corolla** widely funnel-shaped, 3-4cm long, white to pink, spotted; calyx minute, tomentose; ovary densely rufous-tomentose; style glabrous. **Distribution** Taiwan only, 800-2,000m (3,000-6,500ft), in mixed broad-leaved forest.

Picture 1

1. *R. formosanum*, a form with heavily-spotted flowers at Guavis, N. Island, New Zealand.

R. haofui Chun & Fang 1957. H4?

Height 4-12m, a shrub. **Leaves** coriaceous, lanceolate to oblanceolate, *7-22 x 3-7cm*, apex acuminate, upper surface glabrous, lower surface with a fulvous floccose-pannose tomentum; petioles glabrous. **Inflorescence** 5-9 flowered. **Corolla** broadly campanulate, white, sometimes flushed with rose, 4-4.5cm long; calyx minute with long soft hairs; stamens *18-20*, puberulous in lower third; ovary with a dense whitish to pale brown lanate tomentum; style glabrous. **Distribution** Guizhou, Guangxi, Hunan, c. 1,000-1,500m (3,250-5,000ft).

Picture 1

1. *R. haofui* Guiz 75 at Glendoick showing the chlorotic foliage typical of this species in cultivation.

Picture 2

2. A close-up of a truss of *R. haofui* Guiz 75 showing the large number (18-20) of stamens which are the distinguishing feature of this species.

Other species in subsection Argyrophylla usually have 10-15 stamens so *R. haofui* can be readily distinguished in flower by its 18-20 stamens. As cultivated it is characterised by its long hanging leaves and small rounded flower buds. Its closest relative is probably the widespread and variable *R. simiarum*.

This species was introduced in 1985 from Fan Jin Shan, Guizhou, as *R. ririei* Aff. Guiz 75. It has now flowered in cultivation and has been identified as *R. haofui*. So far it is proven rather hard to please, with yellowish foliage and a tendency for its leaves to hang dejectedly. It may prefer a near neutral soil to a very acid one. May-June.

Picture 1

Picture 2

1. A fine form of *hunnewellianum* Wilson 1199 at the Valley Gardens, Windsor, S. England, exhibiting the characteristic narrow hanging leaves and pointed growth buds.
2. *R. hunnewellianum* showing the typically spotted flowers.

R. hunnewellianum makes an interesting foliage plant but the flowers are sometimes rather hidden in the leaves. Introduced 1908-10. February-April.

R. hunnewellianum
Rehder & E.H. Wilson 1913 (*R. leucolasium* Diels.) H4-5

Height 2-7.5m, a usually rounded shrub or small tree. Branchlets whitish tomentose and glandular. **Leaves** 7.6-18 x 1.5-2.5cm, *narrowly oblanceolate*, retained 2-3 years; upper surface glabrous, lower surface with a *bistrate* indumentum, upper layer loose and whitish. **Inflorescence** 4-8-flowered, loose, rachis *very short, 0.6-1.5cm*. **Corolla** widely campanulate, 4-5cm long, white tinged pink to pink, spotted; calyx minute, 1mm, glandular-ciliate; ovary densely tomentose, whitish to fawn; style glabrous or sparsely glandular at the base. **Distribution** N. Central Sichuan, 1,800-3,000m (6,000-10,000ft), woods and thickets.

This species is easily identified by its narrow hanging leaves, pointed at both ends, and its flattened truss with a very short rachis. Ssp. *rockii* which is not in cultivation differs in its yellowish rather than whitish indumentum.

R. insigne Hemsl. & E.H. Wilson 1910 H5

Height 1.5-3.7m, a compact shrub in open situations, leggy in shade. Branchlets thin tomentose; bark grey with age. **Leaves** 7-14 x 1.6-5cm, *thick and rigid,* retained 3-5 years; upper surface *dark green and shiny,* lower surface *with a shiny, plastered* tawny-grey to copper indumentum. **Inflorescence** 6-15-flowered, loose to compact. **Corolla** widely campanulate, to 5 cm long, pale to deep pink, usually with a darker line on each lobe; calyx minute, floccose; ovary densely hairy; style glabrous, capsules ripening *very early* and opening widely, quickly shedding all the seed. **Distribution** apparently only grows on Wa Shan (where it may now be extinct), Central Sichuan, 2,100-3,000m (7,000-10,000ft), in woodlands though it may have been found recently in N.E. Yunnan and S. Sichuan.

With its stiff leaves, shiny on the lower surface, *R. insigne* is very distinctive. *R. thayerianum* differs in its persistent perulae, white, later flowers and the non-shiny indumentum. The leaves of *R. argyrophyllum* ssp. *nankingense* differ in the lower surface being white to grey and matt rather than shiny.

Picture 1

Picture 2

Picture 3

1. A free-flowering *R. insigne* at Glendoick, showing the typical bushy habit of a plant grown in plenty light.
2. A fine pink-flowered form of *R. insigne* showing the shiny plastered indumentum on the leaf lower surface.
3. *R. insigne* showing the typical deeper lines on each corolla lobe and also the good leaf retention characteristic of this species.

R. insigne is horticulturally the best of the subsection with excellent foliage and generally fine flowers which open relatively late in the season. It is quite easy to grow except as a small seedling when it is very sensitive to fertiliser. Young growth rarely appears on flowering shoots until the subsequent year and dead-heading is advisable. Introduced 1908. May-June.

R. longipes Rehder & E.H. Wilson 1913. H4-5?

Height 1-4m, a bushy to erect shrub. Branchlets floccose when young. **Leaves** 5-13 x 1.5-3.5cm, lanceolate to oblanceolate, *apex long, acuminate to cuspidate*, retained 2-3 years; upper surface glabrous, lower surface *with thin pale brown* indumentum. **Inflorescence** *8-15-flowered*, loose. **Corolla** funnel-campanulate, 2.5-4cm long, pale pink to rose and purple, sometimes spotted; calyx 1-2mm; ovary tomentose and glandular; style glabrous.

R. longipes differs from *R. argyrophyllum* in its long, acuminate leaf apex, in the brown indumentum on the lower leaf surface and in the 8-15-flowered truss.

Photographs taken by the Chinese in the wild indicate this recently introduced species to be showy and free-flowering. The species has been divided into 2 varieties which probably differ very little in garden merit:

Var. *chienianum* (W.P. Fang) D.F. Chamb. 1979

Leaves with thicker, more spongy, felted indumentum. **Distribution** N.E. Yunnan, S.E. Sichuan, Guizhou, 1,300-2,000m (4,500-6,500ft), broad-leaved forest and cliffs. Introduced 1995->?

Var. *longipes*

Leaves with a thin compacted indumentum. **Distribution** S.W. Sichuan, 2,000-2,500m (6,500-8,250ft), broad-leaved forest. Introduced 1995->

Picture 1

Picture 2

1. *R. longipes* var. *chienianum* in N.E. Yunnan. Photo Yang Zhenhong.
2. *R. longipes* var. *longipes* near Leibo, S. Sichuan, showing the oblanceolate leaves with cuspidate apices and pale brown indumentum on the lower surface.

R. pingianum W.P. Fang 1939. H5

Height 4-8m, rounded shrub or small tree. Branchlets white tomentose. **Leaves** 8-13.5 x 3-4.2cm, retained 1-3 years: upper surface glabrous, lower surface with white compacted indumentum. **Inflorescence** 8-20-flowered, fairly loose. **Corolla** funnel-campanulate, 2.8-3.5cm long, *pinkish to pale purple*; calyx 1-2mm floccose; ovary *densely rufous-tomentose*, eglandular, style glabrous. **Distribution** C. Sichuan, 2,000-2,700m (6,500-9,000ft), forested mountain slopes.

Closely related to *R. argyrophyllum* and doubtfully deserving separate specific status; apparently differing in its broader leaves, more intensely coloured flowers and its rufous-tomentose eglandular ovary. Recent introductions (1981->) from Emei Shan are free-flowering with fine pink flowers but these plants lack the densely rufous-tomentose ovary and it is uncertain whether they should be classified under *R. pingianum* or not. May.

Picture 1

1. *R. pingianum* KR 150 from Emei Shan. Plants grown from this number lack the densely tomentose ovary, but otherwise seem to match the type description.

R. ririei Hemsl. & E.H. Wilson 1910 H4-5

Height 3.5-13m, an erect but bushy shrub or small tree. Branchlets covered with thin whitish tomentum; bark roughish, brownish-black. **Leaves** 7-16 x 2.7-5cm, retained 2-3 years; upper surface glabrous, lower surface with *a thin compacted silvery-white* indumentum. Growth buds *long, thin and pointed*. **Inflorescence** 4-10-flowered, loose. **Corolla** with 5-7 lobes, campanulate, with *glistening deep violet nectar pouches*, 4-5cm long, *lilac-purple to reddish-purple;* calyx length very variable, sometimes coloured as the corolla and reflexed; ovary whitish-tomentose; style glabrous. **Distribution** S. Central Sichuan, 1,200-2,200m (4,000-7,250ft), forests or thickets.

This distinctive species is characterised by the very thin indumentum in the leaf lower surface, the pointed growth buds, the loose purplish trusses with prominent nectar pouches and the variable calyx. It is rarely confused with its relatives.

R. ririei has fine large but very early flowers which last well if there is no frost; it is often the earliest elepidote species to bloom. Growth also comes early, often getting frosted, causing slight dieback. Otherwise easy to please. Introduced 1904. January-March

Picture 1

Picture 2

Picture 3

Picture 4

1. A free-flowering form of *R. ririei* Tigh Na Rudh form growing at Glendoick.
2. *R. ririei*, Tigh Na Rudh form; *R. ririei* is one of the earliest species to flower.
3. A fine form of *R. ririei* at the Royal Botanic Garden, Edinburgh.
4. Young growth of *R. ririei* at the Royal Botanic Garden, Edinburgh.

R. simiarum
Hance 1884 (*R. fokienense* Franch., *R. fordii* Hemsl.) H2-3

Height 1-6m, a shrub or small tree of variable habit. Branchlets floccose or/and glandular; bark grey. **Leaves** 4-15 x 2-4.5cm, *thick and rigid,* narrowly elliptic to broadly oblanceolate, retained 1-2 years; upper surface usually glabrous, lower surface with a thin, compacted, white to cork-brown indumentum. **Inflorescence** 4-12-flowered. **Corolla** funnel-campanulate, c. 4cm long, white to light rose, spotted; calyx 1-2mm, floccose or glandular; ovary rust to fawn tomentose, sometimes glandular; style glabrous, sometimes hairy or glandular at base. **Distribution** China, Hainan northwards to Guangxi, north-eastwards to Zhejiang, 600-1,650m (2,000-5,400ft), forests, thickets and rocky slopes.

A very variable species, with a widespread distribution. The leaves are characterised by their thickness and rigidity with a thin indumentum on the lower surface.

R. simiarum comes from the lowlands of S. China, so is a plant for warmer areas than Britain, though plants have survived for many years in the south. It has proven useful for hybridising for heat tolerance. The Hong Kong form is exceedingly slow-growing in Scotland but may require more heat than we can offer. Older introductions from mainland China have been grown as *R. fordii*. Introduced 1894. Reintroduced 1971 and 1981-> from Hongkong. April-May.

Picture 1  *Picture 2*

1. Peter Cox looking at *R. simiarum* on Ma-on-shan, Hong Kong New Territories. (Photo Roy Lancaster).
2. Foliage of *R. simiarum* in Warren Berg's garden, Hood Canal, Washington State, U.S.A.

R. thayerianum
Rehder & E.H. Wilson 1913 H5

Height 2-4m, a rigid shrub with stout branches. Branchlets floccose and glandular, *with crowded foliage, persistent perulae,* bark shaggy, rough flaking brown. **Leaves** 7-15 x 1.5-3cm, *thick and stiff, margin recurved,* narrowly oblanceolate, retained 2-6 years; upper surface glabrous but often mottled, lower surface with a compacted fawn indumentum. **Inflorescence** 10-20-flowered, fairly compact to loose. **Corolla** funnel-campanulate, 2.5-3cm long, white or white flushed pink, no markings or a few spots; calyx 2-5mm, glandular; ovary densely glandular, style with *whitish glands to tip.* **Distribution** Central Sichuan, 2,700-3,000m (9,000-10,000ft), woodlands.

This is a very distinct species with its glandular branchlets, stiff crowded foliage, persistent perulae and glandular style. In this subsection, only *R. insigne* and *R. simiarum* have such thick, stiff leaves.

R. thayerianum is an easy, slow-growing species, well worth growing for its unusual foliage and late, though sometimes small, white flowers. Introduced 1910, reintroduced 1991. June-July.

Picture 1

Picture 2

1. A form of *R. thayerianum* from Caerhays in Cornwall growing at Glendoick.
2. *R. thayerianum* showing the characteristic crowded, stiff leaves.

3. Young growth of *R. thayerianum* showing the persistent perulae.

Subsection Auriculata

A monotypic subsection.

R. auriculatum Hemsl. 1899 H4-5

Height 1.8-11m, a large shrub or tree, often umbrella-shaped. Buds *long, conical, outer scales with long perulae*. Branchlets usually *densely glandular-hairy;* bark roughish, greenish-brown. **Leaves** 9.5-32 x 2.8-12cm, oblong to oblong-lanceolate, base *auriculate*, retained 1 or more years; upper surface glabrous or with remains of hairs and glands, lower surface *hairy and glandular,* especially on midrib, veins and on margins. **Inflorescence** 6-15-flowered, fairly loose. **Corolla** funnel-shaped, 6-10cm long, white or occasionally to rose-pink, with greenish blotch, *fragrant;* calyx small, 1-3mm, slightly glandular; stamens *14;* ovary densely glandular; style glandular to tip. **Distribution** E. Sichuan, W. Hubei and E. Guizhou, 500-2,300m (1,600-7,500ft), in forest.

A distinctive species, identified by its long vegetative buds, auricled leaves, general stickiness of branchlets, leaves and flowers, and by its very late flowers and growth. Its closest relations, also with white, fragrant flowers, are in subsection Fortunea.

R. auriculatum is a very fine species at its best but despite its hardiness, it is not suitable for climates where first frosts come early, as the late growth remains soft well into the autumn. Best planted in a situation that will not dry out in late summer when making its growth. Although quite heat-tolerant it does best in some shade and shelter to protect flowers and new growth. Introduced 1901-8, reintroduced 1994->. July-August.

1. *R. auriculatum* at the Royal Botanic Garden Edinburgh, flowering in August. Note that the new growth is only now beginning to elongate. Such late growth is vulnerable to early autumn frosts.
2. *R. auriculatum* at Wakehurst Place, S. England. (Photo A.D. Schilling).
3. Foliage of *R. auriculatum* showing the long, tailed buds and the auriculate bases of the leaves.

Subsection Barbata

Height 1.20-9m, rounded to upright bushes or trees. Branchlets usually *bristly;* bark *peeling, plum, purple to reddish.* **Leaves** dark green, usually *convex,* retained 1-2 years; upper surface glabrous on maturity, often rugulose, lower surface glabrous or with woolly indumentum. **Inflorescence** 10-20-flowered, usually *dense.* **Corolla** tubular-campanulate, with nectar pouches, *scarlet to crimson,* occasionally rose; calyx small to large, often coloured; ovary glabrous, tomentose to glandular; style glabrous. **Distribution** All the species in this subsection come from the Himalaya region from W. Nepal to Arunachal Pradesh and S. Tibet.

Among botanists, there is some disagreement as to which species should be included in this subsection. Davidian places *R. fulgens* and *R. succothii* in their own series. In the Edinburgh revision, Chamberlain placed *R. fulgens* along with *R. sherriffii* and *R. miniatum* in their own subsection but has now placed *R. sherriffii* in subsection Thomsonia. *R. succothii* is placed in subsection Barbata while *R. fulgens* remains in subsection Fulgensia although in our opinion it would be better placed in subsection Barbata.

Species in this subsection make handsome plants with their fine trunks and showy red flowers in early-mid spring. They are particularly good in western Britain. New growth tends to come early so is subject to frost damage.

By far the commonest in cultivation is *R. barbatum. R. argipeplum, R. erosum* and *R. succothii* are fairly widely grown, while *R. exasperatum* remains quite rare.

R. argipeplum Balf.f. & Cooper 1916 (*R. smithii*, Hook.f. *R. macrosmithii* Davidian) H3-4

Height 2-7.5m, a generally upright shrub or small tree. Branchlets bristly or glandular-bristly; bark *reddish to brown, flaking and smooth.* **Leaves** 8-16 x 2.8-6cm, oblong-lanceolate to elliptic-lanceolate, base sometimes cordulate, *convex, rugulose,* retained 1-2 years; upper surface glabrous, lower surface *with a thin, continuous, loose* indumentum, *white at first, turning to buff to rufous or greyish-white.* **Inflorescence** 10-19-flowered, compact. **Corolla** tubular-campanulate, 2.8-4.5cm long, scarlet to crimson; calyx *variable,* 2-10mm, often fleshy, sometimes glandular; ovary dense rufous tomentose and glandular; style glabrous. **Distribution** Sikkim to Bhutan, 2,400-3,700m (8,000-12,000ft), fir and rhododendron forest.

This species has recently (1996) been redefined and now only includes plants formerly known as *R. smithii* Nutt. (*R. macrosmithii* Davidian). All plants with rounded leaves, until recently grown under the name *R. argipeplum,* are now included under *R. erosum. R. argipeplum* differs from the closely related *R. barbatum* in the presence of indumentum on the leaf lower surface. It differs from *R. erosum* in its smaller, narrower, more pointed leaves, lacking a pronounced cordate leaf base, and it usually has a larger, fuller inflorescence of larger flowers.

A fine early-flowering species differing little in garden value or appearance from *R. barbatum* apart from the presence of indumentum. Introduced pre-1859, reintroduced 1980->, mostly from Bhutan. February-April.

Picture 1

Picture 2

1. *R. argipeplum* (formerly *R. smithii* and *R. macrosmithii*), showing the typical non-cordulate leaf base and the large, full truss.
2. Young growth of *R. argipeplum* Cave 6714, at the Royal Botanic Garden, Edinburgh, showing the bristly stems and petioles.

BARBATA

R. barbatum Wallich 1849 (*R. imberbe* Hutch., *R. lancifolium* Hook.f.) H4

Height 2-9m, a generally upright small tree, usually branched from the base. Branchlets *bristly*, sometimes glandular, rarely glabrous; bark peeling, reddish-plum to deep purple, buds sticky, red bud scales. **Leaves** 10-20 x 2.5-7cm, slightly convex, *elliptic to elliptic-lanceolate,* retained 1-2 years; upper surface sea to deep green, glabrous, rugulose, lower surface usually *glabrous*; petiole usually with *bristles*. **Inflorescence** 10-20-flowered, compact. **Corolla** tubular-campanulate, to 5cm long, light scarlet to crimson-scarlet; calyx medium to large, 10-15mm, usually glabrous; ovary densely glandular; style glabrous. **Distribution** Uttar Pradesh (India), Nepal, Sikkim, Bhutan, S. Tibet and W. Arunachal Pradesh, 2,400-3,700m (8,000-12,000ft), mixed forests and thickets.

This species is closely related to *R. argipeplum*, differing in the absence of indumentum on the leaf lower surface. *R. barbatum* is variable in the density and length of its bristles and in some cases these are completely absent (formerly known as *R. imberbe*). As plants with and without bristles sometimes grow together in the wild, the presence or absence of bristles cannot be considered to be of taxonomic significance. Plants in cultivation under the name *R. shepherdii* Nutt. do not agree with the type description and are just forms of *R. barbatum*.

R. barbatum is a fine, easily-grown species, provided it can be protected from wind. Particularly fine are forms with deeper-coloured flowers and those with large leaves. The usually showy bark is an added bonus. Introduced 1829?, 1841 and often since. February-April.

Picture 1

Picture 3

Picture 5

Picture 4

Picture 2

1. *R. barbatum* at the Royal Botanic Garden, Edinburgh.
2. A truss of an excellent form of *R. barbatum* collected by Joseph Hooker during the 1850's.
3. *R. barbatum* at the Royal Botanic Garden, Edinburgh.
4. *R. barbatum* showing the bristles on the petioles which are typical of most forms of this species.
5. The trunk of *R. barbatum* at Baravalla, W. Scotland.

R. erosum Cowan 1937 H3-4

Height 2.5-9m, a fairly compact shrub or small tree. Branchlets with glandular bristles; bark brown, peeling. **Leaves** 8-20 x 3-10cm, convex, *obovate to oval*; base *cordulate*, upper surface glabrous, lower surface with a *sparse to thin continuous pale* *brown floccose and glandular* indumentum. **Inflorescence** 10-15-flowered. **Corolla** tubular-campanulate, with nectar pouches, 2.8-3.5cm long, rose-pink to crimson; calyx medium 2-5mm, coloured, usually glabrous; ovary densely glandular; style glabrous. **Distribution** E. Bhutan to S. Tibet, 3,000-4,000m (10,000-13,000ft), rhododendron, birch and fir forest.

This species is related to *R. argipeplum*, differing in its larger, wider leaves with a pronounced cordulate base and in its usually smaller flowers in less rounded trusses. *R. erosum* has recently been redefined (1996) to include the eastern distribution of what was formerly included under *R. argipeplum*.

R. erosum also resembles *R. exasperatum*, differing in its less bristly branchlets, deciduous perulae and in the presence of indumentum rather than bristles on the lower leaf surface.

A fine foliage plant, especially in young growth which is a copper to plum colour. In most forms, the flowers are rather small in comparison with the size of the leaves. Introduced 1915. March-April.

Picture 1

Picture 2

Picture 3

1. *R. erosum* showing the small, open-topped truss in contrast to the large, wide leaves.
2. *R. erosum* showing the cordulate leaf bases.
3. *R. erosum* showing the attractive young growth.

R. exasperatum Tagg 1931 H3-4

Height 1-4.5m, a compact to open shrub or small tree. Branchlets *with dense glandular bristles* and *persistent shaggy perulae*; bark peeling, reddish-brown. **Leaves** 9-17 x 4.5-10cm, *convex*, broadly obovate to elliptic, retained 1-2 years; upper surface glabrous, *verdigris-green to beetroot when young*, lower surface with scattered bristles, larger bristles on the midrib; petiole *bristly*. **Inflorescence** 10-15-flowered, very compact. **Corolla** tubular-campanulate, with nectar pouches, 3.5-4cm long, brick-red or pink; calyx medium, 3-5mm, coloured; ovary densely glandular; style glabrous. **Distribution** S.E. Tibet, E. Arunachal Pradesh (India), N. Burma, 2,900-3,700m (9,500-12,000ft), on rocks, in thickets, fir forest and on steep slopes.

A distinct species characterised by its large, wide, stiff leaves, glandular-bristly on the lower surface. The persistent perulae and more colourful young growth distinguish it from *R. erosum*.

R. exasperatum is a fine foliage plant, particularly in young growth which unfortunately comes early and is therefore prone to frost damage. Best in a sheltered site, and well worth extra care. The flowers are often small in contrast to the leaves. Forms grown from the original introduction (possibly a single clone) are red but some plants seen recently in the wild had pinkish flowers. Introduced 1926-28, Reintroduced 1996. February-May.

Picture 1

Picture 2

1. *R. exasperatum* Kingdon-Ward 8250 from Arunachal Pradesh, N. India collected in 1928.
2. *R. exasperatum* Kingdon-Ward 8250 at Glendoick.

BARBATA

Picture 3

Picture 4

Picture 5

3. *R. exasperatum* on the south side of the Doshong La, S.E. Tibet, where it always grows on large rocks and boulders.
4. *R. exasperatum* with the characteristic brilliant-coloured young foliage.
5. *R. exasperatum* showing bristles on the midrib as well as on the stem and petiole and also the characteristic persistent perulae.

R. succothii Davidian 1966 (*R. nishiokae* Hara) H4

Height 1-4.6m, a broadly upright to rounded and compact shrub or small tree; bark attractive, *red-brown or purplish grey, peeling*. **Leaves** 5-13 x 2.5-6.5cm, oblong-obovate to elliptic, apex rounded, base cordate, *held in whorls at ends of branches*, often *upturned* in a distinctive manner, retained 1-2 years; upper surface glossy and dark green, lower surface glabrous or rarely hairy; petiole *very short*. **Inflorescence** 10-15-flowered, *compact and rounded*. **Corolla** tubular-campanulate, with nectar pouches at the base, 2.3-3.5cm long, crimson or scarlet; calyx minute, glabrous; ovary and style glabrous. **Distribution** Bhutan, W. Arunachal Pradesh (India), 3,400-4,100m (11,000-13,500ft) amongst rhododendrons and other shrubs, in forest and at forest margins.

This species differs from the other species in the subsection in the combination of its distinctively upturned leaves, the very short petioles, the absence of bristles and the absence of indumentum on the leaf underside. It resembles *R. fulgens* which has a thick woolly indumentum and a longer petiole.

R. succothii was long grown as *R. fulgens* affinity before gaining specific status. It is a distinctive species, recently moved into this subsection on account of its smooth bark and tight truss of red flowers. It is prone to powdery mildew infection and tends to be rather shy-flowering. Introduced 1914-49 and recently reintroduced. February-April.

Picture 1

Picture 2

Picture 3

1. *R. succothii*, Thrumseng La, C.W. Bhutan. (Photo I. Sinclair).
2. *R. succothii*, Glendoick.
3. *R. succothii* foliage showing the characteristic very short petioles, Glendoick.

Picture 4
4. Trunk of *R. succothii* showing the red-brown and grey bark, Glenarn, W. Scotland.

Subsection Campanulata

Height 0.30-11m, compact shrubs to small trees. Branchlets usually glabrous, sometimes floccose, bark grey, fawn to pinkish-brown, smooth to rough with age. **Leaves** 5-18 x 2-9cm, elliptic, obovate, oval to oblong-elliptic; upper surface shiny to glaucous, lower surface with a dense, fawn rusty brown indumentum, or a loose or scattered indumentum or almost glabrous. **Inflorescence** 5-18-flowered, usually compact. **Corolla** white, cream, pink, purplish-red, rosy-purple, blue-lavender, with +/- spotting; calyx 1-3mm, sparsely tomentose to glabrous; stamens 10; ovary and style +/- glabrous. **Distribution** N. India, Nepal, Bhutan, Tibet.

Subsection Campanulata is related to subsection Taliensia but the species usually differ in the the indumentum hair type, the +/-obovate leaf shape and the usually mauve or purple flowers. Also allied to subsection Lanata, the species of which differ in their smaller leaves, white, cream or yellow flowers and the tomentose branchlets, petioles, ovary, and capsule.

It could be argued taxonomically that this subsection consists of a single very variable species. The Edinburgh revision recognises 2 species while Davidian recognises 4. We consider that *R. aeruginosum* would be better maintained at specific rank, despite the taxonomic overlap, as in most of its distribution it is very distinct from *R. campanulatum*. Both *R. campanulatum* and *R. wallichii* are quite common in cultivation.

R. campanulatum D.Don 1821 H3-5

Height 0.30-11m, a low compact to taller bush or small tree. Branchlets usually glabrous, sometimes hairy, bark grey to fawn, rough with age. **Leaves** 7-18 x 2.5-9cm ovate to broadly elliptic, base rounded to cordate, retained 1-2 years; upper surface glabrous when mature or partly glaucous and usually glossy, lower surface *with a continuous, usually dense, fawn to rusty brown* lanate indumentum made up of capitellate to ramiform hairs. **Inflorescence** 6-18-flowered, loose to fairly compact. **Corolla** broadly campanulate, 3-4.5cm long, blue-lavender, mauve, rosy-purple, pink to pale purple or purple and white, with none, many or few spots; calyx minute 0.5-2mm, usually glabrous; ovary usually glabrous; style glabrous.

Ssp. *campanulatum*

Height 2-11m. **Leaves** 9.5-18cm long, *without a glaucous bloom* and usually glabrous on upper surface when mature, lower surface *with a continuous unistrate* indumentum. **Corolla** *white to pale rose or lilac.* **Distribution** Kashmir eastwards to Sikkim, 2,900-4,100m (9,500-13,500ft), forest, thickets and scattered above the tree line.

This subspecies differs from ssp. *aeruginosum* in its unistrate rather than bistrate indumentum, its larger stature and in the absence of a glaucous bloom on the young leaves. The lower leaf surface of *R. wallichii* has discontinuous indumentum or is glabrous.

Ssp. *campanulatum* varies greatly in its ornamental value and hardiness. The western large-leaved forms are the most handsome, often with the best flowers. Selected clones are usually scarce owing to difficulties of propagation, 'Roland Cooper' and 'Knaphill' are amongst the best. The fine S.S. & W. introductions from C. Nepal have medium-sized leaves and white or pale mauve flowers but are relatively tender. Introduced 1825->, reintroduced 1950s->. March-May.

Ssp. *aeruginosum* (Hook. f.) D.F. Chamb. 1979. H4-5.

Height *0.30-1.8m,* (occasionally to 6m), usually of rounded habit and slow-growing. **Leaves** 7-10cm+ long, upper surface *with a glaucous metallic bloom at first* which is often partially persistent, lower surface *with a buff to rusty-brown, bistrate,*

CAMPANULATA

usually *smooth and thick* indumentum. **Corolla** *pink to purple.* **Distribution** Sikkim, Bhutan, (E. Nepal plants are intermediate between ssp. *aeruginosum* and ssp. *campanulatum*), 3,700-4,500m (12,000-14,750ft), about and above tree line on rocky moorland.

Ssp. *aeruginosum* is generally a lower-growing plant than ssp. *campanulatum* with smaller, more convex leaves which have a more intense glaucous bloom on the upper leaf surface especially when young, and a bistrate as opposed to unistrate indumentum on the lower surface. These two subspecies appear to merge in E. Nepal and Sikkim. Ssp. *aeruginosum* may also be confused with some forms of *R. clementinae* which also has convex leaves, often glaucous when young, but differs in its 7 rather than 5-lobed corolla and its usually more upright habit; both are very shy-flowering.

Ssp. *aeruginosum* is a fine foliage plant in its best selections. The flowers take many years to appear and may be of a harsh colour. Introduced 1849?, 1914, reintroduced 1966-> April-May.

Picture 1

Picture 2

Picture 3

Picture 4

Picture 5

Picture 6

Picture 7

Picture 8

Picture 9

1. *R. campanulatum* on the Milke Danda ridge, E. Nepal.
2. *R. campanulatum* Beer, Lancaster & Morris 283 at the Royal Botanic Garden, Edinburgh.
3. *R. campanulatum* Stainton, Sykes & Williams 9106 at Glendoick.
4. *R. campanulatum* 'Roland Cooper' A.M., Cooper 5768, at the Royal Botanic Garden, Edinburgh, named after the collector. This form has particularly large leaves and flowers and is very hard to propagate.
5. *R. campanulatum* 'Waxen Bell' A.M. at the Royal Botanic Garden, Edinburgh.
6. *R. campanulatum* T. Spring Smyth 11 at the Royal Botanic Garden, Edinburgh.
7. *R. campanulatum* ssp. *aeruginosum* at the Rhododendron Species Botanical Garden, Tacoma, W. U.S.A.
8. *R. campanulatum* ssp. *aeruginosum* at Glendoick.
9. *R. campanulatum* ssp. *aeruginosum*, a fine form, showing the glaucous young foliage and convex leaves.

R. wallichii Hook.f. 1849 (*R. heftii* Davidian) H4

Height 2-6m+, a rounded bush to upright small tree. Branchlets glabrous, floccose when young, bark pinkish-brown to grey, becoming rough. **Leaves** 5-14 x 2.2-5.8cm, elliptic to ovate, retained 1-2 years; upper surface dark, often shiny and glabrous when mature, lower surface *glabrous* or with a *discontinuous* layer of black to brown, indumentum consisting of hair tufts. **Inflorescence** 5-10-flowered, fairly compact to loose. **Corolla** funnel-campanulate, 4-5cm long, blue-mauve, rosy-purple, pale mauve fading to near white to white; calyx minute 1-3mm, tomentose or glabrous, irregular; ovary usually glabrous; style glabrous. **Distribution** E. Nepal to W. Arunachal Pradesh (India) and adjacent S. Tibet, 2,700-4,300m (9,000-14,000ft), in forest and above tree line.

R. wallichii is closely related to *R. campanulatum*, differing in the sparse or absent rather than continuous and thick indumentum on the lower leaf surface. *R. wallichii* is generally an inferior foliage plant to *R. campanulatum* but can still be fine in its better forms. *R. heftii* Davidian 1989 refers to the glabrous-leaved forms. The problem is that the description is too narrow: *R. heftii* is described as having 'pure ivory-white' flowers while most of the glabrous leaved *R. wallichii* from E. Nepal to W. Bhutan have pale mauve flowers which fade. It would be more satisfactory to give all the glabrous-leaved forms varietal or subspecific status together. Introduced 1849 or earlier->. March-May.

Picture 1

Picture 2

Picture 3

Picture 4

1. *R. wallichii* at the Royal Botanic Garden, Edinburgh.
2. *R. wallichii* Ludlow, Sherriff and Taylor 6659 at the Royal Botanic Garden, Edinburgh.
3. *R. wallichii* Hruby 16 at the Royal Botanic Garden, Edinburgh, showing the discontinuous leaf indumentum.
4. The glabrous lower surface of the leaf and the white flowers of the *R. wallichii* Heftii Group; this plant from the Arun Valley, Nepal, growing at Glendoick.

Subsection Campylocarpa

Height 0.9-7.6m, shrubs to small trees, compact to open, occasionally leggy with age. Branchlets stipitate-glandular or glabrous. **Leaves** narrowly *obovate to orbicular, glabrous on both surfaces when mature;* lower leaf surface sometimes with minute red punctate glands. **Inflorescence** 4-15-flowered, lax. **Corolla** *5-lobed,* campanulate to saucer-shaped, *nectar pouches absent,* pink, white or yellow; calyx variable; stamens 10; style glabrous or glandular; ovary stipitate-glandular. **Distribution** E. Nepal, Sikkim, Bhutan, Burma, Tibet, Yunnan to Sichuan.

This subsection is related to subsections Thomsonia and Fortunea. Species in subsection Thomsonia differ in their usually crimson to pink flowers (occasionally white and yellow) and in the presence of 5 dark nectar pouches at the base of the corolla, absent in subsection Campylocarpa. The species in subsection Fortunea differ in their sometimes scented white to pink/lavender flowers with 12-25 stamens.

The species in this section are popular garden plants, with *R. wardii* being the most widely grown. They generally require good drainage and cool conditions and are not easy to grow in areas with high summer temperatures.

R. callimorphum Balf.f. & W.W. Sm. 1919, H3-4

Height 0.6-2(3)m, dome-shaped when young, but often straggly when older. **Leaves** 2-7 x 1.5-5cm, broadly ovate to orbicular; upper surface glabrous, lower surface glaucous with minute red punctate glands. **Inflorescence** 4-8-flowered, loose, pedicel glandular. **Corolla** campanulate, 3-4cm long, *pink to deep rose to white* with a large or small deep basal blotch; calyx and ovary glandular; style usually *glabrous.* **Distribution** W. Yunnan/Burma frontier, 2,700-3,400m (9,000-11,000ft) on open stony slopes, margins of thickets, amongst bamboo and on cliffs.

This species is closely related to the yellow *R. campylocarpum* var. *caloxanthum,* differing in the the red punctate glands on the lower leaf surface and in the flower colour. It differs from *R. souliei* in the smaller leaves, the different flower shape and the glabrous rather than glandular style. Some forms of *R. selense* closely resemble *R. callimorphum* but differ in their taller growth habit, the more bristly branchlets and the more funnel-campanulate corolla.

Var. *callimorphum* (*R. cyclium* Balf.f. & Forrest, *R. hedythamnum* Balf.f. & Forrest)

Flowers *pink.*

When well-grown, *R. callimorphum* is free-flowering and attractive. It can suffer bark-split and be cut back in cold areas but often regenerates well from the base. Good drainage is required and it is longer-lived in areas of lower rainfall. Introduced 1912->. April-May.

Var. *myiagrum* (Balf.f. & Forrest) D.F. Chamb. 1978.

Flowers *white,* usually with spotting and/or a crimson blotch.

Rarer in cultivation than var. *callimorphum* and characterised by its very sticky glands on the pedicels and petioles which gave rise to its Latin epithet of fly-catcher. Introduced 1925-31.

Picture 1

Picture 2

1. A magnificent bush of *R. callimorphum* var. *callimorphum* at the Royal Botanic Garden, Edinburgh.
2. *R. callimorphum* var. *callimorphum* at the Royal Botanic Garden, Edinburgh.

Picture 3

Picture 4

3. *R. callimorphum* var. *callimorphum* showing the typical pink corolla with a deep basal blotch.
4. *R. callimorphum* var. *myiagrum*, which differs from var. *callimorphum* in its white flowers.

R. campylocarpum Hook.f. 1851 H3-4

Height 1.5-6m, a shrub or small tree, bushy, often leggy with age. **Leaves** 3.2-10 x 1.5-5.4cm, orbicular, elliptic to oblong-elliptic to ovate, tip obtuse to rounded; upper surface glabrous, sometimes glaucous, lower surface pale glaucous, +/- glabrous. **Inflorescence** 3-10(15)-flowered, loose. **Corolla** campanulate, 2.5-4cm long, pale to bright yellow, with or without a faint basal blotch (rarely white); calyx *minute;* style *glandular at base, to half its length or eglandular;* capsule slender and *curved.*

This species is closely related to *R. wardii* which differs in its widely-campanulate to saucer-shaped flowers, the larger calyx, the style glandular along the whole length, and the straight seed capsule. These two species merge in parts of S.E. Tibet. It is also closely related to *R. callimorphum* which differs in its pink flowers.

Ssp. *campylocarpum* H4

Height to *6m.* Leaf elliptic, *1.6-2.5* x as long as broad. **Distribution** E. Nepal, Sikkim, Bhutan, Arunachal Pradesh and S.E. Tibet 2,900-4,300m (9,500-14,300ft) with other rhododendrons, in forest, bamboo, scrub or rocky slopes.

Differs from ssp. *caloxanthum* in stature, leaf shape and usually in its deeper-coloured flowers.

This is one of the finest yellow-flowered species and it is usually free-flowering and quite easy to grow. Plants grown under Elatum Group tend to be larger and less compact growers with the flowers sometimes tipped orange. Introduced 1849->. April-May.

Ssp. *caloxanthum* (Balf.f. & Forrest) D.F. Chamb. 1978 (*R. telopeum* Balf.f. & Forrest.) H3-4.

Height 0.90-1.8m. **Leaves** *orbicular, 1.1-1.5(1-7)* x as long as broad. **Distribution** Upper Burma, S.E. Tibet, W. Yunnan, 3,400-4,000m, (11,000-13,000ft), rocky slopes, cliffs and forest margins.

Apart from the differences in height and leaf shape, the foliage of ssp. *caloxanthum* is often more glaucous than in ssp. *campylocarpum*, the flowers are usually paler and it comes from further east. Ssp. *caloxanthum* closely resembles the pink *R. callimorphum* when out of flower.

Ssp. *caloxanthum* needs very good drainage and tends to be rather early into growth but it is a fine subspecies in its best forms, especially those with glaucous foliage. Plants under the name Telopeum Group tend to have the smallest and most glaucous leaves. Introduced 1919-1953. April-May.

Picture 1

Picture 2

1. *R. campylocarpum* ssp. *campylocarpum* on the Milke Danda ridge, E. Nepal.
2. *R. campylocarpum* ssp. *campylocarpum* on the Phephe La, C. Bhutan, showing the campanulate corolla. (Photo Roy Lancaster).

Picture 3

Picture 4

Picture 5

Picture 6

Picture 7

Picture 8

3. *R. campylocarpum* ssp. *campylocarpum* at Inverewe, N.W. Scotland.
4. A very fine specimen of *R. campylocarpum* ssp. *campylocarpum* at Glenarn, W. Scotland.
5. A deep yellow form of *R. campylocarpum* ssp. *campylocarpum* at the Royal Botanic Garden, Edinburgh.
6. *R. campylocarpum* ssp. *caloxanthum* at Glenarn, W. Scotland.
7. *R. campylocarpum* ssp. *caloxanthum* showing the orbicular leaves. Blackhills, N.E. Scotland
8. An excellent glaucous-leaved form of *R. campylocarpum* ssp. *caloxanthum* at Glendoick.

R. souliei Franch. 1895 (*R. cordatum* Lév. *R. longicalyx* M.Y. Fang.) H4-5

Height 1.2-5m, a fairly open shrub or small tree which can become leggy with age. **Leaves** 3.5-8.2 x 2.2-5cm, ovate to ovate-elliptic to oblong-elliptic to almost orbicular; upper surface glabrous, glaucous when young, lower surface glabrous. **Inflorescence** 3-8-flowered, fairly loose. **Corolla** *openly cup or saucer-shaped*, 2.5-3.5cm long, pink, pink flushed rose, white, white flushed pink; calyx 3-8mm, lobes rounded, glandular-ciliate; style *glandular to tip*. **Distribution** Central & S.W. Sichuan, 2,700-4,300m (9,000-14,000ft) in scrub, oak and spruce forest.

This species is closely related to *R. wardii* which differs in its yellow flowers. *R. wardii* var. *puralbum* only differs from *R. souliei* in its pure white flowers; it could almost as well be considered a form of *R. souliei*. *R. callimorphum* has smaller leaves, campanulate flowers and a (usually) glabrous style.

R. souliei is one of the most attractive of all species in flower, especially the white and deep pink selections produced by selective seed raising. It needs perfect drainage and protection for its early young growth and it grows best in cool climates with low rainfall such as E. Scotland. Free-flowering from a young age. Hard to root and graft, so usually only available from seed. Recent field studies seem to confirm that *R. longicalyx* Fang is simply a form of *R. souliei* with a slightly larger-than-average calyx. Introduced 1903-10, reintroduced 1990s. May.

Picture 1

Picture 2

Picture 3

Picture 4

1. Two forms of *R. souliei* growing with the yellow-flowered *R. wardii* at Glendoick.
2. *R. souliei* showing the characteristic saucer-shaped flowers.
3. *R. souliei,* a deep pink-flowered selection at Glendoick.
4. *R. souliei,* a deep pink-flowered selection, showing the glaucous young growth typical of many forms.

Picture 5

5. *R. souliei,* a pure white-flowered form at Glendoick.

R. wardii

W.W. Sm. 1914 (*R. astrocalyx* Balf.f. & Forrest, *R. croceum* Balf.f. & W.W. Sm., *R. gloeblastum* Balf.f. & Forrest, *R. litiense* Balf.f. & Forrest, *R. oresterum* Balf.f. & Forrest, *R. prasinocalyx* Balf.f. & Forrest) H4-5

Height 0.90-7.6m, a compact to open shrub or small tree. **Leaves** 3-12 x 2-6cm, more or less orbicular to ovate, oblong-elliptic or oblong; upper and lower surface glabrous, new growth often glaucous, lower surface pale green or glaucous (Litiense Group) with minute punctulations. **Inflorescence** 5-14-flowered, fairly full to loose. **Corolla** *widely campanulate to saucer-shaped,* 2-4cm long, yellow in various shades or white, with or without a crimson blotch, buds sometimes orange-red; calyx 0.4-1.2cm, +/- cupular if developed; style *glandular to the tip;* capsule stout, short or oblong, straight or curved.

This species differs from *R. campylocarpum* in the more saucer-shaped corolla, in the larger calyx (which hangs on after flowering, aiding identification) and in the style which is glandular to the tip. These two species merge in the Tibet/Arunachal Pradesh border area. Differs from *R. souliei* in the flower colour.

Var. *wardii*

Corolla *yellow.* **Distribution** S.E. Tibet, N.W. Yunnan, S.W. Sichuan, 2,700-4,300m (9,000-14,000ft), open hillsides, thickets, bamboo and forest.

This species has a wide distribution and is very variable in appearance and hardiness. Plants grown under Litiense Group have a glaucous leaf underside. *R. wardii* hybridises in the wild with many other species including *R. vernicosum* (*R. x chlorops*) and *R. selense* (*R. x erythrocalyx*-treated under *R. selense*).

R. wardii is in many ways the best of the medium-sized yellow-flowered species. Forms collected under Ludlow and

CAMPYLOCARPA

Sherriff numbers in S.E. Tibet have proved to be amongst the hardiest and the latest into growth while others may grow early so are vulnerable to spring frosts. Most forms flower freely from a fairly young age. There have been many and varied new introductions in recent years from different parts of its range. Introduced 1913->. Reintroduced 1980's->. May-June.

Var. *puralbum*
(Balf.f. & W.W. Sm.) D.F. Chamb. 1978 H4

Corolla *white*, often rose in the bud. **Distribution** disjunct, N.W. Yunnan and S.E. Tibet around the Tsangpo area, open slopes, thickets and forest, 3,400-4,300m (11,000-14,000ft)

Picture 1

Picture 2

Picture 3

This variety was formerly given specific status. Horticulturally it is perhaps closer to *R. souliei* than *R. wardii* differing in its pure white rather than pink or white flushed pink flowers. All plants in cultivation are probably from Forrest 10616 and Yu 14757 from Yunnan, or seedlings of these, and are quite uniform. In hardiness var. *puralbum* is similar to that of the average var. *wardii* but is not as hardy as *R. souliei*. Introduced 1913-17. May.

Picture 4

Picture 5

Picture 6

Picture 7

1. *R. wardii* var. *wardii* by the river which runs down the north side of the Doshong La pass in S.E. Tibet.
2. A pure yellow unblotched form of *R. wardii* var. *wardii* at Napa Hai, Zhongdian, Yunnan.
3. A form of *R. wardii* var. *wardii* with a large red blotch on the Temo La pass in S.E. Tibet. Such fine blotched forms were fairly uncommon.
4. *R. wardii* var. *wardii* Ludlow, Sherriff & Taylor 6586, collected in S.E. Tibet in 1938, at the Royal Botanic Garden, Edinburgh.
5. *R. wardii* var. *wardii,* a form with a prominent blotch, typical of the so-called 'Ludlow and Sherriff' forms.
6. *R. wardii* var. *wardii,* a non-blotched form.
7. *R. wardii* var. *puralbum* Forrest 10616, collected in N.W. Yunnan in 1913, at Glendoick.

Picture 8

8. *R. wardii* var. *puralbum* Yu 14757 at the Royal Botanic Garden, Edinburgh.

Subsection Falconera

Height 3-14m, often tree-like with stout trunks. Branchlets usually tomentose; bark rough or smooth, grey, pink to red-brown, flaking or peeling. **Leaves** thick and leathery, oblanceolate to broadly obovate, to *46 x 17cm;* upper surface glabrous or with a semi-persistent indumentum, lower surface with a *dense white to rufous woolly* indumentum, often of *cup-shaped hairs.* **Inflorescence** 10-30-flowered, fairly full. **Corolla** (5)7-10-lobed, usually *ventricose-campanulate, nectar pouches lacking,* white, cream, yellow, pink to crimson or flushed or lined mixed colours, often blotched or/and spotted; calyx minute, 1-3mm; stamens 12-18; ovary tomentose, glandular or glabrous; style glabrous. **Distribution** East Nepal, S.E. Tibet, Sikkim, Bhutan, Arunachal Pradesh, Yunnan, S. Sichuan, Burma and N. Vietnam.

Subsection Falconera species differ from their nearest relatives in subsection Grandia in the absence of nectar pouches. The woolly indumentum made up of distinctive cup-shaped hairs (more noticeable on some species than others) contrasts with the plastered indumentum of most of the species in subsection Grandia.

Subsection Falconera includes some of the most splendid species in the genus with majestic foliage, usually with flowers to match. All need sheltered, humid, woodland conditions and will not tolerate excessive heat or sudden drops in temperature. Most species take a number of years to start flowering, especially in milder, wetter areas. *R. rex* and *R. rex* ssp. *fictolacteum, R. falconeri, R. arizelum* and *R. hodgsonii* are the most widely grown.

R. arizelum
Balf.f. & Forrest 1920 (*R. rex* ssp. *arizelum* D.F. Chamb.) H3-4

Height 3-7.6m occasionally taller, a flat-topped tree. Branchlets cinnamon to greyish, tomentose; bark *pink to reddish-brown*, buds conical. **Leaves** 7.5-25 x 3-12cm, usually obovate to oblanceolate, retained 1-4 years; upper surface sometimes rugulose, often shiny or with semi-persistent indumentum, lower surface *with a thick, woolly, usually brown to cinnamon* indumentum of *strongly fimbriate, narrowly cup-shaped hairs.* **Inflorescence** 15-25-flowered, fairly compact. **Corolla** 8-lobed, oblique-campanulate, c. 4.5cm long, cream, yellow, cream tipped rose or mauve, apricot, rose, crimson or carmine; calyx densely tomentose; stamens 15-16; ovary densely tomentose. **Distribution** Arunachal Pradesh, N.E. Upper Burma and adjacent S.E. Tibet and N.W. Yunnan, 3,000-4,400m (10,000-14,500ft), conifer or rhododendron forest.

This species is most likely to be confused with *R. falconeri* ssp. *eximium, R. basilicum,* and *R. semnoides.* It differs from *R. falconeri* ssp. *eximium* in its more obovate, less matt-rugulose leaves, the eglandular pedicels and ovary, and usually in its different flower colour. It differs from *R. basilicum* and *R. semnoides* in the non-winged petiole and usually in the deeper-coloured indumentum. Plants intermediate between *R. arizelum* and *R. basilicum* are found in S.E. Tibet.

Picture 1

Picture 2

1. A pale yellow-flowered *R. arizelum* at the Valley Gardens, Windsor Great Park, S. England
2. A form of *R. arizelum* Forrest 25627 with pink-tinged flowers, Royal Botanic Garden, Edinburgh.

FALCONERA

Picture 3

R. *arizelum* is a fine foliage plant and forms with deeply coloured flowers are generally considered the most desirable. Several colour forms have yet to be introduced although some of these may turn out to be natural hybrids. Rubicosum Group has cerise to crimson flowers, often fading to pale pink. Introduced 1917-53, reintroduced 1988 & 1992->. March-May.

R. basilicum

Balf.f. & W.W. Sm. 1916 (*R. megaphyllum* Balf.f. & Forrest, *R. regale* Balf.f. & Kingdon-Ward,) H3-4

Height 3-9m, a flat-topped tree. Branchlets tomentose, bark generally greyer than *R. arizelum*. **Leaves** 13-36 x 7.5-18.6cm, obovate to oblanceolate, with a *winged, flattened, tapering* petiole, retained 2-3 years; upper surface dark green, rugulose, glabrous or with semi-persistent indumentum, lower surface with a bistrate, woolly indumentum, greyish-cream, sometimes becoming rust-coloured. **Inflorescence** 15-25-flowered. **Corolla** obliquely campanulate with 8 lobes, fleshy, 3.5-4.8cm long, white to cream, occasionally to rose, often flushed or lined pink or purple, usually blotched or/and spotted; calyx c. 2mm, tomentose; stamens 16; ovary densely rufous-tomentose. **Distribution**. Yunnan, N.E. Burma (& S.E. Tibet?), 2,700-4,000m (9,000-13,000ft), forests and open hillsides.

This species closely resembles and is often confused with the more northerly distributed *R. semnoides* but differs in its indumentum composed of broadly cup-shaped, but only slightly fimbriate hairs. Both these species and *R. praestans* have winged petioles; the latter differs in its plastered rather than woolly indumentum. It merges with *R. arizelum* in the wild but not with *R. praestans* which does not occur so far south. The very similar *R. rothschildii* also has the flattened petiole, but differs in its granular rather than woolly indumentum. Plants found recently in S.E. Tibet are intermediate between *R. basilicum* and *R. arizelum*

Picture 4

Picture 5

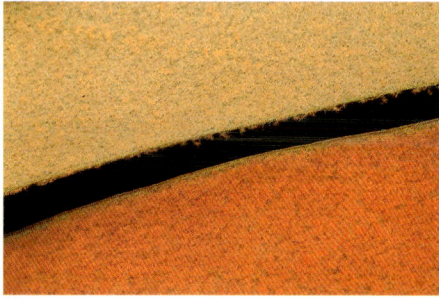

Picture 6

3. An unusual pale apricot-flowered *R. arizelum* at the Younger Botanic Garden, W. Scotland
4. *R. arizelum* Rubicosum Group at Glendoick showing the characteristic deep-coloured flowers of this variety.
5. *R. arizelum* aff. from the south side of the Doshong La pass S.E. Tibet. Intermediate between *R. arizelum* and *R. basilicum* with a winged petiole; flower colour ranged from pink to cream and pale yellow.
6. *R. arizelum* showing indumentum on the lower surface of two leaves, young leaf above, old leaf below. (Photo R.B.G.,E.)

Picture 1

1. *R. basilicum* at the Younger Botanic Garden, W. Scotland. (Photo I. Sinclair).

and usually smaller flowers, but the fine kid-glove-like young growth can be very showy. Some forms have very small flowers and/or small trusses; those with the largest flowers we have seen are from F. 25622 and 25872. Introduced 1918-29. Reintroduced 1988->. April-May.

Picture 2

Picture 3

2. *R. basilicum* at the Younger Botanic Garden, W. Scotland.
3. The typical winged, tapered petioles of *R. basilicum*. (Photo R.B.G.,E.)

R. basilicum can make a handsome specimen, particularly attractive in young growth. It is fine in bloom, but forms with pink-flushed flowers can be somewhat muddy in effect. Introduced 1913-32. March-May.

R. coriaceum

Franch. 1898 (*R. foveolatum* Rehder & E.H. Wilson) H3-4

Height 3-7.6m, a large shrub or fairly flat-topped small tree. Branchlets *relatively thin*, whitish tomentose; bark brownish-grey, flaking, roughish. **Leaves** 10-25 x 3-7.5cm, oblanceolate, retained 3-4 years; upper surface glabrous when mature, lower surface with a fine-textured woolly usually *ash-grey to fawn* indumentum which may wear off. **Inflorescence** 15-20-flowered. **Corolla** (5)-7-lobed, funnel-campanulate, 2-4cm long, white to flushed rose, blotched and sometimes spotted; calyx minute; stamens 10-14; ovary densely tomentose. **Distribution** N.W. Yunnan and S.E. Tibet, 3,000-4,150m (10,000-13,600ft), conifer forest and thickets.

R. coriaceum has paler leaves than *R. rex* and its subspecies, with a lighter-coloured indumentum on the lower surface. The pale-coloured indumentum, thin branchlets and oblanceolate leaves separate this species from other members of the subsection. *R. preptum* is probably a natural hybrid of *R. coriaceum* crossed with *R. rex* or *R. arizelum*.

R. coriaceum is of less garden value and is less hardy than *R. rex* and *R. rex* ssp. *fictolacteum* with less impressive foliage

Picture 1

Picture 2

Picture 3

Picture 4

1. A tree of *R. coriaceum*, Do La Guo, Weixi, Yunnan, China.
2. *R. coriaceum*, Do La Guo, Weixi, Yunnan, China.
3. A particularly fine form of *R. coriaceum* at Glendoick.
4. *R. coriaceum* Forrest 25872 at the Royal Botanic Garden, Edinburgh.

Picture 5

5. *R. coriaceum* showing indumentum on the lower surface of two leaves, young leaf above, old leaf below. (Photo R.B.G.,E.)

Picture 1

R. falconeri Hook.f. 1849 H3-4

Height to 12m, a rounded, rounded-spreading to columnar large shrub or tree. Branchlets *stout* with fawn tomentum, bark *red-brown*, rough but peeling. **Leaves** thick, 15-35 x 7-17cm, elliptic to obovate, retained 1-(3-4) years; upper surface *matt, rugulose*, glabrous or with a scurfy indumentum, lower surface with a thick, rust to brown indumentum of fimbriate narrowly cup-shaped hairs. **Inflorescence** 12-20+ flowered, usually dense. **Corolla** *heavily textured*, lobes 8 (-10), *oblique-campanulate*, 4-6cm long, creamy-white to pale yellow to pink, with blotch; calyx tomentose-glandular; stamens 10-16; ovary densely glandular; style stout, glabrous or tomentose/glandular at base.

Picture 2

This species is usually larger-growing than its closest relative *R. arizelum* and differs in its larger leaves, the more matt upper leaf surface and its glandular ovary.

Ssp. *falconeri*

Taller than ssp. *eximeum*. **Leaves**, upper surface *glabrous* at maturity. **Corolla** white to cream to pale yellow with a purple basal blotch; stamens 12-16. **Distribution** extreme E. Nepal, Sikkim, Bhutan and W. Arunachal Pradesh, 2,700-3,400m (9,000-11,000ft), occasionally higher, in mixed and conifer forest.

Ssp. *falconeri* is one of the grandest of the genus, capable of growing into a magnificent specimen and living well over 100 years in cultivation. The fleshy flowers may last for a full month. Variable in foliage and flower quality. Introduced 1850->, reintroduced 1960s->. Late April-late May.

Picture 3

Ssp. *eximium* (Nutt. 1853) D.F. Chamb. 1979

Generally not as large-growing as ssp. *falconeri* or so tree-like. **Leaves**, *oval or obovate-elliptic*, upper surface with *semi-*

1. Hillside in E. Bhutan with *R. falconeri* ssp. *falconeri*. (Photo I. Sinclair).
2. *R. falconeri* ssp. *falconeri* near Sandakphu, W. Bengal, India.
3. *R. falconeri* ssp. *falconeri*. (Photo A.D. Schilling).

persistent rusty-brown indumentum, lower surface with a *deeper cinnamon-coloured* indumentum. **Corolla** *opens rose to cream flushed rose*, fading out to creamy-pink; stamens 10-14. **Distribution** Bhutan, Arunachal Pradesh, 2,700-3,400m (9,000-11,000ft), forests and ridges.

Ssp. *eximium* is easily confused with some forms of the variable *R. arizelum*. *R. arizelum* generally has smaller, less oval and more obovate leaves with less persistent indumentum on the upper leaf surface. *R. arizelum* has an eglandular ovary and the flowers vary from cream, pale yellow to rose, while they are usually rose fading to cream in ssp. *eximeum*.

A magnificent foliage plant, one of the finest in the genus, especially when in young leaf; the flowers can be a rather harsh colour. Introduced c.1850->, reintroduced 1965 & c.1985->. March-May.

Picture 4

Picture 5

Picture 6

4. *R. falconeri* ssp. *falconeri*. (Photo A.D. Schilling).
5. Young growth of *R. falconeri* ssp. *falconeri*.
6. Old leaves of *R. falconeri* ssp. *falconeri* showing the rugulose matt upper leaf surface and the brown indumentum on the lower surface.

Picture 7

Picture 8

Picture 9

Picture 10

7. *R. falconeri* ssp. *eximium* in Subansiri Division, Arunachal Pradesh, N.E. India.
8. *R. falconeri* ssp. *eximium* Cox & Hutchison 427 when the flowers are newly opened.
9. *R. falconeri* ssp. *eximium* Cox & Hutchison 427 after the flowers have been out for a few days.
10. *R. falconeri* ssp. *eximium* Cox & Hutchison 427 at Baravalla, Argyll, showing the characteristic persistent indumentum on the upper leaf surface.

FALCONERA

Picture 11

11. *R. falconeri* ssp. *eximium* showing the indumentum on the lower surface of two leaves; old leaf above, young leaf below. (Photo R.B.G., E.).

R. galactinum Balf.f. 1926 H4-5

Height 3-9m, a shrub or small tree. Branchlets brown tomentose, bark roughish, grey-brown. Buds *densely tomentose with short scales*. **Leaves** 12-21 x 5-8cm, ovate-lanceolate to oblanceolate, retained 2-4 years; upper surface slightly rugulose, mid-green, glabrous, lower surface with a thinnish, bistrate, buff-grey to pale cinnamon indumentum of fimbriate, narrowly cup-shaped hairs. **Inflorescence** 9-15-flowered, fairly loose; flower buds *ovoid*. **Corolla** white to pale rose with a blotch and spots, campanulate, c. 3cm long; calyx c. 1mm, tomentose, lobes triangular; stamens 14; ovary usually *glabrous*; style glabrous. **Distribution** Central & W. Sichuan, more widespread than previously thought, 2,300-3,300m (7,500-10,800ft), woods and gorges.

A distinct species, very different from the others in this subsection and which might be better placed on its own. Identified by its distinctively shaped, tomentose, short-scaled buds and its glabrous ovary. Leaves tend to be pale and yellowish on cultivated plants.

While lacking the grandeur of the rest of the subsection with its thinner branchlets and lighter-textured leaves, it is valuable in colder areas where most large-leaved species are not hardy. Introduced 1910, reintroduced 1989->. April-May.

Picture 1

Picture 2

Picture 3

Picture 4

Picture 5

1. *R. galactinum*, Wolong-Balang, C. Sichuan, China.
2. *R. galactinum* a pink-flowered form, at Glendoick.
3. A white-flowered form of *R. galactinum* Wilson 4254, at Glendoick, collected on Panlan Shan, Sichuan, China in 1910.
4. *R. galactinum* near Lengqi, C. Sichuan, China.
5. *R. galactinum* showing the distinctive tomentose growth buds. (Photo R.B.G., E.)

R. hodgsonii Hook.f. 1851 H4

Height 3-12m, rounded to erect with age. Branchlets stout, tomentose; bark *creamy to cinnamon, smooth, peeling*. Buds *conical with long tailed bud scales*. **Leaves** leathery, 18-38 x 7-14cm, obovate to oblanceolate to elliptic, retained 2-3 years; upper surface dark green, usually with a thin indumentum, creating an apparent *metallic bloom*, lower surface with a bistrate, dense, silvery to cinnamon indumentum, the *lower layer compacted, the upper layer composed of broadly cup-shaped* hairs; petiole with a thin, floccose, greyish indumentum. **Inflorescence** 15-25-flowered, compact. **Corolla** tubular-campanulate, lobes 7-8, 3-4cm long, pink through purple to deep cherry-red, (liable to fade), with or without a blotch; calyx 1-3mm, tomentose; ovary tomentose. **Distribution** Nepal to W. Arunachal Pradesh and neighbouring S. Tibet, 2,900-4,300m (9,500-14,000ft), ridges, valleys, dominant or in conifer forest.

This is a distinct species, quite easily recognised by its smooth bark, long, tailed conical buds, metallic-looking upper leaf surface, distinctive indumentum and its truss of purplish flowers.

R. hodgsonii makes a very handsome foliage plant with an often outstanding bark. The often 'hot'-coloured flowers are striking but usually fade out to a dirty mauve. One of the hardiest of the subsection and often slow growing. Introduced 1850->, some plants are still in cultivation from that date, reintroduced 1961-> many times. March-May.

R. x decipiens

Lacaita 1916 is the common natural hybrid between *R. hodgsonii* and *R. falconeri* where the two overlap in the wild. In some places there are stable populations. In Bhutan, *R. kesangiae* usually grows between the two aforementioned species, resulting in hybrids between them and *R. kesangiae*. U.C. Pradham and T. Lachungpa in *Sikkim-Himalayan Rhododendrons*, describe what must be a different hybrid under the name *R. decipiens* in Sikkim, probably *R. hodgsonii* x *R. arboreum*.

R. hodgsonii affinity

Tall, to 6m, branchlets densely rufous-tomentose, bark as *R. hodgsonii*. Buds like *R. hodgsonii* with shorter tails. **Leaves** generally smaller and shorter than *R. hodgsonii*, sometimes with a bi-lobed apex; upper surface *a very dark metallic green,* lower surface with a *thick dark chocolate-brown* indumentum, consisting of broadly cup-shaped hairs. **Corolla** colour unknown but opening buds show a similar dark colour range to that of *R. hodgsonii* itself. **Distribution** C. Bhutan on the E. side of the Rudong La, 3,700m (12,000ft), on open hillside.

This uniform population had no other big-leaved species in the vicinity. Hybrids of *R. hodgsonii* x *R. kesangiae* elsewhere had a similar-coloured indumentum so it appears that this plant originated from this cross, but the population has become isolated and stabilised. Seedlings of *R. hodgsonii* aff. Cox, Hutchison and Maxwell MacDonald 3093 appear uniform. It should prove an extremely fine foliage plant in cultivation and it deserves at least varietal status. Introduced 1988. March-May?

Picture 1

Picture 2

Picture 3

1. *R. hodgsonii*, Phokphey-Rudong La, C. Bhutan. (Photo Roy Lancaster).
2. A dark-flowered form of *R. hodgsonii* at the Younger Botanic Garden, W. Scotland.
3. A pink-flowered form of *R. hodgsonii* B.L.M. 323 at Glendoick.

FALCONERA

R. preptum Balf.f. & Forrest 1919 H3-4

Height 2.5-9m, a shrub or small tree. **Leaves** 11-20 x 4-9.5cm, oblanceolate to elliptic; upper surface glabrous and rugulose, lower surface with a bistrate indumentum of *buff*, fimbriate, cup-shaped hairs; petiole *slightly winged or ridged*. **Inflorescence** c.20-flowered. **Corolla** oblique-campanulate, 6-7 lobed, 3.5-4.5cm long, creamy-white, blotched; calyx minute, densely tomentose; stamens 10-14; ovary densely tomentose. **Distribution** N.W. Yunnan, N.E. Upper Burma, 3,400-3,700m (11,000-12,000ft), mixed forest.

R. preptum differs from *R. arizelum* in its usually slightly narrower leaves with paler indumentum on the lower surface, and the slightly winged petiole.

This is probably a hybrid of *R. coriaceum* crossed with *R. arizelum* or with *R. rex* ssp. *fictolacteum*. Rare in cultivation but not usually of particular merit. Introduced 1924. April-May.

Picture 4

Picture 5

Picture 6

Picture 7

Picture 1

Picture 2

4. *R. hodgsonii* showing the distinctive conical bud with long, tailed scales.
5. A trunk of *R. hodgsonii*, at the Royal Botanic Garden, Edinburgh.
6. The indumentum on the leaf lower surface of *R. hodgsonii*, darker on the mature leaf, lighter on the young leaf. (Photo R.B.G., E.)
7. *R. hodgsonii* affinity, Rudong La, C. Bhutan, showing the thick chocolate-brown indumentum and and the bi-lobed leaf apex. (Photo K. Rushforth).

1. *R. preptum*, a form from Portmeirion, showing the pale indumentum on young and old leaves.
2. *R. preptum* at the Royal Botanic Garden, Edinburgh.

FALCONERA

Picture 3

Picture 4

3. Foliage of *R. preptum* at the Royal Botanic Garden, Edinburgh.
4. *R. preptum* with indumentum on the lower surface of two leaves, old leaf above, young leaf below. (Photo R.B.G., E.)

R. rex Lév. 1914. H4-5

Height 2.5-12m, a large erect shrub or small tree. Branchlets stout, *greyish-white* tomentose, bark roughish, grey-brown. **Leaves** 12-37 x 5.5-13.5cm, obovate to oblanceolate, retained 2-4 years; upper surface dark, smooth to rugulose, glabrous on maturity, often shiny, lower surface *with a thick woolly fawn to rufous* indumentum, composed of *slightly* to moderately fimbriate, broadly cup-shaped hairs. **Inflorescence** 12-30-flowered, usually rounded. **Corolla** 8-lobed, with serrated, rather frilled margins, usually oblique-campanulate, 3-4.5cm long, white, through pink to mauve-pink, (pale yellow), usually with a crimson blotch and spots; calyx minute, tomentose; stamens 14-16; ovary densely tomentose, style glabrous. NOTE: this description may not entirely encompass ssp. *gratum*.

R. rex is easily separated from its relatives by the combination of the dark, shiny leaf upper surface, the fawn to rufous, woolly indumentum and the pink or white flowers.

Ssp. *rex.*

Leaves 2.4-3.1 x as long as broad, lower surface with a *fawn* indumentum of only *slightly* fimbriate cup-shaped hairs. **Corolla** 5-7cm long, white flushed pink. **Distribution** S. Sichuan and neighbouring N.E. Yunnan, 3,000-4,300m (10,000-14,000ft), conifer, mixed or pure forests.

Ssp. *rex* is very closely related to ssp. *fictolacteum*, differing in its larger leaves and flowers, and usually paler tomentum on branchlets and leaf indumentum. The two subspecies do merge, making some plants hard to place in one or other. Most mature plants in cultivation are from R. 03800 (18234) and K.W. 4509.

Ssp. *rex* is a very handsome foliage plant with large, well-filled trusses. It is one of the hardiest of the subsection, doing well in relatively cold areas such as E. Scotland and is even worth attempting in milder parts of Scandinavia. Introduced 1913-1932, reintroduced 1992->. April-May

Ssp. *fictolacteum* (Balf.f.) D.F. Chamb. 1979
(*R. fictolacteum* Balf. f., *R. lacteum* var. *macrophyllum* Franch.) H4-5

Branchlets *cinnamon*-tomentose. **Leaves** 10-30 x 3.5cm-11, 2.5-3.8 times as long as broad, retained 2-5 years; upper surface very *dark green,* lower surface *with indumentum, thick buff to rusty-brown* consisting of *moderately* fimbriate cup-shaped hairs. **Corolla** 3-5cm long, usually white, also pale lilac to rose. **Distribution** Widely distributed in W. and N.W. Yunnan and adjoining S.E. Tibet and N.E. Upper Burma, 3,000-4,300m (10,000-14,000ft), in mixed and conifer forest.

Closely related to and merges with ssp. *rex*, differing in its smaller leaves and flowers, and the deeper-coloured and moderately fimbriate indumentum. Also related to and probably hybridising with *R. arizelum* in the wild. Quite variable in foliage and flower.

Ssp. *fictolacteum* is one of the hardiest, most popular and easiest grown of the big-leaved species. The normally white flowers, often freely produced on mature plants, contrast superbly with the dark foliage. Introduced 1886-1937. Reintroduced 1981->. March-May.

Var. *miniforme* Davidian 1989.
Differing from the above in its smaller stature and leaf size. Plants in cultivation so far do not exceed 3.5m, while in the wild it attains 6m. It appears to come from an above average altitude for the subspecies but it has yet to be ascertained if it forms whole populations or just occurs as isolated plants. This will determine whether it really deserves varietal status.

Ssp. *gratum* (T.L. Ming 1981) M.Y. Fang H4?

Young branchlets with greyish tomentum, becoming glabrous.

Leaves obovate-elliptic or obovate-oblong, 18-35 x 7-15cm, lower surface with a bistrate indumentum, upper layer of brown to fawn which is easily rubbed off, revealing a lower layer of +/- agglutinated white indumentum. Petiole *slightly winged*, white tomentose below. **Inflorescence** 15-25-flowered. **Corolla** obliquely campanulate, inflated on one side, 3.5-4cm long, white, milky white, light yellow to rose with purple crimson blotches at base. **Distribution** W. Yunnan, 3,200m, (10,500ft) in coniferous forest and at forest margins.

R. gratum was formerly made synonymous with *R. basilicum* in the Edinburgh revision due to its slightly winged petiole. Chinese botanists have recently classified it instead as a subspecies of *R. rex*.

Ssp. *grantum* seems to be intermediate between *R. basilicum* and *R. rex* or *R. arizelum* and appears to have a very localised distribution. Probably better considered as a form of *R. basilicum* but doubtfully deserving of any more than group status in our opinion as there are so many intermediate forms of many species in this subsection. Introduced 1990's.

Picture 1

Picture 2

Picture 3

Picture 4

Picture 5

Picture 6

1. A very fine form of *R. rex* ssp. *rex* at the Royal Botanic Garden, Edinburgh.
2. *R. rex* ssp. *rex* Rock 18234 from Muli, Sichuan, China, at the Royal Botanic Garden, Edinburgh.
3. Indumentum of *R. rex* ssp. *rex* on two leaves, old leaf above, young leaf below. (Photo R.B.G., E.).
4. *R. rex* ssp. *fictolacteum* (and *R. oreotrephes*), Tian Bao Shan, N.W. Yunnan, China.
5. *R. rex* ssp. *fictolacteum* and E.H.M. Cox at Glendoick.
6. *R. rex* ssp. *fictolacteum* at Glendoick.

Introduced 1929 & 1948. Reintroduced 1988->. April-May.

Picture 7

Picture 8

Picture 1

Picture 2

Picture 3

Picture 4

7. *R. rex* ssp. *fictolacteum* SBEC 0957 from the Cangshan, Yunnan, a form with a very fine large blotch.
8. *R. rex* ssp. *fictolacteum* var. *miniforme*, smaller than the type in all parts. Royal Botanic Garden, Edinburgh.

R. rothschildii Davidian 1972 H4

Height 2.75-6m, a large shrub or tree. Branchlets with thin fawn to brown tomentum; bark rough. **Leaves** 21-36 x 6-14cm, obovate to oblanceolate with *the base tapering into the flat petiole*, retained (1-)-2 years; upper surface glabrous slightly rugulose, lower surface *with a thin granular discontinuous* indumentum of strongly fimbriate hairs. **Inflorescence** 12-17-flowered. **Corolla** obliquely campanulate, 8-lobed, 3.5-4.3cm long, creamy-white, pale yellow to pale pink, sometimes with a blotch; calyx tomentose; stamens 13-15; ovary densely tomentose; style glabrous. **Distribution** Ta-pao Shan between Weixi and Mekong only, W. Yunnan, 3,700-4,000m (12,000-13,000ft), in mixed forest.

This species is very close to *R. semnoides* separated only by its granular, discontinuous indumentum; *R. praestans* differs in its shiny plastered indumentum.

There are few clones in cultivation or satisfactory herbarium specimens of *R. rothschildii* but we have recently seen a large stable population of this species at Weixi, Yunnan (Cox and Cubey 9312). Like *R. semnoides* this species almost certainly originated as a hybrid of *R. arizelum* x *R. praestans* but in this case it has formed a stabilised and seemingly isolated population. A handsome plant with quite a variation in flower colour. Rare.

1. *R. rothschildii*, Xue Long Shan, Weixi, N.W. Yunnan, China.
2. *R. rothschildii,* Xue Long Shan, Weixi, N.W. Yunnan, China.
3. A selection of different forms of *R. rothschildii* at Xue Long Shan, Weixi, N.W. Yunnan, China..
4. The leaf lower surface of *R. rothschildii* showing the granular indumentum, Xue Long Shan, Weixi, N.W. Yunnan, China.

R. semnoides Tagg & Forrest 1926 H3-4

Height 3-6m, a rounded to upright tree or large bush. Branchlets densely tomentose. **Leaves** 10-25 x 4-11.5cm, obovate to oblanceolate, apex sometimes *emarginate*, retained 1-4 years; upper surface usually glabrous, lower surface with a *thin to thick fawn through rufous to brown* indumentum of *strongly fimbriate, narrowly* cup-shaped hairs. Petiole +/- *moderately flattened and winged*. **Inflorescence** 12-20-flowered. **Corolla** oblique-campanulate, c.8-lobed, 3.5-5cm long, white flushed rose, blotched; calyx minute, 1-2mm, tomentose; stamens 16; ovary tomentose; style glabrous. **Distribution** N.W. Yunnan and adjacent S.E. Tibet, 3,700-4,000m (12,000-13,000ft), conifer and mixed forest.

Very similar to *R. basilicum*, only differing in more fimbriate cup-shaped hairs, and to *R. rothschildii* which has granular indumentum.

This species is undoubtedly derived from hybrids of *R. praestans x R. arizelum* and varies in the wild all the way from one species to the other. Further field studies are required to ascertain whether stable populations exist and whether seed of such populations will produce relatively uniform progeny. *R. semnoides* is quite a handsome plant in its better forms but is very similar to *R. basilicum* and *R. rothschildii*. Introduced 1922. Reintroduced 1992. March-May.

Picture 1

Picture 2

Picture 3

Picture 4

3. The leaf lower surface of *R. semnoides*, a young leaf on the left, a mature leaf on the right, showing the winged petioles inherited from *R. praestans*.
4. The lower surfaces of two leaves showing the indumentum of *R. semnoides*, old leaf above, young leaf below. (Photo R.B.G.,E.)

1. *R. semnoides* Rock 25388 at the Royal Botanic Garden, Edinburgh.
2. *R. semnoides* at the Royal Botanic Garden, Edinburgh.

R. sinofalconeri Balf.f. 1916 H2-3?

Height to 20m, tree-like but fairly compact and spreading. Branchlets, tomentose; bark smooth, brown. **Leaves** 17-28 x 11.8-16cm, *broadly obovate to obovate-elliptic*; upper surface glabrous, rugulose, lower surface with a *light brown, woolly* bistrate or unistrate indumentum of moderately fimbriate broadly cup-shaped hairs. **Inflorescence** 10-12-flowered. **Corolla** *8-lobed, oblique-campanulate,* 4.5-5.5cm long, *pale to rich yellow*, pedicel eglandular; calyx small, c.3mm, sparsely tomentose; stamens 16; ovary tomentose, *eglandular*. **Distribution** S.E. Yunnan and adjacent N. Vietnam, 1,600-3,000m (5,250-10,000ft).

R. sinofalconeri differs from its relative *R. falconeri* ssp. *falconeri* in its broadly obovate leaf shape, in its paler indumentum and in its eglandular pedicel and ovary. In flower colour and leaf shape *R. sinofalconeri* resembles *R. macabeanum* which differs in its paler indumentum and more upright stature. The smooth and conical buds resemble those of *R. grande* and *R. protistum*.

As seen in the wild, this is a very handsome species which will hopefully be hardy enough for mildest U.K. gardens and should prove useful for Australia, New Zealand and other similar climates. The yellow flowers may rival those of *R.*

macabeanum. Introduced from N. Vietnam in 1992 and S.E. Yunnan in 1995.

Picture 1

Picture 2

1. *R. sinofalconeri* on Lao Jing Shan, S.E. Yunnan, showing the broadly obovate leaves and smooth conical buds.
2. *R. sinofalconeri*, S.E. Yunnan. Photo Yang Zhenhong.

Subsection Fortunea

Height 1.5-18m, rounded or upright shrubs or small trees. Bark roughish, brown to grey, not peeling (except *R. griffithianum*). **Leaves** to 36 x 12.5cm, +/- *glabrous*. **Inflorescence** 5-30-flowered, often lax. **Corolla** often glandular, *usually more than 5-lobed, often scented*, white to pink to purplish-pink; calyx usually small (not *R. griffithianum*); stamens *12-25*; stipitate-glandular or glabrous; style glabrous or stipitate-glandular to tip. **Distribution** Widely, over China and Himalayan region.

The species in this subsection are characterised by their white to pink flowers and their glabrous leaves. Most species in subsections Thomsonia and Campylocarpa share the glabrous leaves but these tend to be smaller and more rounded than in subsection Fortunea and their flowers are usually red, yellow or deep pink. *R. auriculatum* differs in its later flowering season and in the distinctive long, tailed buds. *R. calophytum* may be related to subsection Grandia, *R. decorum* to *R. auriculatum* and *R. fortunei* and *R. oreodoxa* to subsection Campylocarpa.

Subsection Fortunea contains many useful species for severe climates, *R. fortunei* being the most significant as it is the hardiest species with scented flowers. Some species are also useful for their heat-tolerance. The commonest species in cultivation are *R. decorum, R. oreodoxa, R. fortunei, R. orbiculare, R. calophytum, R. vernicosum* and *R. sutchuenense*, less common are *R. hemsleyanum, R. griffithianum* and *R. praevernum*. Several new species have recently been introduced.

R. asterochnoum
Diels 1921 H5

Small tree. **Leaves** oblanceolate, 18-20 x 5-6cm, apex rounded, base cuneate, lower surface with a *sparse discontinuous whitish stellate indumentum*; petiole 1.5-2.5cm, floccose. **Inflorescence** 15-20-flowered. **Corolla** funnel-campanulate, 7-lobed, c. 4.5cm long, white, tinged with rose, probably with a blotch; stamens c. 20, pubescent at base. Ovary and style glabrous; stigma discoid. **Distribution** C. & S. Sichuan. 3,000-3,660m (10,000-12,000ft).

This species is closely related to *R. calophytum*, differing in the sparse, discontinuous, whitish, stellate indumentum on the leaf lower surface. The indumentum was mostly only on the midrib on specimens we saw at Wolong, Sichuan and on midrib and veins on those at Leibo, Sichuan. Other populations may have indumentum on the whole lower leaf surface.

R. asterochnoum should be a useful hardy species. It may be derived from natural hybrids of *R. calophytum* crossed with a species with indumentum on the lower leaf surface. Introduced 1990-95.

R. calophytum
Franch. 1886 H5

Height 4.6-9m, a wide-spreading large bush or tree. Branchlets with white tomentum. **Leaves** 14-30 x 4-8 cm, *oblong-lanceolate*, coriaceous in texture, usually *flat* without recurved margins, retained 1-2 years; new growth with sparse silver tomentum, lower surface glabrous when mature or with vestiges of juvenile indumentum persisting on the midrib. **Inflorescence** 5-30-flowered, fairly loose, pedicel *very long, red*, 3-7cm. **Corolla** openly campanulate, lobes 5-7, *ventricose*, 5-6cm long, white to pink, pale mauve pink and sometimes purple with deep (red) blotch and spots; calyx c. 1mm, glabrous; stamens 15-20, *very small compared to the style*; ovary glabrous; stigma large, *disk-like*, 5-8 mm across. **Distribution** Common in C.W. and E. Sichuan and N.E. Yunnan, 1,800-4,000m (6,000-13,000 ft), sometimes the dominant species.

R. calophytum has more stamens, a larger stigma and a longer pedicel than its nearest relatives *R. sutchuenense* and *R. praevernum* which also differ in their more oval leaves with more recurved margins. *R. praevernum* has similar flowers to *R. calophytum* but they open earlier. *R. asterochnoum* differs in the presence of stellate indumentum on the leaf lower surface.

R. calophytum is one of the hardiest of the really large-growing species, worth trying in severe climates with adequate shelter. It is long-lived, quite common in cultivation and can form magnificent, free-flowering, umbrella-shaped trees. Introduced 1904-10, reintroduced 1980s->. February-April. Chinese and other botanists have classified this species under several varities; only two are currently in cultivation:

Var. *calophytum*

Leaves *18-30cm* long, apex acuminate. **Inflorescence** *15-30* flowered.

This is the variety commonly cultivated.

Var. *openshawianum*
(Rehder & E.H. Wilson 1913) D.F. Chamb.

Leaves *14-18.5cm* long, apex mucronate. **Inflorescence** 5-10 flowered.

This is probably best considered simply an inferior form of the species with smaller leaves and only 5-10 flowers to the truss, although an isolated population was found in S. Sichuan in 1995. Davidian claims that it has not been in cultivation, but plants have been grown under this name at Dawyck Botanic Garden, Scotland. Introduced 1995.

Picture 1

Picture 2

Picture 3

Picture 4

1. *R. calophytum* on Emei Shan, C. Sichuan, China.
2. *R. calophytum* (*R. sutchuenense* on left) at Glendoick.
3. *R. calophytum* at Glendoick.
4. *R. calophytum* at the Valley Gardens, Windsor Great Park, S. England.

This species differs from *R. oreodoxa* in its long rachis, longer, narrower leaves and more glandular ovary. The apex of the leaf of *R. davidii* is more acuminate than that of *R. oreodoxa*. The narrow leaves distinguish it from *R. sutchuenense* and other members of the subsection. Closely related to the newly introduced *R. huianum* from which it differs in its smaller calyx and generally larger leaves.

Although it is said to have been introduced by Wilson and flowered at Kew, *R. davidii* seems to have been lost to cultivation. The deep coloured flowers in shades of purple make this species a desireable addition to the cultivated species in the subsection. Previous to c. 1993, introductions under the name *R. davidii* turned out to be *R. oreodoxa* or hybrids of it. Apparently, the correct plant has been recently introduced. April-May (in the wild), February-March? (in cultivation).

Picture 5

Picture 6

Picture 7

Picture 1

5. A pink form of *R. calophytum* at the Valley Gardens, Windsor Great Park, S. England.
6. Young growth of *R. calophytum* with the characteristic sparse, silvery tomentum.
7. *R. calophytum* var. *openshawianum* in broad-leaved forest, near Leibo, S. Sichuan, China.

1. *R. davidii* at Mashangping, C. Sichuan, China.

R. davidii
Franch. 1886 H4/5?

Height 3-8m, a large broad shrub. Bark grey. **Leaves** coriaceous, *10-17* x *2-3cm*, oblanceolate, margin recurved; lower surface greenish-white. **Inflorescence** 8-10-flowered, rachis *long, 2.5-6cm*. **Corolla** with 7-8 lobes, openly-campanulate, c. 5cm long, pink, rosy-red, lilac to purplish blue, spotted purple; calyx *1-2mm, rather densely glandular;* stamens 14-16; ovary *densely glandular*; style glabrous or sparsely glandular. **Distribution** Central, S. and W. Sichuan and N.E. Yunnan, 1,800-4,000m (6,000-13,000 ft) in open situations, bamboo and thickets.

R. decorum
Franch. 1886

Height to 9m, a fairy erect and sometimes straggly shrub or small tree. Branchlets with a temporary glaucous bloom. **Leaves** 7-30 x 2.2-11cm, oblanceolate to elliptic, apex and base rounded, retained 1-2 years; upper and lower surfaces glabrous. **Inflorescence** 7-14-flowered, usually open-topped. **Corolla** with 6-8 lobes, openly funnel-campanulate, to 10cm long, *fragrant*, white, pale rose or pale lavender rose, with yellow, green or crimson tinge at base, with or without markings; calyx 1-3mm, stipitate-glandular; stamens *12-20, puberulous at base;* ovary and style stipitate-glandular.

The +/- oblanceolate leaves, fragrant flowers and puberulous stamens distinguish this species from *R. vernicosum* and *R. oreodoxa*. *R. vernicosum* is further distinguished by the dark red glands on the style. *R. fortunei* can be distinguished by its

FORTUNEA

purple petiole and generally more lavender or pink tinged flowers. It is considerably hardier and has a more eastern and northern distribution.

Ssp. *decorum*
(*R. franchetianum* Lév., *R. giraudiasii* Lév., *R. spooneri* Hemsl. & E.H. Wilson) H3-4

Leaves *7-18 x 2.2-7.6cm,* **Corolla** with *6-7* lobes, to 7.5cm; stamens *12-15,* earlier flowering than ssp. *diaprepes*. **Distribution** Sichuan, Yunnan and N. Burma 1,800-4,000m (6,000-13,000 ft) from drier areas, especially pine forests, forest margins, open or cleared areas and limestone cliffs.

This subspecies comes from further north and east than ssp. *diaprepes* but distributions overlap. One of the most plentiful species in the wild, growing over a wide area and considerable altitudinal range. Often found in very dry sites where most other rhododendrons cannot survive. Very vigorous and one of the quickest and most easily-grown species. Hardier clones tend to flower earlier in the season and at a younger age than some of the late flowering, low-altitude forms which are close to ssp. *diaprepes*. Introduced 1887 and many times since. Late April-June

Ssp. *diaprepes*.
(Balf.f. & W.W. Sm.) T.L. Ming (*R. rasile* Balf.f. & W.W. Sm.) H3

Leaves *12-30 x 4.4-11cm.* **Corolla** with *7-8* lobes, 8-10cm, and usually white; stamens *usually 18-20*. Later flowering than ssp. *decorum,* in June-July. **Distribution** Yunnan, Upper Burma and Laos, 1,800-3,400m (6,000-11,000ft), in forests, at forest margins and in thickets.

Ssp. *diaprepes* is now considered to be a low altitude and late flowering form of *R. decorum*. Our recent observations in China bear this out. Larger flowers and leaves than ssp. decorum but less hardy and the early growth can lead to bark split. The clone 'Gargantua' F.C.C. is hardier than the type. Introduced 1913-38 and recently. June-July.

Picture 2

Picture 3

Picture 4

Picture 1

1. *R. decorum* ssp. decorum, Tian Bao Shan, N.W. Yunnan, China.
2. *R. decorum* ssp. *decorum* SBEC 1059, the pinkest-flowered selection from this number, at Glendoick.
3. *R. decorum* ssp. decorum SBEC 1060, Huadianba, Cangshan, W. Central Yunnan, China.
4. *R. decorum* ssp. *decorum* at the Royal Botanic Garden, Edinburgh.

Picture 5

Picture 6

5. A fine white form of *R. decorum*, probably Farrer 979, sometimes considered to be ssp. *diaprepes*.
6. *R. decorum* ssp. *diaprepes* SBEC 1225, Glendoick, showing the leaves larger than in ssp. *decorum*.

R. fortunei
Lindley 1859 H5

Height to 9m, usually much less, a usually upright and fairly rigid shrub or small tree. **Leaves** 7-17 x 3.5-8cm, broadly *oblanceolate to obovate*, apex and base usually rounded, retained 1-2 years; upper surface dark green, lower surface pale; petiole *purplish, bluish or reddish*. **Inflorescence** 5-12-flowered, loose and often pendulous. **Corolla** with 7, often wavy lobes, open to funnel-campanulate, 4-5cm long, 7-9cm across, fragrant, *pale lilac white, pale lilac pink to pink;* calyx 1-3mm, glabrous or glandular; stamens 14-16, filaments glabrous; ovary stipitate-glandular; style slender, stipitate-glandular to tip (white or yellowish).

This species is hardier than *R. decorum* and *R. vernicosum* and is distinguished from them by the leaf shape and purple-red petiole. The distribution of *R. fortunei* is remote from that of these two species.

Ssp. *fortunei*

Leaves *obovate, 1.8-2.5 x as long as broad.* Usually earlier flowering than ssp. *discolor* in cultivated plants. **Distribution** One of the most widely distributed Chinese species: E. Sichuan, Guangxi, Hunan, Guangdong, Jiangxi, Fujian, Anhui, Zhejiang, 600-900m (2,000-3,000 ft).

Ssp. *fortunei* is the hardiest scented species and is also heat resistant and has been much used in breeding. Unfortunately the flowers are often partly obscured by the new growth. The true species is rare in the U.K. but more common in Scandinavia, Germany and E. North America. Introduced 1855 and several times since. May-early June.

Ssp. *discolor*
(Franch.) D.F. Chamb. 1982 (*R. houlstonii* Hemsl. & E.H. Wilson 1910, *R. mandarinorum* Diels, *R. kirkii* Hort, *R. kwangfuense* Chun & Fang.)

Leaves *oblanceolate, 2.8-4 x as long as broad.* **Distribution** Sichuan, Hubei, Guizhou, Guangxi, Hunan, Anhui, Zhejiang, 1,100-2,100m (3,500-7,000ft) in thickets and woodlands.

Morphologically, the only difference between ssp. *fortunei* and ssp. *discolor* is in leaf shape. What is commonly grown in cultivation under the name ssp. *discolor* is normally a late-flowering plant (June-July), around a month later than ssp. *fortunei*. Many such cultivated ssp. *discolor* plants may in fact be *R. auriculatum* hybrids. Introduced 1900. June-July (as cultivated).

Houlstonii Group. Smaller narrower leaves than typical ssp. *discolor* and flowering in May in cultivation. Davidian states that Houlstonii Group plants always have pink flowers but there are cultivated plants under this name with white flowers. Introduced c. 1900. April-May (June).

Picture 1

1. *R. fortunei* ssp. *fortunei* at Glendoick.

FORTUNEA

Picture 2

Picture 3

Picture 4

Picture 5

2. *R. fortunei* ssp. *fortunei* at Glendoick.
3. *R. fortunei* ssp. *fortunei* from Lu Shan at Loch Alsh House, N.W. Scotland.
4. *R. fortunei* ssp. *discolor* Borde Hill form at Glendoick.
5. *R. fortunei* ssp. *discolor* in Vancouver, Canada. (Photo L. Hodgson).

Picture 6

6. *R. fortunei* ssp. *discolor* Houlstonii Group at Glendoick.

R. glanduliferum
Franch. 1886

Height 2m+. **Leaves** 12-16 x 2-4cm (probably larger), oblong-lanceolate, lower surface *pale glaucous green and glabrous,* petioles glabrous. **Inflorescence** 5-7-flowered, loose, with long rachis, pedicels *densely glandular,* **Corolla** 7-8-lobed, funnel-campanulate, *densely stipitate-glandular on outer surface*, 5-6cm long, white, scented; calyx 1-3mm, glandular; stamens 14-16; ovary and style glandular. **Distribution** N.E. Yunnan, only known in three locations, (possibly also N. Guizhou), c. 2,200m (7,250ft), evergreen and deciduous scrub.

A distinctive species in flower with its long, stipitate glands on the pedicels, corolla and calyx.

The handsome scented species *R. glanduliferum* is little-known, and was not introduced into cultivation until 1995. It is evidently rare in the wild and its habitat is seriously threatened by tree felling. Flowering time unknown.

Picture 1

1. *R. glanduliferum* in N.E. Yunnan, China. (Photo Guan Kaiyun).

R. griffithianum

Wight 1850 (*R. aucklandii* Hook.f. *R. oblongum* Griffith) H2-3

Height to 6m in cultivation, 12-15m in wild, an often sparse shrub or small tree. Bark *flaking and peeling,* usually smooth and of several colours. **Leaves** 10-30 x 4-10cm, oblong-elliptic to oblong-ovate, retained 1-2 years; upper and lower surfaces glabrous; petiole long, 2.5-5cm. **Inflorescence** 3-6-flowered, on a *long, stout, rachis to 8cm, usually slightly glandular.* **Corolla**, with 5 lobes, *very large,* widely campanulate, 3-7cm long and to 15cm+ across, sometimes scented, white, white with green base and/or spotting, sometimes tinged or veined red or pink or blush to deep rose-pink or even yellowish; calyx *large,* 1-2 cm, green or pink tinted, irregular; stamens 12-18; ovary and entire style glandular. **Distribution** E. Nepal, Sikkim, Bhutan, N.E. India, 1,800-2,900m (6,000-9,500 ft) in moist forest, oak, rhododendron, magnolia and deciduous forest.

This species is quite easily distinguished from all other species in the subsection by its huge, 5-lobed corolla, well-developed calyx, long rachis and smooth bark.

R. griffithianum has amongst the largest flowers in the genus, but is fairly tender; considerable shade and shelter is required for successful cultivation. There is some variation in hardiness as well as in size of flower and presence or absence of scent. It has been much used as a parent, raising hybrids such as 'Loderi'. Introduced 1850->, reintroduced 1965->. May.

Picture 2

Picture 3

Picture 4

Picture 1

1. *R. griffithianum,* Arun-Tamur divide, E. Nepal. (Photo G.F. Smith).
2. *R. griffithianum,* Arduaine, W. Scotland.
3. *R. griffithianum,* Weesjes's garden, Vancouver Island, B.C., Canada.
4. Trunk of *R. griffithianum,* Chendebji, C. Bhutan. (Photo I. Sinclair).

R. hemsleyanum

E.H. Wilson 1910 (*R. chengianum* Fang) H4

Height 3-6m, a large upright and somewhat spreading shrub or small tree. Bark roughish, flaking. **Leaves** 10-27 x 4-13cm, ovate to ovate-elliptic, base *deeply auricled-cordate,* margins

FORTUNEA

Picture 1

Picture 2

1. *R. hemsleyanum*
2. *R. hemsleyanum* at the Crystal Springs Test Garden, Portland, Oregon, U.S.A.

usually undulated, retained 2 years; upper surface *bright or pale matt-green,* lower surface with a few minute punctulate hair bases and a few glands but otherwise glabrous. Flower buds long and pointed. **Inflorescence** 5-10 or more-flowered. **Corolla** widely campanulate, 6-7 lobes, 4.5-6cm long, fragrant, white, sometimes tinged rose pink, usually with yellow green throat; calyx c. 1mm; stamens c. 14; ovary and entire style glandular. **Distribution** Only found on Emei Shan, W. Sichuan, 1,000-2,000m (3,600-6,550 ft), in woods and thickets.

This species is distinguished by its usually wavy-edged leaves (reminiscent of banana leaves), the auriculate leaf base, its late flowering and its upright, rather sparse habit.

R. hemsleyanum is a distinctive and exotic-looking species which needs shelter from wind, but is otherwise quite robust and easy to please though the flower buds can be damaged by severe frost. Somewhat prone to powdery mildew. Although discovered by Wilson on Emei Shan on 1904, it wasn't introduced into cultivation (USA) until 1937. Reintroduced 1989. June-August.

A taxon found on Lao Jing Shan, W. of Wenshan, S.E. Yunnan in October 1995, resembles *R. hemsleyanum* with the same undulating leaf margins and auricled base. The habit is not as open as *R. hemsleyanum* and the leaves are not as wide.

Picture 3

Picture 4

Picture 5

3. A mature plant of *R. hemsleyanum* at Glendoick showing the open habit.
4. A leaf of *R. hemsleyanum* showing the characteristic auriculate leaf base and wavy leaf margins.
5. *R. hemsleyanum* aff. Cox & Hutchison 7189 on Lao Jing Shan, S.E. Yunnan, China, showing the wavy-edged leaves, narrower than the type.

R. huianum

Fang 1939

Height 2-9m, a shrub or small tree, branchlets glabrous. **Leaves** 8-16 x *1.8-3.5(-5)cm*, oblong-lanceolate to oblanceolate, apex *cuspidate to acuminate*, base cuneate, lower surface glabrous, somewhat glaucous. Petiole 1-3cm, glabrous. **Inflorescence** 6-12 flowered, loose. **Corolla** 6-7-lobed, open-campanulate, 3-5.9cm long, pale or more often *deep rose-purple* or rosy-red, with nectar pouches; calyx *large, 4-10mm,* cupular or with rounded lobes; stamens 12-14; ovary glandular, style *glandular to tip*. **Distribution** S. & S.E. Sichuan, N.E. Yunnan, N.E. Guizhou, 1,000-2,700m (3,250-9,000ft), broad-leaved forest.

R. huianum is separated from its relatives by its acuminate leaf apex, usually deep rose to purple flowers and its large calyx. The more glandular style and larger calyx separate it from its nearest relative R. davidii.

This should prove to be a valuable addition to the cultivated species in this subsection. Introduced 1994-5->. Probably April.

Picture 1

1. R. huianum showing its deep pink flowers with nectar pouches. (Photo Yang Zhenhong).

R. orbiculare

Decaisne 1877 (*R. rotundifolium* David 1973 Nomen nudum)
H4-5

Height to 3m, a shrub or small tree, compact and dense in sun, leggy in shade. **Leaves** 4-12.5 x 4-8cm, *ovate to orbicular,* tip often *rounded,* base *cordate-auriculate;* upper surface bright matt-green, lower surface glaucous. **Inflorescence** 7-10-flowered, loose. **Corolla** 7-lobed, campanulate, 3.5-5cm long, deep pink, rose, deep rose, to purplish pink or purplish rose, with no markings; calyx 0.5mm, glabrous; stamens 14; ovary stipitate-glandular; style glabrous.

This species is most easily recognised by its almost round leaves with long auricles which often overlap. It has no close relatives in subsection Fortunea, in fact more closely resembling *R. williamsianum* which is much smaller in all parts.

Ssp. *orbiculare*

Leaves *orbicular,* 7-9.5cm long. **Distribution** W. Sichuan, N.E. Guangxi?, 2,000-4,000m (6,500-13,000 ft) in woodlands, thickets and on rocks.

This is a fine species in its best forms with striking foliage. Some forms have floppy trusses and/or a harsh flower colour. Requires to be grown in a sunny situation to form a compact bush. Some recent introductions have a pointed leaf apex. Introduced 1904, 1938, reintroduced 1989-> April-May.

Ssp. *cardiobasis* (Sleumer) D.F. Chamb. 1982

Leaves *ovate orbicular,* up to 12.5cm long. **Distribution** Guangxi, 1,700m (5,400 ft)

This subspecies has longer, more heart-shaped leaves than ssp. *orbiculare* with a truss of 6-7 flowers. There are no plants of ssp. *cardiobasis* of known wild origin in cultivation and plants under this name may all be *R. orbiculare* ssp. *orbiculare* hybrids. Further field work will be necessary to ascertain whether ssp. *cardiobasis* is a good taxon or not. May.

Picture 1

Picture 2

1. R. orbiculare ssp. orbiculare at Wolong Panda Reserve, W. Sichuan, China.
2. R. orbiculare ssp. orbiculare at the Royal Botanic Garden, Edinburgh.

FORTUNEA

Picture 3

Picture 4

Picture 5

3. *R. orbiculare* ssp. *orbiculare* at Glenarn, W. Scotland.
4. *R. orbiculare* ssp. *orbiculare*, at the Weesjes's garden, Vancouver Island, B.C., Canada.
5. *R. orbiculare* ssp. *cardiobasis*, University of British Columbia Asian Garden, Vancouver, Canada., perhaps a ssp. *orbiculare* hybrid.

R. oreodoxa
Franch. 1886 H5

Height to 6m or more in cultivation, less in the wild, an upright to bushy shrub or small tree. Young shoots with a thin tomentum; shoots *thin*. **Leaves** 6-8.5 x 2.2-4cm, narrowly elliptic to oblanceolate-elliptic, retained 1 year; lower surface glaucous with very thin indumentum, not noticeable with naked eye. **Inflorescence** 6-12-flowered, usually loose. **Corolla** openly campanulate with 5-8 lobes, 3-4cm long, white, white flushed pale lilac to pink or rose, with or without purple spots; calyx minute, glabrous or glandular; stamens 12-14; ovary glabrous or glandular; style *glabrous*. **Distribution** Sichuan, Gansu, Hubei 2,100-3,000m (7,000-10,000ft) in woodlands. March-April. Var. *shensiense* from Shaanxi only.

This species differs from other members of the subsection in its thin shoots, smaller leaves, lack of scent and early flowering. It differs from *R. vernicosum* in its glabrous rather than glandular style and its significantly thinner shoots.

R. oreodoxa is a fine, easily grown, vigorous and free-flowering species which is frost hardy in the swelling bud, so normally managing to give some display every year, despite its early flowering. Introduced 1901-1904. Reintroduced 1980's.

Var. *oreodoxa* Franch. (*R. haematocheilum* Craib., *R. limprichtii* Diels, *R. reginaldii* Balf.f.)

Ovary *glabrous*. **Leaves** usually narrower. **Corolla** 7-8 lobed, usually paler than in var. *fargesii*.

Var. *fargesii* (Franch.) D.F. Chamb. 1982
(*R. erubescens* Hutch) H5

Ovary *stipitate-glandular*. **Leaves** usually wider than in var. *oreodoxa*. Pedicels glandular. **Corolla** usually 6-7 lobed, lilac pink to deep pink, heavily speckled inside.

Easily grown, even on poor soils. Leaves curl up readily in cold or sunny weather.

Var. *shensiense* D.F. Chamb. 1982

Corolla (always?) 5-lobed. Ovary stipitate-glandular; pedicels sparsely rufous-tomentose. **Distribution** Shaangxi only.

In cultivation in Goteborg Botanic Garden, Sweden. A plant labelled *R. purdomii* at The Younger Botanic Garden, Argyll may be referable to var. *shensiense*, though to our eyes it is closer to *R. maculiferum*.

Picture 1

Picture 2

Picture 3

Picture 4

Picture 5

Picture 6

Picture 7

1. *R. oreodoxa* var. *oreodoxa* Reginaldii Group Farrer 63 from Gansu. A fine form with relatively frost-hardy flowers, at Glendoick.
2. *R. oreodoxa* var. *oreodoxa* Reginaldii Group Farrer 63 from Gansu at Glendoick.
3. *R. oreodoxa* var. *oreodoxa* on Emei Shan, W. Central Sichuan, China.
4. The interior of the corolla of *R. oreodoxa* var. *oreodoxa* showing the glabrous ovary. (Photo R.B.G., E.).
5. *R. oreodoxa* var. *fargesii*, a deep pink-flowered form at the Royal Botanic Garden, Edinburgh.
6. *R. oreodoxa* var. *fargesii,* a pale-flowered form at Glendoick.
7. *R. oreodoxa* var. *shensiense?* (labelled *R. purdomii*) at the Younger Botanic Garden, Benmore, W. Scotland.

R. praeteritum Hutch. 1922 H4-5. This so-called species was described from a single cultivated plant from a batch of *R. wasonii* seed and is almost certainly a natural hybrid, probably of *R. oreodoxa*. The floccose indumentum on the petiole and the 5-lobed corolla with nectar pouches at the base distinguish it from *R. oreodoxa*.

R. praevernum Hutch. 1920 H5

Height to 3.7m or more, an often compact shrub or small tree. **Leaves** 10-18 x 2.5-6cm, elliptic-oblanceolate, retained 1-2 years; upper and lower surface *entirely glabrous,* lower surface midrib prominent. **Inflorescence** 8-15-flowered, open-topped. **Corolla** campanulate, 5-6cm long, white, pale to deep pink, pinkish-purple or pinkish-lilac, *with large purple or crimson blotch at base;* calyx 1-2mm, glabrous; stamens 13-15; ovary and style glabrous. **Distribution** W. Hubei and E. Sichuan, 1,600-2,500m (5,250-8,250ft) in woods, *Quercus* and bamboo.

This species merges with *R. sutchuenense* but in its typical form differs in its smaller leaves, glabrous lower-leaf-surface midrib and the prominent blotch. It is slower and lower growing than *R. sutchuenense* and *R. calophytum.*

R. praevernum is an easy and quite showy species, but rather early flowering. E. H. Wilson, who introduced this species under Wilson 509, considered it and *R. sutchuenense* to be forms of the same taxon and the above number produced plants of both species. Intermediate forms are known as *R. sutchuenense* var. or x *geraldii.* (see under *R. sutchuenense.*) Introduced 1900. February-April.

Picture 1

Picture 2

3. *R. praevernum* showing the red bud scales on the new growth at the Royal Botanic Garden, Edinburgh.

1. *R. praevernum,* a pink-flowered form, showing the characteristic deep blotch, at the Royal Botanic Garden, Edinburgh.
2. *R. praevernum,* a white-flowered form at the Royal Botanic Garden, Edinburgh.

R. sutchuenense Franch. 1895 H5

Height 4.6-10m, usually a large umbrella-shaped shrub or small tree. **Leaves** coriaceous, margins recurved, up to 30 x 7.6 cm, oblong-lanceolate to narrowly oblong-oval, retained 1-3 years; lower surface midrib with +/- *loose indumentum.* **Inflorescence** c.10-flowered, usually open at the top. **Corolla** widely campanulate with 5-6 lobes, 5-7.5cm long, pale pink, pale pinkish-lilac to pale mauve pink *with red spotting but without blotch;* calyx disc-like, +/- glabrous; stamens 13-15; ovary and style glabrous; stigma *reddish.* **Distribution** W. Hubei and E. Sichuan 1,500-2,400m (5,000-8,000ft) in woods among evergreen oaks or bamboo.

The flower without a blotch, with shorter pedicels, fewer stamens, a reddish and smaller stigma and a different corolla shape differentiates this species from *R. calophytum.* In *R. calophytum* the leaves are usually longer and narrower and not recurved at the margins. *R. praevernum* is smaller growing, has a glabrous midrib on the leaf underside and the flower always has a blotch.

This is one of the toughest larger-growing species, with fine trusses of long-lasting flowers, but it grows best with shelter from wind. It starts blooming at a younger age than *R. calophytum. R. planetum* Balf.f. is probably a natural hybrid of *R. sutchuenense* x *R. oreodoxa.* Introduced 1901-1907. February-April.

R. x giraldii Hutch.

Midway between *R. sutchuenense* and *R. praevernum* and originating from seed of W. 509. Wilson considered the two parents species as two extremes of a single taxon and this should perhaps be considered a variety of *R. sutchuenense.*

It has persistent indumentum on the leaf lower surface midrib, a pronounced blotch in the corolla and it is often a spectacular plant in flower with large trusses.

Picture 1

Picture 2

Picture 3

Picture 4

1. *R. sutchuenense* at Glendoick.
2. *R. sutchuenense*, a rich pink-flowered form at Glendoick.
3. A pale flowered form of *R. sutchuenense* at Windsor Great Park, S. England.
4. *R. x giraldii* at Stonefield Castle, W. Scotland.

Picture 5

Picture 6

5. *R. x giraldii* at the Younger Botanic Garden, W. Scotland, clearly showing the blotch inherited from *R. praevernum*.
6. *R. planetum* is a natural hybrid, probably between *R. oreodoxa* and *sutchuenense*.

R. vernicosum

Franch. 1898 (*R. adoxum* Balf.f. & Forrest, *R. lucidum* Franch., *R. araliiforme* Balf.f. & Forrest, *euanthum* Balf.f. & W.W. Sm., *R. hexamerum* Hand-Mazz., *R. rhantum* Balf.f. & W.W. Sm, *R. sheltonii* Hemsl. & E.H. Wilson) H4-5

Height to 7.6m, usually much less, usually forming a rounded, rigid shrub or small tree. **Leaves** 6-12 x 3-5cm, elliptic to ovate, to oblong-oval to oblong elliptic, retained 1-2 years; lower surface pale, with minute punctulate hairs. **Inflorescence** 5-12-flowered, often quite compact. **Corolla** widely funnel-campanulate, with 6-7 lobes, sometimes reflexed, 3.8-5cm long, palest pink, rose-lavender to bright rose, rarely white, +/- crimson spotting; calyx c. 2mm, stipitate-glandular; stamens 14 *glabrous*; ovary and style densely glandular with *dark red glands*. **Distribution** Yunnan, Sichuan, S.E. Tibet, and Gansu, 2,700-4,300m (9,000-14,000ft) in clearings and margins of forests, open and dry situations and by streams.

In subsection Fortunea only *R. vernicosum* has red glands on the ovary and style; all the others have white or yellowish glands or are glabrous. It differs from its closest relatives *R. decorum* and *R. fortunei* in its lack of scent, usually deeper-coloured flowers, glabrous stamens, usually smaller leaves and its more compact truss. It is distinguished from R. *oreodoxa* in its thicker shoots, later flowering, more compact truss and glandular style. The leaf upper surface of R. *vernicosum* becomes sticky, waxy and shiny if heated over a flame.

R. *vernicosum* is one of the most widely-distributed and therefore variable species in China, and this accounts for the many names under which the species was previously known. Very variable in hardiness, partly due to the early growth of some forms. Hybridises with many species in the wild including R. *wardii* (R. *chlorops*) and R. *decorum*. Introduced 1904->, reintroduced 1980s->. April-May.

Picture 1

Picture 2

Picture 3

Picture 4

Picture 5

Picture 6

1. R. *vernicosum* in forest at Shuo Duo Gan, N.W. Yunnan.
2. R. *vernicosum*, Rilong, N. Sichuan, China.
3. R. *vernicosum* 'Sidlaw' P.C. at Glendoick.
4. A pale-flowered form of R. *vernicosum* Yu 13809 at the Royal Botanic Garden, Edinburgh.
5. Young growth of R. *vernicosum*, Bita Hai, Zhongdian, N.W. Yunnan, China.
6. The interior of a flower of R. *vernicosum* showing the ovary and style with the characteristic red glands. (Photo R.B.G., E.).

Subsection Fulgensia

Now considered a mono-typic subsection, though sometimes *R. sherriffii* and *R. miniatum* are placed here.

R. fulgens
Hook.f. 1849 H4

Height 1.5-4.5m, a shrub to small tree, usually fairly compact and tidy. Branchlets glabrous, with *brilliant crimson bracts on young growth*; bark *pinkish-grey to reddish-brown, smooth and peeling*. **Leaves** 5.3-13 x 2.4-7cm, broadly ovate to obovate, apex rounded, base rounded to cordate, retained 1-2 years; upper surface fairly glaucous when young, *dark and shiny* when mature, lower surface with a *unistrate indumentum*, usually thick, fawn to fulvous; petiole usually *1-2cm long*. **Inflorescence** 8-14-flowered, *compact and rounded*. **Corolla** tubular-campanulate, with dark nectar pouches, 2-3.3cm long, deep blood red to bright scarlet, without spots; calyx small, 1-3mm, glabrous; ovary and style glabrous; capsule curved. **Distribution** E. Nepal, Sikkim, Bhutan, S. Tibet and W. Arunachal Pradesh, but usually not common, 3,200-4,300m (10,500-14,000ft), mixed forest and on moorland.

A distinct species, its closest relation being *R. succothii* from which it differs in its long petiole and in the presence of indumentum on the leaf under-surface. *R. fulgens* differs from other species in subsection Barbata in the relatively thick indumentum on the leaf underside and the absence of bristles. *R. sherriffii* has somewhat similar ovate leaves with thick and continuous indumentum but differs in its smaller stature and leaves and smaller trusses of fewer, smaller red flowers. Out of flower it could be confused with *R. campanulatum* but it differs in its smooth bark.

R. fulgens is handsome in foliage and flower, although the truss tends to be rather small and the flowers are early.

Quite easily grown. Introduced 1850, 1915->, reintroduced 1988->. February-April.

Picture 1

Picture 2

Picture 3

1. A compact bush of *R. fulgens*, a form from Maryborough, S. Ireland, growing at Glendoick.
2. *R. fulgens*, a form from the Sherriff garden at Ascreavie, growing at Glendoick.
3. *R. fulgens* showing the characteristic thick, woolly, continuous indumentum on the leaf underside.

Subsection Fulva

Height 2-10m, mostly 3-10m in cultivation, forming large upright shrubs or trees. Branchlets fulvous to greyish-tomentose; bark rough. **Leaves** coriaceous, 6-27 x 2-9cm, elliptic to oblong; upper surface with indumentum at first, later glabrous, lower surface with a dense unistrate or bistrate, *granular*, woolly or silky indumentum. **Inflorescence** 5-30-flowered, *dense*. **Corolla** campanulate to funnel-campanulate, 5-lobed, white to deep rose, usually blotched; calyx minute to small, stamens 10; ovary glabrous, capsule *narrow and curved*. **Distribution** West Yunnan, S. Sichuan, S.E. Tibet.

Related to subsections Taliensia and Argyrophylla but distinguished by the leaves with granular indumentum on the lower surface, the dense trusses of white to pale pink flowers, usually with a blotch and spotting, and the slender, glabrous ovary.

This subsection contains long-lived species with handsome foliage which grow best in sheltered woodland to protect the foliage and rather early flowers. Both are quite commonly cultivated.

R. fulvum

Balf.f. & W.W. Sm. 1917 H4

Height 3-9m, a rounded to upright shrub or small tree. Branchlets whitish to rufous-tomentose, bark moderately rough, greyish to grey-brown. **Leaves** 6-27 x 2-9cm, oblanceolate to elliptic, upper surface dark, shiny to medium matt green, retained 1-3 years; lower surface with a fine to coarsely granular fawn to cinnamon bistrate indumentum, consisting mostly of unusual *capitellate hairs*. **Inflorescence** 8-20-flowered, compact to fairly compact. **Corolla** campanulate, 2.6-4.9cm long, white to pink, usually blotched; calyx minute, 1-2mm glabrous; ovary usually glabrous, style glabrous.

R. fulvum differs from *R. uvarifolium* in its less matted indumentum made up of capitellate rather than dendroid hairs. It somewhat resembles *R. rex* ssp. *fictolacteum* but usually has smaller leaves, a glabrous as opposed to a felted ovary and 10 rather than 14-16 stamens.

Ssp. fulvum

Leaves *10-21cm long*, upper surface *dark shiny green*, lower surface with *finely granular cinnamon* indumentum. **Distribution** Yunnan and adjacent E. Upper Burma, 2,500-4,000m (8,250-13,000ft), fir and deciduous forest.

Although this subspecies merges with ssp. *fulvoides*, the typical forms of each subspecies are easily distinguished; ssp. *fulvum* has a shorter leaf of a dark shiny green and a finer granular cinnamon indumentum.

In foliage ssp. *fulvum* is a fine plant, and it is usually very free-flowering when mature. The leaves curl up readily during frost or bright sunshine. Introduced 1912->. March-April.

Ssp. fulvoides

D.F. Chamb. (*R. fulvoides* Balf.f. & Forrest)

Leaves *10-27cm long*, upper leaf surface *medium matt-green*, lower surface with a *coarsely granular, fawn to brown* indumentum. **Distribution** N.W. Yunnan and S.E. Tibet, 3,200-4,500m (10,500-14,750ft), fir and deciduous forest, generally to the north of ssp. *fulvum*.

The leaves of ssp. *fulvoides* are longer and narrower than those of ssp. *fulvum* with a matt rather than shiny upper surface and an indumentum which is less continuous, coarser, more granular and lighter in colour. It differs from *R. uvarifolium* in its less matted indumentum made up of capitellate rather than dendroid hairs. Morphologically, this plant lies somewhere between ssp. *fulvum* and *R. uvarifolium*.

Ssp. *fulvoides* has recently been reinstated at subspecific status. It is not generally such a fine foliage plant as ssp. *fulvum* but is still very handsome when well-grown. Some introductions have rather small flowers. Introduced 1917-39. Reintroduced 1992. March-April.

Picture 1

Picture 2

1. *R. fulvum* ssp. *fulvum* Farrer 874 (*R. mallotum* in the foreground) at Glendoick.
2. *R. fulvum* ssp. *fulvum* at the Royal Botanic Garden, Edinburgh, showing the fine-textured, continuous indumentum.

FULVA

Picture 3

Picture 4

Picture 5

Picture 6

Picture 7

6. A deep pink-flowered *R. fulvum* ssp. *fulvoides* at the Royal Botanic Garden, Edinburgh.
7. *R. fulvum* ssp. *fulvoides* at the Valley Gardens, Windsor Great Park, S. England, showing the somewhat sparse, granular leaf indumentum.

3. *R. fulvum* ssp. *fulvum*, the best form from Crarae, at Glendoick.
4. The leaf lower surface of ssp. *fulvoides* (above) with coarse, granular, discontinuous indumentum and ssp. *fulvum* (below) with dense, capitellate, indumentum of a darker colour. (Photo R.B.G., E.).
5. *R. fulvum* ssp. *fulvoides* at the Younger Botanic Garden, Benmore, W. Scotland.

R. uvarifolium Diels 1912 H3-4

Height 2-10m, a shrub or tree, fairly compact when young. Young growth *white, suede-like*, branchlets white to grey tomentose; bark rough, greyish to brown. **Leaves** 7-27 x 2.2-7.2cm, oblanceolate to oblong-obovate, retained for 2 years; upper surface *semi-rugulose*, glabrous when mature, usually dark green, lower surface with a very variable indumentum, unistrate or bistrate, usually *white to ash-grey*, matt to woolly, continuous to patchy, hairs dendroid or ramiform. **Inflorescence** 6-20+ flowered. **Corolla** campanulate to funnel-campanulate, 2.7-4cm long, white to rose, usually blotched and spotted, sometimes not; calyx minute 0.5-1mm usually glabrous; ovary usually glabrous; style glabrous.

This very variable species differs from *R. fulvum* ssp. *fulvoides* chiefly in its indumentum hair-type being dendroid or ramiform as opposed to capitellate. *R. uvarifolium* varies in garden value in foliage and flower but is a handsome species as its best, especially in young growth. Introduced 1913->, reintroduced 1990's->.

Var. *uvarifolium*
(*R. dendritrichum* Balf.f. & Forrest, *R. mombeigii* Rehder & E.H. Wilson, *R. niphargum* Balf.f. & Kingdon-Ward)

Indumentum *plastered to woolly, unistrate to bistrate, thick to thin*. E. Arunachal Pradesh, N.W. Yunnan, into S.W. Sichuan, common, 2,100-4,300m (7,000-14,000ft), conifer and mixed forest.

Var. *griseum* Cowan 1953

Leaves often smaller than type, paler above, broadly elliptic to lanceolate with rounded bases. Indumentum *thin, silky, almost matted whitish*. **Inflorescence** 10-15-flowered. **Corolla** white with a crimson blotch at the base. **Distribution** Tsangpo area of S.E. Tibet, 2,300-4000m (7,500-13,000ft), much further west than var. *uvarifolium*.

As cultivated, this variety looks quite distinct from var. *uvarifolium* in its narrower leaves with thin whitish indumentum, and it tends to look rather unhealthy. Wild material we have examined in S.E. Tibet does not appear to match cultivated material, and further fieldwork needs to be done. Introduced 1924, 1947. March-April.

Picture 1

Picture 2

Picture 3

Picture 4

Picture 5

Picture 6

1. *R. uvarifolium* var. *uvarifolium*, Napa Hai, Zhongdian, N.W. Yunnan, China.
2. *R. uvarifolium* var. *uvarifolium* Kingdon-Ward 5660, at the Valley Gardens, Windsor Great Park, S. England.
3. *R. uvarifolium* var. *uvarifolium* 'Yangtze Bend' A.M., Glendoick.
4. *R. uvarifolium* var. *uvarifolium* 'Reginald Childs' A.M., Glendoick. This clone is probably a hybrid.
5. The leaf lower surface of *R. uvarifolium* var. *uvarifolium*, showing indumentum. The indumentum of this species is very variable and may not look like the examples here. (Photo R.B.G., E).
6. *R. uvarifolium* var. *griseum* from Ludlow and Sherriff collection, at the Royal Botanic Garden, Edinburgh.

7. *R. uvarifolium* var. *griseum* on the north side of the Doshong La, S.E. Tibet.

Subsection Glischra

Height 1-9m, bushy to erect shrubs or trees, sometimes open with age. Branchlets often *setose-glandular*; bark roughish, greyish to light brown. Perulae often persistent. **Leaves** to 27 x 8cm, ovate to oblanceolate; upper surface glabrous to bristly, lower surface with *stipitate glands or bristles* or dense ramiform indumentum. **Inflorescence** 6-20-flowered, usually lax. **Corolla** campanulate to funnel-campanulate, without nectar pouches, *white to pink and rose-purple*, often blotched and spotted; calyx *well-developed*; stamens 10; ovary densely glandular; style glabrous or glandular at base. **Distribution** Sichuan, Yunnan, Tibet, Burma.

Subsection Glischra is characterised by the glandular bristles on branches, petioles and sometimes on the pedicels, ovary and leaf midrib or even over the whole leaf surface. Another feature is the prominent calyx. The closely related subsections Glischra and Barbata share the characteristic of prominent bristles but differ chiefly in the flower colour: pink to purplish-pink in subsection Glischra and usually red in subsection Barbata. In the indumentum of *R. recurvoides* and *R. crinigerum* a relationship can be seen with part of subsection Taliensia.

The species in subsection Glischra have distinctive and striking foliage and often showy flowers. The commonest in cultivation are *R. crinigerum* and *R. glischrum*.

R. adenosum
Davidian 1978 (*R. glischrum* var. *adenosum* Cowan & Davidian, *R. kuluense* D.F. Chamb.) H4

Height 2-5m (less in cultivation), a usually rounded shrub. Branchlets usually glandular-hairy. **Leaves** *7-10.5 x 2.4-3.4cm,* ovate-lanceolate to lanceolate, retained 2 years; upper and lower surfaces with *stalked glands (hairs) +/- persistent*. **Inflorescence** *6-8-flowered*. **Corolla** funnel-campanulate, 3.8-5cm long, pale pink or white tinged pink, with crimson spots; calyx *4-5mm*; ovary densely glandular; style glabrous. **Distribution** Kulu, S.W. Sichuan, 2,700-3,500m (9,000-11,500 ft), in *Abies* forest and swamps.

This species differs from the closely related *R. glischrum* in its shorter calyx, in the short, bristly glands on the leaf underside, the fewer-flowered inflorescence and the smaller leaves. It also comes into growth earlier. *R. habrotrichum* has crimson-purple leaf/branchlet glands (hairs) and much wider leaves while *R. crinigerum* has a continuous indumentum on the lower leaf surface.

R. adenosum was only quite recently recognised as a species. It is free-flowering at maturity, relatively easy to please and deserves to be much more widely grown. It seems only to have been introduced under Rock 18226 and 18228 in 1929. April-May.

1. A pink-flowered form of *R. adenosum* Rock 18228 from Kulu, S.W. Sichuan.

GLISCHRA

Picture 2

Picture 3

2. A pink-flowered form of *R. adenosum* Rock 18228 at the Royal Botanic Garden, Edinburgh.
3. A white-flowered form of *R. adenosum* Rock 18228 from Kulu, S.W. Sichuan, at Glendoick.

R. crinigerum Franch. 1908 H3-4

Height 1-5m, usually spreading or compact. Young growth *very glandular with (usually) persistent* bud scales, and sparse woolly indumentum. **Leaves** 7.5-20 x 2-5cm, lanceolate, oblong-oblanceolate to elliptic, very variable, retained 2-4 years; upper surface *usually very shiny*, dark green, sometimes bullate, lower surface with a dense or sparse *indumentum* consisting of ramiform hairs with sparse stipitate glands, ranging from brown, fawn, yellowish, to reddish or whitish; petiole often glandular. **Inflorescence** 10-20-flowered, lax to compact, pedicel glandular and hairy. **Corolla** campanulate, 3-4cm long, white to pink to rose, white striped pink, blotched and/or spotted red; calyx *split* to 1cm long, stipitate glandular; ovary stipitate-glandular; style glabrous or glandular at base. **Distribution** Yunnan, Yunnan-Burma frontier, Arunachal Pradesh and S.E. Tibet, 2,400-4,300m (8,000-14,000ft) in open conifer forests, among bamboo, boulders and on cliffs.

This species differs from the other members of this subsection (except the much lower growing *R. recurvoides*) in its continuous indumentum. Other useful diagnostic features are the split calyx, the bullate, shiny leaves and the sticky young growth.

Var. *crinigerum* (*R. ixeunticum* Balf.f. & W.W. Sm.)

Leaves sparsely glandular beneath, with a dense, matted indumentum.

R. crinigerum var. *crinigerum* has a wide distribution and is very variable in hardiness and garden value. The best forms have fine flowers (especially the striped ones) and attractive bullate leaves but many forms are rather indifferent. Introduced 1914-49. April-May.

Var. *euadenium* Tagg & Forrest 1931.

Leaves more densely glandular beneath, usually with a sparse indumentum.

This rarely cultivated variety appears to be an intermediate or natural hybrid between *R. crinigerum* and *R. glischrum* which occur together on the Burma-Yunnan frontier. Some forms have fine pink flowers with a deeper blotch. It is variable, showing the whole range between the two species but the indumentum usually consists of sparse hairs with a dense covering of stipitate glands.

Picture 1

Picture 2

1. *R. crinigerum* var. *crinigerum*, a pure white-flowered form at the Royal Botanic Garden, Edinburgh.
2. A form of *R. crinigerum* var. *crinigerum* Rock 10983, with pale pink, spotted flowers, at the Royal Botanic Garden, Edinburgh.

Picture 3

Picture 4

Picture 5

Picture 6

3. A pale pink-flowered *R. crinigerum* var. *crinigerum* at Glendoick.
4. A unusually heavily-striped *R. crinigerum* var. *crinigerum* at the Royal Botanic Garden, Edinburgh.
5. New growth of *R. crinigerum* var. *crinigerum* showing a form with dense bristles on the branchlets. Some forms have few or no bristles.
6. *R. crinigerum* var. *crinigerum* showing the thick leaf indumentum (which can vary greatly in colour and thickness), concealing the veins on the leaf lower surface. (Photo R.B.G. E.).

Picture 7

7. The rare *R. crinigerum* var. *euadenium* at the Royal Botanic Garden, Edinburgh. This variety is considered to be a hybrid/intermediate between *R. crinigerum* ssp. *crinigerum* and *R. glischrum*.

R. glischroides
Tagg & Forrest 1930. H3-4

Height 2-4.5m; usually a bushy shrub. Shoots *densely bristly*. **Leaves** 7-17 x 2.5-4.5cm, oblong-lanceolate to oblanceolate; *upper surface very bullate,* lower surface veins with *a thin covering of indumentum,* young bud scales very sticky; petiole with dense, red or purplish bristles. **Inflorescence** 6-10-flowered, loose. **Corolla** campanulate, 3.8-5cm long, white flushed rose to pale, rose or mauve pink with a maroon blotch; calyx to 2cm, crimson purple, glandular, margin with bristles; pedicels *red, setose*; ovary glabrous but covered in bristles. **Distribution** N.E. Upper Burma-Tibet frontier 3,000-3,700m (10,000-12,000 ft), in thickets and on rocky slopes.

The smaller, more bullate leaves with a thin indumentum on the lower surface, the bristly pedicel and the usually more bushy, dense habit distinguish *R. glischroides* from *R. glischrum*. *R. glischroides* is closely related to *R. vesiculiferum*, differing in the absence of white hairs on the leaves etc.

Picture 1

1. *R. glischroides* Forrest 29588, a fine bush in the Valley Gardens, Windsor Great Park, S. England.

Recently regaining specific status, *R. glischroides* has handsome bullate foliage and is free-flowering from quite a young age. Some forms have a tendency to come into growth very early in the season, making them vulnerable to frosted foliage and bark-split. Introduced 1925. February-April.

Picture 2

Picture 3

2. *R. glischroides* Forrest 29588, a fine bush in the Valley Gardens, Windsor Great Park, S. England, showing the deeply veined leaves.
3. *R glischroides* showing the new growth with the characteristic red bristles on the branchlets and petioles.

R. glischrum
Balf.f. & W.W.Sm. 1916 H3-4

Height 1-9m in wild; usually smaller in cultivation, a shrub or small tree. Branchlets *densely bristly-glandular*. **Leaves** 10-25 x 2.5-7cm, oblanceolate to occasionally obovate, retained 1-3 years; upper surface smooth, with or without bristles, slightly rugulose, lower surface usually covered with hairs or bristles; petiole c. 2cm, bristly. Flower and leaf buds *sticky*. **Inflorescence** 6-15-flowered. **Corolla** campanulate, c.4cm long, rose-pink, purple-rose, near scarlet, pink to almost white, usually prominently blotched and spotted crimson; bud scales *usually persistent during flowering*; calyx red, lobes unequal, fringed with hairs; ovary densely glandular; style glabrous or glandular at base; capsule with a persistent calyx.

The bristly branchlets and petioles are the main distinguishing feature of this species, as well as the bristles on the lower leaf surface which are much longer than those of *R. adenosum*. *R. glischroides* differs in its more bullate leaves with a thin layer of indumentum on the veins on the lower surface, and in its generally more dense growth habit. *R. crinigerum* usually has a dense indumentum, although intermediates do occur as *R. crinigerum* var. *euadenium*. *R. vesiculiferum* differs in its white as opposed to green or brown hairs/bristles. Those of *R. habrotrichum* are crimson-purple.

Ssp. *glischrum*.

Leaf upper surface *not rugose* and *lacking* bristles, at least when mature, lower surface bristly but lacking a tomentose indumentum. **Distribution** Yunnan, Upper Burma, S.E. Tibet, 2,100-4,300m (7,000-14,000 ft), common in woodlands and in the open.

Ssp. *glischrum* is very variable and makes a fine and showy plant in its best forms. In Britain the most commonly seen forms are those with flowers of white, flushed rose and rose with a crimson blotch. Introduced 1914 19. Reintroduced 1988->. Late April-May.

Ssp. *rude* (Tagg & Forrest) D.F. Chamb. 1978. H3-4

Height to about 2.7m. Leaf and flower buds very sticky. **Leaves** with bristles/glands on *upper surface* as well as on the branchlets, petiole and lower leaf surface. **Inflorescence** c. 10-flowered. **Corolla** openly campanulate, 2.5-3cm long, pink to cherry-pink, purple-pink, marked crimson. **Distribution** Salween-Kiuchang divide, Yunnan, S. E. Tibet, 3,400-3,700m (11,000-12,000 ft).

Ssp. *rude* differs from ssp. *glischrum* in its bristly/glandular leaf upper surface and its generally paler, yellowish-green leaves, though intermediate forms exist.

Rare in gardens; most cultivated ssp. *rude* originated from a group of plants at Glenarn, Scotland (Kingdon-Ward 10952). Sometimes rather hard to please, but worth the effort as the flowers are very striking. Botanically, ssp. *rude* is doubtfully distinct from ssp. *glischrum* which itself sometimes has persistent bristles on the upper leaf surface; intermediate forms occur in S.E. Tibet. Introduced 1924. Reintroduced 1995-96. May.

GLISCHRA

Picture 1

Picture 2

Picture 3

Picture 4

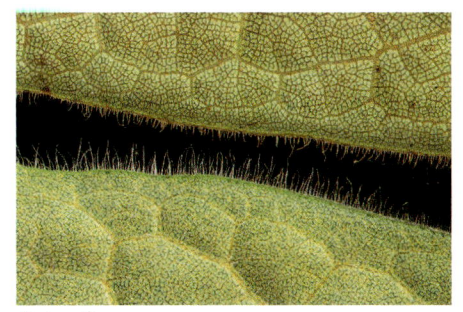

Picture 5

Picture 6

Picture 7

1. *R. glischrum* ssp. *glischrum* at 3,100m, Xue Long Shan, near Weixi, N.W. Yunnan, China, growing with *R. coriaceum* and *R. uvarifolium*.
2. *R. glischrum* ssp. *glischrum*, Ta Pao Shan, near Weixi, N.W. Yunnan, China. It is unusual to find such a well-shaped specimen in the wild.
3. A two-toned pink-flowered form of *R. glischrum* ssp. *glischrum*, Xue Long Shan, near Weixi, N.W. Yunnan, China.
4. *R. glischrum* ssp. *glischrum* Forrest 25619 (previously wrongly identified as *R. crinigerum* var. *euadenium*.)
5. *R. glischrum* ssp. *glischrum* showing the typically deeply veined leaves, loosely covered with hairs on the leaf lower surface. Not all forms have the bristles on the leaf margins. (Photo R.B.G., E.).
6. *R. glischrum* ssp. *rude* at Glendoick.
7. *R. glischrum* ssp. *rude*, one of the original cultivated plants at Glenarn, W. Scotland, which is almost certainly from Kingdon-Ward 10952.

Picture 3

8. Leaf upper surface of *R. glischrum* ssp. *rude* showing the distinctive bristles, usually absent on the upper surface of the leaf of ssp. *glischrum*. (Photo R.B.G., E.).

R. habrotrichum

Balf.f. & W.W. Sm. 1916 H3-4

Height 1.2m, a usually bushy, occasionally leggy shrub. Branchlets *densely clad with reddish-purple, glandular bristles*. **Leaves** 10-18 x 4-8cm, elliptic or rarely elliptic oblong, margin with stalked glands, retained 2 years; upper surface slightly rugose, lower surface glabrous apart from bristles on the midrib and veins; petiole with thick dense *reddish-purple bristles*. **Inflorescence** to 15-flowered, usually compact. **Corolla** funnel-campanulate, to 5cm long, plum-rose to clear rose to white-striped-pink, spotted crimson; calyx irregular with 5 distinct lobes; ovary densely glandular; style glandular in lower third. **Distribution** Yunnan-Burma frontier, 2,700-3,700m (9,000-12,000 ft) in open situations, thickets and among rocks.

This species differs from the closely related *R. glischrum* in its coloured bristles and its wider, shorter leaves which are glabrous as opposed to bristly on the lower surface. It differs from *R. glischroides* in the glandular but not bristly petiole and in the wider, rounder leaf shape.

R. habrotrichum is one of the finest in the subsection for foliage and flower and it blooms quite freely from a surprisingly young age. It requires good drainage. Introduced 1912-38. April-May.

R. diphrocalyx Balf.f. 1919. H3. A natural hybrid of *R. habrotrichum* probably crossed with a member of subsection Neriiflora such as *R. catacosmum*. It has bristly foliage and crimson flowers with a large cup-shaped calyx.

Picture 1

Picture 2

Picture 3

1. *R. habrotrichum* at the Royal Botanic Garden, Edinburgh.
2. *R. habrotrichum* Forrest 15778 at the Royal Botanic Garden, Edinburgh.
3. *R. habrotrichum* showing the characteristic stalked glands on the leaf margins. (Photo R.B.G., E.).

R. recurvoides

Tagg & Kingdon-Ward 1932 H4-5

Height 1-1.5m, a usually dense and compact shrub. *Perulae persistent*. Branchlets setose-glandular. **Leaves** 3-7cm x 1-

3.2cm, lanceolate-oblanceolate, margins recurved; upper surface *rough* and shiny, lower surface with a dense, thick, tawny to cinnamon indumentum of ramiform hairs; Petiole bristly. **Inflorescence** 4-7-flowered, loose. **Corolla** campanulate, 2.6-3cm long, white, flushed pink to rose with crimson spots (no blotch); pedicel bristly; calyx 8-10mm, glandular; ovary densely setose-glandular; style glabrous. **Distribution** N.E. Upper Burma 3,400m (11,000 ft) on steep granite slopes.

A very distinctive species which is the smallest member of the subsection. It is quite easily distinguished from the lower-growing species in subsection Taliensia such as *R. roxieanum* by its bristly stems, pedicels and petioles.

R. recurvoides has only been collected once (Kingdon-Ward 7184) in 1926 and may have originated as a natural hybrid of *R. crinigerum* with a species in subsection Taliensia. One of the finest dwarf species for foliage, and worth growing for this alone. Somewhat variable in width of leaf and thickness of indumentum. In cultivation it is so slow-growing that it rarely reaches 1m. March-May.

Picture 1

Picture 2

Picture 3

Picture 4

3. Young growth of *R. recurvoides* showing the glandular bristles and persistent perulae.
4. *R. recurvoides* showing the typical thick indumentum. (Photo R.B.G., E.).

1. *R. recurvoides* Kingdon-Ward 7184 at the Royal Botanic Garden, Edinburgh.
2. A clone from Keillour, Perthshire, Scotland of *R. recurvoides* growing at Glendoick. This clone has unusually thick indumentum on the lower surface of the leaf.

R. spilotum
Balf.f. & Farrer 1922 H3-4

Height 1-1.5m, usually bushy. Young shoots glandular-setose. **Leaves** 7-11 x 3-4.2cm, upper surface *glabrous on maturity*, lower surface with thin indumentum, *setulose glands only on lower half of midrib*. **Inflorescence** 5-12-flowered, loose. **Corolla** 2.8-4cm long, pink with a basal blotch; calyx c.10mm long, glandular-ciliate; style glabrous, ovary densely stipitate-glandular. **Distribution** N.E. Upper Burma, in alpine woodland.

R. spilotum is probably a natural hybrid between a member of subsection Glischra and a member of subsection Selensia. It was reported to be rare in the wild. Plants in cultivation are not of known wild origin and may be garden hybrids.

This can be a pretty plant in the best selections with neat small heavily blotched flowers. It is not very vigorous. Introduced 1920? April-May.

1. *R. spilotum* at the Royal Botanic Garden, Edinburgh, showing the characteristic deep red blotch. Some forms are pinker than this one.

R. vesiculiferum

Tagg 1931 H3-4

Height to 3m, broadly upright. Branches densely glandular bristly. **Leaves** 8-16 x 2.5-5cm, oblong-lanceolate to oblong-oblanceolate, margin ciliate; lower surface *very hairy*, (hairs vesicular and *white*) especially on the midrib; base of midrib bristly; petiole with bristles and *white* hairs. **Inflorescence** 7-15-flowered, usually loose, pedicels have a few white hairs. **Corolla** campanulate, 3-3.5cm long, almost white through pink to rosy-purple with a deep blotch; calyx purplish with long lobes; ovary long stalked glands and shorter hairs; style with stalked glands and hairy in the lower third; capsule covered with *white hairs*. **Distribution** Yunnan, Burma-Tibet border 2,400-3,400m (8,000-11,000ft) in thickets and *Abies* and *Rhododendron* forests.

This species is distinguished from its closest relative *R. glischroides* by the white hairs on the lower leaf surface, pedicels, ovaries, seed capsule etc.

R. vesiculiferum is probably best considered an extreme form of *R. glischrum* distiguished by its white bristles. Very rare in cultivation. Introduced 1926-38. April-May.

Subsection Grandia

Height 1.20-30m, broadly upright large shrubs or trees. Branchlets stout, glabrous to tomentose; bark rough. **Leaves** to 70+ x 30cm, oblanceolate to broadly elliptic; upper surface glabrous on maturity, lower surface usually with a *plastered,* indumentum (except *R. macabeanum* and *R. balangense*). **Inflorescence** 10-30-flowered, usually large and fairly compact. **Corolla** with *6-10 lobes* (except *R. balangense*), fleshy, tubular-campanulate or campanulate, *oblique or ventricose,* usually without nectar pouches, white, yellow, pink to purple or crimson; calyx minute, 1-2mm; stamens 12-20 (except *R. balangense*); ovary usually tomentose, occasionally glabrous, or glandular (*R. grande*); style glabrous. **Distribution** N.E. India, Nepal, Bhutan, Burma, Tibet, Yunnan, Sichuan, S. Gansu, Vietnam.

A fairly natural grouping of large-growing species, subsection Grandia is closely related to subsection Falconera, differing in the plastered indumentum (except *R. macabeanum* and *R. balangence*) rather than the woolly or spongy indumentum of cup-shaped hairs of subsection Falconera.

This subsection includes some of the giants of the genus with amongst the largest leaves and trusses. They require shelter and only grow to perfection in in a moist mild climate with moderate temperature fluctuations. The commonest in cultivation are probably *R. macabeanum*, and *R. sinogrande*.

R. balangense

W.P. Fang 1983 H4-5?

Height 5-8m, a fairly open shrub or small tree. Leaf bud scales *persistent.* **Leaves** 6-15 x 3.5-8cm, obovate to elliptic-obovate; upper surface glabrous, lower surface with a *loose, floccose greyish-white* indumentum; petiole glabrous, *winged*. **Inflorescence** 13-15-flowered, fairly loose. **Corolla** campanulate, *5-lobed*, 3.5-4cm long, tinged pink in bud, opening to white or white with pink stripes on lobes and a few small spots; calyx 1-2mm long; stamens *10*, puberulous in lower half; ovary and style glabrous. **Distribution** So far only known from Wolong Valley towards Balang, N.W. Sichuan, 2,350-3,400m (nearly 8,000-11,000ft), steep, partially wooded slopes.

The closest relation to this species would appear to be *R. watsonii*, and *R. balangense* may have originated as a natural hybrid of that species perhaps crossed with a subsection Taliensia species such as *R. rufum*. This would account for its having only 10 stamens but not for its persistent bud scales. It has now formed stable populations and has been accepted as a distinct species, though perhaps it should be placed in subsection Taliensia rather than subsection Grandia.

R. balangense has only recently been introduced (1991->), little is yet known of its behaviour in cultivation, though it should make a handsome foliage plant. From its location, it should be bordering on H5 hardiness. May-June in wild, probably April-May in cultivation.

Picture 3

3. *R. balangense* showing the characteristic, loose greyish-white indumentum on the leaf lower surface, at Wolong Panda Reserve, N.W. Sichuan, China.

Picture 1

Picture 2

1. *R. balangense* hugging the cliffs in Wolong Panda Reserve, N.W. Sichuan, China.
2. *R. balangense* with white flowers, flushed pink when opening, Wolong Panda Reserve, N.W. Sichuan, China.

R. grande

Wight 1847 (*R. argenteum* Hook.f., *R. longifolium* Nutt., *Waldemarea argenta* Klotzsch) H2-3

Height 6-15m, a loosely-branched tree, often with a single trunk. Branchlets whitish tomentose. Growth often *very early*. **Leaves** 15-46 x 8-13cm, elliptic to oblanceolate, retained 2-3 years; upper surface slightly rugulose, often medium-green, lower surface with a plastered or occasionally loose, woolly indumentum, *white, silvery to fawn*. **Inflorescence** 15-25-flowered. **Corolla** with 8 lobes, oblique-campanulate, with 8 purple nectar pouches, 5-7cm long, cream, pale yellow to pink with blotch and spots; calyx c. 1mm, glandular; stamens 15-16; ovary densely *glandular*, sometimes eglandular and also tomentose. **Distribution** E. Nepal, through Sikkim, Bhutan into C. Arunachal Pradesh and S. Tibet, 2,100-3,200m (7,000-10,500ft), evergreen and mixed forest.

Picture 1

1. Mossy temperate rain forest, Subansiri Division of Arunachal Pradesh, India, with several trees of *R. grande*.

GRANDIA

This species differs from *R. sinogrande* in its smaller, more lanceolate rather than oval/elliptic leaves, its earlier growth and flowers, and its glandular ovary. It is separated from *R. sidereum* by its larger leaves, the much earlier flowers and the glandular ovary.

A fine plant for relatively mild gardens. In colder gardens it is important to select hardiest forms of *R. grande* that do not grow too early in the season. Only the mildest British gardens are suitable for most introductions. Introduced 1850-1937 and 1965->. February-April.

Picture 5

5. The leaf lower surface of *R. grande* showing the typical pale, plastered indumentum. (Photo R.B.G., E.).

Picture 2

Picture 3

Picture 4

2. *R. grande*, a relatively hardy form which has survived at the Royal Botanic Garden, Edinburgh, for many years.
3. *R. grande* at Glenarn, W. Scotland.
4. Young growth of *R. grande* C.H.&M. 3020, Glendoick.

R. kesangiae D. G. Long & Rushforth 1989 H4

Height 8-12m, a large-headed, upright shrub or small tree. Branchlets often closely covered with tufts of soft hair; bark mid-brown, *not* peeling. Terminal buds usually *rounded*, green, reddish to nearly black. **Leaves** 20-30 x 10-16cm, broadly elliptic to nearly obovate, apex and base usually rounded, retained c. 2 years; upper surface glabrous with *prominent lateral veins*, lower surface with a variable indumentum, woolly to almost plastered, fawn to silvery white on young leaves, *without cup-shaped hairs, uniformly grey-white or sometimes brownish-fawn* on old leaves. Petiole 2-3.5cm, not winged. **Inflorescence** 15-20-flowered, usually compact. **Corolla** campanulate, *with nectar pouches*, 3-4.7cm long, *rose to pink, fading to pale pink* or white; calyx small; ovary densely glandular, sometimes thinly tomentose. **Distribution** Bhutan only, 2,900-3,500m (9,500-11,500ft), fir or hemlock forest, often with other rhododendrons and bamboo.

R. kesangiae differs from *R. hodgsonii* in its non peeling bark, usually rounded buds, rougher upper leaf surface and usually pinker flowers. From *R. falconeri* it differs in its non-peeling brown bark, usually rounded buds, paler indumentum and pink flowers. It hybridises with both the above species in the wild.

This is a newly described species; most herbarium specimens were formerly identified as *R. hodgsonii x falconeri*. It is extraordinary that it was not recognised as distinct by collectors and botanists earlier, as it is the commonest large-leaved species in Bhutan. *R. kesangiae* is a first-rate introduction, likely to be intermediate in hardiness between *R. hodgsonii* and *R. falconeri*. Impressive in foliage and flower, but it requires wind shelter as its long petioles tend be easily broken. It tolerates fairly boggy conditions in the wild. Introduced 1967, reintroduced 1987-> many times. April-May.

Var *kesangiae*.

Corolla rose to pink. **Distribution** W. & C. Bhutan.

Var *album* Namgyel & D.G. Long 1989

Corolla white. **Distribution**: confined to E. Bhutan.

Picture 1

Picture 2

Picture 3

Picture 4

Picture 5

1. *R. kesangiae* on the Yutong La, C. Bhutan. (Photo Roy Lancaster).
2. *R. kesangiae*, C. Bhutan. (Photo I. Sinclair).
3. Variations in flower colour of *R. kesangiae* decorating a bus in central Bhutan. (Photo Roy Lancaster).
4. *R. kesangiae* Bowes Lyon 9703 at Wakehurst Place, S. England, possibly the first truss to open in cultivation. (Photo A. D. Schilling).
5. *R. kesangiae* on the Pele La, W. Bhutan, showing the characteristic rounded buds and the lower surface of a young leaf on the left, mature leaf on the right.

R. macabeanum (Watt ex) Balf.f. 1920 H3-4

Height 3-9m in gardens, taller in the wild, forming an upright tree, often leaving a bare trunk. Branchlets fawn to whitish-tomentose, young growth *white and woolly*. **Leaves** 14-38 x 9-20cm, broadly ovate to broadly elliptic, retained 1-2 years; upper surface, *shiny*, often with +/- vestiges of indumentum, lower surface with a *dense, bistrate* indumentum, the lower layer compacted, the upper layer lanate-tomentose, composed of mostly rosulate and some ramiform hairs; petiole slightly winged, floccose-tomentose. **Inflorescence** to 30-flowered, compact to rather loose. **Corolla** with 8 lobes, tubular to narrowly funnel-campanulate, 5-7.5cm long, *yellow to yellowish-white*, usually blotched purple; calyx c. 1mm, tomentose; stamens 16; ovary rufous-tomentose. **Distribution** N.E. India, Manipur and Nagaland only, 2,400-3,000m (8,000-10,000ft), mixed forest and birch or pure stands.

GRANDIA

The woolly indumentum distintinguishes *R. macabeanum* from other members of subsection Grandia. In appearance, it is perhaps closest to *R. sinofalconeri* (subsection Falconera) which has browner indumentum and which has only recently been introduced. The very upright growth habit, the wide leaves and the yellow flowers usually allow *R. macabeanum* to be quite easily identified.

R. macabeanum is one of the finest, most popular and easiest to grow of the big-leaved species but it is advisable to seek good yellow forms with compact trusses. The most attractive silvery young growth, with contrasting bright scarlet budscales, develops into magnificent foliage. Somewhat hardier than *R. sinogrande*, *R. montroseanum* and *R. sidereum*. Introduced 1927 and 1935. Reintroduced 1994. March-May.

Picture 1

Picture 2

Picture 3

Picture 4

3. *R. macabeanum* leaf lower surface showing the thick woolly indumentum, mature leaf darker, young leaf lighter. (Photo R.B.G., E.).
4. A magnificent plant of *R. macabeanum* at Portmeirion, N. Wales, with Philip Brown and members of the Scottish Rhododendron Society.

1. The best clone of *R. macabeanum* at the Royal Botanic Garden, Edinburgh.
2. The second clone of *R. macabeanum* at the Royal Botanic Garden, Edinburgh.

R. magnificum Kingdon-Ward 1935 H2-3

Height 18m in the wild, less in cultivation. A broadly upright tree. Branchlets tomentose. **Leaves** 20-46 x 6-17cm in cultivation, *oblanceolate to oblong-obovate*, retained 2-3 years; upper surface usually glabrous, lower surface with a *thin, grey to fawn, compacted* indumentum. **Inflorescence** to 30-flowered. **Corolla**, obliquely tubular-campanulate, with nectar pouches, c.7.5cm long, c.6.5cm across, rose to deep reddish-purple, fading; calyx c.1mm, tomentose; stamens c.15; ovary densely rufous-tomentose. **Distribution** N.E. Upper Burma-S.E. Tibet frontier only, 1,500-2,400m (5,000-8,000ft), mixed forest.

R. magnificum is very closely related to *R. protistum* and is likely to be merged with it when further field work has been done. In cultivation, *R. magnificum* is slightly later flowering with narrower leaves and with indumentum which develops on much younger plants. It is distinct from the other big-leaved species in its early reddish-purple flowers in large trusses.

It is a splendid plant where it reaches its full potential but *R. magnificum* and its close relative *R. protistum* are the most tender of the subsection and are only suitable for the mildest gardens, in south west and west Britain, San Francisco, coastal New Zealand and S.E. Australia for example. A hardier plant which used to be grown at Nymans, Sussex, under Kingdon-Ward 13681 from Arunachal Pradesh is probably a distinct species. Introduced 1931,39. February-March.

Picture 1

Picture 2

Picture 3

1. *R. magnificum* at Brodick Castle, Isle of Arran, W Scotland. (Photo J. Basford).
2. *R. magnificum* at Brodick Castle, Isle of Arran, W. Scotland.
3. A pale-flowered clone of *R. magnificum*, Brodick Castle, Isle of Arran, W. Scotland.

Picture 4

4. Young growth of *R. magnificum*, Brodick Castle, Isle of Arran, W. Scotland.

R. montroseanum

Davidian 1979 (*R. mollyanum* Cowan & Davidian) H3

Height to 9m in cultivation so far, to 15m in the wild, an upright or dome-shaped large shrub or tree. Branchlets tomentose. Bud scales red. **Leaves** thick, 20-45+ x 5-20cm, oblong to oblanceolate, retained 1-2 years; upper surface *dark green and shiny, rugulose,* lower surface with a *silvery to buff, plastered and shiny* indumentum. **Inflorescence** 15-20-flowered. **Corolla** with 8 lobes, oblique-campanulate with no nectar pouches, c.5cm long, pale to deep to purplish-pink, crimson-veined and blotched; calyx minute; stamens 15-16; ovary densely tomentose. **Distribution** Tsangpo Gorge, S. Tibet, 2,400-2,700m (8,000-9,000ft), temperate forest.

This is a distinct species, which was formerly known as pink *R. sinogrande*. It is related to *R. pudorosum* from which it differs in its longer leaves, deciduous bud scales and greater stature in cultivation. It differs from *R. sinogrande* in the smaller leaves and in the colour of the flowers, and from *R. grande* and *R. sidereum* in the thicker leaves and pink flowers.

Picture 1

1. *R. montroseanum* 'Benmore' F.C.C., Younger Botanic Garden, Benmore, W. Scotland.

GRANDIA

R. montroseanum is a fine plant attaining a large stature in milder gardens but it is successful only in favourable sites away from the south and west coasts of Britain. It is variable in flower colour, the paler forms generally being considered inferior. A plant of *R. lanigerum* x *R. sinogrande* closely resembling *R. montroseanum*, found in Pemako, S.E. Tibet, in 1996 suggests that this species may be derived from this cross. Introduced 1924 only. February-March.

Picture 2

Picture 3

Picture 4

2. A pale-flowered clone of *R. montroseanum*, Younger Botanic Garden, Benmore, W. Scotland.
3. A rich pink-flowered clone at the Royal Botanic Garden, Edinburgh.
4. *R. montroseanum*, Baravalla, W. Scotland.

R. praestans
Balf.f. & W.W. Sm. 1916 (*R. coryphaeum* Balf.f. & Forrest, *R. semnum* Balf.f. & Forrest) H4

Height 3-9m, a usually rounded and compact large bush or small tree. Branchlets whitish to fawn tomentose. **Leaves** 18-38 x 7-14cm, oblanceolate to obovate, base *tapered*, retained 2-4 years; upper surface glabrous, rugulose, lower surface with a plastered (occasionally scurfy) indumentum *shiny bronzy-brown* when young, *shiny grey-white to fawn* when mature; petiole *flattened and winged*. **Inflorescence** 10-25-flowered. **Corolla** with 8 lobes, oblique-campanulate, without nectar pouches, 3.5-4.5cm long, white through creamy-pink to magenta rose, usually with spots and blotch; calyx minute, c.1mm; stamens 16-18; ovary densely buff tomentose. **Distribution** N.W. Yunnan, S.E. Tibet, just into N.E. Upper Burma, 2,700-4,300m (9,000-14,000ft), conifer, mixed and rhododendron forest.

R. praestans in its type form is easily identified by the combination of a shiny, plastered indumentum, darker in the first year and the flattened and winged petiole. The picture is complicated, however, by its readiness to cross with *R. arizelum* and probably other species giving rise to hybrids which have formed stabilised populations; *R. semnoides* and *R. rothschildii* are almost certainly examples of this. It may also cross/merge with *R. basilicum*.

Picture 1

Picture 2

1. White and pink-flowered plants of *R. praestans* near Zhuzijo, Salween-Mekong divide, N.W. Yunnan, China.
2. An exceptionally deep-coloured *R. praestans*, near Zhuzijo, Salween-Mekong divide, N.W. Yunnan, China.

R. praestans is valuable for its hardiness and it makes a handsome, bold specimen. It is one of the slowest-growing of the big-leaved species and is often slow to bloom. Introduced 1914-49 and reintroduced 1992->. Late February-May.

Picture 3

Picture 4

Picture 5

3. *R. praestans* at the Younger Botanic Garden, Benmore, W. Scotland, showing the typical rounded habit of a plant grown in an open situation.
4. *R. praestans*, Younger Botanic Garden, Benmore, W. Scotland.
5. *R. praestans*, near Zhuzijo, Salween-Mekong divide, N.W. Yunnan, showing the characteristic winged, flattened petiole and the young leaves (left) and mature leaves (right). It is typical of the species to have darker indumentum on the young leaves.

R. protistum
Balf.f. & Forrest. 1920 (*R. giganteum* (Forrest ex Tagg, *R. giganteum* var. *seminudum* Tagg & Forrest) H2-3

Height 6-30m in the wild, to 9m+ so far in cultivation, forming a large domed shrub/tree in an open site, very leggy in shade. Branchlets thin, greyish-tomentose. *Very early growth with bright red bud scales.* **Leaves** 20-56 x 9-25cm, oval, elliptic to oblanceolate, retained c. 2 years; upper surface *matt green* rugulose, lower surface glabrous in juvenile state, developing a buff, continuous, thin tomentum, at least along a marginal band, as the plant matures. **Inflorescence** 20-30-flowered, compact or lax. **Corolla** oblique-campanulate, with nectar pouches, 6-7.5cm long, pale rose to crimson-purple, sometimes two-tone, fading, with or without small blotch; calyx c. 2mm, tomentose, stamens 16; ovary densely rufous-tomentose. **Distribution** W. & N.W. Yunnan, S.E. Tibet, E. Arunachal Pradesh, N.E. Upper Burma, N. Vietnam, 2,700-4,000m (9,000-13,000ft), open and heavy forest.

Formerly described as two species, it is now evident that as originally described *R. protistum* was simply the juvenile form and *R. giganteum* the adult form of the same species. Some authorities maintain these as var. *protistum* and var. *giganteum*. This is often the slowest species to reach full maturity, taking as much as 50-60 years to develop indumentum on the whole leaf underside and sometimes as long to start blooming. The indumentum often develops unevenly on a plant with some leaves remaining glabrous. Closely related to and may merge completely with *R. magnificum* which has a narrower leaf and a thin, compacted indumentum which forms on much younger plants than in *R. protistum*.

Picture 1

1. *R. protistum* (var. *giganteum*), Brodick Castle, Isle of Arran, W. Scotland.

GRANDIA

R. protistum is a splendid species, with deep coloured flowers in full trusses in the best forms, but it is only suitable for the mildest, most sheltered gardens and its very early growth leads to frequent frost damage. Introduced 1918-53. Reintroduced 1990s. January-March.

Picture 2

Picture 3

Picture 4

Picture 5

5. *R. protistum* (var. *giganteum*) with leaf indumentum fully developed on an old specimen plant.

2. *R. protistum* (var. *protistum*), Maryborough, Cork, S. Ireland, showing the flower colour beginning to fade.
3. The early growth of *R. protistum* (var. *giganteum*) Kingdon-Ward 21602 'Pukeiti' at Pukeiti, North Island, New Zealand with the characteristic red bud scales.
4. *R. protistum* (var. *protistum*) Kingdon-Ward 21498, Guavis, North Island, New Zealand. This shows the typical incomplete formation of indumentum on the leaf lower surface, even on a plant nearly 40 years old.

R. pudorosum Cowan 1937 H4

Height to 12m in wild, so far to 2.5m in cultivation, a slow-growing compact to leggy shrub. Branchlets tomentose, with *persistent bud-scales* for 2 to several years, giving a shaggy appearance, buds purple. **Leaves** 8-30 x 3-12cm, oblanceolate, margins recurved, retained 2-3 years; upper surface dark green, slightly rugulose, glabrous, lower surface with a thin, slightly spongy, silvery-fawn, plastered indumentum which is often absent, especially on young plants. **Inflorescence** 15-25-flowered. **Corolla** with 6-8 lobes, oblique-campanulate, without nectar pouches, 3-3.5cm long, pink to mauve-pink, with or without blotch; calyx c.1mm; stamens 16; ovary whitish tomentose. **Distribution** S. Tibet, 3,400-3,800m (11,000-12,300ft), in conifer forest.

A distinct species, especially on account of its persistent bud-scales, perhaps closest to *R. montroseanum* which is larger growing, more tender and lacks the persistent bud scales.

We only know of the 2-3 original seedlings of *R. pudorosum* at the Royal Botanic Garden, Edinburgh, where it is slow-growing and first flowered in 1972. It will probably prove to be more vigorous under damper conditions elsewhere. One of the hardiest big-leaved species. The young growth is often distorted, perhaps due to frost or heat. Introduced 1936. February-March.

GRANDIA

turning to yellow, with blotch; calyx rim-like, 1mm, tomentose; stamens 16; ovary densely rufous-tomentose. **Distribution** Mostly Upper Burma, also E. Arunachal Pradesh, N. and W. Yunnan, 2,400-3,700m (8,000-12,000ft), conifer and mixed forests.

The closest relation to this species is *R. grande* which usually has much larger, wider leaves, earlier growth, flowers which open at least a month earlier, and a glandular ovary. Vegetatively, this species is perhaps most likely to be confused with larger-leaved forms of *R. arboreum*.

R. sidereum is one of the best, large, yellow-flowered species though selection of the better forms is advisable. Often slightly tougher than *R. grande*, though its hardiness is variable, and useful for its late flowering. Introduced 1919-37. Reintroduced 1993. May-June.

Picture 1

Picture 2

Picture 3

1. *R. pudorosum* Ludlow & Sherriff 2752 at the Royal Botanic Garden, Edinburgh.
2. A paler clone of *R. pudorosum* at the Royal Botanic Garden, Edinburgh.
3. Young growth of *R. pudorosum*, Royal Botanic Garden, Edinburgh, showing the characteristic persistent bud-scales.

R. sidereum Balf.f. 1920 H2-3

Height 6-9m, an erect to rounded shrub or small tree. Branchlets *thin*, tomentose. **Leaves** 9-23 x 3-7cm, oblanceolate to narrowly elliptic, retained 2-5 years; upper surface matt, usually glabrous, lower surface with a +/- plastered, silvery-white to fawn indumentum. **Inflorescence** 12-20-flowered. **Corolla** with 8 lobes, oblique-campanulate, c. 5cm long, *cream to clear yellow*, sometimes opening pink

Picture 1

Picture 2

Picture 3

1. *R. sidereum* at the Younger Botanic Garden, Benmore, W. Scotland.
2. *R. sidereum* at the Younger Botanic Garden, Benmore, W. Scotland, showing the relatively small, narrowish leaves. (Photo R.B.G., E.).
3. *R. sidereum*, a form from Maryborough House, Cork, S. Ireland.

GRANDIA

R. sinogrande Balf.f. & W.W. Sm. 1916 H3

Height to 14m in wild, less so far in cultivation, tree-like with wide lower branches. Branchlets stout, thinly pale tomentose. **Leaves** *very large*, 20-70(-90) x 8-30cm, oblanceolate to broadly elliptic, retained 2-4 years; upper surface *dark green, shiny and rugulose*, lower surface with a *plastered silvery to fawn, shiny* indumentum; petiole *rounded*. **Inflorescence** 12-30+ flowered. **Corolla** with 8-10 lobes, oblique-campanulate, 6-7.5cm long, creamy-white to pale yellow with a large crimson blotch; calyx 1-2mm tomentose; stamens 16-20; ovary densely tomentose. **Distribution** N. & W. Yunnan, Upper Burma, S.E. Tibet and E. Arunachal Pradesh, 2,100-3,400m (7,000-11,000ft (higher altitudes claimed are almost certainly exaggerated), in mixed forest.

With the largest leaves in the genus, especially large on young, vigorous plants, this species is usually easily recognised. Its closest relations are *R. grande* and *R. montroseanum*, differing from the former in its larger, darker, more rounded leaves, later growth and flowers and from the latter in its larger, less rugulose leaves, its flower colour and the usually greater number of stamens.

For sheer grandeur of foliage *R. sinogrande* is the supreme species. Hardiness is variable; those introductions with very late growth are especially unsuitable for colder, drier areas. Beware of open-pollinated seedlings as the great majority will be hybrids. Plants under Boreale Group (var. *boreale*) have leaves shorter and narrower than the type and with clear pale yellow unblotched flowers. This is a northern, high-altitude variant and further explorations are necessary to ascertain its botanical standing. Introduced 1912-53, re-introduced 1981->. 1981 SBEC plants may be slightly hardier than most previous introductions. March-May.

Picture 1

Picture 2

Picture 3

Picture 4

1. *R. sinogrande*, Duanqing, Cangshan, C. Yunnan, China.
2. Sir Peter Hutchison with a *R. sinogrande* seedling, Dapingdi, Cangshan, C. Yunnan, China.
3. *R. sinogrande*, Stonefield Castle, W. Scotland.
4. *R. sinogrande* SBEC 0104, Royal Botanic Garden, Edinburgh, exceptionally blooming at c.10 years from seed.

R. watsonii Hemsl. & E.H. Wilson 1910 H4-5

Height 2-6m+, a shrub or small tree, usually fairly compact in cultivation. Branchlets whitish to fawn tomentose. **Leaves** 15-23 x 5-10cm, obovate to oblanceolate, retained 1-2 years; upper surface glabrous *with yellow petiole, colour extending some way down midrib*, lower surface with a plastered silvery-white indumentum (sometimes takes years to develop). Petiole *winged, very short*. **Inflorescence** 10-16-flowered, moderately compact. **Corolla** oblique-campanulate, 7-lobed,

3-5cm long, white or flushed pink, with or without spots and blotch; calyx 1-2mm fleshy; stamens 14, puberulous at base; ovary *glabrous*. **Distribution** N. & W. Sichuan only, 2,600-3,400m (8,500-11,000ft) in birch or conifer forest.

A distinct species, easily identified by the yellow midrib on leaf upper surface, the short flattened petiole and the glabrous ovary. *R. praestans* shares the flattened petiole and plastered indumentum but differs in the size and shape of the leaves. *R. watsonii* also has some resemblance to *R. calophytum* which has longer leaves with little or no indumentum. *R. balangense* differs in its thicker indumentum.

R. watsonii is the hardiest but the least flamboyant of the subsection, with rather small flowers, but it can have quite handsome foliage when well-grown. In cultivation the new growth is often frosted resulting in slow-growing, compact but rather shy-flowering plants. Introduced 1908->, re-introduced 1986->. Rare in cultivation. February-April.

Picture 1

Picture 2

Picture 3

Picture 4

3. A white-tinged-pink form of *R. watsonii*, Balang Mountain, N. Sichuan, China.
4. Seedling of *R. watsonii* showing the characteristic yellow midrib on the upper leaf surface and the typical short, winged petiole. Jiuzhaigou, N. Sichuan, China.

1. A small tree of *R. watsonii*, Huanglongsi, N. Sichuan, China.
2. A white-flowered *R. watsonii*, Huanglongsi, N. Sichuan, China.

Subsection Griersoniana

A monotypic subsection.

R. griersonianum
Balf.f. & Forrest 1924 H3

Height 1.5-3m, a sprawling, spreading shrub. Branchlets *bristly-glandular* and tomentose; bark rough, brown. Buds with *long, elongated bud-scales*. **Leaves** 6-20 x 2-5.3cm, lanceolate; upper surface with indumentum on new growth, glabrous on maturity, lower surface with a continuous or patchy whitish to pale brown indumentum consisting of dendroid hairs; petiole usually *long* 1-3.5cm. **Inflorescence** 5-12-flowered, loose to fairly full. **Corolla** funnel-shaped with *long tube, hairy outside,* 5-8cm long, bright geranium-scarlet to bright rose, with or without spots; calyx minute c.1mm, stamens 10, puberulous in the lower half; ovary densely tomentose, with or without glands; style *deep red*, glandular or floccose or glabrous at base. Distribution W. Yunnan and Upper Burma, 2,100-2,700m (7,000-9,000ft), mixed forest and open glades.

A distinct species with no close relatives, identified by its long, conical tapered flower and growth buds, the narrow leaves with a long petiole and the late, red flowers.

R. griersonianum has fine late flowers but is not very hardy and its late growth is vulnerable to damage from early autumn frosts. It can be pruned into a compact shape in mild areas. Before *R. degronianum* ssp. *yakushimanum* was introduced, *R. griersonianum* was the most popular parent for hybridising. Introduced 1917-31. June-July (August in dense shade).

Picture 1

Picture 2

Picture 3

Picture 4

3. *R. griersonianum* showing the pale brown indumentum on the leaf lower surface.
4. The characteristic pointed flower bud of *R. griersonianum*. (Photo R.B.G.,E.).

1. *R. griersonianum* Forrest 30392, Royal Botanic Garden, Edinburgh, showing the lanceolate leaves and the bright red flowers.
2. *R. griersonianum* and *Cardiocrinum giganteum* at Glendoick.

Subsection Irrorata

Height 1-12m, upright shrubs to small trees. Branchlets stipitate-glandular, sometimes floccose; bark rough. **Leaves** to 19 x 6.5cm; upper surface usually tomentose at first, *glabrous on maturity*, lower surface *glabrous* or with a thin veil of indumentum. *Glands usually over-lie the veins*. **Inflorescence** 4-20-flowered, lax-dense. **Corolla** 5-7-lobed, tubular to open-campanulate, mostly *with nectar pouches*, white, creamy-yellow, pink to mauve and deep crimson; calyx minute to 3mm; stamens 10; ovary glabrous to tomentose or glandular; style glabrous or glandular to tip. **Distribution** Guizhou, Sichuan, Yunnan, Vietnam, Burma, Tibet, Bhutan, Arunachal Pradesh, Malay Peninsula.

This subsection is related to subsection Maculifera which differs in the presence of matted-dendroid tomentum on the petiole, pedicel and usually on the midrib on the lower leaf surface.

IRRORATA

Proving too tender for cultivation in Europe and most of North America, many species in subsection Irrorata are not in general cultivation. *R. irroratum*, *R. aberconwayi* and *R. anthosphaerum* are fairly hardy and quite widely grown, most of the others are rarely cultivated but could be useful in hotter and milder areas.

R. aberconwayi Cowan 1948 H-34

Height to 2.5 m, a usually upright shrub with *rigid* branches. **Leaves** 3-7 x 1.2-3.3cm, oblong-elliptic to broadly lanceolate, apex usually acute, margins recurved, *thick, rigid and leathery*, retained 2-4 years; upper and lower surface glabrous when mature, lower surface with red punctate hair-bases overlying the veins. **Inflorescence** 6-12-flowered, erect. Rachis *often long*. **Corolla** *flatly-campanulate or saucer-shaped, without nectar pouches,* 2-3cm long, white sometimes tinged pink with a few to many crimson spots; calyx c. 1mm, sparsely hairy; ovary densely glandular; style glandular to tip. **Distribution** N.E. Yunnan, Lo Shiueh, under 3,000m (10,000ft).

This species is quite easily recognised; its stiff habit, saucer-shaped, white flowers and its rigid, leathery leaves are unlike those of any other species.

With striking flowers in its best forms, *R. aberconwayi* is one of the finer lower-growing species. Quite variable in flower, it is worth searching out selected clones with large, strongly spotted flowers. Introduced 1937. May-June.

Picture 1

1. *R. aberconwayi* 'His Lordship' A.M., McLaren T41, collected in the Lo Shiueh mountains of N.E. Yunnan, showing the characteristic, stiff, pointed leaves.

Picture 2

Picture 3

2. *R. aberconwayi* 'His Lordship' showing the typical, rather sparse, upright habit.
3. *R. aberconwayi* at the Cecil Smith Garden in Oregon, U.S.A., a form with fainter spotting on the corolla than most.

R. annae Franch. 1898 H3-4

Height to 2.4m (up to 4.6m in ssp. *laxiflorum*), a rather upright, sometimes straggly shrub or small tree. Young shoots thinly floccose +/- glandular. **Leaves** 7-13.5 x 1.5-4cm, narrow elongate-lanceolate to oblong-elliptic, slightly-recurved, retained 2-4 years; upper surface shiny, lower surface glabrous, with red punctate hair bases; petiole *glabrous*. **Inflorescence** 7-12(17)-flowered, fairly loose. **Corolla** *cup-shaped to openly-campanulate*, 2-4.5cm long, white to cream flushed rose to pink, unspotted or spotted pink or purple; calyx 1-2mm, densely glandular; ovary and style densely glandular.

This species differs from *R. aberconwayi* in the shape of its flower, the less rigid, leathery leaves and its taller stature. It is distinguished from *R. araiophyllum*, and *R. lukiangense* in the red hair bases on the lower leaf surface and in the densely glandular ovary.

As there is insufficient material of *R. annae* in western herbaria to make a good evaluation of this species, we have followed the Chinese in recognising two subspecies:

IRRORATA

Ssp. *annae*

Height to 2.4m. **Leaves** 1.5-3.2cm wide; petiole almost glabrous. **Corolla** *2-2.5cm* long. **Distribution** Guizhou, N.E. Yunnan? 1,400-2,400m, (4,600-8,000ft), mixed forest.

A plant under this name with white flowers and quite small leaves is quite widely cultivated. Its origin is unknown. May.

Ssp. *laxiflorum* (Balf.f. & Forrest) Ming 1992
(*R. hardingii* Forrest). H2-3.

Height to 4.6m. **Leaves** oblong-elliptic to oblanceolate; petiole sparsely glandular. **Corolla** *3.4-4.5cm* long, pink to white with red spotting. **Distribution** S.W. Yunnan, 2,440-3355m (8,000-10,000ft).

This subspecies, rare in cultivation, seems to be much more tender than ssp. *annae* and is quite closely related to *R. araiophyllum*. Introduced 1924-25. May-June (in the wild).

Picture 3

Picture 4

3. A pink-flowered form of *R. annae* ssp. *annae*, probably now lost to cultivation. (Photo R.B.G., E.).
4. *R. annae* ssp. *laxiflorum,* a form at Sandling Park, S. England, probably the clone 'Folk's Wood' A.M. (Photo J. Bond)

Picture 1

Picture 2

1. *R. annae* ssp. *annae*, the hardiest of the forms in cultivation, possibly from Guizhou province, China.
2. A view of the hardy form of *R. annae* ssp. *annae*.

R. anthosphaerum

Diels 1912 (*R. chawchiense* Balf.f. & Farrer, *R. eritimum* Balf.f. & W.W. Sm., *R. gymnogynum* Balf.f. & Forrest, *R. heptamerum* Balf.f., *R. persicinum* Hand.-Mazz.) H2-3

Height 1.2-9m, a broadly upright shrub to small tree. Branchlets with rufous glands and hairs. **Leaves** 8-18 x 2.5-5cm, narrowly oblong to obovate to oblanceolate, retained 2-3 years; upper surface *not waxy*, lower surface pale, slightly floccose, papillate, but may appear glabrous and sometimes glaucous. **Inflorescence** 7-15-flowered, fairly compact. **Corolla** with *5-7 (usually 6-7)* lobes, tubular-campanulate, with nectar pouches, 4-5cm long, pale pink to rose, pale lavender through mauve and lilac to deep magenta, white/pink tinged and/or tinged deeper, blotched and sometimes spotted crimson; calyx 1-2mm; ovary *usually glabrous*; style *glabrous*. **Distribution** widely distributed in Yunnan, 2,700-4,000m (9,000-13,000ft), in open or wooded situations.

R. anthosphaerum is unique in subsection Irrorata in that it generally has a 6-7-lobed corolla. *R. lukiangense* differs in its shiny rather than matt upper leaf surface and *R. irroratum* differs in its densely glandular style and ovary.

Very variable in colour, flowering time and hardiness, *R. anthosphaerum* can be quite showy in its best forms but some are of a decidedly 'hot' colour. The very early growth is prone to frost damage. Former *R. eritimum* (now Eritimum Group) tends to have fairly narrow leaves. *R. anthosphaerum* crosses with and may merge with *R. irroratum* in the wild. Introduced 1910-1937, reintroduced 1981->. February to May.

Picture 1

Picture 2

Picture 3

1. A pale mauve-flowered *R. anthosphaerum* north of Zhuzipo, Salween-Mekong divide, N.W. Yunnan, China.
2. *R. anthosphaerum*, a lilac-flowered form at the Royal Botanic Garden, Edinburgh.
3. *R. anthosphaerum*, a cherry red-flowered form from the 1981 SBEC expedition to the Cangshan mountains of Yunnan, China, growing at Maryborough House, S. Ireland.

Picture 4

Picture 5

4. *R. anthosphaerum* Gymnogynum Group Rock 11354, a Tower Court plant now growing at the Valley Gardens, Windsor Great Park, S. England. The deep basal blotch on the corolla is typical of the species.
5. Young growth of *R. anthosphaerum* Forrest 11354 showing the matt upper leaf surface and the glands on the branchlets.

R. araiophyllum
Balf.f. & W.W. Sm. 1917 H2-3

Height to 6m, a slender, upright shrub or small tree. Branchlets +/- *white floccose,* midrib *woolly,* young growth scarlet; bark smoothish and slightly flaking. **Leaves** of thin texture, 5-11.5 x 2-3cm, lanceolate, narrowly elliptic, margin undulate, retained 1-2 years; lower surface shiny, usually glabrous, *midrib sometimes woolly.* **Inflorescence** 6-8-flowered, small, compact to loose. **Corolla** with 5 lobes, *cupular* or openly campanulate, with nectar pouches, 3.5-4cm long, white, white tinged pink or pale pink, with crimson blotch and spots, slightly fragrant; calyx 1-2mm; ovary *pubescent*; style *glabrous,* red. **Distribution** W. Yunnan and N.E. Upper Burma, 2,300-3,400m (7,500-11,000 ft), in and at margins of mixed forest.

This species resembles *R. irroratum* but is distinguished by its woolly lower leaf-surface midrib and narrower leaves. It is distinguished from *R. annae* by its glabrous style and lack of punctate hair bases. Leaves are usually thinly-textured.

IRRORATA

A pretty plant for milder areas, *R. araiophyllum* has flowers with a fine deep blotch and spotting in its better forms but it tends to come rather early into growth. Introduced 1917-25. April-May.

Picture 1

1. The rare *R. araiophyllum* at Arduaine, W. Scotland, showing the characteristic thin leaves and blotched flowers.

Picture 1

Picture 2

Picture 3

Picture 4

R. irroratum

Franch. 1887 H3-4

Height to 9m (usually to 3.7m in cultivation), an erect to rather straggly shrub or small tree. Young shoots thinly tomentose and glandular. **Leaves** up to 14 x 4.4cm, retained 2-3 years, lanceolate, oblanceolate to narrowly elliptic; pedicel *densely or moderately glandular*. **Inflorescence** to 15-flowered, usually loose. **Corolla** campanulate or tubular-campanulate with 5 nectar pouches, 3-5cm long, very variable: pure white, pale yellow, pale to rose pink, unspotted to heavily spotted crimson or purple, sometimes on every lobe; calyx densely glandular, and/or tomentose; stamens 10; ovary and calyx tomentose and/or stipitate-glandular; style *glandular to the tip*.

This is a very variable species and is often hard to distinguish from its relatives such as *R. anthosphaerum*. The best features to look for are the glandular pedicels, ovary and style. *R. irroratum* has 10 stamens and 5 corolla lobes while *R. anthosphaerum* has 10-14 stamens and usually 6-7 corolla lobes.

3 subspecies are recognised, though Chamberlain says that the distinctions between them are rather insignificant and intermediate forms occur.

1. *R. irroratum* ssp. *irroratum* 'Polka Dot' A.M., a heavily spotted clone from Exbury Gardens, growing at Glendoick, showing the 5 nectar pouches at the base of the corolla.
2. A lightly spotted *R irroratum* ssp. *irroratum* SBEC 0064 collected on the Cangshan, Yunnan, China in 1981, growing at Glendoick.
3. *R. irroratum* ssp. *irroratum* 'Carse' a fine form at Glendoick.
4. *R. irroratum* ssp. *pogonostylum* in Yunnan, China. (Photo Kunming Botanical Institute).

Ssp. *irroratum*
(*R. ningyuenense* Hand.-Mazz. ssp. *ningyuenense* T.L. Ming)

Corolla white, cream to pale yellow; ovary and usually calyx *stipitate-glandular, not tomentose*. **Distribution** Yunnan and S. Sichuan, 2,400-3,700m (8,000-12,000ft) in shady rhododendron and conifer forest and at forest margins.

Common in the wild and quite common in cultivation. Some forms are decidedly average, especially those with muddy pink flowers. It is quite easy to please, but its early flowering and growth is vulnerable to frost and it can be cut back or killed in the hardest British winters. Introduced 1886, reintroduced 1981->. Early March-May. Ningyuenense Group; cream to pale yellow flowers. N.E. Yunnan, c. 1,800m. Likely to be tender, suitable for mildest gardens only. This may be reinstated at varietal status. Introduced 1995.

Picture 5

5. *R. irroratum* Ningyuenense Group in N.E. Yunnan. (Photo Yang Zhenhong).

Ssp. *pogonostylum*
(Balf.f. & W.W. Sm.) D.F. Chamb. 1978 (*R. adenostemonum* Balf.f. & W.W. Sm.) H2?

Corolla deeper pink; calyx and ovary *tomentose and glandular*. **Distribution** S. & W. Yunnan, S.W. Sichuan?, 2,100-2600m (7,000-8,500m), forests, rocky slopes.

This subspecies has a more southerly distribution than ssp. *irroratum* and is less hardy but likely to be more heat tolerant. The true plant may not be in cultivation. Introduced 1886-1938, reintroduced 1981? Early March-May.

R. langbianense
A. Chev. ex Dop. 1930 This taxon from Langbian, Vietnam has been introduced during the last few years. This subsection Irrorata species was made synonymous with *R. irroratum* ssp. *kontumense* (from Sumatra) in the Edinburgh revision but recent fieldwork has called this into doubt. Related both to *R. irroratum* and *R. annae*, further work will be required to settle its status. It should be a useful heat tolerant species for mild and warm areas.

Picture 6

6. *R. langbianense* with pale pink flowers and a deeper blotch, photographed in Vietnam. (Photo K. Rushforth).

R. kendrickii
Nutt. 1853 (*R. pankimense* Cowan, *R. shepherdii* Nutt. (type only)) H2-3

Height 3-7.6m, a rather erect, open and sparse shrub or small tree. Branchlets slender. **Leaves** 9-13 x 2-3cm, oblong-lanceolate to lanceolate acuminate, with *wavy edges*, retained 1-2 years; upper surface smooth, lower surface paler and somewhat shiny. **Inflorescence** 8-20-flowered, fairly full. **Corolla** tubular campanulate, *with 5 prominent nectar pouches*, 3-3.8cm long, rose to deep rose to crimson, red and scarlet, sometimes spotted; pedicel c.8mm, *densely or moderately floccose*; calyx 2-3mm with conspicuous rounded lobes; ovary with floccose hairs; style glabrous. **Distribution** Bhutan, Arunachal Pradesh and S. Tibet, 2,100-2,700m (7,000-9,000 ft), common in dense and open evergreen and deciduous forest, even on heavy clay soil.

This species is identified by its stiff, narrow, sharply pointed leaves, the undulate leaf margins (usually recurved) and the pink to red flowers. Very closely related to *R. ramsdenianum* which has wider, less pointed and more leathery leaves. Perhaps these two should be merged. *R. tanastylum* is also very similar; see under this species for differences.

The best forms of *R. kendrickii* are worth seeking out; in many forms the flowers and trusses are rather small. It is also rather tender and its growth comes very early. Foliage and growth habit are quite distinctive and it should have some heat tolerance. The type specimen of *R. shepherdii* fits here, but plants in cultivation under this name are forms of *R. barbatum* from Bhutan. Introduced 1850s, 1915-1938, reintroduced 1965-> April-May.

IRRORATA

Picture 1

Picture 2

Picture 3

1. *R. kendrickii,* a very fine form at Brodick Castle, Arran, W. Scotland.
2. *R. kendrickii* at Mount Congreve, S. Ireland, showing the typical wavy edges to the leaves.
3. *R. kendrickii* showing the typical narrow pointed leaves. The bronzy new growth soon turns green.

R. lukiangense

Franch. 1898 (*R. admirabile* Balf.f. & Forrest, *R. adroserum* Balf.f. & Forrest, *R. ceraceum* Balf.f. & W.W. Sm., *R. gymnanthum* Diels) H2-3

Height 4.5-6m (usually less in cultivation), an upright to rounded shrub or small tree. Young shoots *bronzy, red or purplish*; bark roughish, greenish-grey. **Leaves** up to 19 x 6cm, oblong-lanceolate to oblanceolate, retained 1-2 years; upper surface dark, *waxy and glossy*, lower surface usually glabrous apart from red punctate glands; petiole slightly purplish. **Inflorescence** 8-13-flowered, **Corolla** *with 5 lobes*, tubular-campanulate, 4-4.5cm long, pale or deep rose, magenta-rose to lilac, with or without crimson blotch and spotting; pedicel glabrous; calyx c. 2mm, sparsely ciliate; ovary +/- glabrous; style glabrous. **Distribution** N.W. Yunnan and S.E. Tibet 2,400-4,000m (8,000-13,000ft), in conifer and rhododendron forest, thickets, among rocks and by streams.

This species is usually recognised by its dark leaves with a waxy sheen on the upper surface. It differs from *R. irroratum* in its non-glandular ovary, style and pedicel and from *R. anthosphaerum* in the number of corolla lobes.

R. lukiangense is rare in cultivation but common in the wild. The best forms have quite fine flowers but many are poor. Forms formerly under *R. adroserum* have more rounded leaves while those known as *R. ceraceum* have deeper, more reddish flowers. Rather tender and the early growth is easily frosted. Introduced 1918-1938, reintroduced 1988->. March-April.

Picture 1

Picture 2

1. *R. lukiangense* Forrest 16353 at Royal Botanic Garden, Edinburgh, showing the glossy upper and lower leaf surfaces.
2. The bronzy new growth on *R. lukiangense* which soon turns green.

Picture 3

3. *R. lukiangense* photographed in the wild in N.W. Yunnan. (Photo Kunming Institute of Botany)

Picture 1

Picture 2

1. A fine display from a large old plant of *R. ramsdenianum* at Glenarn, W. Scotland.
2. A truss of *R. ramsdenianum* at a Royal Horticultural Society show, showing the relatively wide leaves compared with *R. kendrickii*. (Photo F. Auckland).

R. ramsdenianum Cowan 1936 H2-3

Height to 12m, less in cultivation, an erect shrub or small tree. Bark light brown. **Leaves** coriaceous, up to 10 x 3.8cm, broadly lanceolate to oblong-lanceolate; upper surface glabrous when mature, lower surface glabrous or with vestiges of brown indumentum with red punctate hair bases overlying the veins. **Inflorescence** 12-15-flowered. **Corolla** tubular-campanulate, with 5 nectar pouches, c.4cm long, rich rose, dark reddish-pink to bright crimson, purplish red or blood red, with or without spots and blotch; pedicel *glabrous or sparsely floccose*; calyx c.2mm, glabrous, lobes broad, rounded; ovary glabrous or with a few rufous hairs; style glabrous. **Distribution** S.E. Tibet 2,400-3,000m (8,000-10,000ft), in deciduous and conifer forest.

This species is very closely related to *R. kendrickii* and these two should probably be merged (probably also including *R. papillatum* which is not in cultivation). The leaves of *R. ramsdenianum* are more leathery, wider and less pointed at the apex than in *R. kendrickii* and the pedicel is glabrous or only sparsely floccose.

R. ramsdenianum only occasionally flowers well with full-sized trusses. This, together with its early growth, makes this a species mainly of interest to collectors and it is suitable for relatively mild gardens only. Introduced 1924. April.

R. tanastylum Balf.f. & Kingdon-Ward H2

Height 6-9m in the wild, forming a loose spindly tree. Branchlets thin, thinly floccose at first. **Leaves** up to *18 x 6.3cm*, broadly lanceolate to oblanceolate; lower surface glabrous or with a thin veil of indumentum; petiole c. 1.5 cm. **Inflorescence** 4-8-flowered. **Corolla** tubular-campanulate, with 5 nectar pouches at the base, 4.5-5.5cm long, rose-purple, magenta or purple-crimson, black-crimson or rich crimson scarlet marked +/- with deeper spots; calyx c. 2mm, glandular or tomentose; style glabrous, ovary glabrous to tomentose and glandular. **Distribution** S.E. Tibet, Upper Burma, W. Yunnan and Arunachal Pradesh (very common), 1,800-3,400m (6,000-11,000 ft), amongst scrub, rocks or in forests, often in valley bottoms.

This species is closely related to and may merge with *R. kendrickii* and *R. ramsdenianum* but the type should have larger leaves and flowers. The floccose and strigose hairs on the pedicel, calyx and ovary of *R. kendrickii* are absent in *R. tanastylum*.

David Chamberlain doubts that true *R. tanastylum* is in cultivation; plants cultivated under this name are closer to *R. kendrickii* or *R. ramsdenianum*. Although rather tender, this species should be more showy in flower that its close relatives and would make a fine garden plant in milder areas. March-May in the wild, April or earlier in cultivation.

Var. *tanastylum* (*R. cerochitum* Balf.f. & Forrest, *R. ombrochares* Balf.f. & Kingdon-Ward)

Leaves at maturity +/- glabrous beneath; pedicel eglandular.

Var. *pennivenium* (Balf.f. & Forrest) D.F. Chamb. 1978

Leaves with persistent, cobwebby, semi-plastered indumentum on lower leaf surface.

A plant under this name is cultivated at the Rhododendron Species Botanic Garden, U.S.A.

R. wrayi

King & Gamble 1905 (*R. dubium* King & Gamble, *R. coruscum* Ridley) H1(-2?)

Height 3-6(12)m, a compact shrub or small, lanky tree. New growth rust-coloured, tomentose. **Leaves** 6-15 x 2-5cm; upper surface glabrous when mature, lower surface with a *thin veil of evanescent white indumentum* and a fine greyish indumentum of cobweb-like hairs in pits, midrib grooved. **Inflorescence** 10-18-flowered; pedicels *rusty-floccose*. **Corolla** with 5 to 7 lobes, widely campanulate, 3-3.5cm long, white flushed pink, blotched, spotted or unspotted; calyx small, floccose-tomentose outside; stamens 10. **Distribution** Malay Peninsula 900-2,100m (3,000-7,000 ft).

The only subsection Irrorata species from peninsula Malaysia and quite easily distinguished from other members of the subsection in the rusty-floccose covering on the leaves and pedicels. It grows among Vireyas such as *R. malayanum* in the wild and has been introduced to Australia, New Zealand and California where it is being used in breeding to impart heat tolerance. December-January (in the wild).

Picture 1

Picture 2

1. *R. wrayi* in the Cameron Highlands, Malaya. (Photo P. Cribb, R.B.G., Kew).
2. *R. wrayi* showing traces of indumentum on the upper leaf surface. (Photo R.B.G., Kew).

Subsection Lanata

Height 0.3-4m, bushy to erect shrubs or small trees. Branchlets and buds *densely tomentose*; bark greyish-purple, slightly rough. **Leaves** 3.5-11cm long, obovate to elliptic; upper surface *often with persistent indumentum*, lower surface with *a dense, unistrate, light brown to rufous, lanate or crisped indumentum composed of dendroid hairs*. **Inflorescence** 3-15-flowered, lax to dense. **Corolla** campanulate to open-campanulate, with no nectar pouches, yellow, white or pink with some to much crimson spotting; calyx minute; stamens 10; ovary densely tomentose; style glabrous; capsule tomentose. **Distribution** Sikkim and S. Tibet eastwards to Arunachal Pradesh and S.E. Tibet.

The species in this subsection are closely related and to some extent merge into one another. The subsection has seen some recent revision; *R. flinckii* and *R. luciferum* have been reinstated at specific status and *R. lanatoides* described as a new species. Subsection Lanata appears to be related to *R. wasonii* of subsection Taliensia and perhaps more distantly to subsection Campanulata.

The species of subsection Lanata are attractive in both leaf and flower but tend to be rather fastidious as to soil conditions. Though hardy, they require some shelter as they flower and grow early in the season. *R. flinckii, R. lanatum* and *R. tsariense* are fairly widely grown, *R. lanatoides* is very rare.

Picture 1

Picture 2

Picture 3

R. flinckii
Davidian 1975 H4-5

Height 1.5-2.5m, a broadly upright shrub or small tree. Branchlets densely tomentose with rusty brown, fawn or whitish cottony tomentum. **Leaves** 4-10 x 2-4.5cm, oblong oval, oblong-lanceolate, elliptic or oblong-elliptic, somewhat *thin*, retained 1-3 years; upper surface pale green, *matt*, hairy or with vestiges of hairs especially in the grooved midrib, lower surface with a *thin*, felted, bright rust-brown, continuous, unistrate indumentum; petiole densely tomentose. **Inflorescence** 3-8-flowered. **Corolla** campanulate, 3.5-5cm long, pale yellow to pale pink? with crimson spotting; calyx minute; ovary tomentose; style glabrous. **Distribution** Bhutan, 3,000-4,150m (10,000-13,500ft), in Abies forest, forest margins, on cliffs.

This species has only recently regained specific status after field studies in Bhutan. It had been reduced to synonymy under the closely related *R. lanatum* which has thicker leaves, shiny and dark green rather than matt on the upper surface, with a thick, dense, indumentum on the lower surface. *R. flinckii* merges with *R. lanatum* in the wild.

R. flinckii seems to be easier to grow than its close relative *R. lanatum*. It has been reintroduced several times in recent years so should gradually become quite well-known. So-called pink forms may be better referable to *R. tsariense* Magnum Group. Introduced 1915, 1936. Reintroduced 1980s and 90s. April-May.

1. *R. flinckii* on the Rudong La, C. Bhutan. (Photo Roy Lancaster).
2. A pale-flowered form of *R. flinckii* at the Royal Botanic Garden, Edinburgh.
3. *R. flinckii* showing the thin felted indumentum on the leaf lower surface. The indumentum on the upper surface soon wears off. Rudong La, C. Bhutan.

R. lanatoides D.F. Chamb. 1982 H4-5?

Height 2-4m, an upright but dense bush or small tree. Branchlets covered in tomentum. **Leaves** 9-11 x 2.1-7cm, *lanceolate*, retained 2-3 years; upper surface *dark* green with some persistent indumentum, otherwise glabrous and shiny, lower surface with a dense, lanate, dark brown to fawn indumentum of dendroid hairs with long straight branches;

petiole 1-1.5cm long, densely tomentose. **Inflorescence** 10-15-flowered, dense. **Corolla** campanulate, 3.5-4cm long, white flushed pink with a few faint flecks; pedicel densely tomentose; calyx sparsely tomentose; ovary densely tomentose; style glabrous. **Distribution** Kongbo, S.E. Tibet among rocks, in conifer forests and gullies, 3,200-3,700m (10,500-12,000ft).

The lanceolate leaf shape differentiates this species from the others in the subsection. It is further distinguished by its very early flowering and thick indumentum. Its closest relative *R. luciferum* differs in its yellow rather than white flowers. It seems to be related to *R. principis* which occurs in the same area and hybridises with it.

R. lanatoides is a recently named species; plants of it have been found in several Scottish and English collections, but it is hard to root and graft and remains very rare. Collected as *R. roxieanum* aff. under Kingdon-Ward 5971 and also found by Ludlow and Sherriff. After much searching, we found a large population of it in S.E. Tibet in 1996. A very fine foliage plant but the flowers are very early and rather small. Introduced 1924-5, reintroduced 1996. February-April.

Picture 1

Picture 2

1. *R. lanatoides* at Tigh-na-Rudha, W. Scotland, with lightly spotted white flowers, showing the characteristic thick, woolly indumentum on the lower leaf surface.
2. *R. lanatoides* showing the lanceolate leaves, the mature ones being dark and shiny on the upper surface.

Picture 3

3. Foliage of *R. lanatoides* above the Rong Chu valley, S.E. Tibet.

R. lanatum Hook.f. 1849. H4-5

Height 0.30cm-3m, erect and loose to bushy shrub or small tree. Branchlets *densely tomentose*, white to brown. Flower and growth buds *heavily tomentose*. **Leaves** 6-12 x 1.8-5cm, narrowly elliptic, ovate, obovate to oblong-obovate, *thick*, apex obtuse to rounded, retained 1-2 years; upper surface of young leaves with reddish brown indumentum which soon wears off, lower surface with a *crisped* unistrate indumentum, grey-fawn, pink-brown, brown or rust-coloured, usually *thick*, petiole tomentose. **Inflorescence** 5-10-flowered. **Corolla** broadly-campanulate, 3-5cm long, cream, pale to sulphur yellow, (white) with red or brown spots; pedicel densely tomentose; calyx tomentose; ovary densely rufous-tomentose; style glabrous. **Distribution** Sikkim, Arunachal Pradesh, Bhutan, S. Tibet?, 3,000-4,300m (10,000-14,000ft), under birches, in bamboo and rhododendron forest, on steep banks and cliffs.

This species is closely related to *R. flinckii* which has thinner leaves and indumentum and a dull rather than shiny upper leaf surface. It differs from *R. tsariense* in its yellow rather than pink or white flowers and in its larger leaves. These three species do merge in the wild. *R. lanatoides* has smaller white flowers and longer leaves while *R. luciferum* differs in its pointed leaf apex and more eastern distribution.

R. lanatum is a very fine species at its best but is hard to grow well on its own roots, requiring much organic matter and excellent drainage. Grafted plants are difficult to obtain but seem to grow better. Introduced 1851 onwards. April-May.

Picture 1

Picture 2

Picture 3

Picture 4

Picture 5

5. A distinct form of *R. lanatum* from Muncaster Castle, N.W. England, with near white flowers and oval leaves with a very thick spongy indumentum. This plant may be given botanical status.

1. & 2. *R. lanatum*, a fine clone at the Royal Botanic Garden, Edinburgh.
3. A pale-flowered form of *R. lanatum*, Royal Botanic Garden, Edinburgh. The characters of this form lie about midway between *R. lanatum* and *R. flinckii*.
4. *R. lanatum* showing a heavily tomentose flower bud. (Photo R.B.G., E.).

R. luciferum Cowan 1953 H4-5.

Height 1.5-7.5m. **Leaves** coriaceous, 8.5-11 x 3-4.5cm, elliptic to ovate, apex *acute to acuminate*, lower surface covered with a thick rusty brown indumentum composed of dendroid hairs. **Inflorescence** 8-10-flowered, dense. **Corolla** funnel-campanulate, 3-4cm long, pale yellow with at least a few red flecks; ovary densely covered with a pale brown tomentum; style glabrous. **Distribution** S.E. Tibet, 3,350-4,000m, 10,750-13,000ft), in conifer or rhododendron forest, bamboo or open hillside.

R. luciferum differs from *R. lanatum* in its larger leaf with a pointed rather than rounded apex and more reddish indumentum on the lower surface, and its generally larger stature. It comes from further east. It differs from *R. lanatoides* in the colour of the flower, the leaf shape and in its distribution.

Recently reinstated as a species, this may no longer be in cultivation. Botanically and geographically it is intermediate between *R. lanatoides* which comes from further east and *R. lanatum* which comes from further south and west. A handsome foliage plant. Introduced 1938. April-May.

R. tsariense

Cowan 1937 (*R. poluninii* Davidian, *R. tsariense* var. *magnum* Davidian, var. *trimoense* Davidian.) H4

Height 1-3m but rarely over 1.5m in cultivation, a fairly compact shrub. Branchlets densely woolly. **Leaves** *2.3-6.2 x 1-3cm*, obovate to oblong; upper surface with *+/- persistent indumentum* through the growing-season and sometimes beyond, lower surface with a dense rufous-tomentose indumentum composed of ramiform hairs; petiole densely tomentose. **Inflorescence** 3-5-flowered. **Corolla** open-campanulate, 2.5-3.5cm long, *white or pale cream, pale pink or white flushed pink, spotted crimson*; pedicel densely tomentose; calyx and ovary densely tomentose. **Distribution** S. Tibet, Arunachal Pradesh, E. Bhutan, 3,000-4,400m (10,000-14,500ft), conifer and rhododendron forest, bamboo, open hillsides and cliffs.

The white or pink flowers and smaller leaves of this species differentiate it from the yellow *R. lanatum* and *R. flinckii*, while *R. lanatoides* and *R. luciferum* have considerably larger and longer leaves.

R. tsariense is a very fine, slow-growing species, suitable for the small garden, well worth growing for its foliage alone, and it is easier to please than its relatives. Davidian has recently described new taxa out of the most extreme forms of this variable species, and we have accorded these Group status. Poluninii Group has larger leaves and more flowers to the truss and may be intermediate between *R. tsariense* and *R. flinckii*. Magnum Group is similar and may be better placed as a pink form of *R. flinckii*. Trimoense Group has paler leaves and pale indumentum and occurs at the northern end of the distribution of the species. Introduced 1914, 1936-37. March-May.

Picture 1

Picture 2

Picture 3

Picture 4

1. *R. tsariense*, a typical small-leaved form at Glendoick.
2. A pinkish-flowered clone, with the typical small leaves and rufous indumentum. Glendoick
3. *R. tsariense* Trimoense Group with slightly larger leaves than the type and paler leaf indumentum. Glendoick.
4. *R. tsariense* Poluninii Group with larger leaves and more flowers to the truss than the type. The Royal Botanic Garden, Edinburgh.

Subsection Maculifera

Height 0.3-10m, low, rounded to upright shrubs or small trees. Branchlets *tomentose or setose-glandular*, sometimes becoming glabrous with age; bark rough. **Leaves** 2.5-17.5 x 1.3-5cm, elliptic, oblong to obovate; lower surface with indumentum or hairs, often confined to midrib. **Inflorescence** 5-20-flowered, loose to dense. **Corolla** narrowly to widely campanulate without nectar pouches or tubular-campanulate with nectar pouches, white through pink to deep red, with or without blotch and spots; calyx minute (except *R. longesquamatum*); stamens 10; ovary tomentose to glandular, rarely glabrous; style glabrous or rarely hairy at base. **Distribution** Jiangxi, Guangxi, Zhejiang, Sichuan, Yunnan, Guizhou, Hubei, Anhui, Taiwan.

This is a fairly diverse subsection from a wide area of China. The tomentose or setose-glandular branchlets and the presence of indumentum or hairs on the midrib are the chief taxonomic characters. Davidian has placed the two species with indumentum over the whole leaf underside (*R. ochraceum* and *R. pachysanthum*) into other series. With the exception of these two species, subsection Maculifera is probably most closely related to subsections Irrorata and Selensia.

This subsection includes several good garden plants including some with excellent foliage. All are fairly easy to grow but the Taiwanese species are prone to leaf-tip burn. *R. morii, R. pachysanthum, R. pseudochrysanthum* and *R. strigillosum* are probably the most widely cultivated.

Picture 1

Picture 2

Picture 3

R. anwheiense
E.H. Wilson 1925 (*R. maculiferum* ssp. *anwheiense* D.F. Chamb.) H5

Height rarely more than 2.4m, a compact, rounded shrub. Branchlets with shedding white hairs. Leaf and flower buds *rounded*. **Leaves** 6.5-15 x 2-3.5cm, oblong-lanceolate, margins recurved, retained 1-3 years; upper surface glabrous, often pale green or yellowish in cultivation, lower surface usually with some indumentum and glands, especially on midrib; petiole sparsely or moderately floccose-tomentose and glandular, 2-3 cm. **Inflorescence** 6-10-flowered. **Corolla** funnel-shaped, 2.4-4cm long, pink to white, flushed pink in bud with purple-red spots; calyx glabrous; ovary +/- glabrous. **Distribution** Jiangxi, Anhui, Zhejiang, 1,200-1,800m (4,000-6,000 ft), in open, rocky or shady sites.

Taxonomically, this species bridges the Maculifera and Irrorata subsections and is usually quite easy to identify in its pale leaf colour, and its more or less glabrous leaves, branchlets, rachis, calyx, ovary and style.

R. anwheiense is a very free-flowering and tough species but the foliage is rather uninteresting and tends to look rather pale and yellowish. The flowers are remarkably frost-resistant. Introduced c.1925. April-May.

1. *R. anwheiense*, the A.M. clone with a fairly full truss, Glendoick.
2. A form of *R. anwheiense* with an open-topped truss, at Glendoick.
3. *R. anwheiense* foliage showing the distinctive rounded buds.

R. longesquamatum
Schneider 1909 (*R. brettii* Hemsl. & E.H. Wilson) H5

Height 2-6m, a shrub or small tree, compact when young, sometimes leggy with age. Branchlets covered with *coarse, shaggy, branched hairs*, fawn, turning to rusty-brown, bark shaggy, rough purplish brown. Bud scales *persistent for several years*. **Leaves** 6-12 x 2-4cm, bunched at ends of the shoots, oblong, oblong-obovate to oblong-oblanceolate,

MACULIFERA

retained 1-2 years; upper surface dark green, lower surface pale green with scattered minute glands and *shaggy, tawny hairs on the midrib;* petiole *stout with dense shaggy hairs.* **Inflorescence** 6-12-flowered, loose. **Corolla** openly campanulate, 4-4.5cm, pink to rose, blotched crimson; calyx *large (up to 10mm), pinkish purple*; ovary and lower half of style stipitate glandular. **Distribution** W. Sichuan 2,300-3,500m (7,500-11,500 ft), in woodland, on cliffs and on grassy slopes.

One of the most easily recognised species with the hairy stems, midrib and petiole and the persistent bud scales being obvious characters. This species also has the largest calyx in the subsection.

R. longesquamatum is a striking foliage plant, quite easy to please and its hardiness makes it useful for fairly severe climates. The flowers are rather small for the leaves in the few forms in cultivation; perhaps new introductions will prove to be better. Quite rare in cultivation. Introduced 1904-10, reintroduced 1990->. May-June in cultivation, as late as July in the wild.

Picture 1

1. *R. longesquamatum* showing loose trusses of blotched flowers and the large calyces, Glendoick.

Picture 2

Picture 3

2. *R. longesquamatum* highlighting the shaggy hairy branchlets and the persistent perulae, Glendoick.
3. *R. longesquamatum* showing the characteristic densely hairy bud and petioles (Photo R.B.G., E.).

R. maculiferum Franch. 1895 H5

Height to 6m, usually less, a fairly compact shrub or small tree. Branchlets with traces of thin tomentum. **Leaves** 6-11 x 2.5-4.5cm, oblong-oval, elliptic to almost oval to obovate, margins sometimes hairy, retained 1-2 years; upper and lower surfaces *glabrous* at maturity, lower surface with a fine-textured, buff-white indumentum on the midrib; petiole with floccose hairs. **Inflorescence** 7-10-flowered, loose; pedicels tomentose. **Corolla** openly campanulate, c. 3.5cm long, white to pale pink, *blotched crimson*, sometimes lined and spotted pink; calyx and ovary *tomentose*. **Distribution** Sichuan, Guizhou and Hubei, 2,100-3,000m (7,000-10,000 ft) in woodlands and on cliffs.

The ovaries, calyx and pedicels, tomentose in *R. maculiferum*, are +/- glabrous in the closely related *R. anwheiense*. *R. pachytrichum* has much denser, shaggier hairs on the stems and leaf midrib than *R. maculiferum* and usually narrower leaves, and their distributions are isolated. See also *R. oreodoxa* var. *shensiense* and *R. purdonii*.

R. maculiferum is a rare but hardy and quite easy species which can be showy, especially in forms with white flowers with large purple-red blotches. Foliage is rather uninteresting. Introduced 1901, reintroduced 1984. March- April.

1. *R. maculiferum* showing the large crimson blotch in the corolla. Royal Botanic Garden, Edinburgh.
2. *R. maculiferum* Guiz 148, introduced in 1985. Glendoick.

R. morii

Hayata 1913 (*R. nakotaisanense* Hayata.) H3-4

Height to 10m, usually less, a fairly erect or spreading shrub or small tree. New growth tomentose, soon becoming glabrous. **Leaves** 6.5-15 x 2-3.5cm, oblong-lanceolate, retained 1-3 years; upper surface glabrous, lower surface glabrous, midrib with +/- indumentum and/or glands; petiole floccose-tomentose and glandular. **Inflorescence** 5-15-flowered, loose to compact. **Corolla** widely campanulate, 3-5cm long, white to white flushed rose, with +/- crimson spots and blotch; calyx small; ovary *glandular and hairy;* style *tomentose near base but otherwise glabrous.* **Distribution** Taiwan 1,650-3,500m (5,500-11,500ft), abundant in central mountain ranges in conifer forests.

R. morii is easily distinguished from *R. pachysanthum* as the latter has thick woolly indumentum over the whole lower leaf surface. *R. pseudochrysanthum* differs in its thick and rigid leaves with more indumentum on the lower leaf surface midrib. These three species do hybridise in the wild and some plants with intermediate characters are occasionally cultivated. *R. morii* superficially resembles *R. irroratum* but is distinguished by the absence of nectar pouches, the mainly glabrous style, the different leaf shape and in the presence of floccose indumentum on the midrib.

R. morii is one of the most free-flowering species, in its best forms setting buds on every shoot every year, from quite an early age. It is quite easy to grow and fairly robust but some forms come very early into growth. Introduced 1918, reintroduced 1970s->. March-May.

1. *R. morii*, D. Dougan's garden, Vancouver Island, Canada.
2. A white-flushed-pink form of *R. morii* at the Younger Botanic Garden, Benmore, W. Scotland.
3. *R. morii* R.V. 71009, a form with a showy blotch, Royal Botanic Garden, Edinburgh.

MACULIFERA

Picture 4

4. *R. morii* R.V., with flowers flushed pink, Glendoick.

Picture 1

1. *R. ochraceum* in N.E. Yunnan. (Photo Yang Zhenhong).

R. ochraceum
Rehder & E.H. Wilson 1913 H4?

Height 2-6m, a bushy to upright shrub or small tree, young shoots covered with glandular bristles. **Leaves** *4-10 x 1-2.5cm*, oblanceolate, apex acuminate, retained 1-2 years; upper surface glabrous when mature, lower surface with *matted woolly yellow-brown to light brown indumentum*, petiole 0.8-2cm long. **Inflorescence** 6-12-flowered, fairly loose to almost full. **Corolla** broadly campanulate to tubular-campanulate, 2.5-4cm long, *rich to dark red*, unspotted, with nectar pouches; calyx 1-1.5mm, glandular; ovary densely glandular, style glabrous. The capsules dehisce very early. **Distribution** Central-S. & S.E. Sichuan, N.E. Yunnan, 1,700-3,000m (5,500-10,000ft), in thickets, forest, cliffs and on boulders.

R. ochraceum is a distinct species, resembling a small-leaved version of *R. strigillosum* with indumentum over the whole lower leaf surface. The indumentum separates this from other red-flowered members of the subsection and it differs from members of the Neriiflora subsection (apart from *R. sperabile*) in its acuminate leaf apex, in the larger number of flowers to the truss and its greater stature. Davidian has transferred *R. ochraceum* to the Neriiflorum subseries on account of its indumentum but geographically it fits much better in subsection Maculifera.

From the photographs in *Sichuan Rhododendron of China* and *Rhododendrons of China Vol. 2*, this looks a splendid species with freely-produced, comparatively large flowers of a striking colour. It may well prove one of the best new introductions of recent years. Introduced 1995. May?

R. pachysanthum Hayata 1913 H4-5

Height up to 3m, a compact and dense shrub, looser with age. **Leaves** up to 11 x 6 cm, lanceolate to ovate, retained 3 years; upper surface with *conspicuous silvery indumentum* (often browner in shade) which is retained, lower surface with a *thick, dense, woolly, bistrate indumentum, pale at first, turning to rich rusty brown at maturity*. **Inflorescence** 10-20-flowered, compact. **Corolla** funnel-campanulate, c. 4cm long, pure white to pale rose pink, usually spotted or flared green or crimson; calyx c.1mm, glandular; ovary densely glandular; style *glabrous*. **Distribution** Central Taiwan, 3,000-3,200m (10,000-10,500ft), in full exposure above the tree-line.

R. pachysanthum is an easily-identified species in the distinctive leaf shape, the thick indumentum on the lower leaf surface, the usually persistent indumentum on the upper leaf surface and the entirely glabrous style. Davidian has placed *R. pachysanthum* in the Taliense series, Wasonii subseries and it certainly does seem to be related to *R. wasonii*, which differs in its unistrate indumentum, eglandular ovary and yellow (or white to pale pink) flowers.

Picture 1

1. *R. pachysanthum* R.V. 72001 at Glendoick.

An outstanding recent introduction, *R. pachysanthum* is a first class foliage plant, compact and dense (especially when young), and is free-flowering once mature, though the flowers open rather early. Natural hybrids of *R. pachysanthum* x *R. morii* with a thin layer of indumentum have appeared from wild seed. Most *R. pachysanthum* in cultivation originate from Glendoick from hand-pollinated seed of the first introduction RV 72001, introduced 1972. March-April.

Picture 2

Picture 3

Picture 4

Picture 5

5. *R. pachysanthum* with persistent silvery indumentum on the upper leaf surface on a plant grown in the sun.

2. A white-flowered *R. pachysanthum*, Glendoick.
3. Flowers of *R. pachysanthum*, a form with a pink flush and a blotch on the corolla, Glendoick.
4. *R. pachysanthum* showing the persistent brown indumentum on the upper leaf surface on a plant grown in some shade. The deeper upturned leaf shows the mature indumentum colour, the paler one that of a young leaf.

R. pachytrichum
Franch. 1886 H4

Height to 12m, a large rounded shrub or small tree. Branchlets *with varying degrees of shaggy, curly, branched, brown hairs*. **Leaves** 6-13 x 2-4cm, oblong to oblanceolate to obovate, margin reflexed, retained 1-2 years; lower surface midrib +/- clothed with *brownish shaggy hairs*; petiole c. 1.5 cm, with shaggy brown hairs. **Inflorescence** 7-17-flowered loose to fairly full and rounded. **Corolla** campanulate, 4-5cm long, very variable, white through pink to rose, rose-magenta with a maroon blotch and usually spotting; calyx c. 1.5 mm, 5 acute lobes; ovary *densely tomentose;* style *glabrous*. **Distribution** Sichuan, 2,100-3,400m (7,000-11,000ft), growing with *R. calophytum* and other species in mixed forest.

R. strigillosum differs from *R. pachytrichum* in its red flowers and usually longer leaves. See under *R. strigillosum* for more information. *R. pachytrichum* resembles and possibly merges with *R. longesquamatum* and it hybridises with many other species in the wild, including *R. strigillosum, R. wiltonii, R. oreodoxa* and *R. calophytum*.

R. pachytrichum is very variable in merit, the best forms being magnificent and the poorest having amongst the least spectacular flowers in subgenus Hymenanthes. Unfortunately selected clones are usually hard to root and graft. It is easy to grow and pretty robust but the early flowers and growth need protection. Introduced 1903-11, reintroduced 1980s->. March-April. Some authorities recognise *R. pachytrichum* var. *monosematum* which Chamberlain says is intermediate between *R. pachytrichum* and *R. strigillosum* although cultivated material under this name looks closer to *R. pachytrichum*.

MACULIFERA

R. pseudochrysanthum Hayata 1908 H4-5

Height 0.5-3m, a low shrub, variable in habit, ranging from very dwarf and compact to upright and occasionally leggy. Young shoots tomentose. **Leaves** 3-8 x 1.5-5cm, ovate to oblong-lanceolate, *thick and rigid*, margins usually recurved, retained c. 2 years; upper surface glabrous or covered in a thin grey indumentum, lower surface glabrous *with a persistent indumentum on midrib only*. **Inflorescence** 5-10-flowered, loose to fairly loose. **Corolla** campanulate, without nectar pouches, 3-4cm long, pink in bud opening white, white flushed pink or pale pink, sometimes striped pink outside, spotted red; calyx c.2mm, glandular-ciliate; ovary densely glandular; style glabrous. **Distribution** Taiwan, 1,800-4,000m (6,000-13,000ft), in conifer forests and above the tree-line.

This species is usually quite easily distinguished by its thick, leathery, pointed leaves, the indumentum on the midrib on the lower leaf surface and the characteristic flowers. Some forms seem quite close to *R. hyperythrum*. *R. morii* is distinguished by its thinner leaves with very little midrib indumentum and *R. pachysanthum* by its thickly indumented leaf lower surface. Natural hybrids have been recorded between *R. pseudochrysanthum* and both the aforementioned.

R. pseudochrysanthum is an excellent and widely-grown species which varies enormously in height; the smallest introductions from RV 72003 have not reached much more than 15cm in 20 years while some forms grow up to 2m or more in cultivation. The dwarfest forms are often slowest to start flowering. It needs light shade and low fertiliser to avoid leaf tip burning. Some authorities recognise var. *nankotaisanense* (Hayata) T. Yamaz which is said to differ from the type in its glandular ovary and longer-than-average pedicels. Introduced 1918, 1938, reintroduced 1970s->. March-May.

Picture 1

Picture 2

Picture 3

Picture 4

1. A particularly fine pink form of *R. pachytrichum* at the Royal Botanic Garden, Edinburgh.
2. A white-flowered *R. pachytrichum* Wilson 1522 or 1525 at the Royal Botanic Garden, Edinburgh.
3. *R. pachytrichum* in young growth at the Royal Botanic Garden, Edinburgh, showing the characteristic hairs on the stem. Other forms can be much more densely hairy with longer hairs.
4. *R. monesematum* (or *R. pachytrichum* var. *monesematum*) which is probably a natural hybrid of *R. pachytrichum*; a clone from Blackhills, E. Scotland.

Picture 1

1. A fine pinkish-flowered form of *R. pseudochrysanthum* from Rowallane, N. Ireland, growing at Glendoick.

MACULIFERA

brownish stellate-tomentose, sometimes glandular; style usually glabrous.

This species was formerly placed in the Parishii series on account of its stellate hairs but otherwise has little connection with the red-flowered species of that series/subsection. It differs from the other species in subsection Maculifera in its more or less glabrous leaves without bristles or indumentum.

Var. *sikangense* (*R. cookeanum* Davidian)

Leaves both surfaces +/- *glabrous at maturity*. **Distribution** S.W. and C.W. Sichuan 3,100-4,400m (10,200-14,500ft), in forest, bamboo and alpine slopes.

R. sikangense var. *sikangense* is quite a showy plant in flower in its best blotched forms, but the foliage is rather uninteresting. *R. sikangense* as found on Erlang Shan makes a fine upstanding vigorous plant which has narrower leaves than plants from Muli. Davidian retains both *R. sikangense* and *R. cookeanum* at specific rank stating that the principal difference is in the former having a rough bark and the latter a smooth bark; further fieldwork should confirm whether this is significant. Introduced 1929, reintroduced 1990->. May-July.

Picture 2

Picture 3

Picture 4

2. A very dwarf clone of *R. pseudochrysanthum* RV 72003 from Mt Morrison, Taiwan, at Glendoick.
3. A medium-sized clone of *R. pseudochrysanthum* RV 72003 from Mt Morrison, Taiwan, at Glendoick.
4. The leaf lower surface of *R. pseudochrysanthum* showing the characteristic persistent indumentum on the midrib only. (Photo R.B.G., E.).

Picture 1

1. *R. sikangense* Rock 18142 (collected as *R. cookeanum*), at the Royal Botanic Garden, Edinburgh, one of the very few plants of this rare species in cultivation before its recent reintroduction.

R. sikangense Fang 1952 II4-5

Height 3-9m, a fairly upright shrub or small tree, branchlets at first rufous to white tomentose and glandular, bark rough or smooth. New growth evanescent tomentose. Buds with unusual, *long, tailed bud scales*. **Leaves** 7-15 x 2.8-6cm, elliptic to oblanceolate; upper surface glabrous at maturity, lower surfaces +/- glabrous at maturity or with some semi-persistent indumentum consisting of stellate hairs along the midrib and onto the petiole. **Inflorescence** 5-15-flowered, loose to fairly compact. **Corolla** campanulate, without nectar pouches, 3.5-5cm long, white to pink to purple, with or without blotch or spots; calyx minute 0.5-2mm; ovary

MACULIFERA

Picture 2

Picture 3

Picture 4

4. *R. sikangense* var. *exquisitum* on Wumeng Shan, N.E. Yunnan, showing the much wider leaves than the Erlang Shan plant and slightly wider leaves than the Muli plant (1. & 2.).

2. *R. sikangense* Rock 18142.
3. Foliage of *R. sikangense* on Erlang Shan, W. Sichuan photographed in 1990, showing the distinctive long, tailed bud scales.

Var. *exquisitum* (T.L. Ming) T.L. Ming 1992

A large rounded shrub to 8m. **Leaves** usually a little wider than those of the Muli plant (formerly *R. cookeanum*) and much wider than the Erlang Shan plant, lower surface with *persistent floccose, rufous indumentum towards the leaf base*. **Corolla** white to pink with a deep red blotch and spotting. **Distribution** N.E. Yunnan, 3,500m (11,300ft).

This was originally described as a new species but has subsequently been reduced to varietal status. The wide leaf and persistent indumentum seem to warrant this status. Seen in the wild this is a handsome foliage plant. Introduced 1994-95. *R. montiganum* T.L. Ming from the same location is probably a natural hybrid of *R. sikangense* var. *exquisitum* x *R. bureavii*. Introduced 1994.

R. strigillosum Franch. 1886 H4

Height 3-6m, a shrub or small tree, *often round and symmetrical in habit*. Branchlets +/- covered with long bristles. Leaf buds *sticky*. **Leaves** 8-18 x 2.5-4.5cm, oblong-lanceolate to oblanceolate, margin often strongly recurved, bristles on young leaves white, retained 1-3 years; lower surface, midrib and petiole with long bristles and +/- shortish stalked hairs. **Inflorescence** 8-12-flowered, usually flat-topped. **Corolla** tubular-campanulate, with depressed, black-crimson nectar pouches, 4-6cm long, *deep red to crimson-scarlet*; pedicel usually covered with hairs; calyx c.1mm; ovary with *long, reddish glandular hairs*; style glabrous. **Distribution** Sichuan and N.E. Yunnan, 2,100-3,400m (7,000-11,000ft), in thickets or in the open on cliffs and slopes.

R. strigillosum is closely related to the newly introduced *R. ochraceum* which differs in its smaller leaves with indumentum over the whole lower leaf surface. *R. pachytrichum* differs in its pink to white rather than red flowers, in the usually shorter leaves, the hairy rather than bristly branchlets and midrib and the tomentose rather than bristly petiole, pedicel and ovary. So-called white *R. strigillosum* reported from the wild is almost certainly *R. pachytrichum*. *R. strigillosum* appears to take the place of *R. pachytrichum* in S. Sichuan, and N.E. Yunnan but the two species do merge/hybridise where their distributions meet.

A fine, early-flowering, red species with distinctive bristly foliage. The best forms of *R. strigillosum* have rich deep red flowers. Early in flower and growth, needing a protected site in most climates. Introduced 1904-10, reintroduced 1990s. February-May.

Subsection Neriiflora

Height 0.05-6m, *creeping, mounded, rounded* to broadly upright shrubs or small trees. Branchlets *mostly thin*, with thin to thick tomentum, sometimes with bristles and/or glands; bark smooth to roughish, often peeling, fawn to cinnamon. **Leaves** narrowly elliptic to orbicular; upper surface usually glabrous on maturity, lower surface glabrous, or with a compacted or woolly, continuous or discontinuous indumentum, whitish to buff or rufous. **Inflorescence** 1-12 (20)-flowered, usually loose, flowers often hanging between the leaves. **Corolla** 5-lobed, *tubular-campanulate* or occasionally campanulate, with depressed nectar pouches, *most often red*, also white, yellow, pink to carmine and mixed colours, usually *fleshy*; calyx minute to large and cup-like, often coloured; stamens *10*; ovary usually *tomentose*, with or without glands; style *glabrous*. **Distribution** Bhutan eastwards to S.E. Tibet, Yunnan and Upper Burma.

A complex subsection whose members often grow together in the wild, merging and hybridising freely. Not surprisingly, this has led to a confused and unwieldy taxonomy and many of the so-called species, subspecies and varieties are simply extreme forms or natural hybrids. The Edinburgh Revision went a long way towards improving the confusion but recent fieldwork has indicated that a further reduction in the number of species is still necessary. We have divided the subsection into alliances following the old subseries divisions; this treatment is implicit in the order of the species in the Edinburgh revision. Subsection Neriiflora is allied to subsection Thomsonia but the species are generally smaller in all parts.

Many species in subsection Neriiflora make good garden plants for cooler areas but they dislike hot summers, require excellent drainage and most dislike fertiliser. The most widely cultivated species in this subsection are *R. haematodes, R. neriiflorum, R. forrestii, R. dichroanthum,* and *R. sanguineum*.

Picture 1

Picture 2

Picture 3

Picture 4

1. *R. strigillosum* an old plant at Glendoick, obtained from Harry White of Sunningdale Nurseries, showing the recurved edges to the leaves.
2. A fine deep red-flowered form of *R. strigillosum* at Crarae. W. Scotland, showing the typical reddish, pointed growth buds.
3. A bush of *R. strigillosum* at Baravalla, W. Scotland.
4. The distinctive new growth of *R. strigillosum* at the Royal Botanic Garden, Edinburgh.

Haematodes alliance

Leaves with thick woolly indumentum on the lower surface. **Corolla** red; ovary usually tomentose,

R. beanianum
Cowan 1938 H3-4

Height to 3m but usually 1.5-1.8m, a fairly compact shrub, often straggly with age. Branchlets *bristly and glandular*; bark greenish-brown to red-brown. **Leaves** 6-9 x 3.2-4.4cm, obovate to elliptic, retained for 2 years; upper surface dark green and bullate, lower surface with a dense unistrate rufous indumentum consisting of dendroid hairs. **Inflorescence** 6-10-flowered, fairly full to very loose. **Corolla** tubular-campanulate, c.3.5cm long, fleshy crimson-scarlet or occasionally pink; calyx c.5mm, sparsely tomentose; ovary tomentose. **Distribution** N.E. Upper Burma and Arunachal Pradesh, 2,700-3,400m (9,000-11,000ft), in bamboo, amongst rocks and at forest edges.

A distinct species characterised by the plentiful bristles on the stems and petioles. Its closest relative is *R. piercei* which differs in the smoother upper leaf surface, the bistrate indumentum on the lower surface and its hairy but not bristly branches.

R. beanianum is a fine foliage plant with good flowers in the best forms. Many forms are rather tender and the early flowers are easily frosted. Introduced 1926. March-May.

Picture 1

1. *R. beanianum*, a form at the Royal Botanic Garden, Edinburgh

Picture 2

Picture 3

2. *R. beanianum*, a form from Maryborough, S. Ireland, growing at Glendoick.
3. *R. beanianum* foliage showing the dark green, bullate upper leaf surface and the rufous indumentum on the lower surface.

R. catacosmum
(Balf.f. ex) Tagg 1927 H4

Height 1.3-3m, a fairly compact low shrub. Branchlets tomentose, with few or no bristles; bark light red-brown. **Leaves** 6-14 x 3-6.5cm, *obovate*, leathery, retained for 1(-2) years; upper surface with a little persistent indumentum, lower surface with a bistrate indumentum, the upper layer tawny, loose, consisting of dendroid hairs, petiole 1-1.5cm, tomentose. **Inflorescence** 5-9-flowered. **Corolla** tubular-campanulate, fleshy, c.4.5cm long, scarlet to rose-crimson; calyx *large, 1.5-2cm*, margins ciliate; ovary densely tomentose. **Distribution** Limited to S.E. Tibet, N.W. Yunnan border, 3,700-4,400m (12,000-14,500ft), forest margins and cliffs.

This species is separated from most of the subsection by its large obovate leaves. It is closely related to *R. haematodes* but is slightly larger in all parts, particularly in its large calyx which gives the flower an almost hose-in-hose effect. *R. haematodes* ssp. *chaetomallum* differs in its more bristly branchlets and petioles.

R. catacosmum is a fine species which makes a shapely bush with good foliage and flowers but it is hard to propagate and remains rare. Introduced 1921-23. March-April.

Picture 1

Picture 2

1. *R. catacosmum* from Wakehurst Place, showing how floriferous a mature plant can be. Glendoick.
2. *R. catacosmum*, a form from Wakehurst Place, showing the obovate leaf shape and large calyx typical of the species.

R. coelicum
Balf.f. & Farrer 1922 H4

Height 1-1.85m, a compact to upright small shrub. Branchlets eglandular or sparsely glandular. **Leaves** 6-8.5 x 3.1-4.4cm, *obovate,* often retained for less than one year and sometimes almost deciduous; upper surface *dark and shiny*, lower surface with a thick fulvous woolly indumentum; petiole 1-1.5cm, sparsely stipitate-glandular. **Inflorescence** c.10-flowered. **Corolla** tubular-campanulate, 3.8-4.5cm long, crimson, fleshy; pedicel stipitate-glandular; calyx 5-7mm, glandular-ciliate margins; ovary tomentose and glandular. **Distribution** W. Yunnan, N.E. Upper Burma border, 3,700-4,400m (12,000-14,500ft), shaded screes, bamboo, cliffs.

This species is very closely related to *R. pocophorum* but is smaller in all parts with eglandular or almost eglandular branchlets and petioles and a more obovate leaf.

The rare *R. coelicum* has fine summer foliage but is inclined to be semi-deciduous and is often rather a weak grower. It is probably an extreme form of *R. pocophorum* or a natural hybrid of it. Introduced 1924. April.

Picture 1

Picture 2

Picture 3

1. *R. coelicum* at Hamish Gunn's garden, Edinburgh (Photo H. Gunn).
2. *R. coelicum* showing the dark green leaves. (Photo H. Gunn)
3. *R. coelicum* foliage in the summer with fresh growth. In winter we find it partly deciduous, Glendoick.

R. haematodes Franch. 1886 H4-5

Height 0.6-1.5m, occasionally more, a low shrub, dense and compact in the open. Branchlets densely tomentose (ssp. *haematodes*) to bristly (ssp. *chaetomallum*); bark light to red-brown. **Leaves** 4.5-11 x 1.8-5.5cm, obovate to oblong, leathery; upper surface smooth, usually glabrous, lower surface with a *dense, bistrate, woolly* indumentum, fawn to rufous. **Inflorescence** 4-12-flowered. **Corolla** tubular-campanulate, 3.5-4.5cm long, fleshy, scarlet to crimson; calyx *minute to large* 3-13mm+, red or green; ovary densely rufous-tomentose.

Ssp. *haematodes*

A generally *low and compact* shrub. Young shoots and petioles predominantly *tomentose* with few bristles. **Leaves** retained 2-3 years; upper surface dark green and *smooth*, lower surface with a *dense* indumentum. **Distribution** W. & N.W. Yunnan, 3,400-4,000m (11,000-13,000ft), fairly widespread at about and above the tree line.

Ssp. *haematodes* is usually easily separated from its relatives by its low, compact habit, the absence of bristles, and its smooth dark leaves, retained for 2-3 years.

Hardier than its relatives, ssp. *haematodes* is one of the finest and most popular low-growing species. It is somewhat variable in denseness of growth and size of flower and it requires very good drainage and a position away from hot sun for best results. Introduced 1910->, reintroduced 1981->. May-June.

Ssp. *chaetomallum* (Balf.f. & Forrest) D.F. Chamb. 1979 (*R. hillieri* Davidian) H3-4.

Young shoots and petioles predominantly *bristly*. **Leaves** usually somewhat rugulose, retained for *1(-2)* years; upper surface often with *some persistent indumentum*, lower surface with a +/- densely woolly indumentum, pale to dark fawn. **Distribution** N.E. Upper Burma, S.E. Tibet, N.W. Yunnan, common, 3,000-4,600m (10,000-15,000ft), forest, bamboo and meadows.

Although the two subspecies do merge in the wild, as cultivated, they are typically quite distinctive: ssp. *chaetomallum* is often taller with a looser habit and has paler and rougher leaves than ssp. *haematodes*. The persistent indumentum on the upper leaf surface and the less dense indumentum on the lower surface also separates ssp. *chaetomallum* from ssp. *haematodes*. Forms with very thin indumentum including the taxon described as *R. x hillieri* Davidian, are almost certainly natural hybrids. Closely related to *R. catacosmum*.

Ssp. *chaetomallum* seldom makes as good a garden plant as ssp. *haematodes* as it is often short-lived or liable to partial die back. Introduced 1917-49. March-May.

Picture 1

Picture 2

Picture 3

1. *R. haematodes* ssp. *haematodes* at the Royal Botanic Garden, Edinburgh. This form has a minute calyx.
2. *R. haematodes* ssp. *haematodes* SBEC 0585 with a large calyx. collected on the SBEC expedition to the Cangshan, Yunnan, China in 1981.
3. A very fine bush of *R. haematodes* ssp. *haematodes* at the Royal Botanic Garden, Edinburgh.

Picture 4

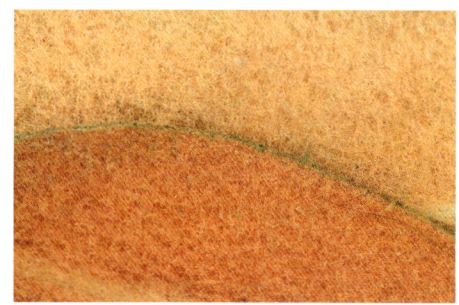

Picture 5

Picture 6

Picture 7

Picture 8

8. *R. haematodes* ssp. *chaetomallum* showing the typically persistent indumentum on the upper surface of the leaf.

R. piercei
Davidian 1976 (*R. beanianum* var. *compactum* Cowan) H3-4

Height 1.2-2.5m, an open, *spreading,* bushy or straggling shrub. Branchlets tomentose; bark light brown to brown. **Leaves** 6-11 x 2.7-5.5cm, ovate to elliptic, retained 2-3 years; upper surface tomentose on the young growth, *shiny and rugulose* at maturity, lower surface with a *bistrate*, woolly, fulvous-tomentose-dendroid indumentum. **Inflorescence** 6-8-flowered. **Corolla** tubular-campanulate, with dark nectar pouches, 2.8-3.6cm long, fleshy crimson; calyx 3-6mm, irregular, glabrous; ovary densely tomentose. **Distribution** S. E. Tibet, only one known location, 3,700-4,000m (12,000-13,000ft).

The distinctive features of this species, which make it one of the most easily recognised in the subsection, are its shiny and rugulose upper leaf surface, its spreading, rather rangy habit and its early flowers. The bistrate rather than unistrate indumentum and the lack of bristles on the branchlets and petioles separate *R. piercei* from its closest relative *R. beanianum*.

R. piercei is an attractive plant with usually fine flowers and is more vigorous, easier to grow and longer-lived than most of its relatives. Introduced 1933 (once only). March-April (-May)

4. *R. haematodes* ssp. *haematodes* foliage showing the deep green leaves with a dense woolly indumentum on the lower surface.
5. The leaf lower surface of *R. haematodes* ssp. *haematodes* showing the dense woolly indumentum, dark on the old leaf, paler on the young one. (Photo R.B.G., E.)
6. *R. haematodes* ssp. *chaetomallum* at Glenarn, W. Scotland.
7. *R. haematodes* ssp. *chaetomallum*, a large bush at the Royal Botanic Garden, Edinburgh.

Picture 1

1. *R. piercei* Kingdon-Ward 11040, collected in 1933 in S.E. Tibet, growing at the Royal Botanic Garden, Edinburgh..

NERIIFLORA

Picture 2

Picture 3

2. *R. piercei* at Glendoick.
3. *R. piercei* foliage, showing the characteristic leaf shape and rugulose upper leaf surface with traces of indumentum.

R. pocophorum (Balf.f. ex.) Tagg 1927 H3-4

Height 1.2-3m, a rather erect shrub. Branchlets *densely glandular*; bark light brown. **Leaves** 10-16 x 4-7cm, oblong to obovate, retained for 1-2 years; upper surface dark, shiny, glabrous, lower surface with a *unistrate, dendroid, woolly* continuous or patchy indumentum. **Inflorescence** *10-20-flowered, dense*. **Corolla** fleshy, tubular-campanulate, 3.6-5.3cm long, light to deep crimson; calyx large 5-10mm, glandular ciliate, otherwise glabrous; ovary densely *glandular*. **Distribution** S.E. Tibet, N.W. Yunnan, 3,700-4,600m (12,000-15,000ft), margins of bamboo and conifer forest, crags and meadows.

Var. *pocophorum*

Leaves, with a *continuous* indumentum on lower surface.

Var. *pocophorum* is the giant of the alliance, taller-growing and with larger, more elongated leaves than its closest relatives *R. haematodes* and *R. coelicum*.

This is a handsome plant when growing well but it can be hard to please and its early flowers are often frosted. Introduced 1921-1928. March-April.

Var. *hemidartum* (Tagg) D.F. Chamb. 1978

Tending to be smaller growing than var. *pocophorum*. **Leaves** with a *patchy discontinuous* indumentum on lower surface.

This variety is most likely to be confused with two subsection Neriiflora species with discontinuous indumentum: *R. floccigerum* has narrower leaves with a papillate lower surface covered in a more speckled indumentum while *R. sperabiloides* has a thin, ramiform rather than dendroid indumentum and a ciliate rather than +/- glabrous calyx.

This taxon is almost certainly a natural hybrid; it can be attractive but is prone to disease and is sometimes lacking in vigour. Introduced 1921-24. March-April.

Picture 1

Picture 2

1. *R. pocophorum* var. *pocophorum*, a very fine form with a large, dense truss.
2. *R. pocophorum* var. *pocophorum* at the Royal Botanic Garden, Edinburgh.

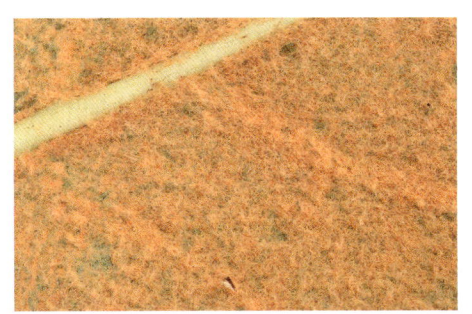

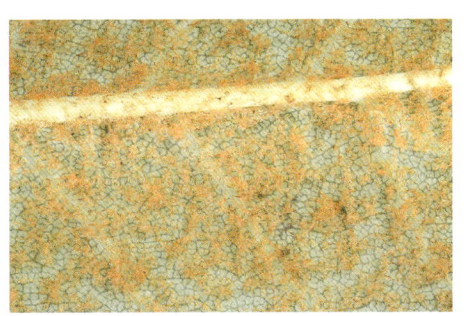

NERIIFLORA

Mallotum alliance

R. mallotum
Balf.f. & Kingdon-Ward 1917 (*R. aemulorum* Balf.f.) H3-4

Height 1.5-4.5m or larger, a shrub or small tree. Branchlets densely rufous-tomentose; bark roughish, flaking, purplish-grey to grey-green. **Leaves** 7-15 x 3.5-7.5cm, *obovate, very thick and stiff,* retained for 1-3 years; upper surface *dark green and rugose,* glabrous on maturity, midrib sometimes tomentose, lower surface *with a unistrate dense woolly cinnamon-brown* indumentum; petiole is sometimes tomentose above. **Inflorescence** 7-15-flowered, *often compact.* **Corolla** tubular-campanulate, with nectar pouches, 4-5cm long, fleshy, usually crimson, also scarlet, cherry-red; calyx small 2-3mm *tomentose*; ovary densely rufous-tomentose; capsule *short c.2cm*; seed *very small*. **Distribution** N.E. Upper Burma – W. Yunnan frontier, 3,400-3,700m (11,000-12,000ft), open rocky slopes, bamboo thickets.

R. mallotum is so distinct from the rest of subsection Neriiflora it might be better placed in its own subsection. The combination of its early red flowers, its tree-like habit and the large, rugose, thickly-indumented leaves make it one of the most easily recognised of all species.

R. mallotum is an excellent foliage plant, especially with the sun shining on the leaf undersides. The flowers are fine but are easily frosted. Introduced 1919-1924. February-April.

3. Leaf lower surface of *R. pocophorum* var. *pocophorum* showing the continuous woolly indumentum. (Photo R.B.G., E.)
4. *R. pocophorum* var. *hemidartum* at Glendoick.
5. The lower leaf surface of *pocophorum* var. *hemidartum* showing the typical patchy, discontinuous indumentum. (Photo R.B.G., E.)

R. x hemigynum Tagg & Forrest. Leaves with thin indumentum or glabrous; Corolla red to magenta. This is considered by Chamberlain to be *R. pocophorum x R. eclecteum.*

1. *R. mallotum* Farrer 815, collected with E.H.M. Cox in Upper Burma in 1919, growing at Glendoick.

NERIIFLORA

Picture 2

A most complex group including some apparently stable species but many varieties and forms in a complete state of flux, often growing together in the wild where they merge and/or hybridise with each other, making identification almost impossible. All are closely related, this being noticeable in both their morphological characters and in their behaviour in cultivation. The sanguineum alliance species form neat, symmetrical garden plants when young but they can become straggly with age and some tend to be shy-flowering.

Picture 3

Picture 4

R. aperantum Balf.f. & Kingdon-Ward 1922 H4

Height to 0.6m, occasionally more in shade, a dense and matted dwarf shrub. Branchlets floccose, perulae *persistent*, covering whole stems, bark reddish-brown. **Leaves** 3-6.5 x 1.4-2.4cm, *crowded at ends of shoots,* obovate to lanceolate, retained 1-2 years; upper surface glabrous, lower surface *glaucous with papillae*, almost glabrous on older leaves, petiole *broad and winged*. **Inflorescence** 4-6-flowered, lax. **Corolla** tubular-campanulate, to 4.5cm long, usually crimson to pink in cultivation but also white, yellow, orange-red and mixed colours in the wild; pedicel *tomentose*; calyx medium 3-6mm, lobes glandular-ciliate; ovary rufous-tomentose with a few glands. **Distribution** N.E. Upper Burma – W. Yunnan frontier, 3,700-4,400m (12,000-14,500ft), open meadows and cliffs.

This is a distinctive species, easily identified by its flat-topped growth habit, the persistent perulae, the leaves in close whorls and the broad, winged petiole.

2. A fine bush of *R. mallotum* at the Royal Botanic Garden, Edinburgh.
3. A *R. mallotum* of this size in full flower is a rare sight. Royal Botanic Garden, Edinburgh.
4. *R. mallotum* showing the dense, woolly, cinnamon-brown indumentum on the lower surface (top) and the sparse indumentum (which usually does not persist) on the upper surface (bottom). (Photo R.B.G., E.)

Sanguineum alliance

Height usually 0.6m-2m. **Leaf** size not over 10cm long, usually less. Indumentum variable, from absent to thick. **Inflorescence** small and loose. **Corolla**, of variable colour, often not as thick and fleshy as that of species in the Haematodes alliance.

Picture 1

1. *R. aperantum* flowering very freely at Glendoick. It rarely performs as well as this.

Although *R. aperantum* is slow to flower and somewhat tricky to grow, it is well worth extra care, as mature, free-flowering specimens are very fine. Best in an open but north-facing, cool situation. Introduced 1919->. April-May

Picture 2

Picture 3

Picture 4

2. A red-flowered form of *R. aperantum* at Glendoick.
3. A pink-flowered form of *R. aperantum* Forrest 27022, Royal Botanic Garden, Edinburgh.
4. Foliage of *R. aperantum* showing the dense habit, leaves crowded at ends of shoots and the glaucous lower surface of the leaf.

R. citriniflorum
Balf.f. & Forrest 1919 H4-5

Height 0.2-1.5m, a compact to leggy dwarf shrub. Branchlets glabrous or with white floccose tomentum; bark dark brown. **Leaves** 4-7.7 x 1.3-2.5cm, obovate to elliptic, retained 1-2 years; upper surface glabrous when mature, lower surface with a *dense, moderate to thick, brown to grey-brown* indumentum. **Inflorescence** 2-6-flowered. **Corolla** moderately fleshy, tubular-campanulate, 3-4.5cm. **Distribution** N.W. Yunnan-S.E. Tibet border, 4,000-4,900m (13,000-16,000ft), moorland, cliffs and screes.

Var. *citriniflorum* (*R. chlanidotum* Balf.f. & Forrest)

Corolla yellow, often flushed pink or red, tubular-campanulate; calyx *small to medium 2-5mm*, pedicel and ovary usually *glandular*.

Var. *horaeum* (Balf.f. & Forrest) D.F. Chamb. 1979 (var. *aureolum* Cowan).

Leaves often with a thicker indumentum than in var. *citriniflorum*. **Corolla** *orangy-red to carmine-red*. Calyx large *0.7-12mm*; pedicel and ovary *eglandular*.

When the indumentum is at its thickest, especially in var. *horaeum*, *R. citriniflorum* is quite a distinct species but as with most members of this alliance, there are a multitude of intermediate forms between it and many of its relatives. Two examples are *R. x hillieri* Davidian with pink to crimson flowers and *R. x xanthanthum* Tagg & Forrest with creamy yellow crimson-flushed flowers. Both probably have *R. haematodes* ssp. *chaetomallum* in their parentage. *R. citriniflorum* and *R. sanguineum* often occur together in the wild, resulting in hybrid swarms between the two species.

Good yellow-flowered forms of var. *citriniflorum* are rare in cultivation but are worth seeking out; other forms with flowers tinged pink may be attractive, but some are muddy and of little merit. The orange-flowered form of var. *horaeum* Forrest 21850 is striking and much sought after. Introduced 1917-32. April-May.

Picture 1

1. A particularly fine form of *R. citriniflorum* var. *citriniflorum* Cox 6538 growing amongst *R. mekongense* var. *mekongense* above Londre, Salween-Mekong divide, N.W. Yunnan, China.

NERIIFLORA

Picture 2

Picture 3

Picture 4

Picture 5

Picture 6

Picture 7

2. *R. citriniflorum* var. *citriniflorum* Cox 6543B with pretty bicoloured flowers growing near to Cox 6538 (No.1); above Londre at 3,700m (12,000ft), Salween-Mekong divide, N.W. Yunnan, China.
3. A cream flushed pink-flowered *R. citriniflorum* var. *citriniflorum*; above Londre at 4,000m (13,000ft), Salween-Mekong divide, N.W. Yunnan, China.
4. A clear pale yellow-flowered *R. citriniflorum* var. *citriniflorum* Rock 108, at Glendoick, with a relatively lax truss.
5. The striking flowers of *R. citriniflorum* var. *horaeum* Forrest 21850, Glendoick.
6. This plant was formerly called *R. sanguineum* ssp. *didymoides* Consanguineum Group but its thick indumentum obviously places it in *R. citriniflorum* var. *horaeum*.
7. Leaf lower surface of *R. citriniflorum* var. *horaeum* showing the dense woolly indumentum, often thicker in var. *horaeum* than in var. *citriniflorum*. (Photo R.B.G., E.)

R. dichroanthum
Diels 1912 H4

Height 0.3-2.3m, a spreading and usually compact shrub. Branchlets with white soft tomentum, sometimes with hairs or bristles; bark light brown to brown. **Leaves** 4-10 x 2-4cm, oblanceolate to elliptic, retained (1-)-2 years; upper surface *glabrous*, lower surface with a continuous +/- loose to compacted *silvery to fawn* indumentum. **Inflorescence** 3-8-flowered, lax. **Corolla** fleshy, tubular-campanulate, 3.5-5cm long, *orange-red, yellow-flushed-red or rarely carmine*; pedicel tomentose or glandular; calyx *3-15mm*, cupular or irregular; ovary rufous-tomentose, sometimes glandular.

This species is characterised by its continuous, rosulate, +/- loose to compacted indumentum. The flowers in orange-red and orange shades distinguish this from all its relatives except the orange forms of R. *citriniflorum* var. *horaeum* which differ in the eglandular ovary and pedicel and the usually smaller leaves.

R. *dichroanthum* is a useful species for its late flowers of often unusual colours, but many forms tend to be rather shy-flowering. It divides into 4 subspecies which differ little botanically or in garden value but have more or less separate distributions.

Ssp. *dichroanthum*

Young shoots without bristles or with non-glandular bristles. **Leaves** longer than in ssp. *apodectum* and not as narrow as in ssp. *septentrionale*. Indumentum silvery, compacted. **Distribution** Only found in Cangshan, W.C. Yunnan, 2,700-3,700m (9,000-12,000ft), meadows and gullies.

Introduced 1910-31. Reintroduced 1981->. May-June.

Ssp. *apodectum* (Balf.f. & W.W. Sm.) Cowan 1940
(R. *jangtzowense* Balf.f. & Forrest, R. *liratum* Balf.f. & Forrest).

Leaves more oval and smaller than in ssp. *dichroanthum*; upper surface *shiny*, indumentum silvery to fawn, non-glandular, often a little *thicker* than in the other subspecies. **Corolla** orange with rose, crimson or scarlet. This is the southern-most occurring sub-species. **Distribution** W. Yunnan into N.E. Upper Burma. 3050-3,660m (10,000-12,000ft).

Introduced 1912-53. May-June.

Ssp. *scyphocalyx* (Balf.f. (and) Forrest) Cowan 1940
(R. *herpesticum* Balf.f. & Forrest R. *torquatum* Balf.f. & Farrer).

Leaves slightly larger than in ssp. *apodectum*, obovate, not shiny on the upper surface, indumentum *thinnish*, glandular, fawn to grey. **Corolla** orange through yellowish-crimson to dull yellow. **Distribution** N.E. Upper Burma into W. Yunnan, 3,050-4,270m (10,000-14,000ft).

Plants under Herpesticum Group are generally smaller than the type in all parts. Introduced 1914-25. May-June.

Ssp. *septentrionale* Cowan 1940

Leaves often *smaller* and narrower than in the other subspecies, indumentum whitish to fawn. **Corolla** yellow or yellow flushed rose. **Distribution** N.W. Yunnan and adjacent N.E. Upper Burma. 3,700-4,300m (12,000-14,000ft).

The rarest subspecies in cultivation. Introduced 1924. May-July.

Picture 1

Picture 2

Picture 3

1. R. *dichroanthum* ssp. *dichroanthum* in a gully on the east flank of the Cangshan, Central W. Yunnan, China.
2. R. *dichroanthum* ssp. *dichroanthum* Forrest 6781, Glendoick. The leaves of this subspecies are larger than those of the other subspecies.
3. R. *dichroanthum* ssp. *dichroanthum* SBEC 0601, Glendoick, collected on the Cangshan, central W. Yunnan, China.

Picture 4

Picture 5

Picture 6

Picture 7

Picture 8

Picture 9

8. *R. dichroanthum* ssp. *scyphocalyx*, Corsock, S. Scotland.
9. Upper leaf surface of *R. dichroanthum* ssp. *septentrionale* showing its small, narrow leaf. (Photo R.B.G., E.)

4. Leaf lower surface of *R. dichroanthum* ssp. *dichroanthum* showing the compacted indumentum. (Photo R.B.G., E.)
5. *R. dichroanthum* ssp. *apodectum* Forrest 18167 at the Royal Botanic Garden, Edinburgh.
6. *R. dichroanthum* ssp. *apodectum*, showing the wider, shorter leaves than in ssp. *dichroanthum*.
7. *R. dichroanthum* ssp. *scyphocalyx* Farrer 1024, Royal Botanic Garden, Edinburgh, showing the leaves are not shiny on the upper surface.

R. eudoxum
Balf.f. & Forrest 1919 H4

Height 0.3m-2m, a fairly dense low shrub. Branchlets tomentose, usually with a few bristles, bark greenish to dark brown. **Leaves** 3.5-7 x 1-3cm, *thin*, elliptic, retained 1-2 years; upper surface glabrous, lower surface with a *thin brownish or whitish* indumentum, often hardly noticeable. **Inflorescence** 2-6-flowered. **Corolla** not fleshy, tubular-campanulate to campanulate, 2.5-4cm long, rose-pink, magenta to bluish-crimson; calyx medium, 2-7mm, sparsely tomentose or glandular; ovary mostly glandular to mostly tomentose. **Distribution** S.E. Tibet and adjacent N.W. Yunnan, 3,400m-4,300m (11,000-14,000ft), on rocky slopes, cliffs and in thickets.

This species is closely related to *R. temenium*, which differs in its glaucous-papillate leaf under surface with no indumentum. Var. *eudoxum* also differs in the presence of at least some glands.

R. eudoxum is almost certainly a series of natural hybrids between various members of the sanguineum alliance crossed with *R. selense*. Usually relatively easy to grow but with rather uninteresting foliage and often harsh-coloured flowers, it is seldom of much merit. Introduced 1917-1950, reintroduced 1995->. April-May.

Divides into 3 varieties which are not of much horticultural consequence:

Var. *eudoxum* (*R. trichomiscum* Balf.f. & Forrest, *R. trichophlebium* Balf.f. & Forrest, *R. temenium* ssp. *albipetalum* Cowan, *R. temenium* ssp. *rhodanthum* Cowan).

Ovary predominantly *glandular*.

Var. *brunneifolium*
(Balf.f. & Forrest) D.F. Chamb. 1979.

Leaves *7-9cm,* indumentum thin brownish below. **Corolla** rose-carmine; ovary tomentose, *eglandular*.

Var. *mesopolium* (Balf.f. & Forrest) D.F. Chamb. 1979 (*R. asteium*, Balf.f. & Forrest, *R. epipastum*, Balf.f. & Forrest, *R. fulvastrum* var. *mesopolium* (Balf.f. & Forrest) Cowan).

Leaves 3.5-7cm, lower surface with a thin *whitish* indumentum. **Corolla** rose-pink; ovary tomentose, with +/- a few glands. Fairly close to *R. sanguineum* and probably a natural hybrid of it.

Picture 2

Picture 3

Picture 4

1. *R. eudoxum* var. *eudoxum* A.M. clone, Glendoick. The rather harsh-coloured, bluish-tinted flowers are a good character for identification.
2. *R. eudoxum* var. *eudoxum* at the Royal Botanic Garden, Edinburgh.
3. *R. eudoxum* var. *mesopolium*, Royal Botanic Garden, Edinburgh. This often has lighter-coloured flowers than var. *eudoxum*.
4. Leaf lower surface of *R. eudoxum* showing the typical very thin indumentum. (Photo R.B.G., E.)

Picture 1

R. microgynum
Balf.f. & Forrest 1919
(*R. gymnocarpum* (Balf.f. ex.) Tagg, *R. perulatum* Balf.f. & Forrest) H4

Height 0.6-1.5m, an umbrella-shaped dwarf shrub, compact when young. Branchlets whitish-tomentose, perulae persistent or deciduous, bark red brown. **Leaves** 5.5-12 x 1.5-4cm, elliptic to oblong-elliptic, *recurved*, retained 1-2 years; upper surface *glabrous, dark* green, lower surface with a *thin, felted, cinnamon to buff* indumentum of rosulate hairs. **Inflorescence** 3-10-flowered. **Corolla** campanulate, 3-4cm long, *deep* *crimson*, rarely to pale rose; calyx medium-large, 2mm-1cm, fleshy to papery; ovary brown-tomentose, *glandular*. **Distribution** Along the N.W. Yunnan-S.E. Tibet border, 3,700-4,300m (12,000-14,000ft), in open forest, bamboo and stony slopes.

This species is quite easily identified by its recurved, deep green leaves which are larger than those of its relatives. It is distinguished from *R. citriniflorum* by its felted indumentum.

R. microgynum is one of the best and most easily-grown of the alliance, flowering freely and relatively young, and it deserves to be more widely grown. Under previous classifications, plants under *R. microgynum* had pinker flowers while those under *R. gymnocarpum* had redder ones. It is probable that only the red forms are currently in cultivation. Introduced 1917-18. April-May.

Picture 1

Picture 2

1. *R. microgynum* Forrest 14242, Glendoick. This species has more recurved, slightly larger leaves and is freer-flowering than most of its relatives.
2. *R. microgynum*, Royal Botanic Garden, Edinburgh.

Picture 3

3. Leaf lower surface of *R. microgynum* showing the characteristic felted, fairly thin indumentum. (Photo R.B.G., E.).

R. parmulatum
Cowan 1936 H4

Height 0.6-2m, a broadly upright or rounded shrub. Branchlets soon glabrous, perulae deciduous. **Leaves** 4.5-8 x 2-3.8cm, obovate to elliptic, retained 1(-2) years; upper surface *rugulose*, lower surface with fine papillae, rather *glaucous and almost glabrous*; petiole broad. **Inflorescence** 4-6-flowered. **Corolla** tubular-campanulate, 4-5cm long, white, yellow flushed pink, pink to red, sometimes bicolour, often heavily spotted; calyx medium c.5mm glabrous; ovary with few scattered hairs. **Distribution** Pemako, S. Tibet, 3,000-3,700m (10,000-12,000ft), steep slopes, rocks and cliffs.

This species is easily identified by its rugulose leaves, almost glabrous and rather glaucous on the lower surface, and by its usually spotted flowers. It is most likely to be confused with *R. temenium* which has smoother leaves and unspotted flowers. The geographic isolation of this species from its closest relatives and therefore the lack of intermediate forms make it taxonomically one of the least problematic in this subsection.

The unusual, spotted flowers of various colours make *R. parmulatum* a distinctive plant and it is easier to please than many of its relatives, though often slow to flower. The huge colour-variation we have seen in the wild has not yet been reflected in cultivated plants. Introduced 1924, reintroduced 1996. March-May.

NERIIFLORA

Picture 1

Picture 2

Picture 3

Picture 4

Picture 5

5. The leaf lower surface of *R. parmulatum*, typically almost glabrous and glaucous with tiny papillae on the veins. (Photo R.B.G., E.).

R. trilectorum Cowan with pale yellow flowers appears to be a natural hybrid between *R. parmulatum* and *R. campylocarpum* which grow together on the south side of the Doshong La. Natural hybrids are common in this area.

1. *R. parmulatum* on the south side of the Doshong La, S.E. Tibet, showing different colour forms and degrees of spotting.
2. The heavily spotted corolla of *R. parmulatum* 'Ocelot' A.M., at Glendoick, showing the typical rugulose upper leaf surface.
3. A pink-flowered *R. parmulatum* on the south side of the Doshong La, S.E Tibet.
4. A exceptionally fine pink *R. parmulatum* on the south side of the Doshong La, S.E. Tibet.

R. sanguineum
Franch. 1898 H4-5

Height 0.3-1.5m. a rounded low shrub, sometimes leggy in shade. Branchlets sparsely white-floccose, perulae deciduous except in var. *didymoides* and ssp. *didymum*. **Leaves** 3-8 x 1.5-3.2cm, elliptic to obovate, retained for 1-3 years; upper surface glabrous, lower surface with a *continuous, compacted, thin, silvery to grey, rosulate* indumentum; petioles floccose when young, usually glabrous with age. **Inflorescence** 3-6-flowered, loose. **Corolla** tubular-campanulate, 2.3-3.5cm long, blackish-crimson, crimson, pink, yellow, white and sometimes a mixture of several colours; calyx medium to large, coloured, glandular, fringed with hairs; ovary tomentose to glandular.

Ssp. *sanguineum* D.F. Chamb. 1979

Perulae usually deciduous (except for var. *didymoides*). Ovary eglandular-tomentose (except var. *didymoides*). Flowering time *March-May*. **Distribution** N.W. Yunnan, S.E. Tibet, 3,000-4,500m (10,000-14,500ft), open slopes, forest margins and clearings and bamboo.

This subspecies is usually distinguished from its relatives by its thin, rather compacted indumentum. Ssp. *didymum* differs in its usually persistent perulae and its later, (dark red) flowers.

Variable in hardiness and freedom of flowering, and as with the other species in this subsection, ssp. *sanguineum* requires good drainage and cool roots. The number of varieties recognised within this subspecies reflects its variability, especially as regards flower colour. Recent fieldwork has confirmed that some of this variation is due to hybridisation with at least three other species, *R. citriniflorum*, *R. forrestii* and *R. temenium*. It seems likely that some of the so-called varieties described here are natural hybrids and it is often very hard to distinguish the parents from the hybrids derived from them. Introduced 1917-49. Reintroduced 1992->. March-May.

Var. *sanguineum*

Corolla *bright crimson*; ovary eglandular.

Var. *haemaleum* (Balf.f. & Forrest) D.F. Chamb. 1979 *R. sanguineum* ssp. *mesaeum* (Balf.f.) Cowan

Corolla *deep black-crimson*. There are intermediates between this and ssp. *didymum* which have been referred to ssp. *atrorubrum* (Cowan) Davidian.

Var. *himertum* (Balf.f. & Forrest) D.F. Chamb. 1979 (*R. nebrites* Balf.f. & Forrest, *R. poliopeplum* Balf.f. & Forrest, *R. sanguineum* ssp. *R. aisoides* Cowan)

Corolla various shades of *yellow*.

Broadly speaking, this differs from *R. citriniflorum* in its non-glandular rachis, pedicel, calyx and ovary. The picture is complicated as not all *R. citriniflorum* are glandular and some intermediates occur with a thin indumentum and a glandular ovary while others have a thick indumentum and a tomentose ovary.

Var. *cloiophorum* (Balf.f. & Forrest) Chamb. 1979 Balf.f. & Forrest, *R. asmenistum* Balf.f. & Forrest)

Corolla *white, yellow flushed pink or pink*. This is almost certainly a hybrid or intermediate form with one of its relatives.

Var. *didymoides* Balf.f. & Forrest 1931 (*R. roseotinctum* Balf.f. & Forrest, *R. mannophorum* Balf.f. & Forrest, *R. sanguineum* ssp. *consanguineum* Cowan)

Perulae usually *persistent*. **Corolla** *white and/or yellow flushed pink to pink, rose or crimson*; ovary *at least partly glandular*.

Some of these have a pretty combination of colours. Some forms are intermediate with *R. citriniflorum*.

Ssp. *didymum* (Balf.f. & Forrest) Cowan 1940

Perulae usually persistent. **Leaves** shiny, 2-6cm long. **Corolla** deep blackish-crimson; ovary at least partly glandular. Flowering time *June-July*. **Distribution** S.E. Tibet, N.W. Yunnan, 4,300-4,600m (14,000-15,000ft), stony meadows and limestone cliffs.

Ssp. *didymum* differs from its relatives in its horticultural requirements, demanding an almost alkaline soil. We had great difficulties in keeping it alive until we applied lime. In cultivation it appears fairly uniform but in the wild it merges with var. *haemaleum*. Introduced 1921-1949. June-July.

Picture 1

Picture 2

1. *R. sanguineum* ssp. *sanguineum* var. *sanguineum* Cox 6521 near the east side of the Doker La, Mekong-Salween Divide, N.W. Yunnan. This grows on part of an avalanche slope, hence the bare foreground and the battered appearance of the plants.
2. Close-up of No. 1. This is a particularly good form of var. *sanguineum* Cox 6521 with large rich red flowers; near the east side of the Doker La, Mekong-Salween divide, N.W. Yunnan, China.

Picture 3

Picture 4

Picture 5

Picture 6

Picture 7

Picture 8

Picture 9

3. *R. sanguineum* ssp. *sanguineum* var. *sanguineum* always has bright crimson flowers, Glendoick.
4. An attractive bicolour-flowered *R. sanguineum* ssp. *sanguineum* var. *didymoides*, Royal Botanic Garden, Edinburgh.
5. *R. sanguineum* ssp. *sanguineum* var. *haemaleum*, Royal Botanic Garden, Edinburgh. This variety, along with its relative ssp. *didymum* have amongst the darkest red flowers of any rhododendron, looking almost black.
6. *R. sanguineum* ssp. *sanguineum* var. *haemaleum* Forrest 20253, Royal Botanic Garden, Edinburgh.
7. *R. sanguineum* ssp. *didymum*, Glendoick.
8. *R. sanguineum* ssp. *didymum* at Arduaine Gardens, W. Scotland.
9. The leaf lower surface of *R. sanguineum* showing the thin greyish indumentum. (Photo R.B.G., E.)

R. temenium
Balf.f. & Forrest 1919 H4-5

Height 0.3-1.5m but rarely over 1m, a usually compact, flat-topped or dome-shaped dwarf shrub. Branchlets glandular to bristly. **Leaves** 3.5-8 x 1.3-3cm, elliptic, leaves retained for 1-3 years; upper surface glabrous or with the remains of a whitish loose indumentum, lower surface *glabrous* (or sometimes with a little whitish indumentum), +/- *glaucous-papillate*; petiole *short*. **Inflorescence** 2-6-flowered, dense. **Corolla** tubular-campanulate, 3.5-4.5cm long, white, pink, yellow, carmine to crimson; calyx small to medium, 2-5mm, ciliate; ovary tomentose. **Distribution** Borders of N.W. Yunnan and S.E. Tibet, 3,700-4,600m (12,000-15,000ft), moorland, scrub, beside water and on cliffs.

This species is quite easily identified by its leaves: relatively pale green and smooth on the upper surface and glabrous to glaucous below. It is closely allied to *R. sanguineum* and *R. eudoxum* and hybridises with these and other members of the subsection, though recent field studies indicate that *R. temenium* is a more distinctly defined species than either *R. sanguineum* or *R. citriniflorum*. *R. fulvastrum* Balf.f. & Forrest is probably a hybrid of *R. temenium x R. sanguineum*.

R. temenium is a neat plant but is often slow to flower and it requires particularly well-drained, cool, but not over-dry soil conditions. Introduced 1917-24. reintroduced 1992->. April-May.

Var. *temenium* (R. pothinum Balf.f. & Forrest.)

Young shoots with bristles. **Inflorescence** *dense*. **Corolla** *pink to crimson*.

Var. *gilvum* (Cowan) D.F. Chamb. 1979
(*R. chrysanthemum* Cowan).

Branchlets *strongly bristly*. **Leaves** often with a little whitish indumentum on lower surface. Truss dense. **Corolla** *cream to yellow*, sometimes tinged pink or red.

Var. *dealbatum* (Cowan) D.F. Chamb. 1979
(*R. glaphyrum* Balf.f. & Forrest).

Branchlets tomentose, usually with only weak bristles. **Leaves** small, lower surface *glabrous at maturity*. **Inflorescence** *lax*. **Corolla** white to deep-rose-pink; pedicel with only weak bristles.

Some plants in cultivation labelled var. *dealbatum* are undoubtedly hybrids.

Picture 1

Picture 2

Picture 3

Picture 4

1. A fine scarlet *R. temenium* var. *temenium* Rock 101, Royal Botanic Garden, Edinburgh.
2. *R. temenium* var. *temenium* growing on the Chunsien La, Salween-Mekong divide, N.W. Yunnan, China.
3. A pretty pink *R. temenium* var. *temenium* Rock 114, Royal Botanic Garden, Edinburgh.
4. *R. temenium* var. *dealbatum* affinity Rock 22070, Royal Botanic Garden, Edinburgh, which is almost certainly a natural hybrid, perhaps crossed with *R. eclecteum*.

R. chamaethomsonii
(Tagg & Forrest) Cowan & Davidian 1951 H4-5

Height 0.1-1m, a dwarf shrub, very compact and dense to rather rangy. Branchlets glandular or sparsely tomentose, perulae +/- deciduous or persistent. **Leaves** *2-9 x 1.8-4cm*, broadly obovate to elliptic; upper surface glabrous, lower surface glabrous or with a sparse, whitish, plastered indumentum. **Inflorescence** 1-5-flowered. **Corolla** campanulate, 2.5-5cm long, carmine or crimson to pink; calyx minute to well-developed, *1mm-1.5cm*; ovary *sparsely hairy*, sometimes glandular. **Distribution** N.W. Yunnan, adjacent S.E. Tibet and Arunachal Pradesh (var. *chamaethauma*), 3,400-4,600m (11,000-15,000ft), steep hillsides, rocks, often with *R. forrestii*.

R. chamaethomsonii is generally taller than *R. forrestii* with larger leaves, a larger calyx, and more flowers to the truss. Some forms of apparent *R. chamaethomsonii* which have indumentum on the leaf underside are probably hybrids.

Recent observations in the wild backed up by examination of cultivated material and herbarium specimens have confirmed that this so-called species needs to be completely rethought. Plants which match *R. forrestii* except for the 2-5 flowered truss should be considered *R. forrestii*. Single-flowered and several-flowered forms grow together in the wild as a cline and cannot sensibly be separated. Most of the cultivated clones and many of the herbarium specimens under the name *R. chamaethomsonii* are in fact natural hybrids of *R. forrestii* crossed with *R. sanguineum*, *R. eclecteum*, *R. selense*, *R. aganniphum*, *R. parmulatum* and several other species. We have seen extensive examples of such hybridisation on the Salween-Mekong Divide and in S.E. Tibet. Nearly all these hybrids have traces of indumentum on the leaf underside. Plants under the name *R. chamaethomsonii* tend to be more vigorous, and often easier to grow and more free-flowering than plants under the name *R. forrestii*. Introduced 1922->. March-May.

Var. *chamaethomsonii*

Leaves very variable in size, glabrous; petioles and young shoots glandular. **Corolla** carmine to crimson; calyx to 7mm; ovary sparsely hairy, sometimes glandular.

Var. *chamaedoron* (Tagg & Forrest) D.F. Chamb. 1979

Branchlets eglandular. Mature leaves with a *thin discontinuous* indumentum on the lower surface; petioles

---

5. *R. temenium* var. *gilvum* Rock 22271, Glendoick. This variety is characterised by its cream to yellow flowers.
6. *R. temenium* var. *gilvum* Rock 22272 'Cruachan' F.C.C. from Tigh-an-Rudha, W. Scotland, at Glendoick. This plant was formerly known as *R. chrysanthemum*.
7. The leaf lower surface of *R. temenium*, glaucous with papillae on the veins. The entire lower surface may also be glabrous. (Photo R.B.G., E.)

Forrestii alliance

Creeping to mound-forming. **Leaves** glabrous or with a little thin indumentum/glands. **Corolla** crimson to (rarely) pink. Essentially there is only one 'good' species here. The number of flowers in the inflorescence distinction said to divide the 2 species often does not hold true and many plants grown under the name *R. chamaethomsonii* are natural hybrids which have undergone some degree of speciation in places.

eglandular. These are probably all natural hybrids.

Var. *chamaethauma* (Tagg) Cowan & Davidian 1951

Leaves *glabrous;* branchlets and petiole *glandular.* **Corolla** *pale to deep pink*; flowers more freely produced than in var. *chamaethomsonii*. Flower buds liable to abort. This appears to be a relatively well-defined, stabilised entity which is probably worthy of taxonomic status.

Picture 1

Picture 2

Picture 3

Picture 4

Picture 5

4. A deep pink clone of *R. chamaethomsonii* var. *chamaethauma,* Glendoick, Plants under the name var. *chamaethauma* should have flowers in shades of pink.
5. A pale pink-flowered *R. chamaethomsonii* var. *chamaethauma* from Exbury, S. England, growing at Glendoick.

1. *R. chamaethomsonii* var. *chamaethomsonii*. A large-leaved and large-flowered clone from Exbury, S. England growing at Glendoick. This clone has an open habit.
2. A broadly upright, fairly compact form of *R. chamaethomsonii* var. *chamaethomsonii* Ludlow, Sherriff & Elliot 13278? from Branklyn, Perth, Scotland, growing at Glendoick.
3. *R. chamaethomsonii* var. *chamaethomsonii* growing on the top of the Doshong La, S.E. Tibet.

R. forrestii
(Balf.f. ex) Diels 1912 (*R. repens* Balf.f. & Forrest) H4-5

Height to 15cm, lateral stems up to 60cm, a creeping to mounding dwarf shrub. Perulae *persistent*. **Leaves** 1-5 x 0.6-3.1cm, *obovate to orbicular*; upper surface glabrous, lower surface glabrous or with a few glands or glaucous-papillate with stipitate glands. **Inflorescence** 1(-2)-flowered. **Corolla** fleshy, tubular-campanulate, 3-3.5cm long, scarlet to *crimson;* calyx *minute c.1mm*; ovary *densely glandular and rufous-tomentose*. **Distribution** N.E. Yunnan and adjacent S.E. Tibet, N.E. Upper Burma and E. Arunachal Pradesh, 3,000-4,600m (10,000-15,000ft), steep often rocky slopes, pastures and even waterside.

R. forrestii differs from *R. chamaethomsonii* in its prostrate habit, smaller leaves, the usually single-flowered inflorescence, the minute calyx and the glandular and rufous-tomentose ovary. The combination of red flowers, small rounded leaves and creeping or mounding habit distinguish this from all other species. The occasional crimson-purple leaf underside, once thought to be botanically significant, is just a form of persistent juvenility.

R. forrestii at its best is a magnificent species but it requires very careful siting for success; a steep bank or wall facing away from the sun but in the open is ideal; too much shade and it will not flower. There is quite a variation in freedom of flowering and quality of flower. Introduced 1914-38. reintroduced 1992->. March-May

Ssp. *forrestii*

Leaf lower surface green or crimson-purple, *usually glabrous or with a few stipitate glands. Leaf ratio 1: 1.1-1.5(-2.2) x as long as broad.*

Two former varieties are now accorded group status. Repens Group includes the lowest growing creeping forms while Tumescens Group has a more dome-shaped habit with creeping outer branches and larger leaves. There are several clones under the name Tumescens which grow well and are relatively free-flowering.

Ssp. *papillatum* D.F. Chamb. 1979

Differs from ssp. *forrestii* in the leaf underside being *glaucous-papillate with conspicuous glands*. Leaf ratio *1: 2.2-3.2*. Several herbarium specimens including the type of Tumescens Group lie between ssp. *forrestii* and ssp. *papillatum*. The latter also apparently intergrades with *R. chamaethomsonii* var. *chamaethauma*. **Distribution** S. Tibet, 3,400-4,000m (11,000-13,000ft).

Doubtfully deserving of subspecific status; it may in fact be a natural hybrid. This subspecies is rare in cultivation; we know of two clones, one much more glaucous on the leaf underside than the other.

Picture 1
1. *R. forrestii* ssp. *forrestii* growing with *Diapensia purpurea* on the Doshong La, S.E. Tibet.

Picture 2

Picture 3

Picture 4

2. *R. forrestii* ssp. *forrestii*, Mekong-Salween divide, N.W. Yunnan, China. Plants with the reddish-purple leaf-underside were formerly distinguished as a separate species but these can have no botanical standing as crimson and green leaved plants grow side by side in the wild.
3. *R. forrestii* ssp. *forrestii*, Baravalla, W. Scotland, showing the typical prostrate, carpeting habit.
4. *R. forrestii* ssp. *forrestii* Rock 59174 Repens Group from Windsor Great Park, S. England growing at Glendoick. This is the best form of this species that we have seen with freely-produced, large, scarlet flowers.

NERIIFLORA

Picture 5

Picture 6

Picture 7

5. *R. forrestii* ssp. *forrestii* Tumescens Group, at Glenarn, W. Scotland. The Tumescens Group has a dome-shaped habit.
6. *R. forrestii* ssp. *forrestii* Tumescens Group from Tower Court, S. England, growing at Glendoick.
7. *R. forrestii* ssp. *papillatum* Kingdon-Ward 5845, Royal Botanic Garden, Edinburgh.

Neriiflorum alliance

Height 0.3-6m, generally more upright and of looser habit than the other alliances. **Leaves** +/- lanceolate, lower surface *glabrous or with a loose woolly indumentum*. Ovary usually *slender, tapered into the base of the style;* capsule *long and curved*.

The species in the Neriiflorum alliance are quite closely related but only the combination of general characters above can separate them from some of the species in other alliances.

R. albertsenianum
Forrest 1919 H4

Height 1-2m, a small shrub with stiff, straight branches. Young shoots floccose-tomentose. **Leaves** 6-9.5 x 1.5-2.5cm, elliptic, retained for 1-2 years; upper surface glabrous, lower surface with a continuous *bistrate*, light brown indumentum. **Inflorescence** 5-6-flowered. **Corolla** tubular-campanulate, c.3cm, bright crimson-rose; calyx medium, 3-4cm, fleshy, sparsely tomentose; ovary densely tomentose, tapering into style. **Distribution** N.W. Yunnan, c.3,000m (10,000ft), open forest.

R. albertsenianum is only known from the type specimen and from plants in cultivation which were part of a mixed seed gathering including *R. sperabile* aff. under Forrest 14195. Doubtfully deserving of specific status, it is probably part of a cline containing *R. euchroum* and *R. sperabile* or may be a natural hybrid of *R. sperabile* or *R. floccigerum*. Cultivated plants of this species have a somewhat discontinuous indumentum and do not seem to match the type description.

The cultivated plants of this species (which may in fact be a single clone) are rather lacking in vigour and prone to mildew. Introduced 1917. April-May.

R. euchroum Balf.f. & Kingdon-Ward. (probably not in cultivation) is very similar, differing in its glandular young shoots, petioles and ovaries, and in its smaller leaves.

Picture 1

1. *R. albertsenianum* from Forrest 14195A, the only time this species has been collected. Royal Botanic Garden, Edinburgh.

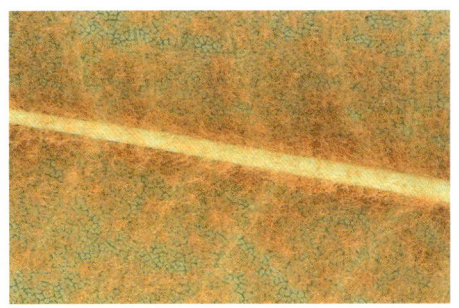

Picture 2

2. The leaf lower surface of *R. albertsenianum*, as cultivated. The indumentum should be thicker and more continuous according to the type description. (Photo R.B.G., E.)

R. floccigerum
Franch. 1898 H3-4

Height 0.3-3m. A shrub, rounded in open, leggy in shade. Branchlets densely tomentose, sometimes bristly and glandular. **Leaves** 5-12 x 1-2.7cm, *narrowly elliptic to oblong* or *elliptic;* upper surface glabrous, lower surface *glaucous-papillate* with a rufous, usually *patchy, speckled* indumentum. **Inflorescence** 4-7-flowered, loose. **Corolla** tubular-campanulate, 3-4cm long, usually rose to deep crimson or scarlet, sometimes a mixture of orange, yellow or pink; calyx small, 1-4mm, slightly hairy; ovary densely tomentose, eglandular, tapering into style. **Distribution** N.W. Yunnan, S.E. Tibet and N.E. Upper Burma, 2,700-4,000m (9,000-13,000ft), scrub, cliffs and open conifer forest.

This species merges with *R. sperabiloides*, which has a similar distinctive discontinuous indumentum, but is a dwarfer plant with shorter leaves lacking the papillate lower leaf surface. *R. neriiflorum* differs in its usually glabrous leaves and *R. sperabile* differs in its continuous indumentum on the lower leaf surface, though it probably merges with *R. floccigerum* in some locations. The patchy indumentum is also a character of *R. pocophorum* var. *hemidartum* which differs in the absence of a glaucous-papillate lower leaf surface and in its glandular ovary.

This species has been introduced many times and is quite widely cultivated in a great variety of colour forms. The red forms are perhaps the most showy while some others are decidedly muddy-coloured. Introduced 1914->. March-May.

Picture 1

Picture 2

Picture 3

Picture 4

1. *R. floccigerum*, Younger Botanic Garden, Benmore, W. Scotland, with the flowers characteristically hanging down between the leaves.
2. *R. floccigerum* Rock 10959, Royal Botanic Garden, Edinburgh.
3. An attractive bicolour *R. floccigerum*, Rock 10, Royal Botanic Garden, Edinburgh. This species is very variable in flower colour; some forms can be termed 'muddy'.
4. *R. floccigerum*, Glendoick.

NERIIFLORA

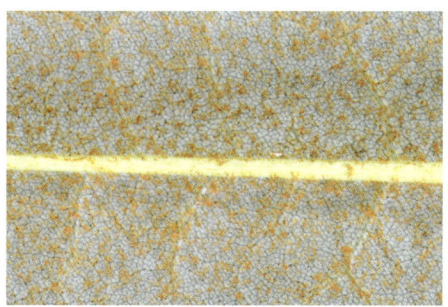

Picture 5

5. The leaf lower surface of *R. floccigerum* showing the characteristic, patchy, speckled indumentum. (Photo R.B.G., E.)

R. neriiflorum
Franch. 1886. H3-4

Height 1-6m. a compact to loose shrub or small tree. Branchlets sparsely tomentose, eglandular or rarely bristly-glandular. **Leaves** 4-9 x 1.9-3.5cm, elliptic to oblong to oblanceolate; upper surface *glabrous,* lower surface *usually glabrous, glaucous-papillate.* **Inflorescence** 5-12-flowered, compact to loose. **Corolla** tubular-campanulate, 3.5-4.5cm long, bright scarlet, crimson, rose, straw yellow and orange; pedicel *eglandular;* calyx *eglandular,* fleshy, *variable 2-15mm;* ovary usually densely tomentose, *eglandular.*

An attractive species of variable hardiness, quite common in gardens. Divides into 3 subspecies:

Ssp. *neriiflorum* (*R. euchaites* Balf.f. & Forrest, *R. phoenicodum* Balf.f. & Forrest)

Leaves 4-9cm long, ratio *1: 1.7-3*; lower surface glabrous, *without* papillae; petiole eglandular. **Corolla** in shades of red; pedicel, calyx and ovary eglandular. **Distribution** N.W. Yunnan, N.E. Upper Burma, 2,100-3,400m (7,000-11,000ft), scrub and forest.

This subspecies differs from the others members of the alliance in its usually glabrous, glaucous leaf underside. It differs from ssp. *phaedropum* in its shorter, wider leaves and in the absence of glands on the pedicel, calyx and ovary.

This subspecies is very variable with a wide distribution in the wild. Plants grown under Euchaites Group are relatively tall and upright. Introduced 1910, reintroduced 1981->. April-May.

Ssp. *agetum* Balf.f. & Forrest

(Probably not in cultivation) Only differs from the above in having pits and papillae on the lower leaf surface. Burma-Tibet border.

Ssp. *phaedropum* (Balf.f. & Forrest) Tagg 1930
(*R. floccigerum* var. *R. appropinquans* Tagg & Forrest, *R. tawangense* Sahni & Naithani) H3(-4)

Leaves 8-11cm long, ratio *1:3-5*; petiole sometimes glandular. **Corolla** red, *rose to straw yellow or tawny orange*; pedicel, calyx and ovary with *some glands.* **Distribution** C. Bhutan eastwards through Arunachal Pradesh, S. Tibet, Upper Burma into N.W. Yunnan, 2,000-3,400m (6,500-11,000ft).

Ssp. *phaeodropum* is most likely to be confused with *R. floccigerum* which has a similar leaf shape and flower colour range but differs in its patchy indumentum. This subspecies differs from ssp. *neriiflorum* in the presence of glands on the calyx and ovary and in its generally longer, narrower leaves and it replaces ssp. *neriiflorum* in the western part of the range of the species.

Often more tender than ssp. *neriiflorum* and therefore not as widely cultivated. Introduced 1922. Reintroduced 1965 and 1980s.

Picture 1

1. *R. neriiflorum*, Cangshan, C. Yunnan, China. This species varies considerably in the size of its truss and flower, this being a decidedly inferior example.

NERIIFLORA

Picture 2

Picture 3

Picture 4

Picture 5

5. *R. neriiflorum* ssp. *phaedropum* with unusual-coloured flowers, at Arduaine Gardens, W. Scotland. This subspecies can be almost as variable in flower colour as *R. floccigerum*.

2. *R. neriiflorum* Ludlow & Sherriff 1352, Royal Botanic Garden, Edinburgh.
3. *R. neriiflorum* Euchaites Group; this group includes the largest growing and finest forms. Blackhills, N.E. Scotland.
4. *R. neriiflorum* ssp. *phaedropum* Ludlow, Sherriff & Taylor 6563, Royal Botanic Garden, Edinburgh. This subspecies is often more tender than ssp. *neriiflorum*, coming from lower elevations.

R. sperabile
Balf.f. & Forrest 1922 H3-4

Height 1-2m. a dense or leggy shrub. Branchlets densely whitish tomentose, *usually glandular*. **Leaves** 5-10 x 1-2.6cm, elliptic; upper surface glabrous, lower surface *with a dense but loose* cinnamon or white indumentum. **Inflorescence** 3-5-flowered. **Corolla** tubular-campanulate, 3.5-4cm long, scarlet to crimson; calyx small, coloured, 2-3mm, glandular-ciliate; ovary densely *rufous-tomentose and glandular*, tapering into style.

Separated from *R. floccigerum* and *R. sperabiloides* by its continuous indumentum. Divides into two varieties, although in some areas such as Weixi, they merge and cannot really be distinguished.

Var. *sperabile* H3

Leaf ratio 1: 2.5-3.5, indumentum *cinnamon* when mature.
Distribution N.E. Upper Burma and near by N.W. Yunnan, 3,000-3,700m (10,000-12,000ft), scrub, screes and cliffs.

A pretty plant in a modest way but we have found it one of the most tender members of the subsection. Rare in cultivation and most forms are less showy than cultivated forms of var. *weihsiense*. Introduced 1919->. April-May

Var. *weihsiense* Tagg & Forrest 1927 H3-4.

Leaf ratio 1:3-4, indumentum whitish when mature.
Distribution N.W. Yunnan, mostly to the north of var. *sperabile*, cliffs and rocky slopes.

NERIIFLORA

Generally larger growing than var. *sperabile* with a more open habit. A little hardier and more showy in flower in its best forms, var. *weihsiense* is more widely cultivated than var. *sperabile*. Introduced 1924, reintroduced 1980s->. April-May

Picture 1

Picture 2

Picture 3

Picture 4

Picture 5

4. *R. sperabile* var. *weihsiense* Kingdon-Ward 20620 at Glendoick.
5. The leaf lower surface of *R. sperabile* var. *weihsiense* showing the fairly dense, loose, pale indumentum. (Photo R.B.G., E.)

1. *R. sperabile* at the Valley Gardens, Windsor Great Park, S. England. Though labelled var. *sperabile*, this is likely to be referable to var. *weihsiense* which has a greater length to breadth leaf ratio.
2. *R. sperabile* at the Valley Gardens, Windsor Great Park, S. England. Although labelled var. *sperabile* it appears closer to var. *weihsiense*.
3. *R. sperabile* var. *weihsiense* Kingdon-Ward 19405, Royal Botanic Garden, Edinburgh.

R. sperabiloides
Tagg & Forrest 1927 H3-4

Height 1-1.5m, a compact and spreading small shrub. Branchlets tomentose, *eglandular*. **Leaves** 3-7 x 1.8-2.5cm, elliptic, upper surface rugulose and glabrous; lower surface with a *discontinuous* indumentum. **Inflorescence** 4-5-flowered. **Corolla** tubular-campanulate, 2.5-3.5cm long, crimson to deep red; calyx *medium 4-7mm* +/- ciliate; ovary densely fulvous-tomentose, eglandular, tending to be slightly tapered into the style. **Distribution** S.E. Tibet and adjacent N.W. Yunnan, 3,700-4,000m (12,000-13,000ft), scrub and rocky slopes.

This resembles *R. albertsenianum* which has thicker indumentum and it differs from *R. sperabile* in its discontinuous indumentum, larger calyx and lack of glands. From *R. floccigerum* it usually differs in its smaller, broader leaves but the two species appear to merge/hybridise.

This is essentially the northern form of *R. sperabile*. Rare in cultivation. Introduced 1922. April-May.

Picture 1

Picture 2

1. *R. sperabiloides*, a plant from Rock's 1949 expedition, probably Rock 125, at Baravalla, W. Scotland.
2. The leaf lower surface of *R. sperabiloides* showing the discontinuous indumentum. (Photo R.B.G., E.)

Subsection Parishia

Height 2-10m, shrubs or small trees. Young shoots with rufous tomentum consisting of stellate hairs. **Leaves** elliptic to broadly obovate, usually rounded at the tip; upper surface *with stellate indumentum at first,* partly glabrous later, lower surface *with a thin indumentum of stellate hairs, mixed with glands, especially on the midrib.* **Inflorescence** 5-15-flowered, compact to lax. **Corolla** with 5 lobes, fleshy, tubular to funnel-campanulate, with pronounced nectar pouches, *red to deep scarlet*; calyx usually small; stamens usually 10; ovary densely tomentose, usually also with glands; style glabrous to glandular. **Distribution** India, Burma, Yunnan.

The species of subsection Parishia are characterised by their late, usually red flowers and by the indumentum of stellate hairs on the leaves. This subsection is allied to Subsection Irrorata and more distantly to subsection Neriiflora.

Subsection Parishia contains beautiful, scarlet, late-flowering species which are unfortunately too tender for most of Continental Europe and North America. Even in milder areas, the typically late growth is often frosted. *R. facetum, R. elliottii* and *R. kyawii* are grown in mild climates such as coastal S.W. England, Scotland, Ireland and New Zealand and have been used in hybridising. The other 3 species in this subsection are probably not in cultivation.

R. elliottii
Watt 1906. H2-3

Height 3-9m, an upright, often rather straggly shrub or small tree. Young shoots tomentose and glandular. Buds rounded and prominent. **Leaves** 8.5-11(-15) x 2.5-6cm, lanceolate, oblong-elliptic to elliptic; upper surface glabrous at maturity, lower surface, glabrous or with vestiges of stellate hairs (indumentum), especially on midrib; petiole c.2.5cm, floccose and glandular at first, glabrous on maturity. **Inflorescence** 10-20-flowered, compact. **Corolla** funnel-campanulate, with nectar pouches, to 5.5cm long and wide, scarlet to crimson, (purplish-red to deep rose-purple) with deeper spots; calyx 3-4mm, glandular sometimes with a few hairs; pedicel *glabrous* or glandular; stamens *glabrous*; ovary and style tomentose and glandular. **Distribution** N.E. India, 2,400-2,700m (8,000-9,000ft), in forests.

Picture 1

1. *R. elliottii* with Graham Smith, Pukeiti, North Island, New Zealand.

R. elliottii is closely related to *R. facetum* and could be considered the most westerly population of this species. *R. elliottii* usually has smaller, more glabrous leaves which narrow abruptly at the base rather than tapering, a more funnel-shaped corolla, usually no hairs on the calyx or pedicel and the stamens are glabrous. *R. kyawii* is larger in all parts and later flowering.

R. elliottii is one of the finest red-flowered species but its tenderness and its frost-vulnerable late growth restrict the extent of its cultivation to mild areas such as the west of Britain and New Zealand. Its hybrids such as R. 'Grenadier' are popular in milder areas. Introduced 1927, 1949. Reintroduced 1994. May-July.

Picture 2

2. *R. elliottii*, Pukeiti, North Island, New Zealand, showing the totally glabrous upper leaf surface.

is some disagreement amongst botanists concerning the distinguishing features. *R. facetum* has a tomentose pedicel and calyx while those of *R. elliottii* are usually glabrous and in addition, *R. facetum* usually has larger leaves, tapering at the base and a tubular-campanulate rather than funnel-campanulate corolla. *R. kyawii* is usually larger in all parts, taller growing and later flowering but it may intergrade with *R. facetum*.

This is a superb red-flowered species for milder climates and it has been much used in hybridising. It is useful for its late flowering but its late growth can be frosted or distorted in heat, drought or by insects. Introduced 1917-38. Reintroduced 1981. June-August.

Picture 1

R. facetum
Balf.f. & Kingdon-Ward 1917 (*R. eriogynum* Balf.f. & W.W. Sm.) H2-3

Height 2.4-4.5m, a compact to leggy shrub or small tree. Branchlets covered in a tomentum, white at first, brown at maturity; bark roughish, flaking. **Leaves** 10-*20* x 5-*7.5*, oblong-elliptic to oblong-lanceolate, retained 2 years; upper and lower surfaces at first with white to light brown indumentum, glabrous on maturity except on the midrib which usually retains some hairs; petiole 2-3cm, tomentose at first, glabrous at maturity. **Inflorescence** 12-20-flowered, compact to fairly compact. **Corolla** tubular-campanulate, with nectar pouches, to 5cm long, clear bright red (also reported to be deep rose in the wild); pedicel *rufous stellate tomentose*; calyx 4-5cm, fleshy, reddish, *stellate tomentose*; ovary tomentose; style floccose with glands. **Distribution** N.E. Upper Burma, W. Yunnan, 2,400-3,400m (8,000-11,000ft), in the open, in forests and scrub.

This species is very closely related to *R. elliottii* which is probably best considered a western variant of *R. facetum*. There

Picture 2

Picture 3

1. *R. facetum*, Baravalla, W. Scotland.
2. *R. facetum*, Arduaine Garden, W. Scotland.
3. *R. facetum* SBEC 1014, Glendoick, showing the evanescent indumentum on the upper leaf surface.

R. kyawii

Lace & W.W. Sm. 1914 (*R. agapetum* Balf.f. & Kingdon-Ward, *R. prophanthum* Balf.f. & Forrest) H2-(3)

Height 4.5-7.5m, a rounded, spreading or broadly upright shrub or tree. Branchlets tomentose or glandular at first. **Leaves** 10-*30* x 3.5-*10*cm, elliptic to oblong; upper surface with stellate hairs at first, later glabrous, lower surface with a cinnamon, stellate indumentum mixed with a few glands, sometimes almost glabrous; petiole *2.5-4cm* (or very short in former *R. agapetum*), tomentose and glandular at first. **Inflorescence** 10-20-flowered. **Corolla** tubular-campanulate, with nectar pouches, 4.5-6cm long, bright crimson to scarlet, without flecks; pedicel with or without hairs; calyx c. 5mm, glandular and hairy; style crimson, usually hairy and glandular. **Distribution** N.E. Upper Burma, W. Yunnan, 1,500-3,700m (5,000-12,000ft) in deep wooded gorges, limestone cliffs, mixed thickets and forests.

This species is similar to *R. facetum* and *R. elliottii* but is less hardy, taller, with broader and longer leaves, usually with a longer petiole, and with larger, later flowers.

A magnificent late-flowering red, *R. kyawii* is unfortunately only suitable for very mild gardens and remains rare. Introduced 1919->. June-July-August.

Picture 1

1. *R. kyawii*, Matthew's garden, North Island, New Zealand, showing the large leaves compared with *R. elliottii* and *R. facetum*. (Photo G. Smith).

Picture 2

2. Young foliage of *R. kyawii*, Fernhill, E. Ireland, showing the very long petioles.

Subsection Pontica

Height to 12m, prostrate, compact and bushy shrubs to open and upright small trees. Branchlets glabrous to densely tomentose or/and glandular; bark roughish. **Leaves** 2.5-30 x 0.8-9cm, oval to linear, leaves retained 1-5 years; upper surface usually glabrous at maturity, lower surface glabrous or with a dense unistrate or bistrate indumentum of radiate, dendroid or ramiform hairs. **Inflorescence** 5-30-flowered, fairly lax to dense, rachis *usually long, resulting in a tall truss*. **Corolla** with 5 (occasionally 6-9) lobes, funnel-shaped to campanulate, white to pale yellow to pink to rosy-purple, often spotted or blotched; calyx 1-9mm long; stamens 10; ovary glabrous or glandular or/and tomentose; style glabrous. **Distribution** This is the subsection with the widest distribution in the genus, stretching from W. & E. North America to Europe and W. & E. Asia and Japan.

The long rachis, the usually tall truss and the erect pedicels and capsules are the most distinctive characters of subsection Pontica. This subsection is most closely related to subsection Argyrophylla through *R. adenopodum*.

Subsection Pontica contains several of the hardiest, most heat-tolerant and most easily-grown members of the genus, capable of succeeding in extreme continental conditions. Several of the species have been much used in hybridising. The commonest species in cultivation are *R. ponticum* and *R. degronianum* ssp. *yakushimanum* in Europe and *R.*

catawbiense, *R. maximum* and *R. degronianum* ssp. *yakushimanum* in E. North America. Other relatively common species are *R. brachycarpum*, *R. degronianum* ssp. *heptamerum* and *R. smirnowii*.

R. aureum

Georgi 1775 (*R. chrysanthum* Pallas 1776, *R. officinale* Salisbury) H5

Height 0.2-1m, forming a creeping to mounding dwarf shrub. Branchlets +/- glabrous, perulae *persistent for up to 4 years*. **Leaves** 2.3-13.5 x 0.8-7cm, very variable, even on one plant, ovate to broadly elliptic; upper and lower surfaces glabrous when mature. **Inflorescence** 5-8-flowered. **Corolla** widely campanulate, thin textured, 2-3cm long, *cream to pale yellow*, usually lightly spotted; pedicel long and erect; calyx small, 2-3mm long, hairy; ovary rufous-tomentose.

This is a distinctive species; its closest relation is *R. caucasicum* which is larger in all parts, more upright and never prostrate, has a persistent indumentum and is later flowering.

R. aureum is a very hardy species, best suited to a fairly continental climate. It grows and flowers early, is very slow growing when young and some forms are shy-flowering. Japanese introductions may be easier to grow than those from mainland Asia. Introduced 1796->. March-April.

Var. *aureum*

Height to 60cm. **Leaves** 2.3-9.5 x 0.8-4cm. **Distribution** Central and E. Siberia and N. China to N. Japan, 1,500-2,700m (5,000-9,000ft), covering huge areas of mountain slopes.

Var. *hypopytis* (Pojark) D.F. Chamb.

Height to 1m. **Leaves** up to 13.5 x 7cm. **Distribution** E. Russia

This variety is larger in all parts and more vigorous. It may not be in cultivation outside Russia.

Picture 1

Picture 2

Picture 3

1. *R. aureum*, 2,000m, Mt. Tokachi, Hokkaido, Japan. (Photo Y. Doi).
2. *R. aureum* and *Phyllodoce caerulea*, Mt. Asahidake, Japan. (Photo T. Takeuchi).
3. *R. aureum*, Glendoick.

R. x *nikomontanum* (Komatsu) Nakai. This is a natural hybrid between *R. aureum* and *R. brachycarpum* which can vary in form all the way from one species to the other.

Picture 4

4. *R x nikomontanum*. This is a natural hybrid between *R. aureum* and *R. brachycarpum*. The full truss and green spots come from *R. brachycarpum*, Glendoick.

R. brachycarpum
D. Don ex G. Don 1834 H5

Height 1.2-4m, forming a rounded and compact shrub in the open, laxer with age in shade. Branchlets at first tomentose, later glabrescent; bark roughish, greyish-brown. Growth buds *pointed*. **Leaves** 7-15 x 3-7cm, oblong to obovate, *usually convex*, retained for 2-3 years; upper surface glabrous when mature, lower surface with a compacted pale fawn to brownish indumentum or glabrous in ssp. *fauriei*. **Inflorescence** 10-20-flowered, upright. **Corolla** broadly funnel-shaped, 2.5-3cm long, white to pink, occasionally deep rose, *often lined and flushed, spotted brown to green*; calyx small c. 2mm long; stamens unequal; ovary densely tomentose; style shorter than longest stamens.

This is a distinct but variable species characterised by its yellow or pale green petiole and midrib and by its pointed vegetative buds. It resembles *R. catawbiense* in foliage but differs in the smaller, pink to white rather than pink-purple flowers, and in addition, the leaves of *R. catawbiense* are completely glabrous when mature.

R. brachycarpum is a useful hardy species but is not suited to areas prone to spring frosts as the growth comes early and is easily damaged. While quite attractive, the flowers are usually small compared with the size of the leaves and they tend to be rather hidden in the young growth. Introduced late 1800s. June-July.

Some authorites maintain two subspecies but further fieldwork may indicate that the two subspecies overlap to such an extent that they should not be maintained.

Ssp. *brachycarpum* (var. *roseum* Koidzumi, var. *rufescens* Nakai., ssp. *tigerstedtii* Nitzelius)

Leaves *with a persistent, compacted grey to fawn indumentum on the lower surface.* **Distribution** Japan, E. Korea, 1,700-2,300m (5,500-7,500ft), among trees or above the tree-line.

Included here is the quite commonly cultivated Tigerstedtii Group. This differs from the type in its greater stature, swifter growth, larger leaves and its wider corolla which is white with green spots. **Distribution** C. and N. Korea and Dagelet Island, 200-1,600m (660-5,300ft), common in forest. Plants under the Tigerstedtii Group are perhaps the hardiest of all rhododendrons, having withstood -45oC (-49oF) at Mustila, Finland. Not very different from the type of *R. brachycarpum* horticulturally.

Ssp. *fauriei* (Franch.) D.F. Chamb. 1979.
(var. *roseiflorum* Miyoshi).

Leaves +/- *glabrous when mature*. **Distribution** Japan, Korea.

Variable in size of flower and freedom of flowering. Ssp. *fauriei* has been used in the breeding of hardy hybrids.

Picture 1

1. *R. brachycarpum* ssp. *brachycarpum*, Royal Botanic Garden, Edinburgh. Typically the young growth is well advanced at the time of flowering.

PONTICA

Picture 2

Picture 3

Picture 4

Picture 5

Picture 6

6. *R. brachycarpum* ssp. *fauriei* only differing from ssp. *brachycarpum* in its glabrous leaf lower surface. Royal Botanic Garden, Edinburgh.

R. catawbiense
Michaux 1803 H5

Height 1-3m+, a rounded shrub, generally wider than tall. Branchlets at first tomentose, later glabrescent; bark roughish, brownish-grey. **Leaves** 7-15 x 3-5cm, elliptic to oval, *convex*, base *rounded*, retained for 2-3 years; upper surface glabrous, lower surface whitish-green, glabrous but with minute hairs; petiole with floccose indumentum. **Inflorescence** 8-20-flowered, usually compact. **Corolla** funnel-campanulate, 3.5-4.3cm long, usually lilac-purple, occasionally pinkish or white with faint spots; calyx minute, c.1mm; ovary *densely tomentose*. **Distribution** North Carolina and Virginia, U.S.A., 1,200-2,000m (4,000-6,500ft), on summits and sides of mountains, often in the open.

R. catawbiense has a more domed growth habit than its relatives *R. macrophyllum*, *R. maximum* and *R. ponticum* and differs significantly in leaf shape. In foliage, *R. catawbiense* most resembles *R. brachycarpum* but the two are usually easily distinguished in flower and by the very pointed growth buds of *R. brachycarpum*.

R. catawbiense has proven to be invaluable for its extreme hardiness which has been used to create a range of 'ironclad' hybrids suitable for severe climates. *R. catawbiense* is usually grown under the cultivar names 'Catawbiense Album', R. 'Roseum Elegans', R. 'Catawbiense Boursault' and R. 'Catawbiense Grandiflorum', selections or hybrids of the species made during the 19th century. There are also white forms such as 'Glass' which have been raised in cultivation. The low elevation form known as Insularis Group with larger leaves and flowers is rare in cultivation. In Britain, *R. catawbiense* does not grow nearly as well as *R. ponticum*. Introduced 1809->. June.

2. *R. brachycarpum* ssp. *brachycarpum* showing the convex leaves and a pink-flushed corolla with a large green blotch. Glendoick.
3. *R. brachycarpum* ssp. *brachycarpum*, Royal Botanic Garden, Edinburgh.
4. A dwarf strain of *R. brachycarpum* ssp. *brachycarpum* Roseum Group about 40 years old, grown as bonsai in a Japanese garden. (Photo Y. Doi).
5. *R. brachycarpum* ssp. *brachycarpum* Tigerstedtii Group, Baravalla, W. Scotland. This is found only in Korea.

5. The white-flowered *R. catawbiense* 'Powell Glass' a cultivated selection, Glendoick.

R. caucasicum
Pallas 1784 H5

Height 0.3-1m, a usually compact shrub, broader than tall. Branchlets sparsely tomentose, perulae *persistent*; bark brownish-grey, fairly rough. **Leaves** 5-10 x 2-4.5cm, obovate to elliptic, retained for 1-2 years; upper surface glabrous, lower surface with a *compacted fawn to brownish indumentum*. **Inflorescence** 6-15-flowered, upright. **Corolla** thinly-textured, broadly campanulate, 2.3-3.5cm long, cream to pale yellow, sometimes flushed pink, usually spotted green; calyx minute, 1-2mm hairy; ovary densely tomentose. **Distribution** N.E. Turkey and adjacent parts of the Caucasus, 1,800-3,000m (6,000-10,000ft), forming large mats above the tree-line.

The closest relation of *R. caucasicum* is *R. aureum* which is smaller in all parts and has a glabrous leaf underside.

In the wild, *R. caucasicum* grows well in a relatively poor, dry soil, low in organic matter. The species itself is rare in cultivation but its many hybrids such as 'Cunningham's White' are very widely grown. Susceptible to bud blast fungus. Introduced 1803->. April-May.

1. *R. catawbiense* growing wild at Craggy Gardens, North Carolina, U.S.A.
2. *R. catawbiense*, flowers showing two different shades of colour, Roan Mt., North Carolina, U.S.A.
3. *R. catawbiense*, Craggy Gardens, North Carolina, U.S.A.
4. Young growth of *R. catawbiense*, Craggy Gardens, North Carolina, U.S.A., showing the convex leaves with rounded bases.

PONTICA

Picture 1

Picture 2

Picture 3

Picture 4

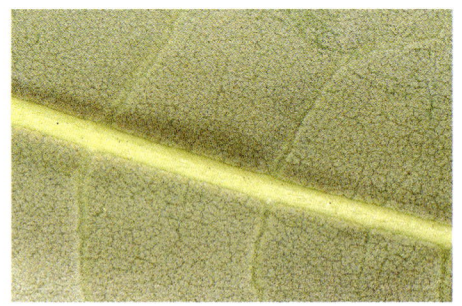

Picture 5

1. *R. caucasicum* in the central Caucasus with Mt. Elbrus behind. In the Baksan valley, this species has uniformly pink-tinged flowers with deeper spots.
(Photo M.J.B. Almond).
2. *R. caucasicum* in the central Caucasus with Mt. Dongusorun. In this valley the flowers are uniformly creamy-yellow. (Photo M.J.B. Almond).
3. *R. caucasicum* A.C.& H. 206 from N.E. Turkey, Glendoick.
4. *R.* 'Cunningham's Sulphur' which appears to be simply a form of *R. caucasicum* with particularly narrow and twisted leaves.
5. The leaf lower surface of *R. caucasicum* showing the compacted indumentum. (Photo R.B.G., E.).

R. degronianum

(Carrière 1869) Hara 1986 (excluding ssp. *yakushimanum* described below) H(4-)5

Height 0.6-2.5m., usually a compact and rounded shrub, often growing wider than high. Branchlets sparsely tomentose; bark light grey-brown, slightly rough. **Leaves** 7-18 x 2-4.6cm, oblong-elliptic to oblanceolate, *recurved*, retained for 2-4 years; upper surface glabrous, deep green, lower surface with a felted, fawn to rufous, bistrate indumentum. **Inflorescence** 6-15-flowered, loose to compact. **Corolla** with *5 or 7 lobes*, funnel-campanulate, 2.8-4.3cm long, soft pink to rose, often lined deeper, occasionally white; calyx small, 2-3mm, glabrous; ovary tomentose.

R. degronianum has wider leaves than *R. makinoi* and much earlier growth.

The classification and reclassification of this species and its near relatives has been a complex saga in recent years but hopefully the current nomenclature will be universally accepted.

R. degronianum is very useful for its extreme hardiness and is widely grown, especially in colder climates. April-May.

Ssp. *degronianum* (R. japonicum (Blume) Schneider, var. *pentamerum* (Maxim.) Hutch., R. metternichii Siebold & Zuccarini, ssp. *pentamerum* Matsumura)

Indumentum present on blade and current-year stem, *sparse or absent on mid-rib and year-old stem.* **Corolla** funnel-shaped, with *5 lobes* **Distribution** N. Honshu, Japan, around and below 1,800m (6,000ft), forming thickets near tree-line.

Distinguished from ssp. *heptamerum* in its 5 rather than 7-lobed corolla and differs from ssp. *yakushimanum* in its less persistent indumentum.

Ssp. *heptamerum* (Hara) D.F. Chamb. & F. Doleshy 1987.

Height *up to 3.7m.* **Leaves** often flat rather than convex, *indumentum thin to velvety.* **Corolla** *6-9 lobed but usually 7-lobed.* **Distribution** S. Japan, 200-1,200m (700-4,000ft), in forest or open slopes.

Differs from ssp. *degronianum* in its frequently greater stature, often flat (not convex) leaves, and in the usually 7-lobed corolla.

A variable subspecies with many different forms from isolated mountains. The low growing var. *micranthum* is now treated as Micranthum Group.

Var. *heptamerum* (R. japonicum (Blume) Schneider, R. metternichii Siebold & Zuccarini.)

Indumentum *thick.* **Distribution** Kyushu, Shikoku, Honsu-Kii Peninsula.

It is sad that the name *R. metternichii*, formerly used for this variety is invalid. This is often a better plant than ssp. *degronianum* with superior foliage and flowers. The indumentum on some forms is striking, especially in forms with deeper coloured midribs. This variety has the most southerly distribution. Introduced 1870 or 1862->. April-May

Var. *hondoense* Hara 1986

Differs from var. heptamerum in its *thin, pale, plastered* indumentum and it has a more northerly distribution. **Distribution** W. Honshu and Kyushu.

Plants under the name var. *intermedium* Sugimoto appear to be intermediate between var. *hondoense* and ssp. *degronianum* and it also merges with ssp. *yakushimanum*.

Var. *kyomaruense* Hara 1986

Differs from var. *hondoense* in its *5-lobed* corolla and from ssp. *degronianum* in its *paler, thinner* indumentum. Doubtfully deserving of varietal status. **Distribution** A single peninsular, C. Honshu.

Picture 1

Picture 2

1. *R. degronianum* ssp. *degronianum*, Mt. Kusatsu Shirane, Japan. (Photo T. Takeuchi)
2. *R. degronianum* ssp. *degronianum*, Royal Botanic Garden, Edinburgh.

PONTICA

Picture 3

Picture 4

Picture 5

Picture 6

Picture 7

Picture 8

3. *R. degronianum* ssp. *degronianum* 'Metternianum', a clone named by the nurseryman K. Wada with pale flowers and striking young foliage with white indumentum on the leaf upper surface. Royal Botanic Garden, Edinburgh.
4. *R. degronianum* ssp. *degronianum*, white-flowered form sometimes known as var. *albiflorum*, at Glendoick.
5. *R. degronianum* ssp. *heptamerum* var. *heptamerum*, Rhododendron Species Botanical Garden, Seattle, U.S.A.
6. *R. degronianum* ssp. *heptamerum* var. *hondoense* in Kaikakedani National Park, Japan. (Photo T. Takeuchi).
7. *R. degronianum* ssp. *heptamerum* var. *hondoense*, Royal Botanic Garden, Edinburgh.
8. Foliage of *R. degronianum* ssp. *heptamerum* var. *hondoense* showing the thin pale-plastered indumentum, Royal Botanic Garden, Edinburgh.

Ssp. *yakushimanum*
(Nakai) Hara 1986

Height 1-1.5(2.5)m, compact and rounded in the typical cultivated plant, more erect and loose in the lowland (var. *intermedium*) forms. Young shoots covered with loose tomentum. **Leaves** 6-21 x 1-3cm, narrowly to broadly elliptic or linear-lanceolate, usually recurved; upper surface glabrous with age but usually with a thin, whitish indumentum persisting for several months, lower surface with a thick, white to tawny, woolly indumentum *which obscures the midrib*. **Inflorescence** 5-10-flowered, compact and rounded to loose. **Corolla** funnel-campanulate, 3-4cm long, pale rose fading to white, spotted or unspotted. Ovary densely tomentose. **Distribution** Yakushima Island, S. Japan, 1,200-1,800m (4,000-6,000ft), taller forms in conifer forest, dwarfer on the exposed tops among rocks.

This subspecies differs from ssp. *degronianum* in the more persistent and thicker indumentum on the stems and midrib and the entire lower leaf surface.

Most gardeners will continue to call this well-known plant *R. yakushimanum* or just 'yak'. One of the most widely grown and versatile of all species, *R. yakushimanum* is hardy, free-flowering and tolerant of sun or light shade. The clones *R.*

'Ken Janeck', R. 'Mist Maiden' and R. 'Pink Parasol' with larger flatter flowers are most likely hybrids, perhaps with ssp. *degronianum* or *R. smirnowii*. Amongst rhododedron breeders, *R. yakushimanum* has been by far the most popular parent in recent years and there are now hundreds of named 'yak' hybrids. Introduced 1934->. May. The plants formerly under the name *R. degronianum* var. *intermedium* (now Intermedium Group) show a complete gradation from tall plants referable to *R. degronianum* to the typical ssp. *yakushimanum* and this is the reason for the inclusion of *R. yakushimanum* as a subspecies of *R. degronianum*.

Picture 9

Picture 10

Picture 11

Picture 12

Picture 13

12. *R. degronianum* ssp. *yakushimanum*, showing the characteristic recurved leaves with semi-persistent indumentum on the upper surface and thick woolly indumentum on the lower surface, Glendoick.
13. A tall-growing clone of *R. degronianum* var. *yakushimanum* growing at the Asian Garden, University of British Columbia, Vancouver, Canada. This tall form resembles an intermediate between *R. degronianum* and *R. yakushimanum*.

9. *R. degronianum* ssp. *yakushimanum*, summit of Mt. Kuromi, Yakushima, Japan. (Photo T. Takeuchi).
10. Different clones of *R. degronianum* ssp. *yakushimanum*, Glendoick.
11. *R. degronianum* ssp. *yakushimanum* 'Koichiro Wada' F.C.C., Glendoick.

R. hyperythrum
Hayata 1913 (*R. rubropunctatum* Hayata) H5

Height 1-2.5m, a compact and rounded shrub, looser in shade. Branchlets usually glabrous; bark fairly smooth, grey-brown. **Leaves** 7-12 x 2.5-4cm, elliptic to elliptic-lanceolate, revolute, margins *recurved* (at least in cultivated plants), retained for 3-4 years; upper surface glabrous, lower surface *superficially glabrous* but with *minute brownish punctulations (dots)*, midrib glabrous or hairy. **Inflorescence** 7-12-flowered, *candelabroid*. **Corolla** funnel-campanulate, 3-5cm long, *white* (in most cultivated plants), also off white or pale pink; calyx small, c.3mm; ovary densely glandular, glandular or hairy near base. **Distribution** Localised in N. Taiwan only, 900-1,200m (3,000-4,000ft), in broad-leaved forest.

This is a distinct species only distantly allied to the others in the subsection. The usually pure white flowers, the recurved leaves (in cultivation) and the punctulations on the lower leaf surface make it quite easily identified.

PONTICA

R. hyperythrum is an attractive, often free-flowering species, best in its larger-flowered, white forms. It prefers a fairly open situation and is both cold and heat tolerant. Apparently the leaves are usually flat in the wild but they always seem to have recurved margins in cultivation. Introduced 1930s. April-May.

Picture 1

Picture 2

Picture 3

1. A fine pure white-flowered clone of *R. hyperythrum* from Windsor Great Park, S. England, growing at Glendoick.
2. *R. hyperythrum*, Royal Botanic Garden, Edinburgh, with white flowers, tinged pink.
3. Young growth of *R. hyperythrum*, Glendoick, showing the strongly recurved leaves.

R. macrophyllum
G. Don 1834 (*R. californicum* Hook.f.) H4

Height 2-9m, a usually erect, vigorous shrub or small tree. Branchlets +/- glabrous; bark fairly rough, dark grey. **Leaves** 7-23 x 3-7.5cm, oblong-obovate to elliptic, retained for 2-3 years; upper surface matt, glabrous, lower surface *glabrous* when mature. **Inflorescence** 10-20-flowered, rachis fairly tall. **Corolla** with often crinkly lobes, broadly campanulate, 2.8-4cm long, pink to rosy-purple, occasionally white, spotted; calyx minute, c.1mm, glabrous; ovary *tomentose*. **Distribution** W. North America, California north to British Columbia, sea-level to 1,200m (4,000ft), forest and forest margins.

This species is related to both the other American members of the subsection, *R. catawbiense* and *R. maximum* and also to *R. ponticum*. *R. catawbiense* is much hardier, has a less upright habit, a different leaf shape and a more purplish flower. *R. maximum* differs in its indumented leaf underside and its generally smaller, paler flowers produced around a month later in cultivation. *R. ponticum* has pinkish-purple flowers, a glabrous ovary and shiny, narrower, glabrous, deep green leaves.

R. macrophyllum is not an easy plant to grow well in Britain or mainland Europe and even in its native habitat it is often scruffy. Coming from an area with little summer rain, it should be invaluable for drought resistance but is not really proving so at Glendoick. Rarely cultivated in Europe. Introduced 1850->. May-June.

Picture 1

1. *R. macrophyllum* showing the typical erect habit.

Distribution A limited area of C. Honshu, Japan, in forest, among ferns and rocks, 180-700m (600-2,300ft).

A distinct species, easily separated from the rest of the subsection by its persistent perulae, narrow leaves, retained for many years, and its late growth. *R. roxieanum* especially in var. *oreonastes* can have equally narrow leaves but *R. makinoi* differs in its persistent perulae, young growth with indumentum on the upper leaf surface, paler indumentum on the lower surface and the usually pink rather than white flowers.

R. makinoi is well worth growing for its unusual foliage but is not the easiest of plants to grow well, being inclined to have chlorotic foliage and a poor root system. It seems to prefer a near alkaline soil and it may perform best as a grafted plant. The wider-leaved forms such as those grown at the Royal Botanic Garden, Edinburgh, are more vigorous and easier to please and may be hybrids or intergrades with *R. degronianum*. Date of introduction unknown, pre 1925. June.

2. A typical pink-flowered *R. macrophyllum*, Glendoick.
3. A pale-flowered *R. macrophyllum*, Glendoick.
4. A dark-flowered *R. macrophyllum* at Britt Smith's garden, Kent, Washington State, U.S.A. The colour of this species can range from pure white to deep pink.

R. makinoi

Tagg 1927 (*R. metternichii* var. *pentamerum* forma *angustifolium* Makino, *R. stenophyllum* Makino, *R. yakushimanum* ssp. *makinoi* D.F. Chamb.) H5

Height 1-2.5m, usually nearer 1m, a usually dense and rounded shrub. Branchlets *densely woolly*; bark slightly flaking, grey-brown; perulae *persistent*. **Leaves** 7-18 x 1-2.5cm, *linear lanceolate, recurved,* retained for 4-5 years; upper surface with a *loose, white* indumentum at first, later glabrous, lower surface with a *dense* tawny, bistrate indumentum. **Inflorescence** 5-8, occasionally to 30-flowered, fairly compact to compact. **Corolla** funnel-campanulate, 3-4cm long, rose to off-white; calyx variable, 1-7mm, hairy; ovary/capsule densely covered with brown woolly tomentum.

1. *R. makinoi* 'Fuju kaku no matsu', Glendoick, with typical narrow recurved leaves.
2. A wider-leaved *R. makinoi*, Royal Botanic Garden, Edinburgh.

PONTICA

Picture 3

3. A fine specimen of *R. makinoi* with Dr. Carl Phetteplace and Patricia Cox in the former's garden near Eugene, Oregon, U.S.A.

its extreme hardiness, it is surprisingly rarely cultivated in Europe and should be more widely grown, especially in selected forms. Introduced 1763->. July.

Picture 1

Picture 2

Picture 3

R. maximum

L. 1753 (*R. ashleyi* Coker, ?*R. latifolium* Hoffmanns. *R. maximum* var. *purpureum* Pursh, *R. procerum* Salisbury, *R. purpureum* (Pursh) G. Don, *R. purshii* D. Don) H5 'Rosebay', 'Great Rhododendron' or 'Great Laurel'.

Height 1-3.7m, occasionally to 12m, a compact, spreading or broadly erect shrub or small tree. Branchlets tomentose and *glandular*; bark slightly rough, dark greyish. **Leaves** 10-*30* x 4-7cm, ovate lanceolate to oblanceolate, retained for 3-6 years; upper surface with some indumentum at first, lower surface with a *film-like indumentum* which usually persists though often it is restricted to a a rim or patch or to near the midrib at maturity. **Inflorescence** 12-30-flowered, upright and compact. **Corolla** campanulate, 2.3-3.1cm long, white, white tinged pink to rose-purple, usually blotched yellow or greenish yellow; calyx 3-5mm, *glandular*; ovary *densely glandular*. **Distribution** E. North America from Georgia to Nova Scotia, from just above sea level to 900m (3,000ft), in conifer or deciduous forest often in dense shade.

This species is easily separated from the others in this subsection by a combination of its large, comparatively narrow leaves, usually with a thin indumentum on the lower surface, its glandular branchlets, petioles and flower parts, and its rather late flowers in small upright trusses.

With its large, dark green leaves, *R. maximum* makes a most handsome evergreen shade plant in its native E. North America but the flowers, by comparison, are rather small. Considering

1. *R. maximum* at Glendoick, showing the typical upright truss with a distinct blotch in the corolla.
2. A pale pink-flowered *R. maximum* with a small blotch, growing wild in North Carolina, U.S.A. (Photo D. Jolley).
3. An unusual almost pure white-flowered *R. maximum*, growing wild on Gregory Bald, Pennsylvania, U.S.A.

Picture 4

4. Foliage of *R. maximum* showing a leaf lower surface partially covered with a film-like indumentum. In some forms, the indumentum on the lower leaf surface may be restricted to a ring or patch. Garden in the Woods, Massachusetts, U.S.A.

R. ponticum

L. 1762 (?*R. adansonii* Pepin, *R. algarvense* Page, *R. baeticum* Boiss. & Reuter, *R. lancifolium* Moench, *R. ponticum* var. *brachycarpum* Boiss, *R. ponticum* ssp. *baeticum* Boiss & Reuter, *R. speciosum* Salis.) H4-5

Height 2-5m, occasionally to 8m, forming a dense to leggy bush, Branchlets usually glabrous; bark rough, brown to dark brown. **Leaves** 10-20 x 3-7.5cm, lanceolate to elliptic, usually *flat*, retained for 2-4 years; upper surface glabrous, lower surface +/- *glabrous*. **Inflorescence** 6-20-flowered, usually upright and compact. **Corolla** campanulate, 4-5cm long, pale to deep lilac-pink to pinkish-purple, spotted or blotched; calyx 1-2mm, glabrous; ovary *glabrous*. **Distribution** Found chiefly in the Caucasus and N. Turkey, at sea-level to 1,800m (6,000ft) but also Lebanon and a few isolated sites in S. Spain and Portugal, in deciduous and conifer forest or more open sites.

This species is easily separated from its American relatives as it is almost completely glabrous, including the ovary. Its narrow, dark green leaves with no traces of indumentum and the lilac-pink rather than pale flowers separate it from *R. maximum*. Some cultivated *R. ponticum* appears to contain some *R. catawbiense* and *R. maximum* blood. There are various selected cultivars including forms with variegated, purple-tinged and extra narrow leaves. Forms with very pale or white flowers are most likely hybrids with *R. caucasicum* (*R. x sochadzeae* Char & Davlianidze 1967).

The ubiquitous *R. ponticum* is the most famous or infamous rhododendron species in the British Isles where it has naturalised itself in many areas and become a serious pest, often spreading through large areas of woodland and covering open hillsides, especially in wetter areas. It spreads by seed or layers and many gardens are now full of *R. ponticum* derived from suckers from grafting understocks (traditionally used for the progagation of hybrids). *R. ponticum* is capable of regrowth from the roots as well as the stems and is therefore extremely difficult to eradicate. In recent years, the invasive nature of the species has been much publicised in the U.K. by environmentalists, farmers and foresters and this has led to rhododendrons in general receiving bad publicity. Outside the U.K., *R. ponticum* is rarely troublesome as it is too tender for large parts of Europe and North America and is also susceptible to *Phytophthora* root rot. Forms from low elevations in the wild may have heat resistance potential but would be rather tender. Introduced to Britain from 1763->. May-July.

Picture 1

Picture 2

1. *R. ponticum* naturalised in the Great Glen, N. Scotland. This species has naturalised itself in many parts of the British Isles.
2. A clone of *R. ponticum* A.C.& H. 205 from N.E. Turkey, growing at Glendoick, with a brownish-yellow blotch on the corolla.

PONTICA

indumentum. **Inflorescence** *6-15*-flowered, loose to fairly compact. **Corolla** funnel-campanulate, 3.1-4cm long, pink to rose-purple, usually with brown to yellow spots; calyx *minute, 1-2mm*, tomentose and glandular; ovary *white tomentose*. **Distribution** N.E. Turkey and adjacent areas of Caucasus, 800-2,300m (2,800-7,500ft), below to just above the tree-line.

This species is closely related to *R. ungernii* from which it differs in its smaller leaves, often with deeper-coloured indumentum on underside, deeper-coloured, earlier flowers, fewer to the truss, a smaller calyx and a tomentose ovary. Otherwise it is distinct with no close relatives although it resembles *R. degronianum* in foliage. Second growth with no leaves is sometimes produced. It hybridises with *R. ponticum* in the wild (*R. x kesselringii* E. Wolf).

R. smirnowii is the hardiest species with a thick indumentum and is satisfactory in many parts of E. North America and Scandinavia. It is inclined to become straggly if grown in a shaded site. Introduced 1886->. May-June.

Picture 3

Picture 4

Picture 5

3. *R. ponticum* A.C.& H. 205 from N.E. Turkey with a yellow blotch on the corolla.
4. A clump of several *R. ponticum* at Glendoick.
5. *R. ponticum* 'Cheiranthifolium', a clone with small twisted leaves and small flowers, Royal Botanic Garden, Edinburgh.

R. smirnowii
Traut. 1886 H5

Height 1-3m, forming a compact bush in the open, more upright in the shade. Branchlets with a *white, felted tomentum*. Perulae *+/- persistent for some years*; bark roughish, brown. **Leaves** *6.5-17.5 x 2-3.5cm*, oblanceolate to elliptic, margins often recurved, retained for 3-5 years; upper surface with white woolly tomentum at first, later dark green and glabrous, lower surface with a dense, pale, fawn to pale brown

Picture 1

Picture 2

1. *R. smirnowii*, near Artvin, N.E. Turkey, with *Picea orientalis* forest in the background.
2. *R. smirnowii* A.C.& H. 118, Glendoick.

Picture 3

3. *R. smirnowii* showing the dark recurved leaves with pale indumentum on the upturned leaf.

R. ungernii
Traut. 1886 H5

Height 2.4-6m, rarely as tall in cultivation, a fairly compact to erect bush. Branchlets not as densely tomentose as *R. smirnowii*; bark brown and slightly flaking. **Leaves** *10-25 x 4-9cm,* oblanceolate to obovate, retained for 2-3 years; upper surface glabrous when mature, lower surface with a dense, woolly, white to fawn indumentum. **Inflorescence** *20-30-flowered, tall and compact.* **Corolla** funnel-campanulate, c.3.5cm long, *white to pale rose* with green spots; calyx, *4-9mm long, glandular;* ovary *densely glandular.* **Distribution** N.E. Turkey and adjacent Caucasus, 800-2,000m (2,800-6,500ft), usually in shade.

This species is closely related to *R. smirnowii* with which it often hybridises in the wild. For differences see *R. smirnowii*. These are no other close relatives.

A fine foliage plant when well grown, *R. ungernii* is useful for its late blooming, though the flowers are rather small in comparison with the leaves. Disliking direct sun, it can thrive in shady conditions where it flowers surprisingly freely. Introduced 1886->. June-July.

Picture 1

Picture 2

Picture 3

1. *R. ungernii* above Artvin, N.E. Turkey. (Photo M.J.B. Almond).
2. *R. ungernii* A.C.&H. 119, from N.E. Turkey, growing at Glendoick, showing the large leaves and the upright, many-flowered truss.
3. *R. ungernii*, probably crossed with *R. smirnowii* A.C.&H. 121 from N.E. Turkey, growing at Glendoick.

Subsection Selensia

Height 0.6 to 6m+, forming compact to straggly shrubs or small trees. Branchlets *bristly-glandular;* bark rough. **Leaves** 2-12.5 x 0.8-6cm, obovate to elliptic; upper surface glabrous, lower surface glabrous or with a loose, thin indumentum. **Inflorescence** 3-10, rarely 1-2-flowered, loose. **Corolla** *funnel-campanulate* to campanulate, *lacking nectar pouches*, white to reddish-purple, with or without spots and blotch, calyx variable, 1mm to 1cm long, stamens 10; ovary glandular and sometimes hairy; style glabrous or glandular at base, capsule slender, often sickle-shaped. **Distribution** S.W. Sichuan, Yunnan, S. E. Tibet.

This subsection is related to subsection Campylocarpa, differing in the longer glandular hairs on the branchlets and petioles and in the more funnel-campanulate corolla shape. It is doubtful if there are really more than two species in this subsection: *R. hirtipes* and the very variable *R. selense*. Recent observations in China and Tibet has confirmed suspicions that several of the so-called species in this subsection are natural hybrids.

The species of Subsection Selensia are not such popular garden plants as the closely related members of subsection Campylocarpa; many forms are rather shy-flowering and not particularly spectacular, though there are some fine ones.

R. bainbridgeanum
Tagg & Forrest 1931 H4

Height 0.6-2.75m, a bushy to upright, open shrub. Branchlets setose-glandular, bud scales +/- persistent; bark roughish, greyish-brown. **Leaves** 6-12 x 2-4.5cm, obovate to elliptic, retained 1-2 years; upper surface glabrous, lower surface with a *loose, scurfy, continuous or discontinuous brownish* indumentum. **Inflorescence** 4-8-flowered, loose. **Corolla** campanulate, 2.8-4cm long, colour very variable, white, white flushed rose to deep pink and reddish-purple, spotted and/or blotched; calyx variable, 2-9mm with unequal rounded lobes, glandular; ovary, densely glandular; style glandular at base. **Distribution** S.E. Tibet, N.W. Yunnan, N.E. Upper Burma, 3,400-4,400m (11,000-14,500ft), thickets near conifer forest and rocks.

R. bainbridgeanum is the only species in subsection Selensia, apart from *R. selense* ssp. *setiferum*, with a +/- continuous indumentum on the leaf lower surface.

Field studies are necessary to confirm suspicions that this species is derived from natural hybrids between *R. crinigerum* and *R. selense* and perhaps with *R. wardii*. There are very few clones of *R. bainbridgeanum* in cultivation, those we know of have reddish-purple or creamy-yellow flowers. Some clones are free-flowering but they tend to flower early and are easily frosted as the buds swell. Introduced 1922->. March-April.

Picture 1

Picture 2

1. A showy plant labelled *R. bainbridgeanum*, at the Younger Botanic Gardens, Benmore, W. Scotland, which may not quite match this species.
2. *R. bainbridgeanum* Forrest 21821, Royal Botanic Garden, Edinburgh.

Picture 3

3. The leaf lower surface of *R. bainbridgeanum* Forrest 21821, showing the discontinuous brown indumentum. (Photo R. B. G., E.)

R. x erythrocalyx

Balf.f. & Forrest 1920 (*R. beimaense* Balf.f. & Forrest, *R. cymbomorphum* Balf.f. & Forrest, *R. docimum* Balf.f., *R. eucallum* Balf.f. & Forrest, *R. panteumorphum* ? Balf.f. & W.W. Sm., *R. truncatulum* Balf.f. & Forrest)

Leaves broadly obovate to oblong; lower surface glabrous or with vestiges of hairs. **Corolla** creamy to pale yellow; calyx fairly large; calyx 3-7mm, stipitate glandular. **Distribution** N.W. Yunnan, S.E. Tibet, 3,355-3,965m (11,000-13,000ft).

A natural hybrid between *R. selense* and *R. wardii* which are often associated in the wild. Rare in cultivation but remarkably common in the wild in disturbed sites, where it often occurs as quite large swarms of variable plants. April-May.

Picture 1

1. *R. x erythrocalyx*, Bei Ma Shan, N. W. Yunnan, China. Both putative parents, *R. wardii* and *R. selense*, were growing nearby.

R. esetulosum

Balf.f. & Forrest 1920 (*R. manopeplum* Balf.f. & Forrest) H4

Leaves 6-12 x 3-4cm, *thick, coriaceous*. **Corolla** 3-5cm long, white flushed rose or purplish usually with a basal blotch and flecks; calyx variable in size, glandular; ovary densely glandular; style *partly glandular* from base. **Distribution** N.W. Yunnan and S.E. Tibet, 3,000-4,300m (10,000-14,000ft), thickets and bouldery slopes.

Plants which exactly fit the type description may not be in cultivation but as it is almost certainly a hybrid, perhaps *R. selense x R. vernicosum*, it is not surprisingly rather variable. Plants under this name are cultivated in Scotland and in S. Scandinavia where they have proved to be hardy. April-May. Plants cultivated under the name *R. calvescens* appear to be hybrids of *R. vernicosum* x *R. beesianum*.

R. hirtipes

Tagg 1930. H4.

Height 0.5-8m, but rarely more than 3m in cultivation, forming an upright bushy shrub. Branchlets with *glandular bristles*; bark roughish. **Leaves** 6-13 x 3.8-7.5cm, broadly obovate to oblong-oval, retained 1-2 years; upper surface glabrous, lower surface with scattered glands and a little loose indumentum, midrib *setose-glandular* usually near base only; petiole setose-glandular. **Inflorescence** 2-7, usually 3-4-flowered. **Corolla** *campanulate*, 3.5-5cm long, pale to deep pink, to white, sometimes flushed, spotted or blotched; calyx *medium to large*, *4-10mm* long, glandular; ovary densely glandular; style glandular at base. **Distribution** S.E. Tibet, well to W. of rest of subsection, 3,000-4,700m (10,000-15,000ft), wet forest, rocks and cliffs.

This is a distinct species, formerly grouped with *R. glischrum* but Chamberlain considers that it is more closely related to *R. selense*. Its isolated distribution from the rest of the subsection and the combination of its setose-glandular branchlets, the sparse indumentum with bristles on the midrib and the campanulate corolla distinguish it from its relatives. *R. hirtipes* differs from the species in subsection Glischra in its broadly obovate leaves with a rounded apex.

F. Kingdon-Ward thought very highly of this species in the wild, and we have seen spectacular forests of it in S.E. Tibet. In cultivation it seldom performs to perfection due to its rather

SELENSIA

frost-vulnerable, early flowers. White forms have yet to be introduced. Introduced 1924-1938, reintroduced 1995->. February-April.

Picture 1

Picture 2

Picture 3

Picture 4

Picture 5

4. Young growth of R. hirtipes Kingdon-Ward 5659 showing the branchlets with grandular bristles.
5. *R. hirtipes* on the Sirchem La, Rong Chu, S. E. Tibet.

1. *R. hirtipes* Ludlow, Sherriff and Taylor 3624 'Ita' A. M. at the Valley Gardens, Windsor Great Park, S. England.
2. *R. hirtipes* Ludlow, Sherriff and Taylor 3624, Royal Botanic Garden, Edinburgh.
3. *R. hirtipes* Kingdon-Ward 5659, Glendoick, showing the characteristic campanulate flowers.

R. martinianum
Balf.f. & Forrest 1919 H4

Height 0.6-2.1m, forming a compact to straggly shrub. Branchlets *thin,* usually setose-glandular; bark grey to greyish-brown. **Leaves** *1.9-5.4 x 0.8-2.5cm*, oblong-elliptic to obovate, retained 1-2 years; upper surface usually glabrous, lower surface glaucous, usually glabrous or with few hairs. **Inflorescence** *1-4*-flowered, loose. **Corolla** funnel-campanulate, 2.5-4cm long, white through pink to purple; calyx small, 1-3mm long, glandular; ovary densely glandular; style sometimes glandular at base. **Distribution** N.W. Yunnan, S.E. Tibet and N.E. Upper Burma, 3,000-4,300m (10,000-14,000ft), margins of forests and thickets to above the tree-line.

This species is closely related to *R. selense*, differing in its thin shoots, smaller leaves and fewer (1-4) flowers per truss. Early introductions are usually rather straggly but Kingdon-Ward 21557 from the Triangle, Burma, has a compact habit and has relatively wide, almost sessile leaves (this taxon may be granted separate status in the future). *R. eurysiphon* Tagg & Forrest may be a hybrid of this species crossed with *R. stewartianum*.

Uncommon in cultivation, *R. martinianum* is essentially a dwarf *R. selense* and is a dainty species suitable for small gardens. Introduced 1914->. April-May.

Picture 1

Picture 2

Picture 3

1. *R. martinianum,* Royal Botanic Garden, Edinburgh. A typically straggly form.
2. *R. martinianum,* from Reuthe's Nursery growing at Blackhills, N. E. Scotland.
3. *R. martinianum* Kingdon-Ward 21557, at Glendoick, which makes a more compact plant than other introductions.

R. selense
Franch. 1898 H4

Height 0.60-6m+, forming a compact to tall and straggly shrub or small tree. Branchlets usually with *short-stalked* glands; bark fairly smooth to roughish, grey. **Leaves** 2.6-9 x 1.5-4cm, ovate, obovate to elliptic, usually retained for only one year; upper surface glabrous and sometimes glaucous, lower surface glabrous, glaucous or with a few small hairs or a thin, sometimes discontinuous indumentum; petiole setose or glabrous. **Inflorescence** 3-8-flowered, loose. **Corolla** funnel-campanulate, 2.5-4cm long, white to deep pink, with or without spots; calyx 1-8mm glandular; ovary densely glandular; style sometimes glandular at base or glabrous.

R. selense now includes a large number of former species which are regarded simply as one variable taxon. Currently divided into 4 subspecies which do undoubtedly merge. Found over a wide area, the picture is complicated by the numerous natural hybrids with *R. wardii, R. vernicosum, R. crinigerum, forrestii, sanguineum, temenium* etc. many of which were previously accorded specific names. So-called *R. selense* from the Tsangpo area of S.E. Tibet are almost certainly natural hybrids between *R. wardii/R. campylocarpum* and other species such as *R. hirtipes*.

Ssp. *selense* (*R. axium* Balf.f. & Forrest,
R. chalarocladum Balf.f. & Forrest, *R. metrium* Forrest, *R. nanothamnum* Balf.f. & Forrest, *R. pagophyllum* Balf.f. & Kingdon-Ward, *R. probum* Balf.f. & Forrest).

Young shoots at maturity *shortly stipitate-glandular*. **Leaves** entirely glabrous, sometimes with a glaucous bloom beneath; calyx normally *small, 1-2mm long*. **Distribution** N.W. Yunnan and S.E. Tibet, thickets, 2,700-4,400m (9,000-14,500ft), open conifer forest, bamboo to above tree-line.

Although this can be quite spectacular in the wild, this subspecies is often disappointing in cultivation, frequently being shy-flowering. Perhaps selections from recent introductions will prove to be more successful. It needs a fairly open situation to avoid stragglyness and to encourage flowers. Introduced 1917->, reintroduced 1992->. April-May.

Ssp. *dasycladum* (Balf.f. & W.W. Sm.) D.F. Chamb. 1978 (*R. dolerum* Balf.f. & Forrest, *R. rhaibocarpum* Balf.f. & W.W. Sm.)

Branchlets with *bristly glands*. **Leaves** often marginally longer than in the other subspecies; lower surface usually not *glaucous*; petiole *with bristly glands*. Pedicel *with bristly glands*; calyx *2-3(-5)mm* long. **Distribution** N.W. Yunnan, S.W. Sichuan, generally to the east of ssp. *selense*.

Ssp. *dasycladum* is related to *R. hirtipes*, differing in its funnel-campanulate corolla. It does not have the same reputation as ssp. *selense* for being shy-flowering but this may be due to a few cultivated clones which flower freely. Introduced 1913-1937.

SELENSIA

R. dasycladoides Hand.-Mazz. 1936 is probably not in cultivation and it is doubtfully distinct from *R. selense* ssp. *dasycladum*. It has a more easterly distribution.

Ssp. *jucundum* (Balf.f. & W.W. Sm.) D.F. Chamb. 1978 (*R. blandulum* Balf.f. & W.W. Sm.).

Branchlets *setose-glandular*. **Leaves** glaucous on the underside. **Corolla** more *campanulate* than that of the other subspecies, usually pink; calyx relatively *large*, (2-)4-6mm long. **Distribution** Confined to Cangshan mountains, Mid W. Yunnan 3,050-3,660m (10,000-12,000ft) where it forms the dominant rhododendron forest.

Ssp. *jucundum* starts to flower at a relatively young age and makes an attractive plant. It requires good drainage. Introduced 1917, reintroduced 1981->.

Ssp. *setiferum* (Balf.f. & Forrest) D.F. Chamb. 1978

This subspecies appears to lie between ssp. *R. selense* and *R. bainbridgeanum* in its *thin, often discontinuous* indumentum; calyx *medium to large* (2-)4-8mm long. It is very likely to be of hybrid origin. **Distribution** N.W. Yunnan, S.E. Tibet.

It is reputed to be free-flowering. Rare in cultivation but not very distinctive.

Picture 2

Picture 3

Picture 4

Picture 5

1. *R. selense* ssp. *selense* on Bei Ma Shan, N. W. Yunnan, China.
2. *R. selense* ssp. *selense* near the east side of the Doker La, Mekong-Salween divide, N.W. Yunnan, China.
3. A selection from different plants of *R. selense* ssp. *selense*, Napa Hai, Zhongdian, N.W. Yunnan, China.
4. Foliage of *R. selense* ssp. *selense*, Mekong-Salween divide, N.W. Yunnan, China, showing the non-bristly branchlets.
5. A tall specimen of *R. selense* ssp. *dasycladum* on Lao Shun Shan, N. W. Yunnan, China.

SELENSIA

Picture 6

Picture 6

10. *R. selense* ssp. *jucundum* SBEC 0544 from the Cangshan, Yunnan, growing at Glendoick.

Picture 7

Picture 8

Picture 9

6. *R. selense* ssp. *dasycladum,* Lao Shun Shan, N. W. Yunnan, China.
7. *R. selense* ssp. *dasycladum* originally from Tower Court, now in the Valley Gardens, Windsor Great Park, S. England.
8. Young growth of *R. selense* ssp. *dasycladum* showing the characteristic granular bristles.
9. Forest of *R. selense* ssp. *jucundum* and *Abies delavayi,* Cangshan, C. W. Yunnan, China.

Subsection Taliensia

Height 0.15-9m, prostrate to rounded and compact shrubs to upright and open small trees. Branchlets glabrous to densely tomentose; bark usually *rough*, brownish with grey flakes. **Leaves** linear to broadly elliptic; upper surface usually glabrous on maturity, lower surface *with a woolly, felted or compacted* indumentum, occasionally sparse or absent; the indumentum made up of radiate, ramiform or fasciculate hairs. **Inflorescence** 5-20-flowered, usually compact. **Corolla** 5-lobed (except *R. clementinae* which has 7 lobes), campanulate to funnel-campanulate, *nectar pouches absent, white to pink to purplish*, occasionally yellow, often spotted, sometimes blotched; calyx variable, 0.5mm-1.5cm long; ovary densely rufous-tomentose to glabrous and/or glandular; style usually *glabrous*, sometimes partly glandular. **Distribution** Bhutan and & S.E. Tibet eastwards to Yunnan and Sichuan and north to Gansu and Shaanxi.

Species of this subsection are generally long-lived and slow-growing with fine foliage but many take years to start flowering freely. All are hardy to H5 but they dislike hot summers, are susceptible to root pathogens and high nitrogen and require good drainage.

The commonest in cultivation are *R. adenogynum, R. balfourianum, R. bureavii, R. phaeochrysum, R. roxieanum, R. traillianum, R. wightii* (hybrid), and *R. wiltonii.*

TALIENSIA

We have divided this subsection into two groups or alliances, considering that *R. lacteum* and its relatives are so distinct from the other species that they could be placed in their own in subsection Lactea.

R. adenogynum
Diels 1912 (*R. adenophorum* Balf.f. & W.W. Sm.) H5

Height 1.3-4m, a broadly upright to spreading shrub, often wider than high. Branchlets tomentose to rarely glabrous; bark grey-brown, rough. **Leaves** leathery, 6-11 x 2-4cm, narrowly elliptic to elliptic, retained for 2-4 years; upper surface usually dark green, rugulose and almost glabrous, lower surface with a unistrate, *dense, woolly, usually spongy* indumentum, *yellowish turning to olive-brown on maturity*. **Inflorescence** 4-12-flowered, *generally loose*. **Corolla** funnel-shaped, 3-4.5cm long, white, white tinged pink, bright rose to rose-purple, sometimes fading to white, with +/- purple markings; calyx *large, glandular, 0.8-1.5cm* long, lobes unequal; ovary *densely glandular*; style *usually glandular in lower half*. **Distribution** S.E. Tibet, W. and N.W. Yunnan and S.W. Sichuan, 3,000-4,300m (10,000-14,000ft), mostly above the tree-line, often in large quantities.

This species and its closest relative *R. balfourianum* have the large glandular calyx and glandular ovary in common but *R. adenogynum* has narrower leaves and a more spongy, darker, olive green-yellow indumentum, whereas the indumentum of *R. balfourianum* is silvery to greyish-pink.

In cultivation, *R. adenogynum* usually makes a compact plant with fine foliage. The paler-flowered forms are often the more desirable as some of the deeper colours can be rather harsh-coloured. It is freer-flowering than many in this subsection and it starts blooming at a younger age. Plants under Adenophorum Group have glands on the branchlets, petioles and the midrib on the leaf underside. Introduced 1910-37, reintroduced 1980s->. March-May.

Picture 1

Picture 2

Picture 3

1. Hillside of *R. adenogynum*, Yulong Shan, N.W. Yunnan, China. (Photo R. McBeath).
2. *R. adenogynum*, from Yulong Shan, N.W. China.
3. A fine clone of *R. adenogynum* Yu 14955 with spotted flowers, Royal Botanic Garden, Edinburgh.

Picture 4

4. A pink-flowered *R. adenogynum*, Royal Botanic Garden, Edinburgh.

R. x detonsum Balf. f. & Forrest 1919.

Leaves larger than in *R. adenogynum* with a much thinner (sometimes absent) indumentum on lower surface. **Corolla** with 5-7-lobes, larger than in *R. adenogynum*. **Distribution** W. Yunnan, 3,000-4,000m (10,000-13,000ft).

This natural hybrid has repeatedly appeared in batches of seedlings of *R. adenogynum*. The other parent is usually *R. vernicosum* but the fact that some clones of *R. x detonsum* are scented would indicate that *R. decorum* is sometimes the involved. Some forms of this are very showy and well worth growing. Introduced 1910->, reintroduced 1986->. May

Picture 5

5. *R. x detonsum* showing the 5-7-lobed corolla. Rhododendron Species Botanical Garden, Washington State, W. U.S.A.

R. aganniphum
Balf.f. & Kingdon-Ward 1917. H5

Height 0.3-3m, a usually compact and rounded shrub. Branchlets glabrous or tomentose; bark rough, grey-brown. **Leaves** 4-12 x 2-5cm, elliptic to broadly ovate-lanceolate, retained for 2-3 years; upper surface glabrous, sometimes with a white 'bloom', lower surface with a *white to greyish-white indumentum which remains pale at maturity or turns red brown*, +/- compacted to spongy, continuous or splitting. **Inflorescence** 10-20-flowered, usually compact. **Corolla** funnel-campanulate, 3-3.5cm long, white to rose, often heavily flushed, spotted; calyx *small 0.5-1mm* long; ovary and style *glabrous*.

This species merges/hybridises with *R. phaeochrysum* but usually differs in having consistently paler indumentum (in var. *aganniphum*) or split indumentum (var. *flavorufum*). Forms with a relatively thick indumentum and larger leaves resemble *R. clementinae* which differs in its 6-7-lobed corolla. It is separated from *R. adenogynum* and *R. balfourianum* in its small, glabrous calyx and glabrous ovary.

Var. *aganniphum* (*R. doshongense* Tagg, *R. fissotectum* Balf.f. & Forrest, *R. glaucopeplum* Balf.f & Forrest)

Indumentum remaining *pale* at maturity and *not splitting into patches*. **Distribution** S.E. Tibet, N.W. Yunnan and W. & N. Sichuan, 3,400-4,600m (11,000-15,000ft), usually above the tree-line.

R. aganniphum is a very common species in the wild, where it covers huge areas of hillside, usually above the tree line. It is not all that common in cultivation, perhaps because it is quite hard to grow as a young plant. It is later flowering than most of its relatives. Introduced 1913-1937, reintroduced 1989->. Several former species are horticulturally distinct and popular with collectors. Glaucopeplum Group has more glandular leaves and petiole and a flatter leaf and seems to be midway between var. *aganniphum* and var. *flavorufum*. Doshongense Group is the extreme western variant, from the Tsangpo gorge area of Tibet, with a thinner, more plastered brown or whitish indumentum. Field observation lead us to conclude that Doshongense would be better considered a form of *R. phaeochrysum* var. *agglutinatum*. April-May (cultivation), June-July (wild).

Var. *flavorufum* (Balf.f. & Forrest) D.F. Chamb. 1978 (*R. schizopeplum* Balf.f. & Forrest) H5.

Indumentum +/- *split (with fissures)*; as the leaves mature, the indumentum turns deep red-brown, with the fissures getting larger, sometimes eventually just leaving spots or small patches of indumentum. **Distribution** S.E. Tibet, N.W. Yunnan and S.W. Sichuan, 3,400-4,600m (11,000-15,000ft).

While other species such as *R. taliense* may show slight indumentum splitting, this variety is unusual in the way the indumentum divides up. The degree of splitting is variable with plants cultivated as Schizopeplum Group having the least split indumentum (these are probably natural hybrids of *R. aganniphum* x *R. phaeochrysum*). Introduced 1917-1937, reintroduced 1992->. April-May (cultivation), June-July (wild)

Picture 1

Picture 2

Picture 3

Picture 4

Picture 5

Picture 6

1. *R. aganniphum* var. *aganniphum* with heavily blotched flowers; above Londre, Salween-Mekong divide, N.W. Yunnan, China.
2. *R. aganniphum* var. *aganniphum* with pale, lightly spotted flowers; above Londre, Salween-Mekong divide, N.W. Yunnan, China.
3. Foliage of *R. aganniphum* var. *aganniphum*, C.C.&H. 4064, Mongbi Mt., N.W. Sichuan, China, showing the typical whitish indumentum on both young and mature leaves.
4. *R. aganniphum* var. *aganniphum*, Glaucopeplum Group, Glendoick.
5. *R. aganniphum* var. *aganniphum*, Doshongense Group, Glendoick. This western variant has a thinner, more plastered leaf indumentum than populations further east and would perhaps be better placed in *R. phaeochrysum* var. *agglutinatum*.
6. The leaf lower surface of *R. aganniphum* var. *aganniphum* showing the typical dense, slightly spongy indumentum of this variety, mature leaf above, young below. (Photo R.B.G., E.)

apex acuminate, retained for 2-3 years; lower surface with a *dense persistent bistrate indumentum*; upper layer made up of long branched hairs, pale to mid-brown, woolly or felted, lower layer whitish and compacted. **Inflorescence** 10-20-flowered, fairly dense. **Corolla** funnel-campanulate, 3-4cm long, white, sometimes flushed rose to lilac-mauve, with +/- blotch or spots; calyx *minute, 0.5-1mm* with a few glands to tomentose; ovary tomentose to glabrous, sometimes with a few glands; style glabrous. **Distribution** S.E. Tibet, W. & N.W. Yunnan and bordering Sichuan, 2,700-4,300m (9,000-14,000ft), open forest to open pasture.

This species is separated from *R. adenogynum* in its small calyx and usually in the colour of indumentum. *R. taliense* has a larger calyx and a glabrous ovary. *R. alutaceum* (as Globigerum group) is said to differ from *R. roxieanum* var. *cucullatum* in its tomentose rather than glandular ovary. (See below for a discussion of this).

This is a confusing and problematic species; recent field-work has revealed that, rather than a stable species, *R. alutaceum* is made up of a multitude of variable plants. Some are extreme forms of *R. roxieanum* but most are natural hybrids between *R. roxieanum* and species such as *R. phaeochrysum, R. beesianum,* and *R. aganniphum*. Current D.N.A. work at the Royal Botanic Garden, Edinburgh, should clarify this. The numerous former species reduced in synonymy under the name *R. alutaceum* are very popular with species collectors, especially in Scandanavia. Introduced 1917-1932. April.

Var. *alutaceum* (*R. globigerum* Balf.f. & Forrest, (*R. roxieanum* var. *globigerum* Balf.f. & Forrest).

Leaves, lower surface with a usually *pale brown* continuous, *woolly* indumentum. Ovary papillate, *glabrous*.

This appears to be an unsatisfactory taxon and it is not clear whether var. *alutaceum* in the strict sense is in fact in cultivation. Most of the cultivated plants (grown under the name *R. globigerum*) have a deeper-coloured indumentum than is described for this variety and are better considered forms of *R. roxieanum*, under which name they were previously classified. Plants with a discontinuous indumentum are referable to var. *russotinctum*.

Var. *russotinctum* (Balf.f. & Forrest) D.F. Chamb. 1978 (*R. triplonaevium* Balf.f. & Forrest, *R. tritifolium* Balf.f. & Forrest).

Leaves, lower surface with a *discontinuous upper layer* of indumentum, mid to rufous brown. Ovary with *sparse rufous tomentum*.

Picture 7

Picture 8

Picture 9

7. *R. aganniphum* var. *flavorufum*, Royal Botanic Garden, Edinburgh.
8. *R. aganniphum* with slightly split indumentum (close to Schizopeplum group). On Bei Ma Shan, N.W. Yunnan, China, there were scattered plants like this amongst typical *R. aganniphum* var. *aganniphum* and *R. phaeochrysum* and this appears to be a hybrid between the two species.
9. Typical *R. aganniphum* var. *flavorufum*, Salween-Mekong divide, N.W. Yunnan, China. This variety occurred here as a uniform population.

R. alutaceum Balf.f. & W.W. Sm. H5

Height 0.60-4.5m, a fairly dwarf and compact shrub or small tree. Branchlets tomentose or glabrous or glandular; bark rough, brown. **Leaves** 5-18 x 2-4cm, oblong to oblanceolate,

TALIENSIA

Var. *russotinctum* merges with var. *alutaceum* and var. *iodes*. The Triplonaevium and Tritifolium Groups tend to have thinner paler indumentum than the type. These two groups resemble *R. phaeochrysum* var. *levistratum* but have narrower leaves, slightly darker indumentum and a non-glabrous ovary. Both are probably derived from hybrids of *R. roxieanum* var. *oreonastes* x *R. phaeochrysum*.

Var. *iodes* (Balf.f. & Forrest) D.F. Chamb. 1982

Leaves, lower surface with a *fine, continuous, mid-brown, felted* upper layer of indumentum. Ovary sparsely tomentose or/and glandular.

Var. *iodes* is intermediate between var. *alutaceum* and var. *russotinctum* and merges with both. As the type specimen is from a mixed gathering of all three varieties, we see very little basis for recognition of this variety at all. Cultivated plants under this name tend to have a greenish-yellow indumentum on the young leaves. Such plants are probably hybrids of *R. roxieanum* x *R. aganniphum*.

Picture 1

Picture 2

1. A plant cultivated as *R. alutaceum* var. *alutaceum* at Glendoick which is probably a *R. phaeochrysum* hybrid.
2. *R. alutaceum* var. *alutaceum* Globigerum Group, Valley Gardens, Windsor Great Park, S. England. This is probably referable to *R. roxieanum* var. *cucullatum*.

Picture 3

Picture 4

Picture 5

3. *R. alutaceum* var. *russotinctum* Rock 33, probably referable to the Triplonaevium Group at the Royal Botanic Garden, Edinburgh.
4. *R. alutaceum* var. *russotinctum*, probably referable to the Triplonaevium Group, Younger Botanic Garden, Benmore, W. Scotland.
5. *R. alutaceum* var. *russotinctum* referable to the Tritifolium Group?, Glendoick.

R. balfourianum

Diels 1912 (*R. balfourianum* var. *aganniphoides* Tagg & Forrest) H5

Height 1-4.5m, a usually rounded and compact shrub. Branchlets glabrous to rarely tomentose; bark rough, greyish-brown. **Leaves** 4.5-12 x 2-4cm, ovate-lanceolate to elliptic, retained for 1-3 years; lower surface with a *shining, dense,* *compacted to spongy* indumentum, *silvery-white when young, turning greyish-fawn to pale pinkish-cinnamon* on maturity. **Inflorescence** 6-12-flowered, loose to fairly compact. **Corolla** funnel-campanulate, 3.5-4cm long, usually *rose* but also white through rose-pink to purplish-pink; calyx *large, 0.6-1cm* long, with uneven *glandular* lobes; ovary glandular; style glandular on lower third. **Distribution** W. & N.W. Yunnan and S.W. Sichuan, 3,000-4,600m (10,000-15,000ft), in forest clearings to above the tree-line.

This species is closely related to *R. adenogynum* with the same large glandular calyx but with a greyish, more plastered indumentum. *R. balfourianum* differs from *R. aganniphum* in leaf shape, in its large glandular calyx and in its glandular ovary. Some authorities recognise var. *aganniphoides* Tagg & Forrest to describe plants with a thicker-than-average indumentum.

R. balfourianum is one of the easiest members of the subsection to cultivate and is free-flowering at maturity. The better forms with rose-coloured flowers in well-filled trusses are very desirable garden plants. Introduced 1910-1937, reintroduced 1992->. March-May.

Picture 1

Picture 2

1. *R. balfourianum* (var. *balfourianum*), Napa Hai, Zhongdian, N.W. Yunnan, China.
2. *R. balfourianum* (var. *balfourianum*) Forrest 16811 at the Royal Botanic Garden, Edinburgh.

Picture 3

Picture 4

Picture 5

3. *R. balfourianum* (var. *balfourianum*), Royal Botanic Garden, Edinburgh, showing the large calyx. The only other subsection Taliensia species with comparably large calyces are *R. adenogynum*, *R. bureavii*, *R. bureavioides*, *R. elegantulum*, *R. faberi* and *R. prattii*. Also the rare *R. mimetes*.
4. *R. balfourianum* (var. *aganniphoides*) which has slightly thicker-than-average indumentum for the species, Royal Botanic Garden, Edinburgh.
5. The leaf lower surface of *R. balfourianum*, young leaf above, mature below. (Photo R.B.G., E.)

R. bathyphyllum Balf.f. & Forrest 1919 H5

0.6-1.5m, a dwarf shrub. **Leaves** elliptic to oblong, 4-7 x 1.5-2cm, lower surface with a dense bistrate indumentum, the upper layer dark rufous brown. **Inflorescence** 10-15-flowered. **Corolla** campanulate, white flushed rose with crimson flecks; calyx glabrous, ovary densely tomentose, style glabrous. **Distribution** S.E. Tibet, N.E. Yunnan 3350-4250m, on

TALIENSIA

bouldery slopes, forest margins, rhododendron moorland etc.

Like a large-leaved *R. proteoides* and differing from *R. roxieanum* in its densely tomentose ovary, this is a natural hybrid of *R. proteoides x R. aganniphum* which we have observed both in the wild and coming up in wild-collected *R. proteoides* seed. Quite a handsome plant, probably only recently introduced into cultivation, though plants have been circulating under this name for several years.

Picture 1

1. *R. aganniphum* and *R. proteoides* growing together on the Salween-Mekong divide near Londre. Occasional plants of *R. bathyphyllum* (hybrids of the above 2 species) grew amongst them and many have come up in seed collected off the *R. proteoides*.

R. bhutanense D. G. Long 1989 H5

Height 0.6-3m, forming quite a compact shrub. **Leaves** 6-12.5 x 3-5cm, broadly elliptic to obovate, base *cordate;* upper surface glabrous *except for some loose hairs at the base of the midrib*, lower surface with a slightly granular felted greyish-brown to orange-brown indumentum. **Inflorescence** 10-15-flowered, pedicels *0.8-1.3cm long*. **Corolla** campanulate, 3-3.5cm long, deep pink buds open medium to pale pink, calyx minute, glabrous; stamens *glabrous*; ovary glabrous. **Distribution** Bhutan (+S.E. Tibet?) 3,800m-4,300m (12,500-14,000ft), amongst scrub and other rhododendrons, above the tree line.

The closest relation to this species is *R. phaeochrysum* var. *agglutinatum* under which old herbarium specimens were placed but it differs from this variety in the hairs on the midrib, the shorter petiole and the glabrous stamens. The upright posture of the foliage, as seen in the wild in autumn, is distinctive.

R. bhutanense is a newly-named species which is the most western-occurring of subsection Taliensia (excluding Lacteum Alliance). So far in cultivation it has proven to be difficult, slow growing and fastidious as to soil conditions. Rather doubtfully distinct from *R. phaeochrysum*. Introduced late 1980's. March-May.

Picture 1

Picture 2

Picture 3

1. *R. bhutanense*, Thampe La, Bhutan. (Photo S. Bowes Lyon).
2. *R. bhutanense*, a recently described and newly-introduced species. (Photo S. Bowes Lyon).
3. Foliage of *R. bhutanense*, Rudong La, C. Bhutan.

R. bureavii
Franch. 1887 (*R. cruentum* Lév.) H5

Height 1.7-7.6m, a compact and rounded shrub in the open, more tree-like in shade, usually not over 3m. Branchlets *densely woolly*, cinnamon-rusty; bark roughish, grey-brown. **Leaves** 4.5-19 x 2-7cm, elliptic, retained for 3-4 years; upper surface usually glabrous, *dark green and shiny* when mature, sometimes with a semi persistent indumentum, lower surface with a *unistrate* indumentum, *dense, deep salmon pink turning to rich, rusty-red* at maturity; petiole *1-2cm long, densely tomentose*. **Inflorescence** 10-20-flowered, usually loose. **Corolla** tubular-campanulate, 2.5-4cm long, white flushed rose, to rose; calyx *large, 0.5-1cm long, densely glandular*; ovary *densely glandular, +/- tomentose*; style glandular at least near the base. **Distribution** N. Central Yunnan and N.E. Yunnan, both in limited areas, 3,000-3,900m (10,000-12,750ft), open forests and thickets.

One of the most distinct and easily identified species in subsection Taliensia, its closest relative is *R. bureavioides* which differs in its shorter, bent petiole, the larger corolla and the presence of glands but absence of hairs at the base of the style. *R. bureavii* differs from *R. rufum* in the thick, rusty, woolly tomentum on the branchlets and the larger calyx.

R. bureavii has amongst the finest foliage of all rhododendron species. As it is easier to please in cultivation than most of its relatives, it has become the most popular garden plant of subsection Taliensia. Cultivated plants from N.C. Yunnan have consistently good foliage but Wumeng Shan plants show a considerable variation in quality of foliage probably caused by natural hybridisation. Introduced 1917-1925 and 1994->. April-May.

Picture 1

Picture 2

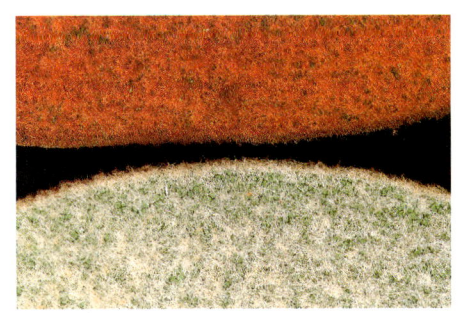

Picture 3

Picture 4

1. *R. bureavii*, Royal Botanic Garden, Edinburgh.
2. Foliage of *R. bureavii* showing the characteristically dark leaves with glabrous to semi-persistent indumentum on the leaf upper surface and the dense rusty-red indumentum on the lower surface. Glendoick.
3. The leaf lower surface of *R. bureavii* showing the mature rusty red indumentum (above) and the pale immature indumentum (below). (Photo R.B.G., E.)
4. *R. bureavii* 'Ardrishaig' A.M. at Glendoick. Although the foliage appears to match that of this species, this clone has much larger flowers than the type and may be a hybrid.

R. pubicostatum T.L. Ming. **Leaf** underside with sparse brown indumentum which is easily rubbed off. **Corolla** light pink. **Distribution** N.E. Yunnan, 3,800-4,000m (12,500-13,000ft) in fir or rhododendron forests. This plant is almost certainly a natural hybrid of *R. bureavii*, probably crossed with *R. sikangense*.

TALIENSIA

R. bureavioides Balf.f. 1920 H5

Height 2-5m, a compact to upright shrub; bark roughish, grey-brown. **Leaves** 10-14 x 4-5cm, oblong-oval; upper surface with indumentum often partly retained, lower surface with a *dense, woolly, indumentum, rusty-red when mature*; petiole *1cm but appearing shorter due to being bent, densely tomentose*. **Inflorescence** c.15-flowered. **Corolla** funnel-campanulate, 4-4.5cm+, white to rose; calyx c.8mm long, floccose and glandular; ovary *glandular, not tomentose*; style glandular at the base, *not tomentose*. **Distribution** W. Central and N.W. Sichuan where it is quite widely distributed, 3,000-3,500m (10,000-11,500ft).

Chamberlain and Davidian state that *R. bureavioides* has bistrate indumentum but it seems to be unistrate in wild and cultivated plants we have observed. It differs from *R. bureavii* in the shorter petiole (giving a very distinctive appearance), larger flowers, and the absence of hairs on the ovary and style. *R. faberi* and *R. prattii* differ in their much thinner indumentum. *R. rufum* has a looser, often thinner indumentum on the lower leaf surface and lacks the thick tomentum on the branchlets and petiole. *R. danbaense* L.C. Hu, with unistrate indumentum, described in *Sichuan Rhododendron of China* is probably a synonym of *R. bureavioides*.

R. bureavioides was apparently not introduced into cultivation before 1986. Plants previously grown under this name do not match the type description or resemble recently introduced plants and are probably *R. bureavii* hybrids. The best forms of *R. bureavioides* should rival *R. bureavii* in foliage and have superior flowers. Natural hybrids of *R. bureavioides* x *R. prattii* with fine foliage have also been introduced. April-May ?

Picture 1

Picture 2

Picture 3

1. *R. bureavioides*, Balang Mountain, N.W. Sichuan. This is quite different to plants erroneously cultivated under this name prior to 1986.
2. Foliage of *R. bureavioides* C. 5076 from central W. Sichuan at Glendoick showing the characteristic short petiole.
3. So-called *R. bureavioides* as cultivated prior to 1986. This does not match wild populations. W. Berg's garden, Hood Canal, Washington State, W. U.S.A.

R. clementinae
Forrest 1915 H5

Height 1-3m. a rounded, usually compact shrub with stiff, stout, usually glabrous branchlets; bark rough, brown to grey-brown. **Leaves** 6.5-14 x 3-8cm, ovate-lanceolate to oval, *usually convex*, retained for 2-5 years; lower surface with a *thick, spongy, whitish to buff, bistrate* indumentum. **Inflorescence** 10-15-flowered. **Corolla** funnel-campanulate, *6-7 lobed*, 4-5cm long, white, usually flushed rose to rose, marked deeper; calyx minute, c. 1mm, stamens *12-14*; ovary and style glabrous. **Distribution** N.W. Yunnan, Zhongdian Plateau southwards and also S.W. Sichuan in Muli area, 3,400-4,300m (11,000-14,000ft), forest margins and moorland.

This is the only member of the subsection with a 6-7 lobed corolla. In the best blue-leaved forms the foliage resembles *R. aeruginosum* which has a 5-lobed corolla and generally lower growth habit. The foliage can also resemble that of some forms of *R. sphaeroblastum* and *R. aganniphum* but these do not have such glaucous young leaves. *R. pronum* also with glaucous foliage is much lower growing with smaller leaves.

Although this species is very slow to flower in cultivation, selected forms of *R. clementinae* are well worth growing for their fine foliage alone. Introduced 1913-1937. March-April.

Picture 1

Picture 2

Picture 3

Picture 4

1. *R. clementinae*, Glendoick, showing the 6-7-lobed corolla.
2. *R. clementinae*, Royal Botanic Garden, Edinburgh.
3. Foliage of *R. clementinae* at Glendoick, showing the convex leaves.
4. The leaf lower surface of *R. clementinae* showing the characteristic thick, spongy indumentum, mature leaf above, young below. (Photo R.B.G., E.)

R. coeloneuron
Diels 1900 H4-5?

Height 3-8m, branchlets *rufous*-tomentose. **Leaves** 7-12 x 2.5-4cm, upper surface somewhat shining and rugulose with impressed veins, lower surface with *dense bistrate* indumentum, upper layer rufous. **Inflorescence** 4-9-flowered, loose. **Corolla** funnel-campanulate, 4-4.5cm long, pink to purplish, spotted; calyx 1-2mm, rufous-tomentose; ovary densely rufous-tomentose, style glabrous. **Distribution** S.E. Sichuan and N. Guizhou, 1,200-2,300m (4,000-7,500ft), in mixed forest.

A confusing species, said to be related to *R. wiltonii* differing in its bistrate rather than unistrate indumentum. In recent years, introductions of what are thought to be typical *R. coeloneuron* have come in from S.E. Sichuan and N. Guizhou. These resemble the recently introduced *R. denudatum* and *R. floribundum* of subsection Argyrophylla rather than *R. wiltonii*. Once seedlings of these introductions are more mature, it should be possible sort out relationships but at present it looks as if *R. coeloneuron* would be better placed in subsection Argyrophylla. Plants found and introduced from Erlang Shan, central-west Sichuan, with thicker indumentum have been referred to *R. coeloneuron* by the Chinese, but these are really nothing more than extreme forms of a variable population of *R. wiltonii*.

Horticulturally this species is likely to require conditions very similar to its probable close relative *R. floribundum*. It should be hardy and make a handsome plant. Introduced 1990s. April-May?

R. elegantulum
Tagg & Forrest 1927 H5

Height 1-1.6m, a usually fairly compact shrub; bark rough, grey-brown. **Leaves** 7-13 x 2.4-3.5cm, elliptic-oblong, retained for 3 years; upper surface soon glabrous, dark green, lower surface with a *dense, fairly woolly, unistrate pinkish-orange* indumentum turning to *rich rufous-brown* in the third year. **Inflorescence** 10-20-flowered, fairly compact. **Corolla** campanulate, 3-4cm long, pale pinkish-purple, sometimes cream, flushed pink with small spots; calyx *large, c. 1.2cm* long, glandular-ciliate; ovary densely glandular; style with a few glands at base. **Distribution** S.W. Sichuan, 3,650-3,950m (12,000-13,000ft), a limited distribution around the tree line.

TALIENSIA

The combination of fairly narrow leaves with a dense woolly indumentum and the freely-produced pink flowers usually make this species fairly easily identifiable.

Free-flowering as a relatively young plant, with fine foliage and fairly easy to please, this makes one of the best garden plants in the subsection. *R. elegantulum* has been rare in gardens until recently but is becoming quite widely grown. Although hand pollinated seedlings are quite uniform, this species is believed to originate as a natural hybrid, perhaps of *R. adenogynum*. Introduced 1922. February-April.

Picture 3

Picture 4

3. Foliage of *R. elegantulum* at Glendoick showing traces of indumentum on the upper surface of the leaves.
4. The leaf lower surface of *R. elegantulum*, showing the dense, woolly indumentum, mature leaf above, young below. (Photo R.B.G., E.)

Picture 1

Picture 2

1. A fine specimen of the best clone of *R. elegantulum* at the Royal Botanic Garden, Edinburgh.
2. The Royal Botanic Garden, Edinburgh's best *R. elegantulum* growing at Glendoick.

R. faberi
Hemsl. 1889 (*R. faberioides* Balf.f., *R. wuense* Balf.f.) H5

Height 1-6m, a dense shrub, sometimes more open with age; bark very rough, greyish-brown. **Leaves** oblong-elliptic, oblong-lanceolate or obovate, 6-11 x 2.8-4.8cm retained for 2-3 years; upper surface glabrous, lower surface with a bistrate indumentum, the upper layer dense, rufous or dark brown, consisting of ramiform hairs, *shedding at maturity* revealing the compacted, whitish lower layer consisting of rosulate hairs; petiole to 2cm, tomentose or woolly. **Inflorescence** 6-12-flowered. **Corolla** campanulate to funnel-campanulate, 3-4cm long, white to pale pink, +/- spotted or blotched; calyx *large 7-10mm* long, moderately-sparsely glandular; ovary densely stipitate-glandular; style glabrous or glandular at base. **Distribution** C. Sichuan with a limited distribution on a few peaks, 3,000-3,400m (10,000-11,000ft).

This is a fairly distinctive species in its dense rounded habit and in the leaf underside with a non-persistent, loose upper layer of indumentum which soon reveals the thin, white, compacted lower layer. It is generally smaller leaved than *R. prattii*.

R. faberi is quite easily cultivated and fairly free-flowering and showy at maturity but the lower-leaf indumentum is slow to develop on young plants and the foliage is not particularly striking. Introduced 1904, reintroduced 1980->. April-May.

Picture 1

Picture 2

Picture 3

1. *R. faberi* from the Valley Gardens, Windsor Great Park, S.E. England. This plant was labelled *R. wuense*.
2. *R. faberi* (*R. wuense*), Glendoick, E. Scotland.
3. *R. faberi* K.R. 193 from Emei Shan, Sichuan, growing at Branklyn, Perth.

R. mimetes Tagg & Forrest 1927 H5

Height 1-2m, a rounded and compact shrub; bark rough, brownish-grey. **Leaves** 6-11 x 3-4.7cm, with *undulating margins*, ovate-lanceolate to elliptic, retained 1-3 years; lower surface with a bistrate indumentum, upper layer matted to loose, fulvous to cinnamon, persistent or evanescent, lower layer compacted, whitish and persistent. **Inflorescence** c.6-7-flowered. **Corolla** funnel-campanulate, 3.5-5cm long, white to rose, spotted; calyx 3-8mm long, lobes narrow, glandular; ovary densely hairy with some glands or glabrous; style glabrous. **Distribution** S.W. Sichuan, 3350-3,650m (11,000-12,000ft), forest margins.

Very doubtfully deserving of specific status, the two varieties of this taxon are almost certainly natural hybrids, not necessarily from the same cross. The name *R. simulans* has been recently found to be an invalid specific name, so at present it remains at varietal status.

Var. *mimetes*

Leaves, lower surface with a *continuous* upper layer of indumentum.

Var. *mimetes* is probably a hybrid of *R. adenogynum*, perhaps with *R. phaeochrysum*. The only clone which we have located in cultivation makes a neat, compact plant but the leaves are inclined to suffer some die-back. Introduced 1921. May.

Var. *simulans* Tagg ex Forrest 1927.

Leaves, lower surface with a *splitting* indumentum.

This is distinguished from var. *mimetes* in the splitting indumentum which resembles that of *R. agganiphum* var. *flavorufum*.

Found in the same area as var. *mimetes*, var. *simulans* seems to be a natural hybrid of *R. sphaeroblastum*, probably with *R. adenogynum*. The only cultivated clone we know of, at the Royal Botanic Garden, Edinburgh, has fine pink flowers.

Picture 1

1. *R. mimetes* var. *mimetes*, Royal Botanic Garden, Edinburgh, showing the undulating leaf margins.

TALIENSIA

Picture 2

2. *R. mimetes* var. *simulans*, Royal Botanic Garden, Edinburgh.

Picture 1

Picture 2

Picture 3

R. nigroglandulosum
Nitzelius 1975 H5

Height 3-5m, upright, spreading shrub. Branchlets with *black glands*; bark grey and light brown, rough. **Leaves** *12-20 x 4-5.5cm* oblanceolate, retained 1-3 years; upper surface midrib and petiole with *black glands*, lower surface with a unistrate lanate, loose woolly tawny indumentum with black glands on the midrib. **Inflorescence** 8-10-flowered. **Corolla** campanulate, 3.5-4cm long, yellowish pink, spotted, opening from carmine buds, or white; calyx small, c.1mm long, tomentose; ovary *glandular and tomentose*; style glabrous. **Distribution** C.W. Sichuan, Tapanshan, N. of Kangding, 3,500m (11,500ft).

Nitzelius stresses that the black glands on the petiole, leaf and branchlets are an important diagnostic feature of this species This characteristic is however also apparent on narrow-leaved *R. phaeochrysum* cultivated at Gothenburg Botanic Garden, also collected by Harry Smith in the same area. This suggests that *R. nigroglandulosum* is a natural hybrid of this species crossed with *R. bureavioides* or *R. rufum* which it also resembles. Only known from H. Smith 13979.

The rare *R. nigroglandulosum* is an interesting plant with quite attractive foliage but rather small flowers. The leaves of some clones are amongst the longest in the subsection. Introduced 1934. March-April.

1. A fine pink form of *R. nigroglandulosum* at Gothenburg Botanic Garden, Sweden. (Photo I. Holmåsen).
2. *R. nigroglandulosum* H. Smith from Gothenburg Botanic Garden, growing at Glendoick, showing the long fairly narrow leaves.
3. *R. nigroglandulosum* H. Smith, growing at Glendoick.

R. phaeochrysum Balf.f. & W.W. Sm 1917 H5

Height 1-4.5m, a compact to upright shrub or small tree; bark rough, shaggy, flaking, greenish-brown. **Leaves** 4-14 x 1-6.5cm, elliptic to ovate-oblong, retained for about 2 years; upper surface dark green, +/- glabrous, lower surface with a dense compacted or felted or agglutinated, pale fawn to rufous-brown indumentum made up of radiate or branched hairs. **Inflorescence** 8-15-flowered. **Corolla** funnel-campanulate, 2-5cm long, white to pink, spotted, sometimes heavily; calyx small c.1mm long, *usually glabrous*; ovary and style *glabrous*.

The extremely variable *R. phaeochrysum* is the commonest subsection Taliensia species in the wild. It merges and/or hybridises with *R. aganniphum*, *R. przewalskii* and with *R. sphaeroblastum*. Hybrids with *R. roxieanum* produce very variable offspring and these appear to be the origin of many of the plants known as *R. alutaceum*. The non-splitting indumentum separates *R. phaeochrysum* from *R. aganniphum* var. *flavorufum* while *R. przewalskii* differs in its thinner or absent indumentum. All these species occupy huge tracts of hillside and are often the only elepidote species present in areas unfavourable to other larger rhododendrons.

Many forms of *R. phaeochrysum* are free-flowering at maturity and make attractive garden plants. Despite sometimes being found in the wild in quite boggy conditions, in cultivation it requires well-drained soil and cool roots as it is particularly susceptible to phytophthora. Introduced 1913->, reintroduced 1980s->.

The species is divided into 3 varieties which are sometimes quite easily distinguishable in the wild but in other populations the varieties merge into one another.

Var. *phaeochrysum* (*R. dryophyllum* Balf.f. & Forrest (part), *R. cupressens* Nitzelius).

Height 1-6m, a rounded to upright shrub or small tree. **Leaves** *8-14.5cm* long, indumentum *felted* not splitting. **Corolla** *larger than in the other varieties, 3-5cm long*. **Distribution** Widespread in N. Yunnan, Central & N. Sichuan and S. Tibet, 3,400-4,700m (11,000-15,500ft), about and above the tree line.

In its best forms, this is perhaps the showiest of the three varieties. Cupressens Group is distinguished by pale indumentum. It is a distinctive plant as seen in Gothenburg and should probably have some status. It appears to be intermediate between *R. phaeochrysum* and *R. aganniphum* or *R. przewalskii*.

Var. *agglutinatum* (Balf.f. & Forrest) D.F. Chamb. 1978 (*R. dumulosum* Balf.f. & Forrest, *R. lophophorum* Balf.f. & Forrest, *R. syncollum* Balf.f. & Forrest.)

Height 1-3m, forming a low, compact and rounded shrub. **Leaves** *4-9cm,* long; lower surface with *agglutinated* indumentum. **Corolla** 2-3.5cm long. **Distribution** Widespread in N. Yunnan, C. Sichuan and S. Tibet.

Completely intergrades with var. *phaeochrysum* and also merges with *R. przewalskii* and *R. aganniphum*. Var. *agglutinatum* can make a tidy and handsome specimen. *R. doshongense*, currently included in *R. aganniphum*, probably fits better under this variety. Introduced 1918->, reintroduced 1990->. March-April.

Var. *levistratum* (Balf.f. & Forrest) D.F. Chamb. 1978 (*R. aiolopeplum* Balf.f. & Forrest, *R. dichropeplum* Balf.f. & Forrest, *R. helvolum* Balf.f. & Forrest, *R. intortum* Balf.f. & Forrest, *R. sigillatum* Balf.f. & Forrest, *R. theiophyllum* Balf.f. & Forrest, *R. vicinum* Balf.f. & Forrest, *R. dryophyllum* Balf.f. & Forrest (part)).

Height 1-4m, a dwarf or more frequently upright shrub, taller than var. *agglutinatum*. **Leaves** 5-14 x 1.4-5.2cm, oblong-lanceolate to lanceolate; lower surface with *felted, not agglutinated, continuous*, usually *fawn through yellowish-brown to light cinnamon* indumentum. **Corolla** 2-3.5cm long. **Distribution** Widespread in N. Yunnan, Central & N. Sichuan and Gansu, 3,000-4,400m (10,000-14,500ft).

Although this variable variety merges with the other two, it is usually separated by the lighter-coloured, felted indumentum and it often makes a looser, more upright plant than var. *agglutinatum*. The flowers are frequently very small, particularly on the forms previously known as *R. dryophyllum*. It also merges with *R. traillianum* and hybridises with *R. roxieanum* producing hybrids which are generally referable to *R. alutaceum* and its varieties. Introduced 1913->, reintroduced 1986->. March-April.

Picture 1

1. *R. phaeochrysum* var. *phaeochrysum*, Big Snow Mountain, N. Yunnan, China, flowering amongst *Larix potaninii*. (Photo I. Sinclair).

TALIENSIA

Picture 2

Picture 6

Picture 3

Picture 7

6. The leaf lower surface of *R. phaeochrysum* var. *agglutinatum* showing the agglutinated indumentum, mature leaf above, young below. (Photo R.B.G., E.)
7. *R. phaeochrysum* var. *levistratum,* Royal Botanic Garden, Edinburgh.

Picture 4

Picture 5

2. *R. phaeochrysum* var. *phaeochrysum* in forest, Tian Bao Shan, N.W. Yunnan. China.
3. A fine form of *R. phaeochrysum* var. *phaeochrysum* with strongly spotted flowers, Tian Bao Shan, N.W. Yunnan, China.
4. *R. phaeochrysum* var. *phaeochrysum*, Tian Bao Shan, Yunnan, China.
5. An exceptionally deep pink *R. phaeochrysum* var. *phaeochrysum* on the Nyima La, S.E. Tibet.

R. prattii Franch. 1895. H5

Height 1-6.5m, a rounded, somewhat compact to rather rangy and open-growing bush or small tree. Branchlets *densely tomentose* and sparsely to moderately glandular. **Leaves**, *broadly-elliptic, elliptic or oblong-elliptic, 8-19 x 5-8cm,* upper surface dark green, usually glabrous, lower surface with a continuous or discontinuous, thin indumentum consisting of long-rayed hairs; petiole to 3.5cm, densely tomentose. **Inflorescence** 6-14-flowered. **Corolla** funnel-campanulate or campanulate, 3.6-5.6cm long, white with red spots; calyx large 5-10mm; ovary glandular and usually hairy; style glabrous, glandular or partly glandular. **Distribution** N. Central to S. Central Sichuan, much more widespread than *R. faberi*, 2,700-4,300m (9,000-14,000ft), usually in forested areas.

The leaves of *R. prattii* are larger and wider than those of its close relative *R. faberi* with a thinner indumentum on the lower leaf surface, and it eventually makes a larger, more open

170

plant. *R. prattii* hybridises freely or merges with *R. bureavioides* in the wild.

The leaves of *R. prattii* are typically the largest in the subsection (excluding the Lacteum Alliance) but the flowers are small compared to the leaf size. It has proven rather shy-flowering in cultivation. Introduced 1904, reintroduced 1989->. April-May.

Picture 1

Picture 2

Picture 3

1. *R. prattii*, Balang Mountain, N.W. Sichuan, China. Apart from in the centre of the picture, the surrounding foliage is all *R. watsonii*.
2. *R. prattii*, Royal Botanic Garden, Edinburgh.
3. *R. prattii* from Corsock growing at Glendoick showing the large leaves compared with the truss size.

R. principis
Bureau & Franch. 1891 (*R. vellereum* Hutch. ex Tagg) H5

Height 2-6m, a bushy shrub or small tree; bark rough, greyish, *vegetative buds pointed*. **Leaves** 6-14 x 1.8-5cm, oblong to ovate-lanceolate, retained for 2 years, margins usually recurved, especially in cultivation; upper surface dark green and shiny, lower surface with a bistrate *thick, spongy, white to fawn* indumentum. **Inflorescence** 10-30-flowered. **Corolla** *more campanulate than most of the subsection*, 2.5-3.8cm long, white through pink, rose to reddish-magenta, spotted or unspotted; calyx *minute, c.1mm long*, glandular-ciliate; ovary and style glabrous. **Distribution** Widespread in S. & E. Tibet, 2,700-4,600m (9,000-15,000ft), in conifer or broad-leaved forest to above the tree-line.

One of the most westerly-occurring subsection Taliensia species, this is a distinctive plant in the combination of its pointed buds, the suede-like, pale indumentum and bell-shaped flowers. It differs from the other paler-indumented species *R. aganniphum* and *R. clementinae* in its pointed buds and narrower pointed leaves and from *R. pronum* in its much greater stature. All plants in cultivation used to be known as *R. vellereum* but the previously described name *R. principis* takes precedence.

R. principis is a fine foliage plant and is pretty in flower, but as both growth and flowers come early, these can be damaged by spring frosts. It tolerates surprisingly dry conditions in the wild. One of the largest-growing species in the subsection. Introduced 1925-1947, reintroduced 1995->. March-April.

Picture 1

1. *R. principis* on the Sirchem La, Rong Chu, S.E. Tibet.

TALIENSIA

Picture 2

Picture 6

6. The leaf lower surface of *R. principis* showing thick spongy indumentum, young leaf above, mature below. (Photo R.B.G., E.)

Picture 3

R. pronum
Tagg & Forrest 1927 H5

Height 15-60cm, a dwarf, prostrate to mounded shrub; branchlets with *persistent perulae*. **Leaves** 4-7.5 x 1.8-2.8cm, elliptic, convex; upper surface glaucous when young, lower surface with a dense greyish to fawn bistrate indumentum. **Inflorescence** 6-10-flowered. **Corolla** funnel-campanulate, 3.5-4.5cm long, white, creamy-yellow to pink, heavily spotted; calyx small 1-2mm long, usually glabrous; ovary and style glabrous. **Distribution** W. Yunnan from a fairly limited area, 3,700-4,600m (12,000-15,000ft), associated with rocks and cliffs.

A distinct species, resembling a dwarf *R. clementinae* or *R. aganniphum*. Its dense, compact habit and bluish or bluish-green foliage make it quite easily recognised.

Picture 4

R. pronum is a much sought-after species for the connoisseur. There have probably only been two clones in cultivation until recently, the R.B. Cooke and Tower Court forms. The former has more glaucous young growth and a more prostrate habit. *R. pronum* is one of the shyest-flowering rhododendron species in cultivation. Introduced 1923-1932. April-May (June-July in the wild).

Picture 5

2. *R. principis*, a group of Ludlow, Sherriff and Elliot 15797 at Glendoick.
3. The best form of *R. principis* Ludlow, Sherriff and Elliot 15797 at Glendoick.
4. *R. principis* on the Sirchem La, Rong Chu, S.E. Tibet.
5. *R. principis* near the Nam La, S.E. Tibet.

TALIENSIA

R. codonanthum Balf.f. & Forrest 1922 (probably not in cultivation). This differs from *R. pronum* in its thinner indumentum and is very likely to be a natural hybrid. The only flowering herbarium specimen has yellow flowers.

R. proteoides Balf.f. & W.W. Sm. 1917
(*R. lampropeplum* Balf.f. & Forrest.) H5

Height 15-30cm (to 1m in the wild), a *slow-growing and very compact* dwarf shrub, eventually forming a low mound; perulae *persistent*. **Leaves** 2-4 x 0.7-1.5cm, elliptic, margins *strongly recurved*; upper surface almost glabrous at maturity, lower surface with a bistrate indumentum, the upper layer, thick, woolly and rufous. **Inflorescence** 5-10-flowered, usually compact. **Corolla** campanulate, 2.5-3.5cm long, white to cream flushed rose, spotted; calyx minute c.0.5mm long; ovary rufous-tomentose; style glabrous. **Distribution** S.E.Tibet, N.W. Yunnan and S.W. Sichuan, 3,700-4,600m (12,000-15,000ft), associated with very steep hillsides, rocks and cliffs.

Easily identified by its small recurved leaves, thick indumentum and low, compact habit. The closest relation of this species is *R. roxieanum* and these two species merge in places where they meet on the mountainside, *R. proteoides* being at the higher elevations. Such plants with narrow leaves up to 8cm long are referable to *R. roxieanum* var. *parvum* Davidian and are also grown under the name *R. proteoides* aff. Hillier. This species also crosses with *R. aganniphum* resulting in *R. bathyphyllum*.

Although very slow to flower, *R. proteoides* is a species prized by the connoisseur for its foliage and habit and is now used in hybridising to produce dwarf hybrids with indumentum. *R. proteoides* comes from steep slopes and cliffs in the wild and appreciates a well-drained soil, rich in organic matter. It may require protection from caterpillars and adult weevils which can destroy the young growth. Introduced 1914-1949, though it is probable that all older plants at present in cultivation date from 1949. Reintroduced 1993->.

Picture 1

Picture 2

Picture 3

Picture 4

1. *R. pronum* on Biluo Snow Mountain, Lushui, Yunnan, at 3,900m. (Photo Kunming Botanical Institute).
2. *R. pronum*, Tower Court (Berg) clone, Glendoick.
3. *R. pronum*, Tower Court (Berg) clone, Glendoick, showing its mound-forming habit.
4. *R. pronum*, Kilbride (Cooke) clone, Glendoick, which has a more prostrate habit than the Tower Court (Berg) clone.

TALIENSIA

Picture 1

Picture 2

Picture 3

Picture 4

Picture 5

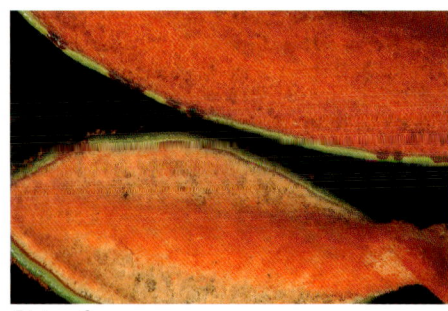

Picture 6

5. The so-called Hillier clone of *R. proteoides* which appears to be intermediate between *R. proteoides* and *R. roxieanum*. Plants resembling this have been seen recently on Bei Ma Shan, N.W. Yunnan, China.
6. The leaf lower surface of *R. proteoides* showing the recurved leaves with thick, rufous indumentum, mature leaf above, young below. (Photo R.B.G., E.)

R. comisteum Balf.f. & Forrest 1919 (probably not in cultivation) is closely related and is almost certainly a natural hybrid between *R. proteoides* and *R. sanguineum* or related species. The deep rose flowers would suggest this cross.

1. *R. proteoides* above Londre, Salween-Mekong divide, N.W. Yunnan, China. (Photo J. Broadus/C. Clark)
2. *R. proteoides* probably Rock 151, Glendoick.
3. A white-flowered form of *R. proteoides* at Glendoick.
4. The attractive young growth of *R. proteoides*, probably Rock 151, Glendoick.

R. przewalskii
Maxim. 1877 (*R. kialense* Franch.) H5

Height 1-2.7m, a slow growing, usually compact shrub. **Leaves** 4.5-10 x 2-4.5cm, broadly elliptic; upper surface sometimes with a trace of juvenile indumentum, giving a metallic, bloom-like effect, lower surface with a *unistrate agglutinated* indumentum, often becoming *glabrous, or always glabrous*. **Inflorescence** 10-15-flowered. **Corolla** campanulate, 1.2-3.5cm long, white (usually) to pale pink, +/- spots; calyx minute c.0.5mm long, glabrous; ovary and style glabrous. **Distribution** N. Sichuan, S.E. Gansu and S.E. Qinghai and perhaps Shaanxi, 3,000-4,300m (10,000-14,000ft), often covering large areas about and above the tree line.

This species is closely related to and merges with *R. phaeochrysum* and also intergrades with *R. aganniphum*. *R.*

przewalskii usually has a distinctive yellow midrib on the leaf upper surface and is distinguished by its very thin or absent indumentum on the lower leaf surface. It is one of the most glabrous members of the subsection.

R. przewalskii is useful for its extreme hardiness but it is one of the least attractive species in the subsection. Introduced 1880, 1904, reintroduced 1986->. April-May.

Ssp. *przewalskii*

Leaves with a thin, compacted whitish to pale brown indumentum on the lower surface. **Corolla** 2.5-3.5cm long.

This variety tends to be the more showy with larger flowers and indumented foliage.

Ssp. *dabanshanense* Fang & Wang 1978

Leaves completely glabrous, leaf lower surface and petiole yellowish green. **Corolla** *very small* for the size of leaf, c. *1-2cm* long.

Picture 1

Picture 2

1. *R. przewalskii*, near Huanglongsi, N. Sichuan, China with *R. capitatum* in the left lower corner.
2. A clone of *R. przewalskii* var. *przewalskii* collected by D. Hummel, from Gothenburg Botanic Garden, Sweden, growing in W, Berg's garden, Hood Canal, Washington State, W. U.S.A.

This variety is well-represented in the collection at Gothenburg Botanic Garden, Sweden. The flowers are so small that it is a plant mainly of interest to collectors.

Picture 3

3. *R. przewalskii* var. *dabanshanense* from the Gothenburg Botanic Garden, Sweden, Glendoick.

R. purdomii
Rehder & E.H. Wilson 1913 H4-5

Height 2-3m?, an upright, fairly compact shrub. **Leaves** oblong lanceolate, 6-9 x 2.5-3.5cm, *margins hairy (in cultivation)*; upper surface *glabrous and shining*, lower surface *glabrous*. **Inflorescence** 10-12-flowered. **Corolla** campanulate, 2.5-3cm long, white to white flushed pink with maroon blotch; calyx c. 1mm, sparsely pubescent; ovary sparsely white-villous; style glabrous. **Distribution** Shaanxi.

The status and validity of this species is very confused, both from a taxonomic point of view due to there being only one poor herbarium specimen and from a horticultural one as there are variable plants labelled *R. purdomii* in cultivation. Chamberlain declines to assign this taxon to any subsection, while Davidian places it in series Taliense, subseries Taliense, admitting it to be aberrant.

Some of the cultivated plants under the name *R. purdomii* have been referred to *R. oreodoxa* var. *shensiense* q.v. and to *R. maculiferum,* two species which also occur in Shaanxi province. Cultivated plants seem to be hardy and quite free-flowering. April-May.

TALIENSIA

Picture 1

Picture 2

1. A fine specimen of a plant labelled *R. purdomii* at Holehird, N.W. England. (Photo P. Bland)
2. *R. purdomii?* at Holehird. (Photo P. Bland)

R. roxieanum
Forrest 1915 H5

Height 0.3-4m but usually 0.6-1.50m, a dwarf and compact to upright and leggy shrub or small tree. Branchlets woolly with +/- *persistent perulae*; bark a little shaggy, brownish-grey. **Leaves** 5-12 x 0.6-4cm, *linear to elliptic, 2-15 times as long as broad*, retained for up to 3 years; upper surface usually dark shiny green, lower surface with a bistrate, thick woolly fawn to rufous indumentum. **Inflorescence** 6-15-flowered. **Corolla** funnel-campanulate, 2-4cm long, white or creamy, often flushed rose, usually spotted; calyx small 0.5-2mm long, *tomentose and glandular*; ovary *densely rufous-tomentose and glandular*; style glabrous or hairy at base. **Distribution** Tibet-Yunnan border, N.W. Yunnan, S.W. Sichuan, 3,400-4,300m (11,000-14,000ft), above and below tree line.

This species is related to the smaller-leaved *R. proteoides* and the two species merge in some locations (var. *parvum*). Plants grown under the name *R. globigerum* or *R. alutaceum* Globigerum Group are better considered forms of *R. roxieanum*. *R. roxieanum* is very variable in habit, and in size and shape of leaf. April-May.

We have recently had the opportunity to study *R. roxieanum* in the wild. In the population at Lao Chun Shan, N.W. Yunnan, forms referable to var. *cucullatum* occur at the highest altitudes. Lower down these gradually evolve into var. *roxieanum* and surprisingly, lower still into the narrowest-leaved form var. *oreonastes*. On Bei Ma Shan, at high altitudes, *R. roxieanum* grows just below *R. proteoides*. Davidian's var. *parvum* appears to cover intermediate forms between these two species. We have seen natural hybrids of *R. roxieanum* crossed with *R. aganniphum, R. balfourinaum, R. selense, R. phaeochrysum,* and *R. vernicosum.*

Var. *roxieanum* (*R. aishropeplum* Balf.f. & Forrest, *R. poecilodermum* Balf.f. & Forrest. *R. globigerum* as cultivated)

Leaves 4-12 x *1-2.3cm*, lanceolate, oblanceolate, ratio length to breadth *1:4-8*.

These include forms intermediate between the other two varieties. Classifying the very variable *R. roxieanum* into varieties is largely for the convenience of gardeners as so many intermediate forms exist. Some forms bloom relatively young while others take many years. What is cultivated under the name *R. globigerum* probably best fits here. Introduced 1913-1937, reintroduced 1989->. April-May.

Var. *cucullatum* (Hand.-Mazz). D.F. Chamb. 1978 (*R. coccinopeplum* Balf.f. & Forrest, *R. porphyroblastum* Balf.f. & Forrest) H5

Leaves *oblong-lanceolate to oblanceolate, 2-4cm wide*. ratio length to breadth *1:2.2-4*. Indumentum *thick and spongy*.

This variety covers the wider leaved variations of *R. roxieanum*. In some locations the plants are quite uniform but recent field studies in N.W. Yunnan have revealed large, very variable populations. Shy-flowering in the wild and in cultivation. Introduced 1918-1932?, reintroduced 1991->. April-May.

TALIENSIA

Var. *oreonastes* (Balf. f. & Forrest) Davidian 1992
(*R. recurvum* Balf.f. & Forrest, *R. roxieanum* var. *parvum* Davidian)

Leaves *linear or linear-lanceolate, 2-8mm wide*, ratio length to breadth *1:8-15*.

This variety which includes the narrowest leaved-forms is a very popular garden plant, well worth growing for its foliage alone. Sometimes free-flowering from a young age but often takes many years to flower. Plants under the name var. *parvum* or Parvum Group are intermediate between *R. roxieanum* and *R. proteoides*. Introduced 1914->. Reintroduced 1994->.

Picture 1

Picture 2

Picture 3

Picture 4

Picture 5

Picture 6

Picture 7

1. *R. roxieanum* var. *oreonastes* at Lao Shun Shan, N.W. Yunnan, China.
2. *R. roxieanum* var. *oreonastes* Rock 59222, Glendoick.
3. *R. roxieanum* var. *oreonastes* Rock 59221, ex Branklyn, growing at Glendoick.
4. *R. roxieanum* var. *roxieanum* on Lao Shun Shan, N.W. Yunnan, China.
5. *R. roxieanum* var. *roxieanum* on Lao Shun Shan, N.W. Yunnan, China.
6. *R. roxieanum* var. *cucullatum*, Muli, S.W. Sichuan, China, showing the rich cinnamon-coloured indumentum on relatively wide leaves. (Photo W. Berg)
7. A wide-leaved form of *R. roxieanum* var. *cucullatum* on Lao Shun Shan, N.W. Yunnan, China.

TALIENSIA

Picture 8

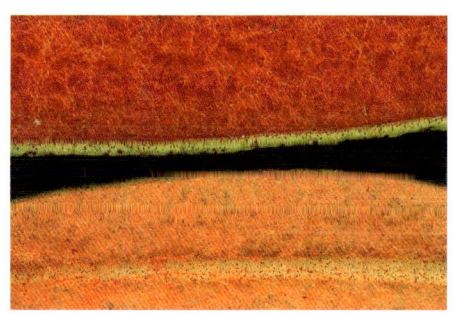

Picture 9

8. Leaf variation of all three varieties of *R. roxieanum* from Lao Shun Shan, N.W. Yunnan, China.
9. The leaf lower surface of *R. roxieanum* var. *roxieanum* showing the thick woolly indumentum, mature leaf above, young below. (Photo R.B.G., E.)

R. rufum

Batal. 1891 (*R. weldianum* Rehder & E.H. Wilson). H5

Height 1.3-6m, a usually fairly compact and rounded shrub, or small tree. Branchlets *glabrous or thinly* tomentose; bark greyish, slightly rough. **Leaves** 6.5-11 x 2.3-5cm, narrowly obovate to elliptic, retained for c. 2 years; upper surface shiny, not rough, lower surface with a *bistrate indumentum*, thin to dense reddish-brown, whitish when young; petiole moderately tomentose at first, *later almost glabrous*. **Inflorescence** 6-12-flowered, fairly compact. **Corolla** funnel-campanulate, 2-3.2cm long, white to deep pink, sometimes striped; calyx *very small c.0.5mm* long; ovary densely reddish-tomentose with a few glands below style; style usually glabrous. **Distribution** N. Sichuan and S. Gansu, 2,400-4,000m (8,000-13,000ft), below and around the tree line.

This is a distinct species, possibly most closely related to *R. nigroglandulosum* which differs in its unistrate indumentum. (*R. nigroglandulosum* may in fact be a natural hybrid of *R. rufum*). Plants under Weldianum Group tend to have a relatively dense indumentum and oval or almost orbicular leaves. *R. rufum* bears a resemblance to *R. bureavii* and *R. bureavioides*, differing in the thin or absent tomentum on the branchlets, the thinner bistrate indumentum on the leaf lower surface, the almost glabrous petiole and the smaller calyx. It hybridises in the wild with *R. watsonii* and *R. przewalskii*.

R. rufum is very hardy and makes a fine foliage plant, though perhaps not as good as *R. bureavii* and *R. bureavioides*. It takes up to 6 years from seed to develop its indumentum and the flowers can be rather small. Introduced 1917-1926, reintroduced 1986->. April.

Picture 1

Picture 2

Picture 3

1. A bank of *R. rufum*, Huanglongsi, N. Sichuan, China.
2. *R. rufum*, Huanglongsi, N. Sichuan, China.
3. Foliage of *R. rufum* showing the almost glabrous branchlets. Huanglongsi, N. Sichuan, China.

Picture 4

Picture 5

4. *R. rufum* Weldianum Group, Mongbi Mountain, N.W. Sichuan, China. This group differs from the type in the thicker indumentum and in the typically oval leaves.
5. The leaf lower surface of *R. rufum* showing the bistrate, somewhat loose indumentum, white on the young leaf, reddish-brown on the mature leaf. (Photo R.B.G., E.)

R. sphaeroblastum
Balf.f. & Forrest 1920 H5

Height 1-3, rarely to 7m, a compact rounded shrub which can grow into an upright and straggly small tree with age; bark very rough, brown to greyish-brown. **Leaves** 6-12 x 3.6-6.2cm, narrowly obovate to elliptic, retained for 2 years; upper surface dark shiny green, lower surface with a bistrate indumentum, upper layer *dense, felted,* reddish-brown when mature, usually white when young, lower layer compacted; petiole 1-1.5cm long. **Inflorescence** 10-20-flowered. **Corolla** funnel-campanulate, 3-4cm long, white, sometimes to pink, spotted or lined; calyx quite small, 1.5-2mm long, *glabrous*; ovary, stamens and style *glabrous*. **Distribution** Mostly S.W. Sichuan, also N.W. Yunnan, 3,000-4,500m (10,000-14,500ft), below and above tree line.

This species is most likely to be confused with *R. phaeochrysum*, from which it differs in its more felted, bistrate indumentum, and with *R. taliense* which has narrower leaves. *R. sphaeroblastum* can also look similar to some forms of *R. clementinae* which differs in its 7-lobed corolla and which usually has more convex leaves and a paler indumentum.

R. sphaeroblastum is a fine species in both foliage and flower and although rare in gardens, it deserves to be more widely grown. Introduced 1904-1932. April-May

Var. *wumengense* K.M Feng 1992

Recently described by the Chinese, differing from the above in the following characters: **leaves** *elliptic to oblong-obovate*, lower surface with soft, light yellow-brown, woolly indumentum, petiole *2-2.5cm* long. **Inflorescence** 8-16-flowered. **Distribution** N.E. Yunnan at 3,750-3,900m (12,250-12,750ft)

This variety has recently been introduced from N.E. Yunnan, where it is plentiful and variable with some leaves almost round. It may not warrant varietal status. Introduced 1994,95.

Picture 1

Picture 2

1. *R. sphaeroblastum* at Glendoick, showing the typical neat trusses of white flowers.
2. *R. sphaeroblastum*, Blackhills, N.E. Scotland.

TALIENSIA

Picture 3

Picture 4

3. *R. sphaeroblastum*, Blackhills, N.E. Scotland, showing the characteristic dark and shiny leaves.
4. *R. sphaeroblastum* var. *wumengense* foliage, Wumeng Shan, N.E. Yunnan.

R. taliense
Franch. 1886 H5

Height 0.8-4m, a dwarf and compact to upright shrub which can get leggy with age; bark brown to grey, rough. **Leaves** 5-11 x 2-4cm, oblong-ovate to broadly lanceolate, retained for 3-4 years; upper surface dark green, usually glabrous, lower surface with a *dense, sometimes very thick, tawny-brown to rufous,* bistrate indumentum, sometimes divided but not split into sections. **Inflorescence** 10-20-flowered. **Corolla** funnel-campanulate, 3-3.5cm long, creamy to pinkish-white, spotted and often flushed deeper; calyx small, 0.5-2mm long, *glabrous*; ovary and style *glabrous*. **Distribution** W. Yunnan, 3,000-4,000m (10,000-13,000ft), mostly above the tree line.

This species is characterised by its tomentose branchlets, continuous bistrate indumentum and glabrous, eglandular ovary. It differs from *R. sphaeroblastum* in its narrower leaves and from *R. alutaceum* in its generally wider leaves as well as its glabrous ovary.

R. taliense has been reintroduced several times in recent years and is becoming much more widely grown. The best forms are those with creamy flowers and those with the thickest rufous indumentum. The form of this species which has been in cultivation for many years has a fairly dense tawny-brown indumentum. Another form, which grows amongst the former, has a very thick rufous indumentum and was introduced under SBEC 0350 in 1981 (as *R. roxieanum* var. *cucullatum*). This taxon appears to be intermediate between *R. roxieanum* and *R. taliense* particularly as it has persistent perulae. When it eventually flowers its ovary should indicate where it should be placed. Introduced 1910-1932, reintroduced 1981->. April-May.

Picture 1

Picture 2

1. *R. taliense*, Cangshan, C.W. Yunnan, China, with view looking down to Dali Plain.
2. *R. taliense*, Cangshan, C.W. Yunnan, China.

TALIENSIA

Picture 3

Picture 4

3. *R. taliense* SBEC 0555 with a thicker-than-average indumentum, Glendoick.
4. The leaf lower surface of *R. taliense* showing the dense tawny-brown indumentum, darker on the mature leaf than on the young. (Photo R.B.G., E.)

R. traillianum
Forrest & W.W. Sm. 1914 H5

Height (0.6-)1.20-8m, a stiff, broadly upright shrub or small tree; bark rough, grey to greyish-brown. **Leaves** 7-13 x 3-6.5cm, obovate to elliptic, retained for 3-4 years; upper surface usually dark green, lower surface with an unistrate indumentum consisting of long or short-rayed radiate hairs. **Inflorescence** 6-15-flowered. **Corolla** funnel-campanulate, 2.5-4.5cm long, white, sometimes flushed rose to rose, with or without spots or blotch; calyx small c.1mm long, glabrous; ovary glabrous or sparsely red-brown tomentose; style glabrous.

This species is closely related to the variable *R. phaeochrysum* but is far more uniform in leaf shape and general appearance and usually differs in its fine, powdery rather than agglutinated or felted indumentum (var. *traillianum* only). *R. traillianum* usually has wider leaves with a deeper-coloured indumentum than *R. phaeochrysum* var. *levistratum*.

Var. *traillianum* (*R. aberrans* Tagg & Forrest)

Leaf apex *apiculate*; lower surface with indumentum consisting of *short and pear-shaped* hair arms, which gives a a powdery appearance to the naked eye. **Corolla** *2.5-3.5cm long*; ovary tomentose or *glabrous*. **Distribution** S.W. Sichuan, S.E. Tibet and W. Yunnan, 3,000-4,600m (10,000-15,000ft), in the open and on margins of conifers, common.

Var. *traillianum* is a handsome foliage plant which is fairly easy to cultivate but the flowers may be small in contrast to the leaves. Introduced 1910-1937, reintroduced 1986->. April-May.

Var. *dictyotum* (Balf.f. & Tagg) D.F. Chamb. 1978 H5

Leaf apex apiculate to *acuminate*. Lower surface indumentum with *long ribbon-like hair arms* giving a thin, felted layer. **Corolla** *3.5-4.5cm long*; ovary *always tomentose*. **Distribution** S.E. Tibet, W. Sichuan and N.W. Yunnan, 3,400-4,300m (11,000-14,000ft).

Further fieldwork needs to be done to ascertain the status of this plant as most cultivated material under this name is incorrectly labelled. It may not actually be directly related to *R. traillianum* being perhaps derived from natural hybrids between *R. beesianum* and *R. phaeochrysum*. Introduced 1917-1937. April-May.

Picture 1

1. *R. traillianum* var. *traillianum*, Glendoick.

Picture 2

Picture 3

Picture 4

2. *R. traillianum* var. *traillianum* Rock 18444, with an unusually heavily blotched corolla, Royal Botanic Garden, Edinburgh.
3. A pure white, unspotted from of *R. traillianum* var. *traillianum*, Glendoick.
4. The leaf lower surface of *R. traillianum* var. *traillianum* showing the dense, powdery indumentum, mature leaf above, young below. (Photo R.B.G., E.)

R. wasonii
Hemsl. & E.H. Wilson 1910 H5

Height 0.6-2m, a compact shrub when young, becoming open with age; bark roughish, grey-brown. **Leaves** 5-10 x 2.5-4.5cm, *ovate-lanceolate*, retained 2-3 years; upper surface dark green, lower surface with a unistrate, sparse to dense, pale to dark, reddish-brown woolly indumentum. **Inflorescence** 8-15-flowered, fairly loose. **Corolla** open-campanulate, 3-3.5cm long, *yellow* to white to pink, lightly spotted, of thin texture; calyx small 0.5mm long; ovary *densely rust-red tomentose*; style glabrous or villose. **Distribution** C. Sichuan only, 2,300-3,700m (7,500-12,000ft) and probably higher, in forest and around the tree line.

This is a distinct species with no close relatives, characterised by the leaf shape, the colour and texture of the indumentum, the papery corolla and the tomentose ovary. Some of the pink and pink-fading-to-white-flowered forms have a less reddish indumentum (Rhododactylum Group).

R. wasonii is a fine plant, particularly in its yellow-flowered forms, and it is relatively easy to grow in cool climates. Plants under Rhododactylum Group (usually Wilson 1876) have pink or white-tinged-pink flowers. Introduced 1904-1911, reintroduced 1989. April-May.

Var. *wenchuanense* L.C. Hu

Leaves lower surface with a thin, compacted and sub-agglutinate indumentum. **Corolla** funnel-campanulate, 2-3.5cm long, pink fading to white. style villose at the base. **Distribution** C.W. Sichuan, Wenchuan County, 2,300-3,520m (7,500-11,500ft).

This is a confusing taxon as the description of the indumentum as thin, compacted and sub-agglutinate does not match any wild or cultivated material we have seen or the photographs which claim to illustrate this taxon in *Sichuan Rhododendrons of China* which clearly show a thick woolly indumentum. The type specimen was probably a *R. wasonii* x *R. phaeochrysum* natural hybrid.

R. wasonii affinity.

Leaves smaller than the type, with a thicker, darker and rougher indumentum. **Corolla** creamy-pink fading to white. **Distribution** C.W. Sichuan (known from a single mountain) 3500-3,700m (11,000-12,000ft), above the tree-line.

Deserving botanical recognition at least at varietal status, this taxon from Kangding County, Sichuan, long cultivated as *R. wasonii* aff. or *R. bureavii* aff. McLaren AD 106, introduced in 1932, is a fine, slow-growing plant with a striking indumentum. It is well-illustrated on p. 172-3 of *Sichuan Rhododendron of China* but wrongly under the name var. *wenchuanense* whose description does not match the pictures. Introduced 1932. Reintroduced 1990. April-May.

TALIENSIA

Picture 1

Picture 2

Picture 3

Picture 4

4. *R. wasonii* affinity Cox 5046, with indumentum thicker than the type, collected in the same locality as AD 106, above Kangding, C.W. Sichuan, China.

R. inopinum Balf.f. 1926.

Leaves with small patches of unistrate indumentum on the lower surface. **Corolla** white, cream or pale yellow, flushed pale pink with a crimson blotch and spotting.

The type specimen was raised from *R. wasoni* Rhododactylum Group Wilson 1876 seed and is a natural hybrid, probably crossed with *R. pachytrichum*.

Picture 5

5. *R. inopinum*, a natural hybrid, probably of *R. wasonii* x *R. pachytrichum*. This form from the Royal Botanic Garden, Edinburgh, growing at Glendoick.

1. *R. wasonii*, a pale yellow-flowered form, Royal Botanic Garden, Edinburgh.
2. *R. wasonii* Rhododactylum Group W. 1876, with white flushed pink flowers, Glendoick.
3. *R. wasonii* affinity McLaren AD 106, Glendoick. The flower colour fades when fully out.

R. wiltonii
Hemsl. & E.H. Wilson 1910 H5

Height 1-4.5m, a broadly upright shrub; bark light to dark brown roughening with age. **Leaves** 5-12 x 1.5-4cm, oblanceolate to broadly elliptic, retained 2-3 years; upper surface *bullate* with deeply impressed veins, lower surface with a unistrate indumentum, *fairly dense, cinnamon to rusty-red*, consisting of fasciculate to ramiform hairs. **Inflorescence** c.10-flowered, fairly loose. **Corolla** campanulate, 3-4cm long, white to pink, spotted and blotched; calyx minute, 1mm long,

TALIENSIA

tomentose; ovary *densely rusty-red tomentose*; style glabrous or hairy at base. **Distribution** W. Sichuan, 2,400-3,400m (8,000-11,000ft), in forests, often confined to cliffs and edges above cliffs.

This species is most easily identified by its deeply-veined leaf upper surface. *R. wiltonii* is such a distinct species with little affinity with the rest of the subsection it would perhaps be better placed in its own subsection. Most plants found on Erlang Shan, C. Sichuan, have relatively obovate leaves with a thicker indumentum and have been (incorrectly) referred to *R. coeloneuron* by the Chinese.

R. wiltonii is an attractive plant with unusual foliage, free flowering with age and quite easy to grow. Most of the recently introduced Erlang Shan plants should have a thicker indumentum. Introduced 1904-1910, reintroduced 1980s->. Erlang Shan form 1990-1992. April-May.

Picture 1

Picture 2

Picture 3

Picture 4

3. *R. wiltonii* a form from Blackhills, N.E. Scotland, showing the bullate leaves.
4. *R. wiltonii* Cox 6150, Erlang Shan, C. Sichuan, China.

R. paradoxum is a natural hybrid of *R. wiltonii*, almost certainly crossed with *R. pachytrichum*, raised in Edinburgh from seed of Wilson 1353. Many seedlings of recent introductions from Erlang Shan appear to be the same cross as *R. paradoxum*. Introduced 1908, reintroduced 1990->.

Picture 5

5. *R. paradoxum*, a natural hybrid between *R. wiltonii* and *R. pachytrichum*. Glendoick.

1. A fine specimen of *R. wiltonii*, W. Berg's garden, Hood Canal, Washington State, W. U.S.A.
2. *R. wiltonii*, Glendoick.

Lacteum Alliance

This is a group of allied species which was part of the old Lacteum series and which shows relationships with subsections Fulva and Campanulata. These species are rather unsatisfactorily placed in subsection Taliensia and there is a good case for assigning them to a separate subsection.

Height 1.8-9m, broadly erect shrubs or small trees with stout stiff branches. **Leaves** thick and leathery with a thin compacted, grey-brown indumentum. **Inflorescence** compact. **Corolla** widely-campanulate, pink, white or yellow; ovary tomentose; style glabrous.

In cultivation these four species are notoriously difficult to grow; small seedlings are especially tricky and mature plants are more-than-averagely susceptible to root pathogens and require a very acid soil. They undoubtedly grow best in relatively dry, cool gardens and, where scion material is available, grafting them certainly helps make them easier to grow and longer-lived.

R. beesianum

Diels 1912 (*R. colletum* Balf.f. & Forrest, *R. emaculatum* Balf.f. & Forrest, *R. microterum* Balf.f.) H4-5

Height 1.8-9m, a broadly erect, stiff shrub to small tree. Branchlets *stout;* bark roughish, greenish-brown. Flower and growth buds *sticky*. **Leaves** *9-29 x 2.6-8.2cm*, oblanceolate to elliptic, retained 1-2 years; upper surface +/- glabrous, lower surface with a *unistrate, thin, compacted, grey, fawn to brown* indumentum of *radiate* hairs. **Inflorescence** 10-20-flowered, flat-topped to rounded. **Corolla** broadly campanulate, 3.5-5.3cm long, *white through pink, rose, purplish to nearly red*, with or without spots or blotch; calyx 0.5-1mm long, glabrous; ovary densely white to brown tomentose; style glabrous. **Distribution** N.W. Yunnan, S.W. Sichuan, S.E. Tibet and N.E. Upper Burma, 3,000-4,400m (10,000-14,500ft), preferring conifer forest but also sheltered rocky slopes, often as the only species present.

The flower colour separates this species from *R. wightii* and *R. lacteum. R. lacteum* has a more elliptic leaf shape and does not have sticky buds. Separated from *R. dignabile* by its continuous indumentum.

R. beesianum is common in the wild but rare in cultivation, due to difficulties of culture. It grows best when grafted and is a handsome plant when well grown, preferring shelter from wind and some shade. It is variable in the colour of flower; pure white and deep pink forms are especially fine. Introduced 1913-1949, reintroduced late 1980s->. April-May.

Picture 2

Picture 3

Picture 4

1. *R. beesianum*, Tian Chi, Zhongdian, N.W. Yunnan, China, typically growing in conifer forest.
2. A fine deep pink *R. beesianum*, near Napa Hai, Zhongdian, N.W. Yunnan, China.
3. *R. beesianum*, Tian Chi, Zhongdian, N.W. Yunnan, China.
4. A fine pure white-flowered *R. beesianum*, Tian Chi, Zhongdian, N.W. Yunnan, China.

TALIENSIA

Picture 5

5. *R. beesianum* showing the typical sticky bud of this species. The petioles are only sometimes coloured. (Photo R.B.G., E.)

R. dignabile
Cowan 1937 H5?

Height 1.8-9m, shrub or small tree. **Leaves** 7.5-18 x 4-6.5cm, elliptic to obovate-lanceolate, base cordate to rounded, upper surface +/- glabrous, lower surface with a *thin, barely visible layer of brown indumentum of hairs and glands*. **Inflorescence** 8-15-flowered. **Corolla** campanulate, 2.5-4.5cm long, pink, cream, lemon-yellow or white, with or without blotch or spots; calyx 0.5-3mm, usually glabrous; ovary glabrous or floccose; style glabrous, occasionally glandular near base. **Distribution** S.E. Tibet, 3,350-4,550m (11,000-14,750ft), rhododendron thickets, conifer forest and rocky slopes.

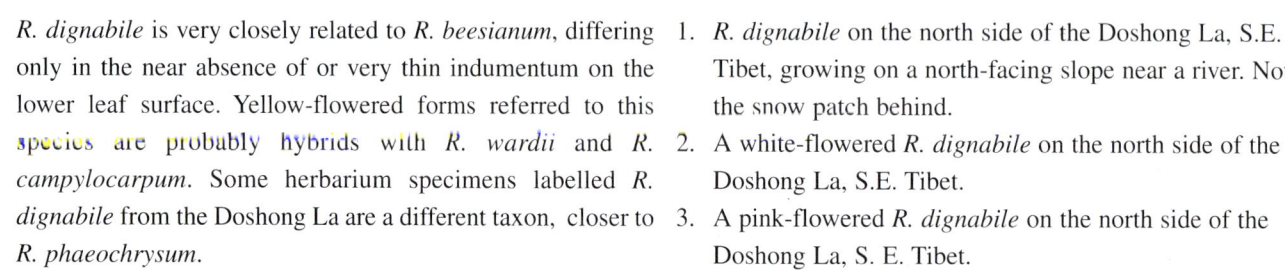

R. dignabile is very closely related to *R. beesianum*, differing only in the near absence of or very thin indumentum on the lower leaf surface. Yellow-flowered forms referred to this species are probably hybrids with *R. wardii* and *R. campylocarpum*. Some herbarium specimens labelled *R. dignabile* from the Doshong La are a different taxon, closer to *R. phaeochrysum*.

As this species was only introduced in 1995, nothing is known about its behaviour in cultivation but, as it is botanically so close to *R. beesianum*, it is likely to prove just as difficult to grow. The best forms seen in the wild were quite showy. April-May?

Picture 1

Picture 2

Picture 3

1. *R. dignabile* on the north side of the Doshong La, S.E. Tibet, growing on a north-facing slope near a river. Note the snow patch behind.
2. A white-flowered *R. dignabile* on the north side of the Doshong La, S.E. Tibet.
3. A pink-flowered *R. dignabile* on the north side of the Doshong La, S. E. Tibet.

R. lacteum
Franch. 1886 (*R. mairei* Lév.) H4-5

Height 2-7.5m, usually considerably less in cultivation, a stiff, often sparsely-branched shrub or small tree. Branchlets *stout*; bark slightly rough, greyish-brown. **Leaves** leathery, dull green, 8-17 x 4.5-10cm, elliptic to obovate, *base rounded to cordate*, retained 1-2 years; upper surface glabrous when mature, lower surface with a thin, suede-like, grey-brown unistrate indumentum. **Inflorescence** 15-30-flowered, usually full. **Corolla** broadly campanulate, 4-5cm long, c. 6.3cm across, *pale to clear canary-yellow*, with or without a blotch; calyx minute, 0.5-1mm long, usually glabrous; ovary densely tomentose; style glabrous. **Distribution** W. Central and N.E. Yunnan, 3,000-4,000m (10,000-13,000ft), just below and above tree line.

The only close relations to this species appear to be *R. beesianum*, *R. dignabile* and *R. wightii*. *R. lacteum* differs from the first two species in the leaf shape and flower colour and from the last in its non-sticky buds, stouter branchlets, the larger truss and the flowers of a deeper colour. *R. lacteum* collected in N.E. Yunnan in 1994-95 had larger-than-average, apple-green leaves and usually had dark red buds and petioles.

When healthy and flowering freely, *R. lacteum* is one of the finest yellow-flowered species. It is hard to grow well and is most likely to be longer-lived as a grafted plant. It prefers light shade and a very acid soil, even being successful in pure sphagnum as long as it doesn't become waterlogged. The depth of flower colour varies from year to year and it seems to be deeper-coloured in the wild. Introduced 1910-1931, reintroduced 1981->. April-May.

Picture 1

Picture 2

Picture 3

Picture 4

1. *R. lacteum*, Cangshan, C.W. Yunnan, China.
2. *R. lacteum*, a form with a small corolla blotch, Cangshan, C.W. Yunnan, China.
3. *R. lacteum*, an unblotched form. Royal Botanic Garden, Edinburgh.
4. *R. lacteum* SBEC 0235, with a large blotch on the corolla, Glendoick.

R. wightii
Hook.f. 1851 H5

The following description is for the true species only. Height 2-6m, a fairly open and rigid shrub or small tree, very slow growing; branchlets fairly stout; bark greenish-brown to brown; leaf and flower buds *sticky*. **Leaves** 5-18 x 5-7cm, *usually flat,* broadly elliptic to obovate, retained 1-2 years; upper surface dark green and glabrous, lower surface with a somewhat dense, *compacted* indumentum consisting of radiate hairs, pale fawn, turning brown with age. **Inflorescence** 12-20-flowered. **Corolla** campanulate, 5-lobed, 3-4cm long, white, cream to lemon-yellow; calyx minute, 0.5mm long, glabrous; ovary densely red-brown tomentose; style glabrous. **Distribution** Bhutan, Arunachal Pradesh, S. Tibet, 3,400-4,500m (11,000-14,750ft), at about and above the tree line.

This species is geographically isolated from its relatives (except *R. bhutanense*), sharing the sticky buds of *R. beesianum* but differing in the flower colour. *R. wightii* differs from *R. lacteum* in leaf-shape and in the smaller flowers.

True plants of *R. wightii* are very rare in cultivation and it is apparent that there are no plants older than those from the 1971 B.L.M. introduction. It is very slow-growing and liable

TALIENSIA

to die off, particularly as a small seedling, and for the best hope of success it should be grafted if material can be obtained. It is reported to be variable in flower quality in the wild, those with clear yellow flowers being the most attractive. March-April.

Picture 1

Picture 2

Picture 3

Picture 4

4. The indumentum on lower leaf surface and the characteristic sticky bud of the 'true' *R. wightii*.

R. wightii hybrid, H4

Height 2-6m, forming an open, *ungainly* shrub; bark grey. Buds *not* sticky. **Leaves** 9-25 x 3.5-6.5cm, with *convex* margins, obovate, indumentum *much looser* than the type, and of the *same light rusty* colour on young and mature leaves. **Inflorescence,** *lax, one-sided.* **Corolla** *5-7-lobed, slightly ventricose*, pale yellow.

There is now no doubt that the plant we have known in cultivation as *R. wightii* is a hybrid originating at Littleworth in Surrey. The other parent is probably *R. falconeri*, which was introduced from the Himalaya at the same time. The slightly bullate leaf and 5-7-lobed ventricose corolla indicate this parentage. Whether this hybrid originates from wild seed (unlikely owing to altitudinal difference between the putative parents) or cultivated seed, nobody knows but it is a much easier plant to grow than the true species and is found in many gardens. It should probably be given a cultivar name.

Picture 5

1. The 'true' *R. wightii*, on the Sikkim-Nepal border. (Photo G.F. Smith).
2. *R. wightii*, Rudong La, C. Bhutan. (Photo Roy Lancaster).
3. *R. wightii*, B.L.M. 92, Glendoick, showing the full truss.
5. The so-called *R. wightii* of cultivation which is a hybrid, perhaps *R. wightii* x *R. falconeri*, showing the lax one-sided truss, Glenarn, W. Scotland.

Subsection Thomsonia

Height 0.8-14m, rounded to upright shrubs or small trees; bark *smooth or peeling*. **Leaves** orbicular, broadly elliptic, obovate to oblong to oval, mostly *glabrous* at maturity; lower-surface sometimes strongly papillate, or with a thin dendroid indumentum, or with red, punctate hair bases which appear as small red flecks when viewed with a hand lens. **Inflorescence** 1-12-flowered, usually loose. **Corolla** campanulate to tubular campanulate, *with 5 nectar pouches*, crimson to pink, white and yellow; calyx *usually large*; stamens 10, ovary glabrous to glandular or tomentose; style usually glabrous; capsule straight. **Distribution** E. Nepal eastwards to just into Sichuan.

This subsection has been found to be biochemically distinct from subsection Campylocarpa which was previously included in the Thomsonii Series. The species in subsection Campylocarpa lack the nectar pouches of the species in subsection Thomsonia and none have red flowers. *R. sherriffii* is aberrant in this subsection due to its dense indumentum and it could be more closely related to *R. fulgens* with which it was formerly associated.

Most of the species in this subsection have showy flowers and often fine bark but some are unfortunately susceptible to powdery mildew infection. *R. thomsonii* itself is widely grown and most of the others are commonly found in good collections.

R. cerasinum Tagg. 1931. H4

Height 1.8-3.6m, an upright but rounded shrub. **Leaves** 5-10 x 2-4cm, oblong to oblanceolate to oblong-elliptic, +/- rounded at both ends, retained for one year; upper and lower surfaces glabrous; petiole 0.7-2cm long with a few hairs. **Inflorescence** 3-5-flowered, loose, sometimes hanging down between the leaves. **Corolla** *thick, fleshy,* campanulate, to 5cm long, with *deep purple nectar pouches*; cherry red or creamy white to pale pink with a broad red band around the edge of the lobes, rarely scarlet or pink; calyx 1-5mm long, wavy edged, +/- glandular; style glandular to the tip. **Distribution** S.E. Tibet, Arunachal Pradesh, Upper Burma, 3,000-3,700m (10,000-12,000ft), in dense thickets, coniferous forest and forest margins.

In flower, this is a very distinctive species, both in colour (especially the bicoloured forms) and in the dark nectar pouches. The leaves hang downwards characteristically in winter.

In flower, particularly in the two-toned forms, *R. cerasinum* is a striking species. Kingdon-Ward described the bicolour forms as 'Cherry Brandy' and the pure red forms 'Coals of Fire'. The two forms seem to come from Burma and S.E. Tibet respectively. The flowers are long-lasting but can be partly hidden by the foliage and it often partially blooms in autumn as well as in spring. Introduced 1924, reintroduced 1980s, 1995->. May-June.

Picture 1

Picture 2

1. *R. cerasinum*, Glendoick.
2. *R. cerasinum*, Glendoick, a fine bicoloured clone. Such forms were referred to by F. Kingdon-Ward as 'Cherry Brandy'.

Picture 3

3. *R. cerasinum* Kingdon-Ward 11011, such forms were named by the collector 'Coals of Fire', Royal Botanic Garden, Edinburgh.

R. cyanocarpum

W.W. Smith 1914. (*R. thomsonii* var. *cyanocarpum* Franch., *R. hedythamnum* Balf.f. & Forrest var. *eglandulosum* Hand.-Mazz., *R. cyanocarpum* var. *eriphyllum* Balf.f. & W.W. Sm.) H4.

Height 1.2-7.6m, a upright shrub or small tree with *very stiff branches*. Branchlets and petioles *yellowish-green*. **Leaves** 5-12.6 x 4-9cm, broadly elliptic to orbicular, retained for 1 year; upper surface glabrous and often glaucous, lower surface glabrous or minutely hairy; petiole 1.5-3cm eglandular. **Inflorescence** 6-11-flowered, loose or open-topped. **Corolla** campanulate to widely-funnel-campanulate with dark nectar pouches, 4-6cm long, pure white to creamy white, pink to rose-pink, flushed rose to purplish-rose; calyx cup-shaped, greenish or coloured, glabrous; ovary usually glabrous; capsule *with a bluish-glaucous bloom*. **Distribution** C.W. Yunnan (probably Cangshan only) 3,000-4,000m (10,000-13,000ft) common in rhododendron thickets, ravines, bamboo and forest margins.

This species is related to *R. thomsonii* and perhaps more closely related to *R. meddianum* but differs from both in the paler flowers and its isolated distribution.

R. cyanocarpum is quite impressive in flower, especially in its pure white and deep rose-pink forms. It is still rare in cultivation and is not always easy to please, perhaps growing better as a grafted plant. Susceptible to powdery mildew. Introduced 1910-22. Reintroduced 1981->. March-April.

Picture 1

Picture 2

Picture 3

Picture 4

1. *R. cyanocarpum*, Cangshan, C.W. Yunnan, China, the last truss of the season.
2. *R. cyanocarpum*, Royal Botanic Garden, Edinburgh.
3. *R. cyanocarpum* SBEC 0951, Glendoick, showing the somewhat glaucous leaves and the nectar pouches in the corolla.
4. A pure white *R. cyanocarpum* from Dawyck Botanic Garden growing at Glendoick.

R. eclecteum
Balf.f. & Forrest 1920 (var. *brachyandrum* Balf.f. & Forrest)
H3-4

Height 0.6-2.4m, a bushy shrub, sometimes sparse with age; bark of variable colours, fairly smooth and flaking. New growth with vivid red bud scales. **Leaves** 5-15 x 2-6cm, obovate to oblong-obovate to oblong, retained for 1-2 years; upper surface glabrous and often glaucous, lower surface usually glabrous, midrib often with at least some straight hairs on either side; petiole *0.4-1cm* long, *narrowly winged*. **Inflorescence** 6-12-flowered, loose to compact. **Corolla** fleshy, tubular-campanulate, with nectar pouches, 3-5cm long, very variable in colour: white, pale pink, rose or purple, salmon pink, red or yellow, unmarked, heavily spotted or flushed, or stained giving a bicoloured effect such as cream to yellow edged pink; calyx glabrous; ovary glandular; style eglandular. **Distribution** N.W. & W. Yunnan, S.E. Tibet, S.W. Sichuan, Upper Burma and ? Arunachal Pradesh, 3,000-4,300m (10,000-14,000ft), usually in the open on bouldery slopes, in thickets and bamboo.

This species is closely related to and may merge with *R. stewartianum* which usually has a little indumentum on the lower leaf surface and a longer petiole. *R. eclecteum* differs from *R. meddianum* in the winged petiole and only rarely has red flowers. Brachyandrum Group (formally var. *brachyandrum* Balf.f. & Forrest) was originally described as having smaller flowers than the type.

R. eclecteum is extremely variable in its flower colour; some forms are very attractive while others are only fit for the bonfire. Unfortunately this species is often susceptible to powdery mildew. In the wild, it hybridises with many other species including *R. pocophorum* (known as *R. hemigynum*), *R. forrestii* and *R. sanguineum*. Introduced 1917 and many times subsequently. Reintroduced 1992. January-April.

R. eclecteum var. bellatulum (Balf.f. ex) Tagg.
Recent field studies have confirmed that is a hybrid between *R. eclecteum* and *R. selense*. *R. x bellatulum* differs from the type in its longer petiole and usually has white to yellow flowers.

Picture 1

Picture 2

Picture 3

Picture 4

1. A bank of *R. eclecteum* with mixed flower colours on the west side of the Chunsien La, Salween-Mekong divide, N.W. Yunnan, China.
2. *R. eclecteum* with rose-coloured flowers on the west side of the Chunsien La, Salween-Mekong divide, N.W. Yunnan, China.
3. An almost pure white *R. eclecteum* from Borde Hill, growing at Glendoick.
4. *R. eclecteum* with slightly pink-tinged flowers, Royal Botanic Garden, Edinburgh.

THOMSONIA

Picture 5

Picture 6

Picture 7

Picture 8

8. *R.* x *bellatulum* (*R. eclecteum* var. *bellatulum*) Rock 11027, Royal Botanic Garden, Edinburgh.

5. A salmon-pink-flowered *R. eclecteum*, Glendoick.
6. A fine yellow-flowered *R. eclecteum* from Rowallane, growing at Glendoick.
7. The distinctive foliage of *R. eclecteum*, taken on the Salween-Mekong divide, N.W. Yunnan, China, showing the characteristic short petiole which distinguishes this species from *R. stewartianum*.

R. faucium
D.F. Chamb. 1978 H3-4

Height 1.5-6.5m, an erect shrub or small tree; bark smooth, peeling. **Leaves** 7-12 x 2.5-3.5cm, oblanceolate, apex rounded, retained for 2-3 years; upper surface glabrous, lower surface greenish with red punctate hair bases and a few scattered hairs towards the midrib; petiole *0.7-1.5cm* long, often winged for part of its length. **Inflorescence** 5-10-flowered, compact to open-topped. **Corolla** campanulate, 3.5-4.5cm long, *pink* to white tinged pink, (?rarely yellow), spotted purple. Ovary *densely glandular.* **Distribution** S. Tibet, 2,600-3,350m (8,500-10,850ft), forest margins, rock faces.

R. faucium differs from *R. hookeri* in the absence or near absense of tufts of hairs on the lower leaf surface and from *R. subansiriense* in the pink flower colour and the glandular ovary. Most plants grown in cultivation as *R. hylaeum* fit better into *R. faucium*. Davidian considers these two species synonymous while Chamberlain states that *R. faucium* differs in the smaller leaves which taper towards the base, the shorter petiole and the glandular ovary.

Picture 1

1. A dark-flowered *R. faucium* on the south side of the Doshong La, S.E. Tibet.

A fine plant for sheltered and milder gardens, with especially showy bark. *R. faucium* is a recently named species from Tibet which is still rare in cultivation but common in the wild. Introduced 1947, reintroduced 1995. March-May

Picture 2

Picture 3

Picture 4

Picture 5
5. Trunk of *R. faucium* showing the peeling bark.

R. hylaeum
Balf.f. & Farrer 1922 H3

Height 2.5-12m, an erect shrub or small tree; bark smooth, peeling. **Leaves** 8.5-14.5 x 3.4-5.7cm, oblong to oblanceolate; upper surface glabrous, lower surface greenish, epapillate, with scattered bunched hairs from red bases on the veins; petiole *1.5-2cm, narrowly winged*. **Inflorescence** 10-12-flowered. **Corolla** tubular-campanulate, 3.5-5cm long, rose-pink, spotted. Ovary *glabrous*. **Distribution** N.E. Upper Burma, S.E. Tibet, 2,700-3,700m (9,000-12,000ft), open mixed forest.

This species differs from *R. faucium* in its larger leaves, the longer petiole and the glabrous ovary. It differs from *R. subansiriense* in flower colour and in the glabrous ovary.

Although plants are cultivated under the name *R. hylaeum*, most are better placed in *R. faucium* and true *R. hylaeum* may not in fact be in cultivation. The two species should probably be merged. Introduced 1921->? March-May.

2. Young growth of *R. faucium* of particularly good colour, in the Rong Chu valley, S.E. Tibet.
3. *R. faucium*, Glendoick.
4. *R. faucium*, Younger Botanic Garden, Benmore, W. Scotland.

R. hookeri
Nutt. 1853 H3(-4)

Height 3-6m, a usually upright shrub or small tree; bark smooth, grey and maroon. **Leaves** 6.3-17 x 3-7.5cm, oblong to oblong-oval, retained 1-2 years; upper surface glabrous, lower surface with *isolated tufts of fasciculate hairs (like hooks) on lateral veins*; petiole eglandular, slightly winged.

Inflorescence 8-18-flowered, loose to compact. **Corolla** fleshy, tubular-campanulate to funnel-shaped, with dark nectar pouches, 3.5-4.5cm long, bright red, blood red, cherry red, deep crimson, or deep rose; calyx *eglandular*, greenish-yellow; ovary and style *eglandular*. **Distribution** W. Arunachal Pradesh, 3,000-3,700m (10,000-12,000ft). Abies and rhododendron forest on sheltered slopes.

The tufts of hair or 'hooks' on the leaf underside are the best diagnostic feature of this species, distinguishing it from its close relatives *R. faucium*, *R. subansiriense* and *R. hylaeum* which sometimes have sparse small hairs. The eglandular calyx and ovary further distinguish it from *R. faucium* and *R. subansiriense*.

R. hookeri is very fine in its best red forms but is not very hardy and is vulnerable to bark split and damaged growth after spring frosts. An added bonus is the showy new growth resembling red candles. The pinkish-flowered forms are closely related to *R. faucium*. Introduced 1849, 1928. March-April.

Picture 3

Picture 4

3. The inferior pinkish-purple *R. hookeri* Kingdon-Ward 8238, Royal Botanic Garden, Edinburgh.
4. The leaf lower surface of *R. hookeri* with the typical, easily-seen, isolated tufts of hairs. (Photo R.B.G., E.)

Picture 1

Picture 2

1. A splendid clone of *R. hookeri*, Tigh-na-Rudha, W. Scotland.
2. *R. hookeri*, Glenarn, W. Scotland.

R. meddianum
Forrest 1920 H3-4

Height 1-2.3m, an upright shrub or small tree with stiff and rigid branches; bark slightly rough. Young shoots glabrous. **Leaves** 8-11 x 4.5-5.2cm, obovate to broadly elliptic, retained 1-2 years (usually 1 in cultivation); upper and lower surfaces glabrous; petiole 1-3cm. **Inflorescence** 5-10-flowered. **Corolla** tubular campanulate, with nectar pouches, 4-6cm long, deep rose to scarlet, crimson to deep blackish-crimson; calyx cup-shaped, fleshy, red to 2cm long, lobes unequal; style *glandular*. **Distribution** W. Yunnan (var. *meddianum* only), and N.E. Upper Burma, 2,400-3,700m (8,000-12,000ft).

This species is an eastern variant of *R. thomsonii*, differing in its less glaucous leaves, non papillate lower surface, more elongated leaf shape, more tubular flower shape and the glandular style. In cultivation, *R. meddianum* characteristically drops its previous year's leaves as soon as the new year's leaves unfurl, and thus has a rather sparse appearance. In foliage, *R. meddianum* is very similar to the pink or white flowered *R. cyanocarpum*.

Var. *meddianum* H4

Ovary +/- *glabrous*.

The rare var. *meddianum*, which can be very fine in flower, is typically slow-growing and rather lacking in vigour, but is hardier than var. *atrokermesinum*. Introduced 1917-25. April-May.

Var. *atrokermesinum* Tagg 1930. H3

Ovary densely *glandular*.

In cultivation this variety is not as hardy as var. *meddianum*, but is more vigorous, and the best forms have amongst the finest red flowers in the genus. Introductions from the Triangle in Upper Burma in 1953 are later into growth and therefore more robust than those introduced earlier. Introduced 1925 and 1953. March-April.

Picture 1

Picture 2

1. *R. meddianum* var. *meddianum* Forrest 24104, Valley Gardens, Windsor Great Park, S. England.
2. A lovely clone of *R. meddianum* var. *meddianum*, Royal Botanic Garden, Edinburgh.

Picture 3

Picture 4

3. A first rate introduction of *R. meddianum* var. *atrokermesinum* Kingdon-Ward 21006A from the Triangle, Upper Burma. Glendoick.
4. *R. meddianum* var. *atrokermesinum* Kingdon-Ward 21006A, Glendoick.

R. sherriffii
Cowan 1937 H4

Height 2-3m in cultivation, 4.5-6m in the wild, a bushy to rather erect shrub or small tree. Branchlets thin, *tomentose and glandular*; bark *red-brown to pinkish-grey, peeling*. **Leaves** 4.5-7.5 x 2.5-4cm, obovate, retained 1-3 years; upper surface *dark and shiny* on maturity, lower surface with a *thick chocolate-brown* indumentum. **Inflorescence** *4-6-flowered, loose*. **Corolla** funnel-campanulate, with nectar pouches, 3.5-4cm long, *rich deep carmine red*; calyx 3-5mm long, glabrous; ovary glabrous; style glabrous, carmine. **Distribution** S. Tibet in a limited area, 3,700m (11,500ft), near edge of conifer forest.

This is an easily recognised species with neat rounded leaves with distinctively-coloured indumentum on the lower surface and small trusses of pendant flowers, reminiscent of those of *R. cinnabarinum* Roylei Group. Apart from *R. miniatum* which has never been introduced, its closest relative is undoubtedly *R. thomsonii* ssp. *lopsangianum* which mainly differs in its absent or occasionally thin indumentum. The thick indumentum makes *R. sherriffii* aberrant in this subsection and suggests an affinity with *R. fulgens* with which it was formerly associated. *R. sherriffii* is rather hard to please, lacking in vigour but is worth growing for its neat appearance and pendulous dark red flowers. Introduced 1936. February-April.

Picture 1

Picture 2

1. *R. sherriffii*, Glendoick.
2. *R. sherriffii*, from Glenarn, growing at Glendoick, showing the thick chocolate-brown indumentum on the leaf lower surface.

R. stewartianum

Diels 1912 (*R. nipholobum* Balf.f. & Farrer, *R. aiolosalpinx* Balf.f. & Farrer., *R. stewartianum* var. *tantulum* Cowan & Davidian) H3-4.

Height 0.6-2.7m, a usually a thin-branched, upright small shrub; bark fawn, peeling. **Leaves** 5-12 x 2.5-6.5cm, obovate to *elliptic* to oblong-obovate, retained 1-2 years; upper surface bright green and glabrous, lower surface often with a *thin veil of hairs*, sometimes persistent and sometimes only found on fresh young growth; petiole *0.5-2cm* long. **Inflorescence** 3-7-flowered, loose. **Corolla** tubular-campanulate, 3.6-5.4cm long, colour very variable; white, cream, yellow, pink, rose, red and crimson, often flushed a different colour on the margins or on the outside of the corolla, unmarked to heavily spotted; calyx variable, 5-15mm, cup-shaped, margins glabrous or fringed with small hairs; ovary usually densely glandular; style glabrous. **Distribution** Upper Burma, S.E. Tibet, N.W. Yunnan, 3,000-4,300m (10,000-14,000ft), on hillsides, in bamboo and occasionally in forests.

This species is very similar to and merges with *R. eclecteum* which typically differs in its glabrous lower leaf surface and its winged and shorter or almost non-existent petiole. In addition, *R. eclecteum* tends to have larger leaves and thicker, sturdier branches.

The early flowering *R. stewartianum* is very variable in flower colour; the best forms are most attractive but some of the combinations of colours look muddy from a distance. It is quite rare in cultivation, often short-lived and lacking in vigour, and it is rather susceptible to powdery mildew. Introduced 1919->. February to April.

R. eurysiphon Tagg & Forrest 1931 is probably a natural hybrid between *R. stewartianum* and *R. martinianum*. Corolla creamy-white flushed rose. Some of the specimens used to describe this species are in fact referable to *R. stewartianum*.

Picture 1

1. *R. stewartianum*, Royal Botanic Garden, Edinburgh.

THOMSONIA

9,200ft), common in mossy forest with other rhododendrons.

The combination of red flowers, the glands on the lower leaf surface and the densely tomentose ovary distinguish this species from its closest relatives *R. hylaeum*, *R. faucium* and *R. hookeri*. In cultivation its extremely early growth, which is usually frosted, is a distinguishing feature.

R. subansiriense has smallish but bright red flowers and a fine bark but is so early into growth that the young shoots invariably get frosted, even twice in one spring. Introduced only once under Cox and Hutchison 418 in 1965. March.

Picture 2

Picture 3

Picture 4

2. *R. stewartianum*, Glenarn, W. Scotland.
3. *R. stewartianum* from Tigh-na-Rudha, W. Scotland, growing at Glendoick, showing the thin branches and the way the flowers typically hang down between the leaves.
4. *R. stewartianum* Forrest 26980, Valley Gardens, Windsor Great Park, S. England.

R. subansiriense

D.F. Chamb. & Cox 1978 H3(-4)

Height to 14m (in the wild), much less in cultivation so far, an erect small tree; bark *smooth or peeling*, grey to red. **Leaves** 7-13 x 2-4cm, retained 1-2 years; upper surface glabrous, lower surface with *numerous unstalked glands over-lying the veins, with a vestige of hairs*, otherwise glabrous. **Inflorescence** 10-15-flowered, fairly dense. **Corolla** tubular-campanulate, with nectar pouches, c.4cm long, *crimson-scarlet*, with a few spots; calyx 4-5mm long, cup-shaped, margins hairy; ovary *densely tomentose*. **Distribution** Subansiri Division, Arunachal Pradesh, 2,550-2,750m (8,400-

Picture 1

Picture 2

1. *R. subansiriense* C.&H. 418, Younger Botanic Garden, Benmore, W. Scotland.
2. *R. subansiriense*, Stonefield Castle, W. Scotland, showing the growth buds elongating, coinciding with the early flowers.

Picture 3

3. The smooth trunk of *R. subansiriense*. This was the only plant in the rain forest of N.E. India that was not smothered in moss.

R. thomsonii
Hook.f. 1851 H(3-)4

Height to 7m, an upright shrub or small tree; bark *smooth and peeling, usually of mixed colours*. **Leaves** 3-11.3 x 2-7.5cm, thick and leathery, *orbicular, ovate to broadly elliptic*, retained 1-2 years; upper surface dark to medium green and glabrous, often very glaucous when young, lower surface glaucous-papillate, glabrous; petiole glabrous or sparsely glandular. **Inflorescence** 3-12-flowered, loose. **Corolla** fleshy, campanulate, with 5 nectar pouches at the base, 3.5-6cm long, *deep blood-red to dark wine-red*, +/- spotted, often with a bloom, (occasionally white or white-flushed-pink in the wild); calyx cup-shaped 0.6-2cm long, whitish-green, greenish-scarlet, purple or red; ovary glabrous or glandular; style glabrous.

Ssp. *thomsonii*

Height 1.3-7m,. **Leaves** glabrous, *5-11cm long*. Calyx *large (6)-10-20mm*. **Distribution** E. Nepal, Sikkim, Bhutan, S. Tibet, Arunachal Pradesh, 2,400-4,300m (8,000-14,000ft), in rhododendron scrub and Abies forest.

This subspecies differs from ssp. *lopsangianum* in its greater stature, longer leaves and larger calyx, and is easily distinguished from other species by its peeling bark, orbicular leaves, blood-red flowers and the persistent calyx which can be of several colours. *R. meddianum* can be considered the western occurring equivalent of *R. thomsonii* but it has more elongated leaves and more tubular flowers, and is rare in cultivation, partly because it is hard to grow. *R. viscidifolium* is also related but has smaller leaves, sticky on the underside, a less smooth bark and orange-brown flowers.

Widely grown and very attractive with fine flowers, foliage and bark, *R. thomsonii* ssp. *thomsonii* is variable in hardiness and performs best with shelter from winds and late frosts. It is rather susceptible to powdery mildew. Whitish forms have recently been found in Sikkim and Nepal. Introduced 1850 and often since. March-May.

Picture 1

Picture 2

1. *R. thomsonii* ssp. *thomsonii*, Jaljale Himal, E. Nepal.
2. Trusses of *R. thomsonii* ssp. *thomsonii* taken off different plants, Milke Danda, E. Nepal, showing both red and whitish-green calyces.

Picture 3

Picture 4

Picture 5

Picture 6

3. A very floriferous *R. thomsonii* ssp. *thomsonii*, Glenarn, W. Scotland.
4. A particularly good *R. thomsonii* ssp. *thomsonii* B.L.M. 228, Younger Botanic Garden, Benmore, W. Scotland.
5. *R. thomsonii* ssp. *thomsonii* Ludlow & Sherriff, Glendoick, E. Scotland. Plants from this introduction flower earlier in the season than most other collections of this species.
6. A natural hybrid of *R. thomsonii* ssp. *thomsonii* x *R. campylocarpum* ssp. *campylocarpum*, Glendoick. This plant was formerly known as *R. thomsonii* var. *candelabrum*.

Picture 7

7. A trunk of *R. thomsonii* ssp. *thomsonii* showing the multi-coloured peeling bark, Glendoick.

Ssp. *lopsangianum*
(Cowan) D.F. Chamb. 1982. H3-4

Height 0.6-1.8m, an upright but compact shrub. **Leaves** *3-4.5cm* long, lower surface glabrous, occasionally with a thin layer of indumentum. **Inflorescence** *3-5*-flowered, loose. **Corolla** 3-4.2cm long; calyx *small, 2-4mm long*. **Distribution** S. Tibet, 2,600-4,300m (8,500-14,000ft), rocky slopes, open hillsides.

This subspecies differs from ssp. *thomsonii* in its smaller stature, leaves, flowers and calyx and has an isolated distribution in S.E. Tibet, ssp. *thomsonii* occurring further south and west. Some forms of ssp. *lopsangianum* have a thin layer of indumentum on the lower leaf-surface and this suggests that this subspecies may originate as a hybrid between ssp. *thomsonii* and *R. sherriffii*. The closely related *R. sherriffii* differs in its thick, felted indumentum on the lower leaf surface.

Picture 8

8. *R. thomsonii* ssp. *lopsangianum*, Royal Botanic Garden, Edinburgh.

This subspecies is not as showy as its larger relative but it is slow-growing with a neat upright habit and is suitable for a sheltered, small garden. Introduced 1936. April.

Picture 9

Picture 10

9. *R. thomsonii* ssp. *lopsangianum*, a clone with no indumentum on leaf lower surface. Glendoick.
10. *R. thomsonii* ssp. *lopsangianum*, a clone from Keillour Castle growing at Glendoick, with a thin indumentum on leaf lower surface. The calyces are much smaller than in ssp. *thomsonii*. This form could be an intermediate or hybrid between *R. thomsonii* var. *lopsangianum* and *R. sherriffii*.

R. thomsonii hybridises in the wild with several of the other Himalayan species. *R. x candelabrum* Hook.f. is a hybrid of *R. thomsonii* x *R. campylocarpum* with pink flowers. This hybrid is common in the wild and there may be some stable populations. *R. sikkimense* Pradhan & Lachungpa 1990 is almost certainly a hybrid of *R. thomsonii* x *R. arboreum* which may have stabilised in some areas. *R. thomsonii* var. *pallidum* Cowan seems to be a natural hybrid between ssp. *lopsangianum* and *R. campylocarpum*.

Picture 11

11. *R. sikkimense* in N. Sikkim, India. Both the *R. thomsonii* and *R. arboreum* influence are easily observed in this photograph. (Photo K. White)

R. viscidifolium
Davidian. 1966. H3-4

Height 0.6-2.4m, usually a rounded and compact shrub but can become leggy with age. Branchlets slender; bark brownish, fairly smooth, tending to split. **Leaves** 4-10 x 3-6.5cm, oval to rounded, retained for c.2 years; lower surface *whitish and sticky*; petiole 1-2.5cm. **Inflorescence** 1-3-flowered, often pendant. **Corolla** tubular-campanulate, with crimson nectar pouches, 3.5-4.5cm long, *copper-red to coppery-orange*, spotted crimson; calyx cup-shaped, 4-9mm, green or copper-red, margin hairy; ovary *densely tomentose*; style crimson, glabrous. **Distribution** S. Tibet, 2,700-3,400m (9,000-11,000ft), cliff faces near waterfalls.

R. viscidifolium is quite unlike any other rhododendron species in flower and is easily recognised. Out of flower it resembles *R. thomsonii*, differing in the slender branchlets, absence of a smooth bark and in the presence of characteristic sticky glands on the white lower leaf-surface. *R. viscidifolium* is a most unusual species which has only recently become more widespread in cultivation. D. Hobbie in Germany and the Gibsons at Glenarn, Scotland, were among the very few to raise this from the original Ludlow, Sherriff and Taylor seed collection in 1938. May.

Subsection Venatora

A monotypic subsection.

R. venator
Tagg 1934 H3-(4)

Height 1.5-2.7m, a fairly compact to rather straggly shrub. Branchlets white setose-glandular. Bark grey-brown, roughish. **Leaves** 6-15 x 1.7-4.5cm, *oblong lanceolate to oblanceolate*, recurved, retained for 2-3 years; upper surface *rugulose,* lower surface glabrous, apart from thin *stellate indumentum and folioliferous hairs* on the midrib. **Inflorescence** 4-10-flowered, loose. **Corolla** tubular-campanulate with nectar pouches, 3-3.5cm long, scarlet to reddish-orange; calyx medium, 3-4mm long, glandular; stamens 10; ovary tomentose and sometimes glandular; style hairy at base or glabrous. **Distribution** S.E. Tibet, Tsangpo Gorge, 2,400-2,600m (8,000-8,500ft), in thickets, swamps and rock faces.

This is a very distinct species with no close relatives. It superficially resembles some members of subsection Neriiflora, particularly those in the Neriiflorum alliance. Davidian retains it in the Parishii Series, due to its red flowers and stellate hairs.

R. venator is quite easily grown but is not hardy enough for the coldest parts of Britain. It may require pruning to redress stragglyness. Introduced 1924. May-June.

Picture 1

1. *R. viscidifolium* Ludlow, Sherriff and Taylor 6567, seed collected under one number only. Glenarn, W. Scotland. The 1-3-flowered trusses look sparse on an old specimen.
2. *R. viscidifolium*, Glenarn, W. Scotland.
3. *R. viscidifolium*, Glendoick. This species is unique in sub-genus Hymenanthes for its small coppery-orange flowers.

Picture 1

1. *R. venator* Watt 32, Royal Botanic Garden, Edinburgh.

WILLIAMSIANA

Picture 2

Picture 3

2. *R. venator* Watt 32, Royal Botanic Garden, Edinburgh, showing the loose hanging trusses.
3. *R. venator* Kingdon-Ward 6285, Royal Botanic Garden, Edinburgh, with less pendulous trusses than the Watt introduction.

Subsection Williamsiana

R. williamsianum
Rehder & E.H. Wilson 1915 H4-5

Height 0.6-1.5m, a *dense and spreading shrub, becoming dome-shaped*. Branchlets thin and bristly-glandular. Young growth *bronzy*. **Leaves** 1.5-4.5 x 1.3-3.5cm, *ovate-orbicular, base cordate*, retained for 2-3 years; upper surface glabrous, lower surface glabrous except for red sessile glands. **Inflorescence** 1-3-flowered, rarely to 5 in selected forms. **Corolla** with 5-6 lobes, campanulate, 3-4cm long, pale pink to rose or pale pink fading to white, spotted; calyx minute, c.1mm long; stamens 10; style *glandular to tip*; ovary glandular. **Distribution** Confined to two mountains in C. Sichuan, 2,400-3,000m (8,000-10,000ft), on isolated places on cliffs.

With its dense habit, almost round leaves and campanulate pink flowers, this species is usually easily identified. Its closest relations are probably *R. martinianum, R. selense, R. callimorphum* and *R. souliei,* all of which have larger or less ovate-orbicular leaves, a less compact growth habit and greater stature.

A first rate garden plant for foliage and flowers. Despite its early growth which is easily frosted, this species is best grown in the open where it forms a neat, dense bush. The bronzy new growth is most attractive and this is inherited by many of its numerous, widely-grown hybrids. Introduced by Wilson in 1908. April-May.

Picture 1

Picture 2

1. A typical dome-shaped *R. williamsianum* growing at Glendoick.
2. *R. williamsianum* showing its ovate-orbicular leaves with cordate bases, and the campanulate flowers, 1-3 per inflorescence.

MUMEAZALEA

Height 0.6-3m, a bushy, fairly erect shrub with irregularly whorled, slender, spreading branches. Branchlets pubescent and glandular. Branches grey to dark brown. **Leaves** *deciduous, thin*, 2-5 x 0.7-2.5cm, elliptic to ovate, edges *serrated*, sometimes hairy; petiole densely pubescent. **Inflorescence** *single-flowered from lateral buds*; flowers appear below the expanded leaves. **Corolla** *rotate-funnel-shaped, 1.5-2cm* across, white, yellowish-white to white flushed pink, spotted red; calyx 1-2mm long; stamens *5, very unequal*; ovary setose-glandular; style glabrous. **Distribution** Japan, thickets and forests, in mountain localities.

This species is barely recognisable as a rhododendron, more closely resembling a Menziesia. It is characterised by its thin, deciduous leaves and its insignificant, solitary flowers, with 5 unequal stamens, which hang down below the expanded leaves. The branchlets, petioles, pedicels, calyx and ovary are densely glandular.

R. semibarbatum is grown mainly for its foliage which is often attractively tinted through the summer, and colours to red, orange or yellow in the autumn. The flowers are very small and are almost invisible below the foliage. Introduced 1914. June.

Picture 3

Picture 4

3. *R. williamsianum* 'Special', which has slightly larger flowers and leaves than normal, showing the characteristic bronzy young growth. Glendoick.
4. *R. williamsianum*, a clone with flowers which open pale pink and soon fade to white. Glendoick.

Picture 1

1. *R. semibarbatum* showing the pubescent branchlets and petioles and the rotate-funnel-shaped, red-spotted corolla. (Photo A.D. Schilling)

SUBGENUS MUMEAZALEA

A monotypic subgenus.

R. semibarbatum

Maxim. 1870 (*Azalea semibarbata* Kuntze, *Azaleastrum semibarbatum* Makino, *Mumeazalea semibarbata* Makino)
H4

Picture 2

Picture 3

2. *R. semibarbatum* at Arduaine, W. Scotland, showing the insignificant flowers hanging below the already developed, serrated leaves.
3. Autumn colour of *R. semibarbatum* at Glendoick.

SUBGENUS PENTANTHERA

Height 1-10m, spreading to upright, *deciduous* shrubs or small trees (sometimes evergreen when juvenile), sometimes stoloniferous. Branchlets bristly, hairy or glabrous. **Leaves** alternate or occasionally in whorls, oblong-lanceolate, oblong-oblanceolate to obovate, elliptic to orbicular, usually covered wholly or partially in small or/and large hairs. **Inflorescence** *terminal*, 1-15-flowered. **Corolla** tubular-campanulate, broadly-rotate or rotate-campanulate, narrowly to broadly funnel-form or slightly to distinctly two-lipped, white, yellow, pink, purple, orange, scarlet or crimson; calyx 5-lobed, glabrous or covered with hairs; stamens 5-10; ovary usually hairy; style glabrous or pubescent or glandular in lower half. **Distribution**. N. America, E. Europe & the Caucasus, China, Japan, Korea, Manchuria.

This subgenus contains most of the species referred to as deciduous azaleas and it includes the former Canadense, Luteum, Nipponicum and (part) Schlippenbachii subseries. This subgenus is characterised by its deciduous leaves, covered in hairs, and its terminal rather than axillary inflorescence.

SECTION PENTANTHERA

Leaves consistently alternate, though often close together. **Corolla** *pubescent* on outside; stamens 5; testa (outer seed coating) loose.

The species in this section are quite closely related to one another and have been much hybridised by breeders. The eastern north American species provide considerable taxonomic problems as they hybridise and intergrade frequently in the wild. Such intermediates or hybrids can form dominant populations, particularly in areas where the habitat has been much disturbed. The presence or absence of hairs and glands on the flower buds is a significant diagnostic feature for the species of this subsection. The most commonly cultivated species are probably *R. calendulaceum, R. molle* ssp. *japonicum, R. luteum, R. occidentale* and *R. viscosum.* This section was revised in 1993 by K.A. Kron of the University of Florida and this revision is followed here.

Although not as well known as the Mollis and Exbury hybrids derived from them, many of the species in this subsection make fine garden plants, some being especially useful for their scent and/or late flowering. Many cultivated plants under species names, grown from wild seed, are in fact natural hybrids.

Subsection Pentanthera

Corolla *narrowly* funnel-shaped, sometimes blotched but *not* spotted; style *much exserted beyond* the corolla.

R. alabamense
Rehder & E.H. Wilson 1921. Alabama Azalea H3-4

Height to 3m, usually a non-stoloniferous shrub. Branchlets hairy when young, occasionally glandular. **Leaves** 3-6.3 x 1.3-3cm, obovate to elliptic to oblong-obovate; lower surface often glaucous; leaves have *a distinct spicy odour when crushed*. Flower buds glabrous. **Inflorescence** 6-12-flowered; flowers appearing before or with the leaves. **Corolla** tube *narrow, glandular outside*, 2.2-3.5cm long, white, usually with a *yellow blotch, lemon*-scented; calyx small, bristly; stamens twice as long as tube; ovary hairy, sometimes glandular. **Distribution** Alabama and adjacent Georgia, Tennessee and Florida, intermediates/hybrids Georgia, Mississippi and Florida, sea level-500m (0-1,500ft), stream sides, dry rocky hillsides and tops.

R. alabamense is characterised by the yellow blotch on the corolla which separates it from the other eastern North American species. Out of flower, it is hard to separate it from *R. canescens* with which it merges and hybridises, but the two generally enjoy different habitats: *R. alabamense* in dry woods, *R. canescens* in stream beds and moister sites. *R. alabamense* is lower growing and earlier flowering than *R. arborescens* and *R. viscosum*.

R. alabamense is a dainty species with good-sized, delicately sweet-scented flowers. It is rare in cultivation and is unlikely to succeed in areas without considerable summer heat. Introduced to Europe 1922. May-June.

Picture 2

1. *R. alabamense* showing its relatively large flowers, with the characteristic yellow blotch and narrow tube, appearing with the leaves. (Photo M. Beasley).

Picture 2

2. *R. alabamense* at the Rhododendron Species Botanical Garden, Washington State, U.S.A. (Photo L. Fickes).

R. arborescens
Torrey 1824 (*Azalea fragrans* Raf.) Sweet Azalea H5

Height to 6m, a usually non-stoloniferous shrub. Branchlets *yellow-brown and usually glabrous*. **Leaves** 3.8-10.5 x 1.3-3cm, ovate, obovate to elliptic, *lustrous*, upper surface *glabrous* except for the usually glandular midrib. The leaves smell of mown grass when bruised or dried. Flower buds chestnut-brown, usually glabrous. **Inflorescence** 3-7-flowered; flowers appearing with or after the leaves. **Corolla** with tube longer than lobes, pubescent and glandular, 3.8-5.5cm long, white, often tinged pink, sometimes with a yellow blotch; heliotrope-cinnamon-scented; calyx 1-8mm long, glandular-ciliate; stamens twice as long as tube, *purplish-red*; ovary with reddish glands; style with long hairs at base, *purplish-red*. **Distribution** Widespread from Pennsylvania south to Georgia and Alabama, 300-1,500m (1,000-5,000ft), stream sides, moist woods and mountain tops.

R. arborescens is most closely related to *R. viscosum*, differing in its usually glabrous, yellow-brown, smooth branchlets, its smooth leaves, the distinct heliotrope-like scent and the purplish-red stamens and style. These two species occasionally hybridise and *R. arborescens* also hybridises with *R. cumberlandense* producing offspring in a wide range of colours.

R. arborescens is a fine, fairly late-flowering, scented species which is not cultivated as much as it should be. The coloured stamens and style contrast well with the white flowers. Introduced to Europe pre. 1814. May-July.

PENTANTHERA

Picture 1

Picture 2

Picture 3

1. *R. arborescens* in Braxton Co. West Virginia, U.S.A. showing the characteristic, long, purplish-red stamens and style. (Photo D. Jolley).
2. A pinkish-flowered form of *R. arborescens* at the Royal Botanic Garden, Edinburgh.
3. A pure white-flowered form of *R. arborescens* at the Arnold Arboretum, Massachusetts, U.S.A. showing the long, reddish purple style and stamens.

R. atlanticum

Rehder & E.H. Wilson 1921 (*R. neglectum* Ashe). Coastal, Dwarf or Atlantic Azalea H5

Height *0.3-1.5m*, a low, spreading, *stoloniferous* shrub, developing into thickets, with arching branches. Branchlets usually glabrous. **Leaves** 3-5.2 x 0.8-2cm, ovate, obovate to elliptic; upper surface often *glaucous*. Flower buds with small hairs or glabrous. **Inflorescence** 4-13-flowered; flowers appearing before or with the leaves. **Corolla** tubular-funnel-shaped, with hairs and *stalked glands* outside, 2.8-4cm long, white to white flushed pink, tube usually deeper-coloured, fragrance very musky; calyx 2-4mm long, usually glandular-ciliate; stamens longer than tube; ovary glandular and hairy; style long. **Distribution** Coastal plains from Pennsylvania and Delaware south to Georgia, sea level-150m (500ft), understorey species of pines, at forest margins and along streams, tolerating grazing, burning and strimming.

This species hybridises/intergrades with the related *R. periclymenoides* in the northern part of its range (sometimes known under the names *R. x pennsylvanicum* and 'Choptank River Hybrids') and with *R. canescens* in the south. *R. atlanticum* typically differs from *R. periclymenoides* in its low habit of growth, its often glaucous leaves and its eglandular corolla tube. From *R. canescens* it differs in its lower habit of growth and in the often glaucous, almost glabrous leaves. *R. atlanticum* differs from *R. arborescens* and *R. viscosum* in its lower growth habit, smaller leaves and earlier flowers.

R. atlanticum is a most attractive azalea, capable of making large, relatively low, spreading clumps. It enjoys a moister soil than most of its relatives. Introduced to Europe 1922. May-June.

Picture 1

Picture 2

1. A pinkish-flowered *R. atlanticum* at Glendoick.
2. A white flushed pink-flowered form of *R. atlanticum* at the Royal Botanic Garden, Edinburgh.

Picture 3

Picture 4

3. The thin branchlets with the characteristic glaucous leaves separate *R. atlanticum* from other species of section Pentanthera.
4. *R. atlanticum* showing the hairs and stalked glands on the outside of the corolla. (Photo R.B.G., E.).

R. austrinum

Rehder 1917, Florida Azalea H3-4

Height to 5m, usually much less, a usually non-stoloniferous shrub. Branchlets reddish-brown, *pubescent*. **Leaves** 3-8.8 x 2-3.8cm, ovate, obovate to elliptic, margins ciliate; upper and lower surface *pubescent and sometimes glandular*. Flower buds *greyish pubescent, with glandular margins*. **Inflorescence** 10-24-flowered; flowers appearing *before or with* the leaves. **Corolla** tubular-funnel-shaped, tube pubescent and glandular outside, 2.5-3cm long, *yellow to orange-yellow*; slightly fragrant; calyx c.2mm long, pubescent and glandular; stamens long; ovary with whitish hairs, partly glandular; style slightly longer than stamens. **Distribution** N.W. Florida, adjacent Georgia and Alabama to S.E. Mississippi, sea level to 100m (to 330ft), bluffs, woods and stream sides.

In its pure yellow forms, *R. austrinum* is a distinctive species, differing from the yellow-flowered *R. calendulaceum* in its pubescent flower buds, the greater number of flowers to the inflorescence the light fragrance and the smaller corolla. It is separated from *R. canescens* in its glandular flower bud scales, leaf margins, petioles and pedicels. Forms with eglandular flower bud scale margins and yellow flowers with a dark pink-reddish tube are probably hybrids of *R. austrinum* x *R. canescens*. *R. austrinum* is perhaps most closely related to *R. luteum* and *R. occidentale* although their wild habitats are thousands of miles away.

This is a pretty azalea but which performs poorly in cooler and more northerly climates where it will not ripen its young growth or flower freely. In more southern climates, it is quite hardy and free-flowering. Introduced to Europe 1916. May.

Picture 1

Picture 2

1. *R. austrinum* with orange-yellow flowers at the Rhododendron Species Botanical Garden, Washington State, U.S.A. The reddish tube suggests that this is possibly a natural hybrid with *R. canescens*. (Photo K. White).
2. A fine specimen of *R. austrinum*. (Photo W. Berg).

PENTANTHERA

Picture 3

3. *R. austrinum* 'Moon Beam' at the Rhododendron Species Botanical Garden, Washington State, U.S.A. (Photo K. White).

R. calendulaceum

(Michx.) Torrey 1824 (*Azalea aurantiaca* Dietr. *A. coccinea* Lodd., *A. crocea* Hoffsgg., *A. flammea* Bartr.) Flame or Fiery Azalea, Flame Flower, H5.

Height to 10m, usually much less in cultivation, a *non-stoloniferous*, upright, spreading shrub or small tree. Branchlets hairy and bristly. **Leaves** 3-9.1 x 1.3-3.8cm, ovate, obovate to elliptic; upper surface pubescent or glabrous, lower surface sometimes covered with hairs. Flower buds usually glabrous, bud scale margins *glandular*. **Inflorescence** 5-9-flowered; flowers appearing *before or with* the leaves. **Corolla** funnel-shaped, 3.8-6.3cm long, 4.5cm across, glandular and pubescent outside, *orange to red or yellow* with a blotch, with a little acrid fragrance or no fragrance; calyx 1-3mm long, glandular and/or hairy; stamens up to nearly *three* times length of tube; ovary glandular and hairy. **Distribution** S. New York (? cultivated) south through Carolinas to N. Georgia, 180-1,500m (600-5,000ft), on open dry sites in woods, on cliffs and hillsides.

R. calendulaceum is closely related to *R. cumberlandense* but is not stoloniferous, often comes from slightly lower altitudes, tends to be larger in all parts, and has larger, generally paler-coloured flowers before or with leaves. There are relatively few reports of natural hybrids of *R. calendulaceum*, probably due to the fact that it is said to be consistently tetraploid while *R. cumberlandense* and *R. arborescens* are diploid, thus creating a genetic barrier. Late-blooming, deep-coloured *R. calendulaceum* tend to have glandular hairs on the pedicel and calyx margins.

R. calendulaceum is more vigorous and easier to grow than other orange to red-flowered species and can make a very fine, showy plant. Some of the named Ghent azaleas such as 'Coccinea Speciosa' may simply be selections of this species. Plants from high elevations in the wild tend to have slightly smaller, deeper-coloured flowers which open relatively late in the season. Introduced to Europe 1750?, 1804. May-June.

Picture 1

Picture 2

1. An orange-yellow-flowered form of *R. calendulaceum* on Roan Mountain, North Carolina, U.S.A.
2. A salmon pink and orange-flowered form of *R. calendulaceum* from Flatwood, West Virginia, U.S.A., showing the characteristic, long style and stamens, almost 3 x the length of the tube. (Photo D. Jolley).

PENTANTHERA

R. canescens

(Michx.) Sweet 1830 (*R. bicolor* Pursh. *R. candidum* Small). Florida Pinxter, Hoary or Piedmont Azalea. H3-4

Height to 6m, a usually non-stoloniferous shrub. Branchlets hairy and slightly bristly. **Leaves** 4.5-9.8 x 1.3-3.6cm, ovate, obovate to elliptic, lower surface usually *pubescent* or hairy. Flower buds *greyish-pubescent*. **Inflorescence** 6-19-flowered. Flowers before or with leaves. **Corolla** tubular-funnel-shaped, 2.5-3.5cm long, pale to deep pink, rarely white to rose, usually with no blotch, tube glandular, longer and sometimes deeper-coloured than the lobes, *covered with small hairs*; calyx minute 1-2mm, *ciliate, rarely glandular*; stamens *3 x as long* as tube; ovary densely covered with hairs; style pubescent in lower third. **Distribution** Coastal plains from North Carolina to Florida westwards to S.E. Texas and Oklahoma, usually in moist woodlands and beside streams.

R. canescens is very similar to and hybridises/merges with *R. periclymenoides* and *R. prinophyllum* and these three species cause considerable taxonomic problems. *R. canescens* has a more southerly distribution than its relatives and it usually differs from *R. periclymenoides* in the small hairs which cover the corolla tube. *R. prinophyllum* is more distinctive, differing in its broader, more gradually expanded corolla tube, its +/- glandular calyx margins, pedicel and ovary and the glandular hairs on the capsule.

R. canescens is horticulturally similar to *R. periclymenoides* and *R. prinophyllum* but is not generally as hardy as either and is therefore rarely cultivated in Europe. It is well worth growing the better forms of this species in mild areas with warm or hot summers. So called var. *candidum* Small, now Candidum Group, has a whitish-glaucous leaf underside and densely pubescent branches. Cultivated c. 1750 or 1810 in Europe. April-May.

Picture 3

Picture 4

Picture 5

Picture 6

Picture 1

3. Mrs P.A. Cox standing beneath an orange-yellow-flowered form of *R. calendulaceum* in the Blue Ridge Mountains, Virginia, U.S.A.
4. A fiery orange-flowered form of *R. calendulaceum* at Glendoick.
5. *R. calendulaceum* near the summit of Roan Mountain, N. Carolina. U.S.A.
6. The flower of *R. calendulaceum* showing the glandular hairs on the pedicels and calyces. (Photo R.B.G., E.)

1. A form of *R. canescens* with an orange tube. (Photo G. McLellan).

Picture 3

2. An almost pure white-flowered *R. canescens* showing the long tube and stamens. (Photo M. Beasley).

Picture 1

Picture 2

Picture 3

R. cumberlandense
E.L. Braun 1941 *(R. bakeri* Lemmon) Cumberland Azalea
H4-5

Height to 2m but often *much lower, even below 0.6m*, a low, often stoloniferous shrub. Branchlets usually bristly. **Leaves** 3-8 x 1.3-3.5cm, ovate or obovate to elliptic; upper surface nearly glabrous, lower surface glabrous to moderately pubescent. Flower buds +/- glabrous. **Inflorescence** 3-7-flowered; flowers appearing *after* the leaves. **Corolla** tubular-funnel-shaped, pubescent and slightly glandular outside, 3-4.5cm long, usually orange to red, also yellow (not as brilliant-coloured in gardens as in the wild), with a slight, rather acrid fragrance; pedicels and calyx *usually eglandular*; stamens *twice* as long as tube. **Distribution** Adjoining areas in Kentucky, Virginia, Tennessee to Georgia, Alabama and W. North Carolina, mostly over 900m (3,000ft), in open woods and on ridge tops at high elevations.

The diploid *R. cumberlandense* is very closely related to the tetraploid *R. calendulaceum* and the two species are often very hard to separate by visual characters alone. *R. cumberlandense* tends to differ in its lower, more compact, less vigorous growth, in the smaller and later flowers, produced after the leaves have developed, and usually in the eglandular pedicels and calyx. In several areas it hybridises with *R. arborescens* or *R. viscosum* giving a range of hybrids in pink and other shades.

This species has long been grown under the name *R. bakeri* but the type specimen has recently been identified as a probable natural hybrid rendering the specific name *R. bakeri* invalid. *R. cumberlandense* is an excellent, low-growing, late orange to red-flowered azalea, suitable for the small garden. Introduced to Europe 1950s? June-July.

1. The deep red *R. cumberlandense* 'Camps Red' selected by H. Skinner from Big Black Mountain, Kentucky.
2. *R. cumberlandense,* an orange-red-flowered form on Gregory Bald, Tennessee, U.S.A.
3. A red-flowered form of *R. cumberlandense* on Gregory Bald, Tennessee, U.S.A.

R. flammeum
(Michx.) Sargent 1917 *(R. speciosum* (Willd.) Sweet). Oconee Azalea. H3-4

Height 0.6-2.5m, a usually non-stoloniferous shrub. Branchlets pubescent and bristly. **Leaves** 3-8.2 x 1.3-2.7cm, ovate, obovate to elliptic; upper surface usually bristly, lower surface glabrous or pubescent, sometimes bristly. Flower buds *usually pubescent*. **Inflorescence** 6-11-flowered; flowers appear *before or with the leaves*. **Corolla** funnel-shaped, c.4.4cm long, tube finely pubescent and hairy, *longer than lobes, rarely glandular, orange to scarlet*, with a large blotch, with a slight, acrid fragrance; calyx 1-3mm long, setose; stamens more than twice as long as tube; ovary bristly, *eglandular*; style pubescent on lower third. **Distribution** Lower Piedmont, Central Georgia and South Carolina, sea level to 500m (to 1,500ft), open woods, clay bluffs and sandhills.

R. flammeum is related to *R. cumberlandense* and *R. prunifolium* differing in its earlier flowering. It differs from *R. calendulaceum* in its pubescent rather than glandular flower buds and eglandular rather than glandular corolla tube. *R. flammeum* hybridises with *R. canescens* and other species, giving a range of fragrant, colourful offspring. According to Galle, such hybrids have floral buds which are often glabrous at the top, but with pubescent scales at the base.

Like many of the southern most eastern North American species, this azalea is unsatisfactory in cooler and northern climes, where it fails to ripen its wood. It may not be currently in cultivation in Europe but is quite highly rated in S.E. U.S.A. Introduced to cultivation before 1789. May.

Picture 1

Picture 2

1. *R. flammeum* showing the typical yellowish-orange flowers produced with the leaves. (Photo M. Beasley).
2. A fine red-flowered clone of *R. flammeum* known as 'Harry's Red'. (Photo M. Beasley).

Picture 3

3. *R. flammeum* at the Rhododendron Species Botanical Garden, Washington State, U.S.A. (Photo L. Fickes).

R. luteum

Sweet 1830 (*R. flavum* G. Don, *Azalea pontica* L.)
Common yellow azalea, Pontic azalea H5

Height 1.8-3.6m, an upright, spreading, sometimes suckering shrub. Branchlets glandular-hairy when young. **Leaves** 5-14.6 x 1.3-4.2cm, linear-oblong to oblong-oblanceolate; upper surface glaucous, upper and lower surfaces pubescent at first, becoming *glabrous* apart from some glandular bristles. Flower buds usually glabrous. **Inflorescence** 7-17-flowered; flowers appear before or with the leaves. **Corolla** tubular funnel-shaped, c.3.8cm long, 4.5-5cm across, glandular-pubescent outside, *bright yellow with darker blotch*; fragrance *strong, sweet*; calyx 1-6mm long, glandular-pubescent; stamens as long or longer than corolla; ovary sparsely hairy; style slightly longer than stamens. **Distribution** Caucasus; Turkey and adjacent areas, with isolated populations in Poland and Slovenia, sea-level to 2,300m (to 7,500ft), in dense forest and open grassland and swamps.

R. luteum has an extremely isolated distribution, being the only azalea species from the Caucasus region. It is characterised by its +/- glabrous leaves (at maturity), and its sweetly scented yellow flowers with a relatively short corolla tube compared to its closest relation the American *R. austrinum*. Yellow-flowered forms of *R. molle* differ in their wider leaves and larger, less scented or unscented, less tubular flowers.

R. luteum is the most commonly cultivated deciduous azalea species in Europe and it has naturalised in parts of Britain (though unlike *R. ponticum*, it is not proving troublesome). In

PENTANTHERA

much of Europe it is a first rate garden plant, popular for its ease of cultivation, its showy, yellow, scented flowers and its splendid red, orange to purple autumn colour, but it is seldom grown in North America. The long-flowering season is due to the flowers in each inflorescence opening in succession. Many *R. luteum* in old gardens have originated from suckers of the species which was formerly used as an understock to graft deciduous azalea cultivars. Introduced to U.K. 1793. May-June.

Picture 1

Picture 2

Picture 3

Picture 4

4. *R. luteum* showing the fine autumn colour typical of this species.

1. *R. luteum* showing the deep yellow blotch on the corolla which distinguishes this species from *R. austrinum*.
2. A mass of flower on *R. luteum* at the Royal Botanic Garden, Edinburgh.
3. *R. luteum* growing with *R. ponticum* near Artvin, N.E. Turkey. (Photo P. Nagy).

R. occidentale

A. Gray 1876 (*R. sonomense* Greene, *Azalea californica* Torr. & A. Gray ex Durand). Western or Pacific Azalea. H4-5

Height 1-10m, a usually non-stoloniferous shrub or small tree. Branchlets pubescent and glandular, or glabrous. **Leaves** 1.5-10.8 x 0.5-3.6cm, usually oblanceolate, occasionally linear to almost orbicular; upper surface often glossy, upper and lower surface slightly pubescent or glandular. Flower buds pubescent, rarely bristly or glandular, green or red. **Inflorescence** 3-54-flowered, usually 5-25; flowers appear with or after the leaves. **Corolla** broadly funnel-shaped to occasionally tubular-salverform, to *10cm across, glandular outside*, usually white to white flushed pink, with a yellow blotch, occasionally almost pure white, pale pink, yellow or reddish, usually with a sweet fragrance; calyx 1-4+mm long, ciliate or glandular; stamens c. equal to length of corolla; ovary usually glandular and hairy; style usually pubescent near base. **Distribution** W. U.S.A., S. California north to S. Washington State, mostly coastal, with some inland colonies (with small flowers), sea-level to 2,700m (to 9,000ft), shrubby areas to forest, near moist places.

R. occidentale is a very variable species, with a distribution completely isolated from its relatives. It is characterised by its relatively large stature, stout branches, its often glossy leaves and its late, scented flowers which are usually white with yellow markings.

R. occidentale is less hardy than its eastern North American relatives but in moderate climates it is an excellent species, particularly in its numerous selected clones. These clones are still relatively little cultivated as they are not very easy to propagate from cuttings and are somewhat difficult to grow as young plants, partly due to vulnerability to frost damage. *R.*

occidentale (and its hybrids) is susceptible to azalea powdery mildew in its native habitat and in cultivation in some parts of the world. Introduced c.1851, selected clones 1960s->. June-July.

Picture 1

Picture 2

Picture 3

Picture 4

Picture 5

Picture 6

Picture 7

1. *R. occidentale* Smith and Mossmann 127, a pink-flowered selection, at Glendoick.
2. One of the finest *R. occidentale* selections, Smith and Mossmann 28-2 'Crescent City Double' at Glendoick.
3. *R. occidentale* Smith and Mossmann 136, a fine white-flowered selection, at Glendoick.
4. A view of the *R. occidentale* Smith and Mossmann collections at Glendoick.
5. The curious unstable sport *R. occidentale* Smith and Mossmann 502 'Humboldt Picotee' which is unfortunately hard to propagate and often lacks vigour.
6. *R. occidentale* Smith and Mossman 252 at Glendoick.
7. Flowers of *R. occidentale* covered with the characteristic glandular hairs on the outside of the corolla. (Photo R.B.G., E.).

R. periclymenoides

(Michx.) Shinners 1962 (*R. nudiflorum* (L.) Torr.). Pinxterbloom or Honeysuckle Azalea. H5

Height to 2.7m in cultivation, to 5m in wild, a usually non-stoloniferous shrub. Branchlets sparsely hairy. **Leaves** 3-10.9 x 1.3-3cm, elliptic to oblong-obovate, margins ciliate; upper surface *bright* green, upper and lower surfaces sparsely hairy. Flower buds *usually glabrous*, chestnut-brown, margin *usually* *ciliate*. **Inflorescence** 6-15-flowered; flowers appear before or with the leaves. **Corolla** funnel-shaped, 2.5-3.8cm long, tube longer than lobes, *pubescent outside*, sometimes glandular, deep pink, also white to rose-pink, tube often deeper-coloured, usually with a light, sweet fragrance; calyx 1-2mm long, ciliate; stamens *nearly 3 x* as long as tube; ovary hairy; style pubescent on lower third. **Distribution** From N. Georgia and N. Alabama along Atlantic seaboard and inland to Appalachians, north to Ohio, New York, Massachusetts and Vermont, near sea-level to 1,150m (3,800ft), in moist or dry open woods.

This species most closely resembles *R. canescens*, differing in its more glabrous bud scales and leaves and more pubescent, and often eglandular corolla. It differs from *R. atlanticum* in its bright green leaves, deeper-coloured and usually non-glandular corolla and from *R. prinophyllum* in its glabrous buds. Firm identification of *R. periclymenoides* is often difficult as it merges/hybridises with all three of the above species.

Quite pretty and worth growing in its better forms but *R. periclymenoides* is probably not as fine a garden plant as either *R. atlanticum* or *R. prinophyllum*. It is easily grown, tolerating both moist and dryish situations. Introduced to Europe 1734. May.

Picture 1

Picture 2

1. *R. periclymenoides,* a white-flowered form with a reddish tube at the Royal Botanic Garden, Edinburgh.
2. *R. periclymenoides,* a pink-flowered form we obtained from K. Gambrill, showing the characteristic long stamens.

R. prinophyllum

(Small) Millais 1917 (*R. roseum* (Lois.) Rehder & E.H. Wilson) H5

Height to 3m, usually less in cultivation, a rarely stoloniferous shrub. Branchlets hairy when young. **Leaves** 3-8.7 x 1.2-3.7cm, elliptic to obovate-oblong, margins *ciliate*; upper surface *dull lightish-green*, usually pubescent, lower surface *usually pubescent*. Flower buds *usually pubescent*. **Inflorescence** 4-13-flowered; flowers appear before or with the leaves. **Corolla** tubular-funnel-shaped, c. 4.5cm across, 2-4cm long, tubes short, *glandular*, deep to rose pink, rarely white, with a *spicy, clove-like* fragrance; calyx 1-2mm long, usually glandular-hairy; stamens *twice* as long as tube; ovary bristly glandular; style pubescent towards base; capsule *covered with glandular hairs*. **Distribution** Widely scattered distribution, E. Oklahoma, Arkansas to Virginia and north to S.W. Quebec, c.150-1,500m (c.500-5,000ft), usually in mountains, thickets, open woods and sphagnum bogs.

R. prinophyllum is closely related to *R. periclymenoides*, differing in its pubescent buds and dull, light green leaves. It differs from *R. canescens* in its longer pedicels, broader, glandular corolla tube, stamens only twice as long as the tube and less pubescent, more glandular capsules. It merges/hybridises with *R. canescens* and *R. periclymenoides* in the wild.

This is one of the toughest E. North American azalea species, hardier and more showy than most *R. canescens* and usually superior to *R. periclymenoides*. It has proven to be a good doer in Scotland and S. Scandinavia and is generally easy to grow although it is a particular favourite of hungry rabbits. Galle states that many wild-populations of this species contain extensive hybrid swarms. Introduced to Europe before 1812. May.

Picture 1

1. *R. prinophyllum* Spruce Knob, West Virginia, U.S.A., showing the stamens, twice as long as the tube. (Photo D. Jolley).

Picture 2

Picture 3

Picture 4

Picture 5

5. A view of *R. prinophyllum* in the wild at Pittsfield, Massachusetts, E. U.S.A.

R. prunifolium
(Small), Millais 1917. Plumleaf Azalea H3-4

Height to 5.5m in the wild, much less in cultivation, a non-stoloniferous shrub. Branchlets *glabrous and smooth*. **Leaves** 3-15.2 x 2-4.2cm, usually elliptic, margins ciliate; upper and lower surfaces glabrous apart from hairs on the midrib. Flower buds greenish to light brown, *glabrous apart from the ciliate margins to the bud scales*. **Inflorescence** 4-7-flowered; flowers appear after the leaves. **Corolla** funnel-shaped, 3.8-5cm across, tube nearly glabrous, *usually orange-red to vivid red*, occasionally light orange, not fragrant; calyx c.1mm long, hairy; stamens over twice the length of tube; ovary hairy; style glabrous, longer than stamens. **Distribution** Georgia-Alabama border only, c.90-200m (300-660ft), only growing in moist shady ravines.

R. prunifolium has a restricted distribution which is geographically isolated from its nearest relatives *R. calendulaceum*, *R. cumberlandense* and *R. flammeum*. It is characterised by its flowers opening after the leaves have expanded, by the +/- absence of a blotch on the corolla and in the near absence of hairs on the branchlets, buds, leaves and the outside of the corolla.

This species is highly rated in the U.S.A. but is not really suitable for most of N. Europe, requiring more summer heat. It is sometimes successful in England and it should be good in warm, damp areas elsewhere. Introduced to Europe 1918. July-August.

2. A deep pink-flowered *R. prinophyllum* at Glendoick.
3. *R. prinophyllum* at the Royal Botanic Garden, Edinburgh. This form may be a hybrid with *R. atlanticum* or *R. periclymenoides*.
4. *R. prinophyllum* in the wild at Pittsfield, Massachusetts, E. U.S.A.

PENTANTHERA

Picture 1

Picture 2

1. *R. prunifolium,* a fine red-flowered form, showing the absence of hairs on all parts which distinguishes this species from *R. cumberlandense*. (Photo S. McDonald).
2. *R. prunifolium* foliage showing the absence of hairs on the leaves and branchlets which distinguishes this species.

R. viscosum

(L.) Torrey 1824 (*R. coryi* Shinners, *R. hispidum* (Pursh) Torr., *R. nitidum* (Pursh) Torr. *R. oblongifolium* (Small) Millais, *R. serrulatum* (Small) Millais). Swamp Azalea, Swamp Honeysuckle. H5

Height to 6m, usually much less, a sometimes stoloniferous shrub. Branchlets *pubescent and bristly*. **Leaves** 2.1-7.9 x 1.3-3.1cm, ovate to oblong-lanceolate; upper surface largely glabrous, lower surface bristly on the midrib. Flower buds glabrous or finely pubescent. **Inflorescence** 3-14-flowered; flowers usually appear after the leaves. **Corolla** funnel-shaped, c.3.6cm across, tube long and slender, glandular outside (hence its name), white, occasionally tinged pink or purplish, with a sweet spicy fragrance; calyx 1mm+ long, glandular and hairy; stamens somewhat longer than tube; ovary usually glandular; style usually pubescent in lower half and *not* coloured. **Distribution** Maine and Vermont, southwards to S.E. Texas to Florida, sea level-1,500m (5,000ft), marshes, stream sides to mountain forests.

R. viscosum differs from *R. arborescens* in its pubescent and bristly branchlets, sticky flowers and non-coloured stamens and style. *R. viscosum* is very widespread in the wild, and variable, and several former species have recently been made synonyms of it. We have retained these as groups. Oblongifolium Group differs from the type in its larger leaves and earlier flowers and comes from the south-western part of *R. viscosum*'s distribution. Serrulatum Group differs from the type in being largely stoloniferous, with smaller leaves and later flowers and comes from the south-eastern end of the distribution of the species. Aemulans Group from south central Georgia is slow growing, earlier-flowering and has densely pubescent winter buds. Montanum Group from North Carolina's Blue Ridge is particularly low-growing with densely pubescent buds with few scales.

R. viscosum is generally very hardy, apart from the southerly Oblongifolium and Serrulatum Groups which require hot summers to ripen their growth. Late and well-scented but not very showy, *R. viscosum* enjoys a very moist soil and is quite widely cultivated. Introduced to Europe c.1680 and 1734. Early June-August.

Picture 1

Picture 2

1. *R. viscosum* showing the long corolla tube and white stamens which distinguish this species from *R. arborescens*.
2. *R. viscosum,* a pure white-flowered form.

Picture 3

Picture 4

3. *R. viscosum* Serrulatum Group in S.E. U.S.A. which is later flowering than the type. (Photo S. McDonald).
4. *R. viscosum* Oblongifolium Group in S.E. U.S.A. (Photo M. Beasley).

Subsection Sinensia

Corolla *funnel-shaped, red, orange to yellow only,* upper lobe *spotted*; stamens and style *included in or only slightly exserted from the corolla.* **Distribution** China and Japan only.

R. molle
(Blume) G. Don 1834 H3-5

Height 1-2m, a usually non-stoloniferous shrub. Branchlets usually hairy when young and often bristly. **Leaves** 4.1-15 x 1.7-5cm, oblong to oblong-elliptic; upper surface pubescent (at least on young leaves) or with scattered hairs, lower surface with white pubescence over the whole surface or with hairs on the midrib only. Flower buds sparsely or densely covered with hairs, occasionally glabrous. **Inflorescence** 3-13-flowered; flowers appear before or with the leaves. **Corolla** *broadly funnel-shaped,* pubescent outside, 5-8.8cm across, yellow pink, orange or red with greenish blotch; sometimes with a sweet fragrance but usually barely noticeable; calyx minute to small, pubescent, sometimes bristly; ovary pubescent and bristly; style glabrous or pubescent on lower third.

R. molle, the only species in subsection Sinensia, differs from subsection Pentanthera in the more broadly funnel-shaped corolla and shorter stamens and style which are included or only just exserted from the corolla.

Ssp. *molle* (*R. sinensis* Sweet & also Lodd, *Azalea mollis* Blume). Chinese Azalea. H3-4?

Leaves, lower surface usually with dense white pubescence over the whole surface, at least on young leaves. **Corolla** *yellow only*; capsule with *sparse hairs, only 1-4 per mm^2*. **Distribution** E. China, especially Zhejiang, also to Hubei, probably naturalised into Yunnan, sea level-2,400m (8,000ft), thin woods and open hillsides.

Ssp. *molle* is rarely seen in cultivation outside China (where it is a popular garden and pot plant). It appears to be less hardy than ssp. *japonicum* and is perhaps less easily-grown and less vigorous. Introduced 1823, reintroduced 1995. April-June.

Ssp. *japonicum* (A. Gray) K.Kron 1993 (*R. glabrius* Nakai, *Azalea japonica* A. Gray.) H4-5.

Leaves, lower surface usually with hairs only on the midrib. **Corolla** *yellow, orange, red, salmon-pink, pink*; capsule with *5-60 hairs per mm^2*. **Distribution** Kyushu north to S.W. Hokkaido, Japan, 100-2,100m (330-7,000ft), hot grassy plains, bog edges, cliffs, scrub woodland and moorland.

Ssp. *japonicum* is easily grown, versatile and very hardy, particularly from its northern locations. Most Azalea Mollis so-called hybrids are simply forms of this subspecies or crosses between the two subspecies. It is tolerant of less acid conditions than the American species in subsection Pentanthera. Introduced to Europe 1830s, U.K. 1860s. May.

SINENSIA

Picture 1

Picture 2

Picture 3

Picture 4

Picture 5

Picture 6

1. *R. molle* ssp. *molle* with the typical yellow flowers. Cultivated in Yunnan, China. (Photo T. Takeuchi).
2. *R. molle* ssp. *japonicum* Doleshy 14 from Kusatsu, Japan, collected in 1965 and growing at Glendoick.
3. *R. molle* ssp. *japonicum* in the wild at Nikko Heights, Japan. (Photo T. Takeuchi).
4. A bright red-flowered form of *R. molle* ssp. *japonicum* at the Royal Botanic Garden, Edinburgh.
5. A form of *R. molle* ssp. *japonicum* with salmon-pink striped flowers at the Royal Botanic Garden, Edinburgh, showing the short stamens which distinguish this species from those in subsection Pentanthera.
6. *R. molle* ssp. *japonicum,* a yellow-flowered form from S. Hokkaido, Japan. This form has pale green leaves.

SECTION RHODORA

Corolla +/- *rotate-campanulate, two-lipped*, white to pink or rose purple.

This section contains two species which share a two-lipped corolla but are otherwise not all that closely related.

R. canadense

(L.) Torrey 1839 (*Rhodora canadensis* (L.), *R. rhodora* S.F. Gmelin). Rhodora H5

Height to *1.20m*, an upright, often stoloniferous shrub. Branchlets puberulous at first, turning greyish-brown with age. **Leaves** 1.9-5.7 x 0.8-1.9cm, elliptic to oblong; upper surface *dull bluish-green*, slightly hairy, lower surface *greyish-tomentose*. Autumn colour *bluish-purple* on rosy-purple-flowered plants, *yellowish* on white-flowered plants. Flower buds very small, ovoid, pinkish. **Inflorescence** 3-6-flowered; flowers appear *before* the leaves. **Corolla** rotate-campanulate, *two-lipped; upper lip with 2 short lobes, lower lobes divided into 2 distinct lobes*, c.1.9cm long, *tube essentially absent*, pale to deep rosy-purple or white, spotted; calyx minute; stamens 10; ovary densely covered with hairs, slightly glandular; style usually glabrous; capsule *covered with hairs*. **Distribution** Labrador and Newfoundland, Canada south to N.E. Pennsylvania and N. New Jersey, U.S.A., sea level to 1,900m (6,300ft), stream banks, swamps and moist woods.

This very distinct species is unlikely to be confused with any other azalea, having a low, upright, stoloniferous habit, and a rosy-purple or white corolla, two-lipped with the lower lip divided, more or less lacking a tube.

R. canadense is a tough, dainty azalea for a moist, acid site with small but attractive flowers which open earlier in the season than most of its relatives. If isolated, the white form comes true from seed. Introduced 1767. April-May.

1. A fine dark-flowered *R. canadense*, Royal Botanic Garden, Edinburgh.

2. The pretty white-flowered form of *R. canadense* at Glendoick.

3. Foliage of a typical rosy-purple-flowered *R. canadense* showing the dull bluish-green upper surface. Royal Botanic Garden, Edinburgh.

R. vaseyi

A. Gray 1880. Pink Shell Azalea. H5

Height to 4.6m but usually much lower in cultivation, an upright, spreading shrub. Branchlets hairy at first, becoming glabrous, older branches roughish. **Leaves** 5-12.5 x 2-5cm, elliptic, tapering at each end, margins often *undulate*; upper surface dark green, largely glabrous, lower surface lighter green, almost glabrous, with a few scattered hairs. Flower buds *ovoid, almost appearing stalked from the shedding of lower scales*. **Inflorescence** 5-8-flowered; flowers appear before the leaves. **Corolla** widely funnel-shaped to rotate-campanulate, 2.5-3cm long, to 5cm across, appearing 2-lipped, *tube very short with 5 wing-like lobes*, pale to deep pink or white, with greenish throat, with reddish-orange spots at the base, not scented; calyx variable, 0.3-8.5mm long, oblique, glandular; stamens (5)-7, unequal, glabrous; style usually glabrous; ovary glandular-hairy. **Distribution** North Carolina only, 900-1,100m (1,000-3,500ft), in ravines and swamps.

R. vaseyi is not all that closely related to *R. canadense* and differs in its larger stature, later flowers, smaller number of stamens and distinctive leaf and flower bud shape. As far as we know. R. vaseyi has not been crossed with any other azalea species.

R. vaseyi is an excellent, hardy garden plant, preferring a relatively moist soil. The flowers are very pretty and the autumn colour can be striking with the willow-like leaves turning yellow and red. 'White Find' is a lovely white-flowered clone. Introduced early 1880s. April-May.

Picture 3

Picture 2

Picture 3

Picture 4

4. The light green leaves of *R. vaseyi* 'White Find' showing the typical tapering leaf shape of this species.

1. A fine pink-flowered *R. vaseyi* showing the 5 wing-like corolla lobes. Glendoick.
2. A pale-flowered *R. vaseyi*, Glendoick.
3. The lovely pure white *R. vaseyi* 'White Find' showing the 5-7 unequal stamens. Glendoick.

SECTION SCIADO-RHODION

Shrubs or small trees. **Leaves** deciduous, +/- obovate, usually with 5 leaves (occasionally 3-9) clustered in a whorl at the shoot apex, or alternate (*R. albrechtii*). **Inflorescence** 1-6 flowered from terminal buds; flowers open before or with the leaves. **Corolla** +/- zygomorphic, broadly rotate and funnel-form to rotate campanulate, tube much shorter than the limb, and gradually expanding into it, white, pale to deep pink or purplish-pink, outside glabrous, inside sparsely pubescent; stamens *10*, included to slightly exserted; style declinate; seeds *non-winged*. **Distribution** Japan, Korea and adjacent Russia.

The species in section Sciadorhodion are characterised by the +/- obovate leaves, lacking in marginal hairs, the corolla with 10 stamens and the seeds which are neither tailed or fringed. This section is quite closely related to section Brachycalyx (subgenus Tsutsusi) but differs in the 5-leaved rather than 3-leaved whorls of leaves and the obovate rather than rhombic leaf shape.

R. albrechtii
Maxim. 1871. Albrecht's Azalea. H5

Height 1-3m, a usually upright, twiggy shrub. Vegetative shoots *from below* terminal flower bud; branchlets at first glandular-hairy, becoming glabrous and purple-brown. Young foliage *bronzy*. **Leaves** *alternate, lacking a terminal whorl of 5 leaves* (although they sometimes appear pseudowhorled), 3.8-11.3 x 1.3-6.3cm, obovate to oblong-lanceolate, margins and upper and lower surface, *hairy*. **Inflorescence** 2-5-flowered; flowers before or with leaves. **Corolla** rotate-campanulate, 3-4cm across, *rose to deep rose-purple*, tube much shorter than the limb; calyx small, hairy, sometimes pubescent; stamens 10, *dimorphic*; style glabrous to pubescent near base; ovary glandular and hairy. **Distribution** Mid Honshu to Central Hokkaido, c. 1,000m (3,300ft), in sub-alpine areas.

R. albrechtii differs from the other species in this section in that the leaves are alternate, although the leaves can appear to be whorled at the ends of branchlets. It is also characterised by its dimorphic stamens.

R. albrechtii is a fine, hardy species, especially in its deeper-coloured forms. It grows best in woodland conditions where its early growth can be protected from frost. It tends to be relatively slow growing and is generally not very successful in areas with high summer temperatures. Probably introduced c. 1914. April-May.

Picture 1

Picture 2

1. An excellent form of *R. albrechtii* with deep-coloured flowers. Glendoick.
2. A fairly dark-flowered *R. albrechtii* at Valley Gardens, Windsor Great Park, S. England.

Picture 3

Picture 4

3. A rich pink-flowered *R. albrechtii* showing the rotate-campanulate corolla. June Sinclair's garden, Hood Canal, Washington State, W. U.S.A.
4. Foliage of *R. albrechtii* showing c.5 leaves at the end of each shoot. Royal Botanic Garden, Edinburgh.

R. pentaphyllum
Maxim. 1887 (*R. nikoense* Nakai). Fiveleaf Azalea. H4-5

Height 1-2m in cultivation, to 5m+ in the wild, a broadly upright shrub. Vegetative shoots from *below the terminal bud*; branchlets slightly hairy when young; becoming glabrous; new growth *often tinged red to maroon at edges or over whole leaf*; bark shiny, red-brown turning to greyish in 2nd year, *not corky*. **Leaves** deciduous, *5 in whorls at ends of branchlets, 3-6.3 x 1.5-3cm,* elliptic to obovate; petiole with a *conspicuous fringe of eglandular hairs*. **Inflorescence** 1-2-flowered; flowers appear before leaves. **Corolla** rotate-campanulate, 4-6cm across, pale pink or mauve to deep rose, rarely white, +/- glabrous; calyx small to medium 2-5mm, *glabrous* or fringed with hairs; stamens 10, variable in length; ovary and style glabrous. **Distribution** Mid Honshu south to S. Kyushu and Shikoku, Japan, upper deciduous woodland in mountains.

R. pentaphyllum is most closely related to *R. quinquefolium*, differing in the shoots arising from below rather than within

SCIADORHODION

the terminal bud, the absence of corky bark, the flowers appearing before rather than with the leaves and in the usually pink as opposed to white corolla. Kron reports that in cultivation these two species are often misidentified.

This is a graceful species with most attractively poised foliage which colours well in autumn and which can be magnificent in its native Japan, rivalling Magnolia as a spring-flowering tree. It is rare in cultivation outside Japan and has proved difficult to please as it resents transplanting, is vulnerable to spring frosts and tends to be slow-growing and shy-flowering. Introduced to U.S.A. and Europe c. 1910. March-April.

Picture 1

Picture 2

Picture 3

1. Mt. Gozaishodake, Japan, showing *R. pentaphyllum* scattered in the forest. (Photo T. Takeuchi).
2. *R. pentaphyllum* in a Japanese garden showing the dainty pink flowers, produced before the leaves. (Photo H. Suzuki).
3. *R. pentaphyllum* in Gora Botanic Park, Hakano, Japan. (Photo H. Suzuki).

R. quinquefolium
Bisset & Moore 1877. Cork Azalea. H4-5

Height to 3m in cultivation, taller in the wild, a shrub or small tree. Vegetative shoots from *within the terminal bud*; branchlets glabrous; bark *corky* and fissured. **Leaves** in whorls of 4-5 at ends of branchlets, 3.8-5 x 2-3cm, broadly elliptic to obovate, margins hairy; upper surface hairy, often with reddish-purple markings towards the margins, lower surface hairy. **Inflorescence** 1-3-flowered; flowers appearing *with the leaves*. **Corolla** rotate-campanulate to funnel-shaped, to 5cm across, *white*, with olive-green spotting or eye; calyx 1-3mm, margin usually fringed with hairs; stamens 10, variable in length; style +/- glabrous; ovary glabrous or densely pubescent. **Distribution** N. Honshu to Shikoku, c. 1,000m (3,300ft), in large thickets, often with *R. degronianum* and *R. pentaphyllum*. April-May.

This species is closely related to *R. pentaphyllum*, differing in its corky bark, the vegetative shoots coming from within the terminal bud, and the white flowers which open as the leaves develop. Out of flower, the two species look almost identical and are often confused.

R. quinquefolium is a beautiful, dainty azalea with lovely flowers, foliage and autumn colour which performs better in S. England than in Scotland, suggesting that it requires some summer heat. The flowers can be somewhat hidden in the foliage and it is a particular favourite of rabbits. Introduced to Europe in 1896.

Picture 1

1. The beautiful, evenly-spaced flowers of *R. quinquefolium*, Glendoick.

flowers. Although it prefers a fairly continental climate it does well in the maritime climate of Britain provided its rather early young growth is not badly frosted. It prefers a less acid soil than most rhododendrons. It is hard to root from cuttings but selected clones are now commonly available through micropropagation. Introduced 1893 and 1917-18>. May.

Picture 2

Picture 3

2. *R. quinquefolium*, showing the leaves typically margined red-purple. Glenarn, W. Scotland.
3. A free-flowering *R. quinquefolium* in Japan. (Photo T. Takeuchi).

Picture 1

Picture 2

Picture 3

R. schlippenbachii
Maxim. 1871. Royal Azalea. H5

Height 1-4.6m, a shrub or small tree. Vegetative shoots appear from *within* the terminal bud. Branchlets pale brown and glandular at first, turning grey and glabrous. **Leaves** in whorls of 5, 2.5-11.7 x 0.9-7.2cm, *obovate, apex rounded or notched*, largely glabrous when mature. **Inflorescence** 3-6-flowered; flowers appear before or with leaves. **Corolla** broadly rotate-funnel shaped, 5.5-8.8cm across, pale to rose-pink, occasionally white, spotted red-brown, lightly fragrant; calyx *large* c.7mm, glandular and fringed with hairs; stamens 10; ovary glandular-pubescent; style +/- glabrous, with a few hairs in lower half. **Distribution** Korea and N.E. Manchuria, common in open woodlands.

The distinctive *R. schlippenbachii* is larger in all parts than the other species in the section and is unlikely to be confused with any other species. In foliage it most resembles *R. nipponicum* in the large, obovate leaves but *R. nipponicum* differs in its papery, shredding bark and its very different, small, pendulous, tubular flowers.

R. schlippenbachii is one of the finest azalea species with attractive foliage, good autumn colour and large handsome

1. A fine, clear pink-flowered *R. schlippenbachii*, Rhododendron Species Botanical Garden, W. U.S.A.
2. *R. schlippenbachii* with near-white flowers, Royal Botanic Garden, Edinburgh.
3. Foliage of *R. schlippenbachii* showing the whorls of leaves.

VISCIDULA

Picture 4

4. Autumn colour of *R. schlippenbachii*. The neat rounded habit is typical of this species. Royal Botanic Garden, Edinburgh.

Section Viscidula

A monotypic section.

R. nipponicum
Matsumura 1899. H4-5

Height to 2m, a fairly compact shrub. Vegetative shoots coming from *below* the terminal bud; branchlets rigid, hairy and glandular when young; bark *papery, shredding, revealing polished brown stems*. **Leaves** alternate, 5-18 x 3.8-9cm, obovate to oblong-obovate, margins densely hairy, upper and lower surface with scattered glandular hairs; petiole *very short*. **Inflorescence** 6-15-flowered; flowers with or after leaves. **Corolla** *tubular-campanulate with short, sub-erect lobes, pendant*, 1.5-2.5cm long, c.1cm across, white, lacking spots, glabrous; calyx to 6mm with scattered hairs; stamens 10 not exserted from the tube; ovary glandular-hairy; style *straight*, glabrous; capsule moderately covered with glandular hairs; Seeds small, tailed. **Distribution** N. Honshu, Japan, 9,00-1,300m (3,000-4,500ft), deciduous forest and hillsides.

This is an extremely distinct species in its small, pendant, tubular, *Menziesia*-like flowers which are often hidden beneath the leaves. It resembles *R. schlippenbachii* in foliage, but *R. nipponicum* can be distinguished by its peeling bark, usually larger leaves and its vegetative shoots appearing from below the terminal bud.

While quite a fine foliage plant with attractive bark and good autumn colour, the flowers are small and often barely noticeable and *R. nipponicum* is grown as much as a curiosity as for its ornamental value. Introduced 1914. June-July.

Picture 1

Picture 2

1. The very distinct *R. nipponicum* showing the large, reticulate leaves and small tubular-campanulate yellowish-white flowers with short corolla lobes. Glendoick.
2. Autumn colour of *R. nipponicum*.

SUBGENUS RHODODENDRON

Plant bearing *scales on the young growth and undersurface of the leaves*, often also on the leaf upper surface, pedicel, calyx, corolla, ovary, style.

Section Pogonanthum

Height to 2.5m, dwarf, prostrate, mounded or erect evergreen shrubs. Branchlets lepidote; leaf-bud scales deciduous or persistent. **Leaves** evergreen, *strongly aromatic*; upper surface dark green, lower surface densely lepidote with characteristic *overlapping, lacerate* scales. **Inflorescence** dense, many-flowered, often resembling that of a Daphne. Flower bud scales *always fringed with large, branched, dendroid hairs*. **Corolla** *narrowly tubular with spreading lobes* (hypocrateriform) with a ring of hairs in the throat, white, cream, pink, purplish-pink, lavender, near red, yellow; calyx 0.5-7mm, conspicuously lobed; stamens 5-10, *not exserted from corolla tube*; ovary small, usually lepidote; style very short, thick, straight, not exserted from the corolla tube; capsule short; seeds unwinged, with obscure fins. **Distribution** Afghanistan, Kashmir eastwards through Nepal to N.W. Yunnan, Tibet, Sichuan and S. Gansu.

Species within this section are easily recognised by their highly aromatic leaves covered with overlapping scales and by the small, daphne-like flowers with non-exserted stamens and style.

The species in this section vary in horticultural merit; some are straggly and shy-flowering while others are amongst the élite of all dwarfs. The aroma of the leaves has been variously described as pineapple, Thuja, Friar's Balsam, bog myrtle and even human sweat. Not always easy to cultivate.

Section Pogonanthum species dislike hot dry conditions, are sensitive to fertiliser and prefer a less acid soil than most rhododendrons. Chlorosis can often be corrected by applications of lime.

R. anthopogon
D. Don 1821 H4-5

Height to 1.5m, a fairly compact shrub. Leaf-bud scales persistent or not. **Leaves** *very aromatic*, 1-3.8 x 0.8-2.5cm, ovate, elliptic to orbicular; upper surface dark green, sometimes glaucous, shiny, lower surface usually *reddish-brown* with 2-3 tiers of overlapping scales. **Inflorescence** 5-9-flowered, open-topped. **Corolla** narrowly tubular with spreading lobes, 0.8-1.6cm long, white, pale yellow, pink to deep rose, *texture like limp tissue-paper*; calyx 2-5mm lepidote and hairy; stamens *(5-)6-8(-10)*, ovary lepidote.

This species is distinguished by the reddish-brown leaf under-surface, the thinly-textured corolla and the 5-10 but usually 6-8 stamens. It differs from *R. collettianum* in the darker leaf lower-surface, (usually) the flower colour and usually in the fewer number of stamens.

Ssp. *anthopogon*
(*R. haemonium* Balf.f. & Cooper.)

Leaf bud-scales *deciduous*. **Distribution** E. Nepal through Bhutan to N.W. Arunachal Pradesh and S. and S.E. Tibet, 2,700-4,900m (9,000-16,000ft), common on moorland and rocks.

While attractive in its more deeply coloured and yellow forms, it takes up to 10 years to flower and is liable to die off in cultivation as it matures to flowering age. Flowers are very prone to weather damage. Introduced 1820 and many times recently in several colour forms. April-May.

Ssp. *hypenanthum* (Balf.f.) Cullen 1979

Leaf bud scales *persistent*. **Distribution** Mostly to the W. of ssp. *anthopogon* from Kashmir east to E. Nepal, also E. Bhutan, 3,400-5,500m (11,000-18,000ft).

Some selections from Stainton, Sykes and Williams 9090 have much smaller leaves, resembling those of *R. sargentianum*, and flower freely from a young age. April-May.

Picture 1

1. A pale pink-flowered form of *R. anthopogon* ssp. *anthopogon* growing with *R. bhutanense* at Thompe La, Bhutan. (Photo S. Bowes Lyon)

POGONANTHUM

Picture 2

Picture 3

Picture 4

2. *R. anthopogon* ssp. *anthopogon*, a yellow-flowered form with glaucous foliage on the Milke Danda, E. Nepal. The flower colour was uniform in this area.
3. *R. anthopogon* ssp. *anthopogon* 'Betty Graham' A.M. Ludlow & Sherriff 1091 collected in E. Bhutan on the Orka La in 1934. Glendoick.
4. *R. anthopogon* ssp. *hypenanthum* 'Annapurna' A.M. from Stainton, Sykes & Williams 9090 collected in 1954 in Nepal and raised and selected at Glendoick. Most seedlings of this number had much larger leaves and a looser habit and were inferior.

R. anthopogonoides Maxim. 1877 H4-5?

Leaf lower surface with scales, borne at +/- *one level*. **Inflorescence** *very dense*, 10-20-flowered. **Corolla** white or greenish-white, rarely flushed pink; pedicel *pubescent, elepidote*; calyx *erose (irregularly toothed)*. **Distribution** Qinghai, Gansu, possibly N. Sichuan, 3,000-3,400m (10,000-11,000ft).

Although plants are cultivated under this name, none that we have seen are correctly identified and this species is almost certainly not currently in cultivation.

R. cephalanthum Franch. 1885. H4-5

Height 0.1-1.2m, an almost prostrate to upright shrub; leaf-bud scales *persistent*. **Leaves** aromatic, 1-4.7 x 0.5-2.3cm, oblong-elliptic to oval, margins recurved; upper surface dark glossy green, lower surface with fawn to rusty scales overlapping in 2-3 tiers. **Inflorescence** 5-10-flowered, flat-topped to *rounded*. **Corolla** narrowly tubular with spreading lobes, tube 0.6-1.4cm long, white to deep rose (yellow in Nmaiense Group,); calyx 3-7mm long with ciliate margins; stamens 5(-8); ovary lepidote.

This species very much overlaps in distribution with the often taller *R. primuliflorum*. The essential difference between them is the bud scales, persistent in *R. cephalanthum*, deciduous in *R. primuliflorum* and in addition the inflorescence tends to be fuller in *R. cephalanthum*. There are intermediate forms with very few persistent bud-scales (currently classified under *R. cephalanthum*), and maintaining the two taxa as separate species is rather unsatisfactory.

Ssp. *cephalanthum* (*R. chamaetortum* Balf.f. & Kingdon-Ward, *R. crebreflorum* Hutch. & Kingdon-Ward, *R. nmaiense* Hutch. & Kingdon-Ward)

Leaves 1-2.6 x 0.5-1.5cm. **Corolla** tube 0.6-1.3cm long. **Distribution** E. Arunachal Pradesh, N. Burma, N., N.W., Central Yunnan, S., S.E. Tibet, 2,700-4,900m (9,000-16,000ft), moorland, rocky slopes, cliffs and on limestone.

Good, compact, pink-flowered forms can make some of the best garden plants in the section while taller, more open growers are of limited merit. Usually superior to *R. primuliflorum*. Crebreflorum Group includes the smallest growing forms, 5-25cm height, with a low or prostrate, often stoloniferous habit and pale to clear pink flowers. Nmaiense Group from S.E. Tibet and Upper Burma has yellow, pale pink to creamy-white flowers. Introduced 1908, reintroduced 1981->. April-May.

Ssp. *platyphyllum*
(Franch. ex Balf.f. & W.W. Sm.) Cullen 1979

Leaves *2.5-5 x 1.5-2.8cm*. **Corolla** tube *1.3-2cm* long, white to pale rose. **Distribution** N.E. Burma, N., N.W. and Central Yunnan, cliffs and ledges, 3,000-4,400m (10,000-14,500ft).

This subspecies differs from ssp. *cephalanthum* in its larger leaves and consistently larger corolla, and from *R. anthopogon* in its persistent leaf-bud scales and usually in the smaller number of stamens. It differs from *R. collettianum* in its persistent leaf-bud scales, slightly wider leaves and in the number of stamens. It usually differs from both the aforementioned species in its shorter petiole.

Introduced 1914-1929 but lost to cultivation. Reintroduced 1992->. Possibly deserves to be reinstated at specific rank as it seems to be consistently larger than ssp. *cephalanthum* and looks fairly distinctive.

Picture 1

Picture 2

Picture 3

Picture 4

Picture 5

Picture 6

1. *R. cephalanthum* ssp. *cephalanthum* with *R. fastigiatum* growing on the west flank of the Cangshan, Yunnan, China.
2. *R. cephalanthum* ssp. *cephalanthum* SBEC 0751, a very fine pure white-flowered form which has proven to be rather difficult to grow in cultivation.
3. A fine specimen of *R. cephalanthum* ssp. *cephalanthum* growing on a very steep bank near Zhuzipo, Salween-Mekong divide, N.W. Yunnan, China.
4. An excellent pink-flowered form of *R. cephalanthum* ssp. *cephalanthum* near Zhuzipo, Salween-Mekong divide, N.W. Yunnan, China, showing the characteristic rounded, condensed inflorescence.
5. *R. cephalanthum* ssp. *cephalanthum*, a pale pink-flowered clone growing at Glendoick.
6. *R. cephalanthum* ssp. *cephalanthum* Crebreflorum Group, a good pink-flowered selection, at Glendoick.

POGONANTHUM

Picture 7

Picture 8

in the paler-coloured lower surface of the leaf. It differs from *R. cephalanthum* ssp. *platyphyllum* in its deciduous bud scales and larger number of stamens.

Although the handsome foliage and quite showy flowers make this rare species an attractive plant when well grown, it is not easy to please, requiring a sheltered site and very good drainage. Introduced 1880 but lost to cultivation; reintroduced 1969. May.

Picture 1

Picture 2

Picture 3

7. Close-up of branchlets of *R. cephalanthum* ssp. *cephalanthum* showing the persistent bud scales which may be as plentiful as shown here or very few. (Photo R.B.G., E.)
8. *R. cephalanthum* var. *platyphyllum* SSNY 350 photographed on cliffs on the Cangshan mountains, Yunnan, China.

R. collettianum
Aitchison & Hemsl. 1881 H4

Height to 1m, a bushy, fairly upright shrub. Leaf-bud scales deciduous. **Leaves** 3-6*cm, lanceolate* to ovate, slightly recurved; upper surface pale to bright green, usually elepidote, lower surface light yellowish-buff with overlapping scales. **Inflorescence** 5-12-flowered. **Corolla** narrowly tubular funnel-shaped, 1.5-2.4cm long, white, sometimes pink in bud; calyx 3-5.5mm, hairy; stamens (8)-10; ovary lepidote. **Distribution** Afghanistan-Pakistan frontier area, 3,000-4,000m (10,000-13,000ft), stony slopes, often amongst juniper.

R. collettianum differs from *R. anthopogon* in its larger, more lanceolate leaves, the usually greater number of stamens and

1. *R. collettianum* on Mt. Sikaram, Afghanistan. (Photo P. Wendelbo)
2. *R. collettianum* Hedge and Wendelbo 8975 collected on Mt Sikaram, Afghanistan, showing the characteristic narrowly tubular flower shape. Royal Botanic Garden, Edinburgh. (Photo P. Wendelbo)
3. *R. collettianum* Hedge and Wendelbo 8975 at the Royal Botanic Garden, Edinburgh, showing its typical yellowish leaves.

POGONANTHUM

4. Shoot of *R. collettianum* showing the absence of persistent bud scales. (Photo R.B.G., E.)

Picture 4

R. kongboense
Hutch. 1937 H4-5

Height to 1.5m, rarely to 2.5m, a dwarf, compact to broadly upright shrub. Leaf-bud scales deciduous. **Leaves** usually *strongly aromatic*, 0.9-2.8 x 0.4-1.2cm, oblong-lanceolate to oblong-elliptic; upper surface usually matt, dark to greyish-green, lepidote, lower surface fawn to brown with +/- overlapping, plastered scales, pale brown, most with well-developed domed centres. **Inflorescence** 6-12-flowered. **Corolla** variably pilose inside and out, narrowly tubular with spreading lobes, *0.8-1.1cm long, deep strawberry-red*, (?white flushed pink); calyx 3-5mm; stamens *5*; ovary lepidote; capsule sparsely lepidote. **Distribution** S. Tibet, 3,200-4,700m (10,500-15,500ft), forest, rocky slopes, cliffs and moorland.

R. kongboense appears to merge into the very variable *R. primuliflorum*, but typically differs in its smaller and usually deeper-coloured flowers and in the distinctive scales on the lower surface of the leaf. The foliage is strongly aromatic with a pungent, strawberry-like odour.

Deep-coloured forms are quite striking although the flowers are small, and the aromatic scent of the foliage is one of the strongest and most unusual in the genus. White and pink-flowered forms are better referred to *R. primuliflorum* Cephalanthoides Group. *R. kongboense* is a challenge in cultivation, hard to establish and difficult to please. Introduced 1924, reintroduced 1995->. March-May.

Picture 2

Picture 3

2. *R. kongboense* with deep-coloured flowers at Glendoick.
3. Previously known as *R. primuliflorum* 'deep pink', this plant is undoubtedly closer to *R. kongboense*. At Glendoick.

Picture 1

1. *R. kongboense* growing in open woodland below the Tra La, Rong Chu, S.E. Tibet.

R. laudandum
Cowan 1937 H4-5

Height to 0.6m, occasionally more, a fairly compact to leggy small shrub. **Leaves** 1.1-1.8 x 0.6-1cm, oblong to ovate to almost orbicular, retained 2-3 years; upper surface *matt to shiny dark to greyish-green,* lower surface *dark chocolate brown with dense, overlapping scales* borne in 2-3 tiers (lower surface green with cream-fawn scales on cultivated hybrid clones). **Inflorescence** dense, 5-8-flowered. **Corolla** narrowly tubular with spreading lobes, tube 0.5-1.2cm long, white or pink, rarely yellowish (pale lavender-pink as cultivated); calyx 5-6mm; stamens 5-6; capsule very small. **Distribution** S.E. Tibet, 2,900-4,700m (9,500-15,500ft), moraines and open slopes.

This species is a confusing one horticulturally and taxonomically. The few clones (probably two) long cultivated under *R. laudandum* var. *temoense* are almost certainly hybrids, as the scales on the lower leaf surface are green rather than dark brown. These clones have distinctive, relatively early lavender-pink flowers and they often flower in autumn too. Introduced 1924. Botanists have made life even more confusing as the type specimen of var. *temoense* was from the Doshong La while var. *laudandum* was first collected by Kingdon-Ward from the Temo La! After studying this species in the wild, we have concluded that var. *temoense* is a distinctive taxon while var. *laudandum* is probably better considered a form of the very variable *R. primuliflorum*.

Var. *laudandum*

Leaves more than 2 x as long as broad. **Corolla**, usually pink, tube densely pilose outside. Appears to merge with *R. primuliflorum* and may be better considered an extreme form of this species. Often leggy in habit.

This variety is probably not currently in cultivation.

Var. *temoense*
(Kingdon-Ward ex) Cowan & Davidian 1947.

Leaves retained for several years, very dark green, less than 2 x as long as broad, lower surface scales very dense and dark. **Corolla** usually white, tube laxly pilose outside.

Picture 1

1. A magnificent plant of the true *R. laudandum* var. *temoense* on the Temo La, S.E. Tibet.

The small, dark green leaves, retained for several years, with dark brown, almost black scales densely covering the lower surface distinguish the true var. *temoense* from its relatives.

The true var. *temoense* with white flowers, occasionally tinged pink, promises to be one of the best of the subsection. Introduced 1995>. This is very distinct and uniform as seen on four passes in S.E. Tibet. March-April.

Picture 2

Picture 3

Picture 4

2. *R. laudandum* var. *temoense* on the Nyima La, S.E. Tibet.
3. A pink tinged *R. laudandum* var. *temoense* on the Nyima La, S.E. Tibet.
4. This plant has been quite widely grown under the name *R. laudandum* var. *temoense*. It lacks the dark scales on the leaf underside and is probably a natural hybrid. At Glendoick.

R. primuliflorum

Bureau & Franch. 1891 (*R. acraium* Balf.f. & W.W. Sm., *R. cephalanthoides* Balf.f. & W.W. Sm., *R. clivicolum* Balf.f. & W.W. Sm., *R. cremnophilum* Balf.f. & W.W. Sm., *R. gymnomiscum* Balf.f. & Kingdon-Ward, *R. lepidanthum* Balf.f & W.W. Sm., *R. praeclarum?* Balf.f. & Farrer, *R. tsarongense* Balf.f. & Forrest) H4-5

Height 0.05-1.80m and over, a low and compact to erect and leggy shrub. Branchlets densely lepidote. Leaf-bud scales *quickly deciduous*. **Leaves** 0.8-3.5 x 0.5-1.4cm, lanceolate, narrowly ovate to oval; upper surface dark green, shiny, lepidote or elepidote, lower surface with fawn to dark brown, overlapping scales, *often in 2-3 distinct tiers*. **Inflorescence** dense, 5-10-flowered. **Corolla** narrowly tubular with spreading lobes, 1-1.6cm long, white to pink, often yellowish-orange at base; calyx 2.5-5mm; stamens 5. **Distribution** N. & N.W. Yunnan, S. & S.E. Tibet, S.W. to N. Sichuan, probably Gansu, 3,400-4,600m (11,000-15,000ft), forest margins, cliffs to alpine meadows.

R. primuliflorum is closely related to *R. cephalanthum*, differing in its completely deciduous bud scales. These two species do merge and they often grow on the same mountainsides though sometimes in differing habitats. In the deepest pink, upright forms, it also appears to merge with *R. kongboense*. Probably also merges with *R. laudandum* var. *laudandum*.

In habit, this is a very variable species from tall and rangy to dwarf and compact. Some of the taller forms take many years to bloom and the flowers can be small. Plants under Cephalanthoides Group are characterised by the hairy outer surface of the corolla tube; some of these with deep-coloured flowers are close to *R. kongboense*, while others such as the excellent clone 'Doker La' A.M. seem closer to *R. trichostomum*. Introduced 1910, reintroduced 1986->. April-May.

Picture 1

Picture 2

Picture 3

Picture 4

Picture 5

1. A pink-flowered *R. primuliflorum* on the Nyima La, S.E. Tibet.
2. *R. primuliflorum*, a pink-flowered form at Huanglongsi, N. Sichuan, China.
3. *R. primuliflorum* SSNY 148, a white-flowered form on Bei ma Shan, N.W. Yunnan, China.
4. *R. primuliflorum* Cephalanthoides Group 'Doker La' A.M. which originally came from Gigha, W. Scotland. This clone seems to be an intermediate form between *R. primuliflorum* and *R. trichostomum*.
5. A superb pink form of *R. primuliflorum* on the Dokar (Showa) La, Pome, S. E. Tibet.

POGONANTHUM

Picture 6

6. A shoot of *R. primuliflorum* showing the densely lepidote leaves and branchlets and the non-persistent bud scales. (Photo R.B.G., E.)

R. fragrans (Adams) Maxim. This species differs from *R. primuliflorum* in its pubescent corolla lobes inside and the uniform pale yellow scales on the leaf underside. Siberia and Mongolia. Apparently recently introduced.

R. sargentianum
Rehder & E.H. Wilson 1913 H5

Height to 0.6m but usually not over 0.3m in cultivation, a usually *dense and symmetrical* low shrub. Leaf bud scales *persistent*. **Leaves** 0.9-2.3 x 0.5-0.8cm, elliptic to oval; upper surface dark green and shiny, elepidote, lower surface pale brown to rust, 2-3 tiers of overlapping scales. **Inflorescence** 5-12-flowered. **Corolla** narrowly tubular with spreading lobes, 1.2-2.3cm long, lepidote and pubescent outside, *pale yellow to cream*; calyx 2-4mm, margins ciliate; stamens 5; capsule sparsely lepidote. **Distribution** Central Sichuan, 3,000-4,300 (10,000-14,000m), on exposed rocks and cliffs.

This is a distinct species in the combination of its dense, low habit, its persistent leaf-bud scales and the yellow to cream corolla, lepidote on the outside.

Although *R. sargentianum* is not always easy to grow well, at its best it is a first class plant, free-flowering from a young age and one of the showiest species in the section. Introduced 1903-4. April-June.

Picture 1

Picture 2

Picture 3

Picture 4

1. A yellow-flowered *R. sargentianum* at Glendoick.
2. *R. sargentianum* at Glendoick.
3. *R. sargentianum* 'Whitebait' A.M. at Glendoick, a cream-coloured clone which originated at the Royal Botanic Garden, Edinburgh.
4. A close-up of *R. sargentianum* showing the typical elliptic leaves, elepidote above, with densely lepidote branchlets and buds. (Photo R.B.G., E.)

R. trichostomum

Franch. 1895 (*R. fragrans* Franch. (not Adams) Maxim., *R. ledoides* Balf.f. & W.W. Sm., *R. radinum* Balf.f. & W.W. Sm., *R. sphaeranthum* Balf.f. & W.W. Sm.) H(3-)4

Height 0.3-1.2m, a dwarf shrub, rangy when young or in shade, upright, but wide and compact with age in sun. Branchlets usually with some hairs; leaf-bud scales usually deciduous. **Leaves** 0.8-3.4 x 0.2-0.8cm, *linear, oblanceolate to linear-lanceolate*, margins recurved; upper surface dark green and matt, lower surface usually pale brown with overlapping scales in 2-3 tiers. **Inflorescence** 8-20-flowered, *rounded*. **Corolla** narrowly tubular with spreading lobes, 0.8-1.6cm long, white to rose; calyx 0.5-2mm; stamens 5. **Distribution** N. & N.W. Yunnan, S.W. & Central Sichuan, 2,400-4,600m (8,000-15,000ft) but usually between 3,000-4,000m (10,000-13,000ft), forest, scrub or open slopes, habitat often surprisingly dry.

R. trichostomum is characterised by its narrow leaves with recurved margins and its rounded truss, and in cultivation it is later flowering than its nearest relatives. It usually has a smaller calyx, larger truss and is freer-flowering from a younger age than most forms of *R. primuliflorum*.

Variable in hardiness and in quality of flower. In its best clear pink and pure white-flowered forms this is a first rate, free-flowering species. Some strong-growing introductions may be hybrids, perhaps with *R. primuliflorum*. Introduced 1908-21 reintroduced c.1990->. May-June.

R. radendum Fang 1939. Only known from type collection, this seems to be just an extreme form of *R. trichostomum* with particularly hairy branchlets and it should therefore probably be treated as a synonym.

Picture 1

Picture 2

Picture 3

Picture 4

Picture 5

1. *R. trichostomum* N.E. of Zhongdian, N.W. Yunnan, China, showing some of the colour variations in this species.
2. *R. trichostomum*, N.E. of Zhongdian, N.W. Yunnan, China, on a dry bank in pine forest.
3. *R. trichostomum*, a very fine pink-flowered form, at Glendoick.
4. A white-flowered form of *R. trichostomum* Kingdon-Ward 3998 collected in the Yunnan-Sichuan border region of China in 1921 (as var. *radinum*.)
5. A shoot of *R. trichostomum* showing the characteristic lanceolate leaf shape and densely lepidote branchlets. (Photo R.B.G., E.)

Section Rhododendron

Scales entire, crenulate or undulate, bud scales fringed with simple hairs. Capsule valves hard and woody on dehiscence.

Subsection Afghanica

A monotypic subsection.

R. afghanicum Aitch. & Hemsl. 1880 H3

Height to 0.5m, a low shrub, with ascending prostrate branches, rather straggly. Branchlets lepidote and sometimes pubescent. **Leaves** thick, 3-8 x 1.-2.5cm, narrowly elliptic to elliptic; upper surface olive to bluish-green, usually elepidote, lower surface pale green, with unequal scales 1-2 x their own diameter apart. **Inflorescence** 8-16-flowered, terminal, rachis *elongated* 1.3-5cm long. **Corolla** campanulate, 0.8-1.3cm long, base tubular, white to greenish-white; calyx irregular, 1-5mm; stamens 10; ovary densely lepidote; style bent to straight, elepidote. **Distribution** E. Afghanistan and Pakistan border, 2,100-3,000m (7,000-10,000ft), usually in fairly dense forest, on rocks and cliffs.

This species was formerly placed with *R. hanceanum* but, due to its very distinctive inflorescence, Cullen has placed it in its own monotypic subsection.

This rarely cultivated species has rather small flowers and is of little horticultural value. It flowered at Glendoick but was killed by a hard winter. In its native habitat it is known to be particularly poisonous to livestock. Introduced 1880 then lost, reintroduced 1969. June-July.

Picture 1

Picture 2

1. The natural habitat of *R. afghanicum* in E. Afghanistan, showing armed guards with two plants. (Photo P. Wendelbo).
2. *R. afghanicum* in the wild. (Photo P. Wendelbo).

Subsection Baileya

A monotypic subsection.

R. baileyi
Balf.f. 1919 (*R. thyodocum* Balf.f. & R.E. Cooper) H(3-)4

Height 0.5-2m, a straggly to fairly compact shrub. Branchlets lepidote. **Leaves** 2.1-7 x 0.8-3.3cm, narrowly elliptic, elliptic to obovate; upper surface with mostly persistent scales, lower surface usually dark brown, with *overlapping, notched or* *toothed scales.* **Inflorescence** *4-9(18)-flowered.* **Corolla** rotate, flattish, to 3cm across, 0.8-1.6cm long, purple to reddish-purple, with or without spots; calyx 2-4mm, lepidote, +/- ciliate; stamens 10; ovary lepidote; style *sharply deflexed,* short, usually glabrous. **Distribution** Sikkim, Bhutan, S. Tibet, 2,400-4,300m (8,000-14,000ft), conifer forest to open hillsides.

R. baileyi is undoubtedly quite closely related to *R. lepidotum* and probably ought to be returned to subsection Lepidota. *R. baileyi* differs from *R. lepidotum* in its greater stature, its larger and usually more evergreen leaves, in the dense covering of notched or toothed scales on the leaf underside, the many-flowered inflorescence and the sharply deflexed style. *R. cowanianum*, with a similar upright habit, is completely deciduous and is also differentiated by the sparse rather than dense covering of scales on the leaf lower surface.

This is a rather neglected species with striking, usually freely-produced flowers. Some forms are sparse and straggly, while recent introductions have proved to be relatively compact. Introduced 1913, 1930s and 40s, reintroduced 1985->. April-May.

Picture 1

Picture 2

1. A good clone of *R. baileyi* from the Rhododendron Species Botanical Garden, Glendoick.
2. *R. baileyi* K. Rushforth 911 from W. Bhutan. This introduction is more compact than others we have seen. Glendoick.

Picture 3

3. *R. baileyi* A.M. clone, Glenarn, W. Scotland, showing the many-flowered inflorescence (compared to that of *R. lepidotum*).

Subsection Boothia

Low to medium-sized shrubs, often epiphytic in the wild. Bark sometimes smooth, reddish or brown. Young growth often bristly. **Leaves** evergreen, lower surface whitish-papillose with rimmed or vesicular scales deeply sunk in pits. **Inflorescence** 1-10-flowered. **Corolla** campanulate to almost rotate, *yellow or white;* calyx well-developed, lobed; stamens 10; ovary densely lepidote, *tapering into style;* style lepidote and *sharply deflexed.* **Distribution** Arunachal Pradesh, S.E. Tibet, Upper Burma and W. & N.W. Yunnan.

The species of this subsection are epiphytic and are characterised by their leaves with rimmed or vesicular scales and their yellow to white corollas. Subsection Boothia is related to subsections Edgeworthia and Maddenia in general morphology and in the epiphytic habit and winged seeds. The vesicular scales suggest a relationship with subsection Trichoclada.

All species in this subsection are rather tender, only suitable for milder gardens. Where conditions are suitable, they make moderate to good garden plants with the exception of the small-flowered *R. leptocarpum*. All require good drainage.

R. boothii
Nutt. 1853 (*R. mishmiense* Hutch. & Kingdon-Ward) H1-2?

Height to 2m+, a usually an epiphytic shrub, inclined to be straggly. Branchlets lepidote, *covered with twisted bristles*. **Leaves** leathery, *7.8-12.6 x 3.5-6.2cm*, ovate to elliptic; upper surface usually with *bristles on midrib and margins,* lower surface with brown scales their own diameter apart. **Inflorescence** 3-10-flowered. **Corolla** campanulate, 2-3cm long, dull to bright yellow, sometimes spotted; calyx *large*, 7-15mm, lepidote and ciliate. **Distribution** Arunachal Pradesh, India and S. Tibet, 1,800-3,000m (6,000-10,000ft), usually epiphytic, occasionally on rocks.

R. boothii differs from *R. chrysodoron* in its large calyx and it differs from *R. sulfureum* in its bristly leaves and branchlets.

The flowers of *R. boothii* are small compared with the leaf size. We have been unable to locate this species in cultivation in the British Isles but it used to be successfully grown out of doors in Cornwall. It is reported to be in Australia and California but we have not been able to confirm if these plants are correctly identified. Plants formerly under the name *R. mishmiense* (now Mishmiense Group) were separated on account of their spotted flowers. Introduced 1852, 1928 (as *R. mishmiense*). April-May.

Picture 1

Picture 2

1. *R. boothii*, Royal Botanic Garden, Edinburgh. (Photo R.B.G., E.).
2. *R. boothii* (Mishmiense Group) in U.S.A. (Photo A.P. Dome).

R. chrysodoron
(Tagg ex) Hutch. 1934 (*R. butyricum* Kingdon-Ward) H2-3

Height to 1.80m, an upright or bushy shrub. Branchlets bristly at first; bark reddish, peeling. **Leaves** 4.5-8.8 x 2-4.5cm, broadly to oblong elliptic, hairy when young; lower surface with golden yellow scales 1-2 x their own diameter apart. **Inflorescence** 3-6-flowered. **Corolla** campanulate, 3-4cm long, bright canary yellow; calyx *minute 1-3mm, rim undulate*, lepidote and bristly. **Distribution** N. Burma and W. Yunnan, 2,000-2,600m (6,500-8,500ft), epiphytic or on rocks.

R. chrysodoron differs from *R. boothii* in its smaller leaves and calyx, and from *R. sulfureum* in its smaller calyx.

This rare species has the finest yellow flowers in the subsection but unfortunately it is also one of the least hardy and with very early growth, is only suitable outdoors for mildest gardens. Introduced 1924, reintroduced 1953. February-April.

Picture 1

1. *R. chrysodoron* at Hamish Gunn's garden, Edinburgh. (Photo H. Gunn).

R. leptocarpum
Nutt. 1854 (*R. micromeres* Tagg) H3-4?

Height 0.9-1.8m, a lax to spreading shrub. Bark dark mahogany-brown on older wood, peeling. **Leaves** 3-8 x 1.5-3.6cm, narrowly elliptic to elliptic; upper surface bright green, lower surface greenish-grey, with scales ½-1½ x their own diameter apart, unequal, sunk in pits. **Inflorescence** 3-8, perhaps to 13-flowered. **Corolla** rotate-campanulate, *0.8-1.4cm* long, pale yellow to white, pedicels *long, 2-4.2cm* ;

calyx 2-4mm, lobes *usually reflexed*. **Distribution** Widely distributed and common from E. Bhutan, S.E. Tibet, through Arunachal Pradesh to W. Yunnan, 2,400-4,300m (8,000-14,000ft), epiphytic on other rhododendrons and conifers and on rocks.

R. leptocarpum was moved by Cullen from subsection Glauca. It differs from other members of subsection Boothia in its disappointingly small, +/- pale yellow flowers with a usually reflexed calyx. It resembles *R. brachyanthum* in its flowers and *R. auritum* in its reflexed calyx lobes.

This species tends to be of little horticultural value as the flowers are generally rather small. Variable in leaf shape and size. It is usually epiphytic in nature so requires very good drainage. Introduced 1922->, reintroduced 1965, 1990. May-June.

Picture 1

Picture 2

1. *R. leptocarpum*, E. Nepal, showing the small pale yellow flowers. (Photo R. McBeath).
2. *R. leptocarpum* in bud, growing epiphytically on the south side of the Doshong La, S.E. Tibet.

R. leucaspis
Tagg 1929 H3

Height to 1m, a low shrub, usually growing wider than high, straggly in shade. Branchlets *hairy/bristly*. **Leaves** 3-6 x 1.8-3cm, broadly elliptic; upper surface and margins *densely hairy/bristly*, lower surface with vesicular scales, sunken in pits; petiole lepidote and hairy/bristly. **Inflorescence** 1-2 but usually 2-flowered. **Corolla** *almost rotate*, c. 5cm across, white, sometimes tinged pinkish; calyx *large, hairy, green or reddish*; stamens hairy with *very dark anthers*. **Distribution** S.E. Tibet, 2,400-3,000m (8,000-10,000ft), cliff faces and grassy banks, sometimes epiphytic.

As cultivated, this species is easily recognised by its almost rotate white flowers with chocolate-brown anthers. As seen in in S.E. Tibet, it appears to merge with the closely related to *R. megeratum* which as cultivated generally has yellow or cream flowers and smaller, less hairy leaves.

At its best this is a showy species but its early flowers are rather frost-vulnerable and it is somewhat prone to powdery mildew. Taxonomically, it should probably be considered a subspecies or variety of *R. megeratum*. Introduced 1924-6. March-April.

Picture 1

Picture 2

1. *R. leucaspis* at Arduaine Garden, W. Scotland, showing its typical rounded habit in an open situation.
2. *R. leucaspis*, Glendoick.

BOOTHIA

Picture 3

Picture 4

3. *R. leucaspis* growing on mossy rocks on the south side of the Doshong La, Pemako, S.E. Tibet.
4. Young growth of *R. leucaspis* showing the characteristic densely hairy leaves. Glendoick.

R. megeratum
Balf.f. & Forrest 1920 (*R. tapeinum* Balf.f & Forrest) H3

Height 0.3-0.9m, a bushy, often compact, dwarf shrub. Young shoots hairy. **Leaves** 1.5-4 x 1-2cm, elliptic to orbicular; upper surface glabrous except for a few hairs at the base of the midrib, margins usually hairy, lower surface *glaucous-whitish* with vesicular scales sunken in pits; petiole hairy. **Inflorescence** 1-3-flowered. **Corolla** broadly campanulate to rotate-campanulate, 2-3cm long, yellow, cream, off white with a yellow blotch, sometimes spotted deeper; calyx 7-10 mm; style short, sharply deflexed. **Distribution** S.E. Tibet, Arunachal Pradesh, Upper Burma, N.W. Yunnan, 2,400-4,100m (8,000-13,500ft), on mossy rocks and cliffs or epiphytic.

As cultivated, *R. megeratum* is smaller-leaved and less hairy than *R. leucaspis*, with yellow rather than white flowers. As seen in the wild in S.E. Tibet, it has white flowers with a yellow blotch and may merge with *R. leucaspis*. *R. valentinianum* differs in being generally larger in all parts, with funnel-campanulate, yellow flowers and hairs on the upper leaf surface.

At its best, *R. megeratum* is a fine dwarf species with deep yellow flowers and attractive foliage but many forms tend to be tender and/or shy flowering. It requires good drainage. Introduced 1914->. March-April.

Picture 1

Picture 2

Picture 3

1. An exceptionally well-flowered specimen of *R. megeratum*. (Photo A. Reid).
2. *R. megeratum* Bodnant form at Glendoick.
3. A form of *R. megeratum* with cream-coloured flowers with yellow markings, growing epiphytically on the south side of the Doshong La, Pemako, S.E. Tibet.

R. sulfureum
Franch. 1887 (*R. cerinum* Balf.f. & Forrest, *R. commodum* Balf.f. & Forrest, *R. theiochroum* Balf.f. & Forrest.) H2-3

Height 0.6-1.6m, a fairly compact to straggly shrub. Young growth often hairy at first; bark brown, smooth and often peeling. **Leaves** 2.6-8.6 x 1.3-4.2cm, obovate to elliptic; upper surface dark green, quite shiny, margins often hairy, lower surface +/- glaucous, scales unequal, *with upturned rims*, ½ to their own diameter apart, sunk in pits. **Inflorescence** 3-8-flowered. **Corolla** campanulate, *1.3-2cm* long, greenish to bright or deep yellow, unspotted; calyx *3-6mm*. **Distribution** S.E. Tibet through Upper Burma to W. & N.W. Yunnan, 2,100-4,000m (7,000-13,000ft), cliffs, rocks or epiphytic.

This variable species is characterised by its scales with upturned rims. It differs from *R. chrysodoron* in its smaller flowers and larger calyx.

Although the flowers tend to be rather small in comparison to the foliage, they are of a deep colour and are long-lasting, and the peeling bark is an added bonus. In its hardier forms, this species is tougher than its relatives and has succeeded well out of doors in a sheltered site at Glendoick. Introduced 1910->, reintroduced 1981. March-April.

Picture 1

Picture 2

1. A fine plant of *R. sulfureum* SBEC 0249 at Portmeirion, W. Wales.
2. *R. sulfureum* SBEC 0249 from Cangshan at Glendoick.

Picture 3

3. *R. sulfureum* SBEC 0249 showing the characteristic large calyx (in the top left hand corner of the inflorescence). Glendoick.

Subsection Camelliiflora

A monotypic subsection.

R. camelliiflorum

Hook.f. 1853 (*R. cooperi* Balf.f., *R. sparsiflorum* Nutt., *R. lucidum* Nutt., H3-4

Height 0.6-3m, a loose to fairly compact, usually epiphytic shrub. Branchlets densely lepidote; bark reddish, peeling. **Leaves** 3.8-12.3 x 1.5-3.8cm, narrowly to broadly elliptic, retained 1-2 years; upper surface dark green, partly lepidote at first, lower surface pale green to brownish with *dense, usually touching or slightly overlapping*, brown scales. **Inflorescence** 1-3-flowered. **Corolla** *broadly tubular with spreading lobes*, c. 3.8cm across, white, cream, white marked pink, pink or deep wine-red; calyx 5-11mm, with some scales; stamens *12-16*; ovary lepidote; style usually shorter than stamens, *sharply bent*. **Distribution** E. Nepal to Bhutan and probably Arunachal Pradesh, 2,700-3,700m (9,000-12,000ft), plentiful on many trees, sometimes rocks.

Variable but very distinct, showing some relationship with subsections Boothia and Maddenia, *R. camelliiflorum* is quite easily identified by its densely lepidote lower leaf surface, and the small, almost flat flowers with 12-16 stamens.

An interesting species, of somewhat limited garden value due to its small flowers which do resemble those of a camellia. It requires perfect drainage and, while reasonably hardy, it is liable to be short-lived out of doors owing to its preference for being perched in a tree. Introduced 1851, reintroduced 1970s->. Uncommon in cultivation. June-July.

Picture 1

Picture 2

Picture 3

1. *R. camelliiflorum*, a white, pink-tinged form in E. Nepal showing the camellia-like shape of the corolla and the large number of stamens (12-16). (Photo G.F. Smith)
2. *R. camelliiflorum* K. Rushforth 1289 showing the characteristic sharply bent style. Glendoick.
3. *R. camelliiflorum* Sinclair & Long 5696. Wine-red-flowered forms are evidently quite plentiful in Bhutan. This shows the sharply bent style and the scales on the outside of the corolla. (Photo R.B.G., E.)

Subsection Campylogynum

A monotypic subsection

R. campylogynum

Franch. 1884 (*R. caeruleo-glaucum* Balf.f. & Forrest, *R. campylogynum* var. *celsum* Davidian, *R. cerasiflorum* Kingdon-Ward, *R. charopoeum*, Balf.f. & Forrest, *R. cremastum* Balf.f. & Forrest, *R. damascenum* Balf.f. & Forrest, *R. glauco-aureum* Balf.f. & Forrest, *R. myrtilloides* Balf. & Kingdon-Ward, *R. rubriflorum* Kingdon-Ward). H3-5

Height 0.05-1.3m, prostrate, mound-shaped to erect small shrubs. Branchlets +/- lepidote. **Leaves** evergreen, 0.7-3.7 x 0.3-1.2cm, obovate to narrowly elliptic; upper surface dark to mid-green, *often shiny*, rarely lepidote, lower surface *glaucous* or pale green (Cremastum Group), with *deciduous, distant, vesicular scales, 1-6 x their own diameter apart*. **Inflorescence** 1-3-flowered. **Corolla** campanulate, 1-2.3cm long, very variable, claret, red, purple to black-purple, pink, cream, often with a bloom, *usually held above the foliage on long pedicels*; calyx 1-6mm; stamens (8-)10; ovary lepidote; style glabrous, *thick and sharply bent*. **Distribution** Arunachal Pradesh, S. and S.E. Tibet, N.E. Upper Burma, N.W. to S.W. and N.E. Yunnan, 2,400-4,900m (8,000-16,000ft), among larger rhododendrons, rocks, cliffs to open hillsides.

This species is easily recognised in flower by its thimble-shaped flowers on long stalks. Horticulturally the closest relations are in subsection Glauca which are larger in all parts, especially leaf size. There are cultivated plants with pink to purplish flowers which may be natural hybrids with members of subsection Glauca such as *R. charitopes* var. *tsangpoense*.

R. campylognum is very variable and was formerly divided into several species. It is an extremely popular garden plant and it is worth differentiating between the various former species and varieties from a horticultural point of view.

R. campylogynum type form. Leaves dark and shiny. Corolla claret to plum-coloured with a bloom.

Celsum Group, the tallest form with an erect habit. Corolla plum-purple.

Charopoeum Group, compact and spreading with the largest flowers, rose to plum-purple. This may be a natural hybrid.

Cremastum Group, habit compact to erect, leaves green on the lower surface. Flowers of the commonly-cultivated clones are cerise or red.

Leucanthum (clone), low and compact with cream-coloured flowers.

Myrtilloides Group, very small leaves and purple to deep purple flowers.

R. campylogynum is one of the finest dwarfs and a popular species where it grows well. We have collected different forms over the years and now have at least 15. It is variable in hardiness, some forms suffering in a hard winter at Glendoick, and it dislikes high summer temperatures and dry soil. Introduced 1912->, reintroduced 1981->. May-June.

Picture 1

Picture 2

Picture 3

Picture 4

Picture 5

Picture 6

1. A selected clone of *R. campylogynum* SBEC 0587 from Cangshan, C.W. Yunnan, China, showing the bloom on the outside of the flowers. Glendoick.
2. A form of *R. campylogynum* known as 'Claret', at Glendoick.
3. A form of *campylogynum* with salmon pink flowers, Royal Botanic Garden, Edinburgh.
4. An unusual-coloured seedling of *R. campylogynum* with 'Claret' behind, Glendoick.
5. *R. campylogynum* Charopoeum Group with the largest flowers of the species, Glendoick.
6. *R. campylogynum* Cremastum Group with cerise flowers, Glendoick.

CAMPYLOGYNA

Picture 7

Picture 8

Picture 9

7. *R. campylogynum* Cremastum Group 'Bodnant Red' A.M. at I. & M. Young's garden, Aberdeen, E. Scotland. This is one of the very few lepidotes with red flowers.
8. *R. campylogynum* 'Leucanthum', a cream-flowered clone, a selection made by the great plantsman Collingwood Ingram. Glendoick.
9. *R. campylogynum* Myrtilloides Group which has the smallest leaves and flowers of the species. Glendoick.

Subsection Caroliniana

Now considered a monotypic subsection.

R. minus Michaux 1792 H3-5

Height 1-6m, generally much less in cultivation, a compact to straggly shrub. Branchlets sparsely lepidote, green or purplish. **Leaves** evergreen, dark to light green (particularly in white-flowered forms), 5-11.8 x 2-6.5cm, broadly elliptic to lanceolate; upper surface +/-lepidote, midrib hairy, lower surface with small-rimmed, brownish scales, *touching to their own diameter apart*. **Inflorescence** 4-12-flowered. **Corolla** narrowly to openly-funnel-shaped, 2.1-3.8cm long, purplish-pink, pink to white; calyx 1-2mm, *sparsely ciliate, fairly deeply-lobed*; stamens 10; ovary lepidote; style +/- lepidote at base. **Distribution** Tennessee and N. Carolina south to Georgia and Alabama, U.S.A., mountains to plains.

This species is morphologically closely related to the members of subsection Heliolepida, the chief differences being in the calyx. Horticulturally and geographically they are quite distinct.

Var. minus (*R. carolinianum* Rehder, *R. cuthbertii* Small, *R. punctatum* Andrews)

Leaf apex *acute or acuminate*; branchlets usually not erect and rigid.

R. minus has proven invaluable as a parent of hardy hybrids such as 'P.J.M.', but the species itself often has poor chlorotic foliage and a weak root system. Typical var. *minus* is a useful plant in S.E. U.S.A. for its hardiness and heat tolerance, but does not grow well in cooler northern climates where it tends not to ripen its wood. Plants formerly known as *R. carolinianum* now Carolinianum Group from the northern end of the distribution of *R. minus* are generally smaller growing than typical var. *minus* with a more openly funnel-shaped corolla. It has proved easier to grow in northern areas with cool summers (such as Scotland), where it flowers quite freely. Introduced 1786->. May-June.

Var. *chapmanii* (Gray) Duncan & Pullen 1962 H3

In habit, *more erect* than var. *minus*, partly stoloniferous. **Leaves** *smaller, oval to oblong-oval, rugose with a more rounded apex*. **Distribution** Florida gulf country, sand dunes, open pinelands and dry creek banks, rare.

While quite hardy, this variety is a very poor performer in northern areas which lack sufficient summer heat. Some plants labelled var. *chapmanii* in cultivation with non-rugose leaves are referable to var. *minus*. Var. *chapmanii* is rarely cultivated outside the U.S.A. April-May?

Picture 1

Picture 2

Picture 3

Picture 4

Picture 5

1. *R. minus* var. *minus* ex Georgia, U.S.A., growing at the Arnold Arboretum, Massachusetts, E. U.S.A. This shows the long corolla tube of typical var. *minus*.
2. *R. minus* var. *minus* Carolinianum Group, an unusual form with a blotched corolla, Gregory Bald, Pennsylvania, U.S.A.
3. *R. minus* var. *minus* Carolinianum Group, a pink-flowered form, Glendoick, showing the typical widely-funnel shaped corolla typical of this group.
4. *R. minus* var. *minus* Carolinianum Group, a white-flowered clone, Glendoick.
5. *R. minus* var. *chapmanii* ex Florida, growing at the Arnold Arboretum, Massachusetts, E. U.S.A. Note the rugose leaves with an almost rounded apex.

Subsection Cinnabarinum

Medium to tall, upright to umbrella-shaped shrubs. **Leaves** usually evergreen, often glaucous, lower surface densely lepidote. **Inflorescence** terminal or axillary, flowers usually pendulous. **Corolla** fleshy, tubular to campanulate, very variable in colour, yellow, orange, red to plum-crimson or purple, *nectar held in 5 drops in the base of the corolla*; calyx small; stamens 10; ovary lepidote; style straight, sometimes hairy near base. **Distribution** E. Nepal to S.E. Tibet and Upper Burma.

The species in this subsection are usually easily identified by their pendulous flowers with nectaries at the base of the corolla. This subsection is related to subsection Tephropepla but is horticulturally very

CINNABARINA

distinct. *R. cinnabarinum* ssp. *tamaense* and ssp. *xanthocodon* Purpurellum group are related to and may even merge with *R. oreotrephes* in subsection Triflora.

The species in subsection Cinnabarina and their many hybrids were extremely popular for their striking fleshy, often tubular, semi-pendant to pendant flowers but alas the majority are highly susceptible to powdery mildew, resulting in a decrease in their use in the garden.

R. cinnabarinum Hook.f. 1849. H3-4

Height to 7m, usually less, an upright shrub, sometimes spreading with age. Branchlets often purplish, lepidote; bark shaggy, light brown with age, peeling. **Leaves** evergreen (except ssp. *tamaense*), 3-11.5 x 1.8-5.5cm, broadly to narrowly elliptic to lanceolate; upper surface deep to medium green, *usually elepidote, often glaucous*, especially when young, sometimes shiny when mature, lower surface often glaucous, with +/- unequal scales, touching to their own diameter apart. **Inflorescence** 2-9-flowered, *terminal*. **Corolla** *fleshy* tubular to tubular-campanulate, semi-pendant, elepidote outside, 2.6-5cm long, yellow, orange-yellow, orange, pink, red, purple, plum-crimson or mixed colours; calyx minute, 1-2mm, lobed or just a rim; stamens 10; ovary densely lepidote; style +/- hairy at base.

R. keysii, the closest relative of *R. cinnabarinum*, differs in its tubular flowers from mostly axillary buds. The waxy, bell-shaped, pendulous flowers separate *R. cinnabarinum* from all other non-vireya lepidote species.

This species is much loved on account of its pendant, tubular, fleshy flowers of many colours and its often fine, glaucous foliage. Many forms are particularly susceptible to powdery mildew which may result in defoliation and even death but it is otherwise fairly easy to grow. All forms take a few years to flower freely and young plants are rather vulnerable to bark split from unseasonable frosts. Introduced 1849->, reintroduced many times. April-July.

Ssp. *cinnabarinum*
(*R. blandfordiiflorum* Hook.f., *R. roylei* Hook.f.)

Leaves narrower than in ssp. *xanthocodon*, usually *elepidote* on upper surface. **Corolla** +/- tubular-campanulate, red to orange to yellow or a combination of colours.

Blandfordiiflorum Group. Corolla often narrowly-tubular, usually bi- or tricoloured. Some forms are late-flowering. Particularly susceptible to powdery mildew.

Roylei Group. Leaves glaucous. **Corolla** red to plum-crimson with a bloom. Moderately susceptible to powdery mildew. Distribution E. Nepal to Bhutan and S.E. Tibet, 2,100-4,000m (7,000-13,000ft), common in thickets, forests and on hillsides. May-July

Picture 1

Picture 2

1. An orange-flowered clone of *R. cinnabarinum* ssp. *cinnabarinum*, Glendoick.
2. A yellow-flowered *R. cinnabarinum* ssp. *cinnabarinum* on the Milke Danda. E. Nepal.

Ssp. *xanthocodon*

(Hutch.) Cullen 1978 (*R. concatenans* Hutch., *R. cinnabarinum* var. *pallidum* Hook.f., var. *purpurellum* Cowan)

Leaves *broader* than var. *cinnabarinum*; upper surface *usually persistently lepidote*. **Corolla** *tubular-campanulate to campanulate*, yellow, orange, apricot or purple. Intermediates between the two subspecies occur. **Distribution** Arunachal Pradesh, Bhutan, S. Tibet, 3,050-3,950m (10,000-14,000ft).

Plants under the name ssp. *xanthocodon* typically have medium-green leaves and smallish, pale yellow flowers. In our observation, ssp. *xanthocodon* as cultivated does not necessarily have the relatively wide leaves or lepidote upper surface which Cullen describes and we call into question the validity of ssp. *xanthocodon* as it presently stands. It is only moderately susceptible to powdery mildew. Introduced 1924->. May-June.

Concatenans Group (likely to be reinstated as var. *concatenans*). Habit rounded. **Leaves** large, broad, very glaucous. **Corolla** apricot-yellow. **Distribution** Pemako, S.E. Tibet.

Only plants from K.W. 5874 and recent reintroductions from Pemako, S.E. Tibet, should be included here. After fairly extensive study of large populations of this species in Pemako, it is clear that this is a stable and uniform taxon with no overlap with typical *xanthocodon*. Indeed it is probably more closely related to ssp. *cinnabarinum* Roylei group and should be given at least varietal status. Introduced 1924, reintroduced 1995->.

Purpurellum Group (likely to be reinstated as var. *purpurellum*). **Leaves** relatively small and dark. **Corolla**, short campanulate, plum-purple to pinkish-mauve.

This seems to be related to *R. oreotrephes* and some plants seen in S.E. Tibet seem to be intermediate between the two taxa. Somewhat resistant to powdery mildew. S. Tibet. Introduced 1936.

Ssp. *tamaense* (Davidian) Cullen 1978

Leaves *almost deciduous*, lower surface scales usually *2-5 x their own diameter apart*. **Corolla** *purple, lepidote outside*. **Distribution** N. Triangle, N. Burma, forest margins and fir forest, 2,700-3,200m (9,000-10,500ft), to E. and S. of the remainder of the subsection.

This rare subspecies is not of much garden merit having a rather leggy habit and not particularly showy flowers. Not as susceptible to powdery mildew as most of its relatives. Introduced 1953. May-June.

Picture 3

Picture 4

Picture 5

3. *R. cinnabarinum* ssp. *cinnabarinum* Blandfordiiflorum Group with typical bicoloured flowers, in June Sinclair's garden, Hood Canal, Washington State, W. U.S.A.
4. *R. cinnabarinum* ssp. *cinnabarinum* Roylei Group, Glendoick.
5. *R. cinnabarinum* ssp. *xanthocodon* showing the typically broader leaves than ssp. *cinnabarinum* and smallish pale yellow flowers.

CINNABARINA

Picture 6

Picture 7

Picture 8

Picture 9

6. *R. cinnabarinum* (ssp. *xanthocodon*) Concatenans Group on the south side of the Doshong La, Pemako, S.E. Tibet, in its type location.
7. *R. cinnabarinum* (ssp. *xanthocodon*) Concatenans Group showing the characteristic wide, glaucous leaves.
8. *R. cinnabarinum* (ssp. *xanthocodon*) Purpurellum Group with its small leaves and plum-purple flowers. Royal Botanic Garden, Edinburgh.
9. *R. cinnabarinum* ssp. *tamaense* Kingdon-Ward 21003, Glendoick. Nearly all the leaves shown are young growth, the plant being almost deciduous.

R. keysii
Nutt. 1855 (*R. igneum* Cowan) H3-4

Height 1.2-6m, an often straggly or leggy, vigorous shrub. Branchlets lepidote; bark fairly smooth to roughish, fawn to brown. **Leaves** evergreen, 5-15 x 1.5-3.6cm, elliptic to lanceolate; upper surface *light to medium green, not glaucous*, moderately lepidote above, lower surface densely lepidote, scales half to their own diameter apart. **Inflorescence** terminal *and axillary in top 1-3 leaves*, 2-6 flowers per bud often forming dense clusters. **Corolla** *narrowly tubular, slightly oblique*, 1.5-2.5cm long, usually bicolored in shades of deep red to salmon or tinged orange to green with yellow or red lobes, sometimes all red (Unicolor Group), rarely yellow; calyx 1mm +/- lepidote and ciliate; style usually pubescent towards base. **Distribution** Common in Bhutan, W. Arunachal Pradesh and S. Tibet, 2,400-3,700m (8,000-12,000ft), thickets, conifer and mixed forest.

This is a distinct species, differing from *R. cinnabarinum* in its more willowy habit, and in the clusters of tubular flowers from both terminal and axillary buds.

The massed tubular flowers of *R. keysii* are curiously attractive and quite unlike those of any other species. Some forms are less hardy than typical *R. cinnabarinum*. Introduced 1851->, reintroduced 1987->. May-July.

Picture 1

1. *R. keysii*, Gante Gompa, C. Bhutan, with its characteristic narrowly tubular flowers in dense clusters. (Photo C.R. Lancaster)

EDGEWORTHIA

Picture 2

Picture 3

2. A very rare yellow-flowered *R. keysii*, Gante Gompa, C. Bhutan; alas this has not yet been introduced into cultivation. (Photo C.R. Lancaster)
3. *R. keysii* from Blackhills, growing at Glendoick, showing the typical lax growth habit.

Subsection Edgeworthia

Height to 2.5m, compact to straggly shrubs. **Leaves** often bullate on upper surface, lower surface covered in *dense indumentum made up of curled hairs*, usually obscuring the yellow scales. **Inflorescence** 1-4(-6)-flowered. **Corolla** funnel-campanulate to campanulate, white, pale pink to yellow, scented or not scented; calyx well-developed, conspicuous, 5-lobed; stamens 10; style either long or short and deflexed; capsule tomentose. **Distribution** Himalaya, S. & S.E. Tibet, N.W. Yunnan.

Within section Rhododendron, only subsections Edgeworthia and Ledum have an indumentum on the lower surface, which more or less obscures the scales.

All species in this subsection are epiphytic in the wild or grow on cliffs and rocks and therefore require excellent drainage in cultivation. Indoors, pots should have extra drainage holes added and plants should not be over-potted. Outdoors in wet areas, plant on decaying wood or moss-covered rocks; in dry areas in a loose well-drained soil. *R. edgeworthii* is a beautiful and showy plant with an excellent scent and is quite common in cultivation, *R. pendulum* is less common and *R. seinghkuense* is extremely rare.

R. edgeworthii

Hook.f. 1849 (*R. bullatum* Franch. *R. sciaphilum* Balf.f. & Kingdon-Ward) H2-3

Height 0.3-3.3m, a leggy to fairly compact shrub. Branchlets *densely woolly*. **Leaves** 6-15 x 2.5-5cm, oblong-ovate, oblong-lanceolate or rarely elliptic; upper surface strongly bullate, lower surface scales obscured by a dense, woolly indumentum, white when young, turning rust-coloured to fawn at maturity. **Inflorescence** 1-5-flowered. **Corolla** funnel-campanulate, 3-7.5cm long, white, sometimes flushed pink or rose, with or without a yellow blotch at the base, usually strongly fragrant; calyx large and deeply lobed; style *long and straight*; capsule covered in downy hairs. **Distribution** Sikkim eastwards through Bhutan into Upper Burma, N.W. & Central Yunnan and S.E. Tibet, 1,800-4,000m (6,000-13,000ft), epiphytic, and on rocks and cliffs.

This species is very easily recognised in its heavily indumented, bullate leaves. *R. seinghkuense* and *R pendulum* both have smaller leaves, unscented flowers and a deflexed style.

R. edgeworthii is one of the most impressive scented species with fine foliage and usually showy flowers. It needs excellent drainage and grows well on tree stumps, mossy rocks and similar sites. Plants originally from Yunnan, China, often circulate under the name Bullatum Group but these cannot be distinguished from the type of *R. edgeworthii*. It is variable in hardiness, with the hardier clones succeeding outdoors even in eastern Scotland. It makes a fine indoor plant in colder areas, as do its many hybrids. Introduced 1851-1953, reintroduced 1965->. April-May.

Picture 1

1. *R. edgeworthii* SBEC 0607 on the Cangshan, Central W. Yunnan, China.

EDGEWORTHIA

Picture 2

Picture 3

Picture 4

2. *R. edgeworthii*, a relatively hardy form which grows outdoors at the Royal Botanic Garden, Edinburgh.
3. *R. edgeworthii* showing the bullate upper leaf surface and woolly lower surface typical of this species.
4. Foliage of *R. edgeworthii* C. 6021 from the Mekong-Salween divide, N.W. Yunnan, China, growing at Glendoick.

beige indumentum. **Inflorescence** 2-3-flowered. **Corolla** openly funnel-campanulate, 1.5-2.2cm long, white, white flushed pink, cream; calyx red tinged; style *short and deflexed*. **Distribution** E. Nepal, Sikkim, Bhutan and S. Tibet, 2,300-3,700m, (7,500-12,000ft), epiphytic or on steep slopes or cliffs.

This is a distinctive species differing from *R. edgeworthii* and *R. seinghkuense* in the near oval leaf shape and the openly funnel-campanulate white flowers. *R. leucaspis* has more rotate flowers and has hairy rather than indumented leaves and stems.

R. pendulum is not as showy as its two relatives but is quite attractive at its best. Poor drainage causes leaf spot and dieback, often resulting in death. Introduced 1850-1938, reintroduced 1987->. April-May.

Picture 1

Picture 2

1. *R. pendulum* Ludlow, Sherriff and Taylor 6660 (collected in Bhutan in 1938), growing out of doors at the Royal Botanic Garden, Edinburgh.
2. *R. pendulum* K. Rushforth 1232 showing the oval leaf shape which differentiates this species from its relatives.

R. pendulum Hook.f. 1849 H3-(4)

Height 0.3-1.2m, a straggling to fairly compact shrub. Branchlets covered with dense, woolly indumentum. **Leaves** 2.3-5 x 1.2-3cm, *oblong-elliptic, convex*; upper surface *smooth*, glabrous, lower surface with a thick, woolly, whitish to

R. seinghkuense Hutch. 1930. H2-3?

Height 0.3-0.9m, an erect small shrub, sometimes sprawling with age. Branchlets covered with dense or moderately dense indumentum. **Leaves** 3-8 x 1.6-4cm, elliptic or narrowly ovate; upper surface *not convex*, somewhat bullate, lower surface with a woolly, pale to dark brown indumentum. **Inflorescence** 1-2-flowered. **Corolla** rotate-campanulate, c.2cm long, *bright yellow*; ovary *densely tomentose all over*; style *short and deflexed*. **Distribution** N.W. Yunnan, S.E. Tibet, 1,800-3,000m (6,000-10,000ft), epiphytic or on rocks.

In appearance, this species resembles a dwarf, small-leaved *R. edgeworthii* with yellow, unscented flowers and a deflexed, short style.

R. seinghkuense is extremely rare in cultivation. Plants under the name *R. seinghkuense* with larger leaves are most likely forms of *R. edgeworthii*. This species is not easy to grow well and needs perfect drainage. Introduced 1922. May.

Picture 1

Picture 2

1. *R. seinghkuense* Kingdon-Ward 9254 at Glendoick. Until recently, there was probably only one clone in cultivation and possibly only one plant which survived at Glendoick. Seedlings of this clone have now been grown.
2. *R. seinghkuense* Kingdon-Ward 9254 collected in 1931 on the Tibet-Burma frontier, showing the characteristic woolly indumentum on the lower leaf surface.

Subsection Fragariflora

A monotypic subsection.

R. fragariflorum
Kingdon-Ward 1929 H4-5

Height 0.1-0.4m, a prostrate, compact, mounding to straggly shrub. Branchlets *lepidote and puberulent on the young growth.* Leaves 0.5-1.7 x 0.3-0.9cm, oblong-elliptic, margins recurved, sometimes ciliate, minutely toothed; upper surface rough, glossy, with shiny yellow scales, lower surface paler, with golden or brown scales, 1½-6 x their own diameter apart. **Inflorescence** 2-6-flowered. **Corolla** widely funnel-shaped to almost rotate, 1-1.8cm long, strawberry-pink, through purple to purplish-crimson; calyx conspicuous, reddish, 3-8mm, lepidote, +/- ciliate; stamens 10; ovary densely lepidote; style red, glabrous, longer than stamens. **Distribution** N.E. Bhutan, S. and S.E. Tibet, 3,500-4,600m (11,500-15,000ft), common on open hillsides, rocks and swamps.

This species closely resembles *R. setosum*, differing in its dwarfer stature, and puberulent but not bristly branchlets, petioles and leaves. *R. fragariflorum* would be more appropriately placed in a subsection together with *R. setosum* as it appears to be an eastern, less hirsute extension of that species. Both species seem to lie botanically between subsections Lapponica and Saluenensia.

Picture 1

1. *R. fragariflorum* growing on the Temo La, S.E. Tibet.

FRAGARIFLORA

Until recently there have probably only been 1-2 rather shy-flowering clones of this rare species in cultivation, from Ludlow, Sherriff and Elliot 15828. Very free-flowering and showy in the wild. Not easy to root or grow, but hopefully new introductions will prove more amenable to cultivation. Introduced 1924, 1947. Reintroduced 1995->. May-June.

Picture 2

Picture 3

Picture 4

2. *R. fragariflorum* on the Temo La, S.E. Tibet, showing the conspicuous reddish calyces.
3. *R. fragariflorum* on the Temo La, S.E. Tibet, with *Cassiope selaginoides*.
4. *R. fragariflorum* showing the shiny yellow scales on the upper leaf surface. (Photo R.B.G., E.)

Subsection Genestieriana

A monotypic subsection

R. genestierianum

Forrest 1920 (*R. mirabile* Kingdon-Ward). H2-3

Height 1.2-4.6m, an erect and slender shrub. Bark red, dark to light mahogany, peeling. *Young growth reddish-purple*. **Leaves** 5-15.3 x 1.4-4.5cm, narrowly elliptic-oblanceolate, with a *very pointed tip*; lower surface *very glaucous*, with minute and very *scattered* scales, 4-10 times their own diameter apart. **Inflorescence** 4-15 flowered, terminal. **Corolla** tubular-campanulate, c. 1.5cm long, fleshy, reddish-purple with a *distinctive glaucous bloom*; pedicel *very long, thin, with a bloom*; calyx also with a glaucous bloom; style short and sharply deflexed, glabrous; stamens 8-10. **Distribution** S.E. Tibet, Upper Burma, Yunnan, 2,400-3,400m(+?) (8,000-11,000ft(+)?), in scrub, thickets and forest margins.

This is a very distinctive species, characterised by its peeling bark, very early, coloured young growth, glaucous leaf underside and most unusual, pruinose or 'bloom'-covered corolla, pedicel and calyx.

A species for very mild gardens only, partly due to the vulnerability to frost of its very early growth. Farrer thought it ugly while Forrest liked it and it certainly is a most unusual species. Further exploration may reveal hardier forms more suited for general cultivation. Introduced 1919-1953. April.

Picture 1

1. *R. genestierianum* Kingdon-Ward 20682, Brodick Castle, W. Scotland, showing the very early young growth.

Picture 2

Picture 3

2. *R. genestierianum* Kingdon-Ward 20682, Brodick Castle, W. Scotland, showing the characteristic glaucous bloom on the outside of the flower buds.
3. Foliage of *R. genestierianum* showing the very glaucous leaf lower surface and very pointed leaf apex.

Subsection Glauca

Height 0.3-2m, evergreen, upright to compact shrubs. **Leaves** 2-9cm long, *aromatic*, lower surface white or greyish or bluish *with dimorphic scales, the smaller milky or golden, the larger brown*. **Inflorescence** 3-10-flowered. **Corolla** campanulate to tubular-campanulate, pink, reddish, purple, yellow, white; calyx deeply 5-lobed; stamens 10; style impressed, usually glabrous, usually deflexed. **Distribution** Nepal to S.E. Tibet, Upper Burma and W. to N.W. Yunnan.

This is quite a distinctive subsection in its aromatic leaves, glaucous on the underside and with dimorphic scales. All except *R. glaucophyllum* var. *tubiforme* have a deflexed style.

Most species are quite widely represented in large collections but have never been the most popular garden plants. Several flower rather early and/or are on the tender side.

R. brachyanthum
Franch. 1886 H3-4

Height 0.3-2m, a fairly dense shrub, leggy in shade. Bark, peeling. **Leaves** 2-6.5 x 1-2.6cm, elliptic to narrowly obovate; upper surface dark green, lower surface glaucous, scale distribution variable (see below), the smaller scales clear or milky. **Inflorescence** 3-10-flowered. **Corolla** campanulate, 1-2cm long, pale or greenish yellow; pedicel long; calyx with rounded lobes. **Distribution** N.E. Burma, N.W. Yunnan, S.E. Tibet. 2,700m-4,400m (9,000-14,500ft), dry open situations in scrub and forest.

The late yellow 'thimbles' on long stalks are unlike the flowers of any other species. Out of flower, the glaucous leaf underside and peeling bark are distinctive. *R. luteiflorum* has more scattered scales than *R. brachyanthum* ssp. *hypolepidotum* and the flowers open much earlier.

The flowers are small but quite attractive, and it is useful for its late blooming. The leaves are very aromatic, sometimes unpleasantly so. The two subspecies are indistinguishable from a horticultural point of view. Introduced 1914-1917->, reintroduced 1981->. May-July.

Ssp. *brachyanthum*

Lower leaf surface with very *scattered*, sometimes entirely deciduous scales. **Distribution** Only found on Cangshan, Yunnan.

Picture 1

1. *R. brachyanthum* ssp. *brachyanthum* SBEC 0641, Glendoick.

Ssp. *hypolepidotum*
Cullen (*R. charitostreptum* Balf.f. & Kingdon-Ward.)

Lower leaf surface with scales their own diameter to 2-3 x their own diameter apart. Much more common than ssp. *brachyanthum* in the wild and in cultivation.

Picture 2

Picture 3

Picture 4

5. Leaf lower surface of R. brachyanthum ssp. *hypolepidotum* showing the scales 1-3 x their own diameter apart. (Photo R.B.G., E.)

2. *R. brachyanthum* ssp. *brachyanthum*, Royal Botanic Garden, Edinburgh. The little yellow 'thimbles' are produced late in the season.
3. *R. brachyanthum* ssp. *brachyanthum* showing the characteristic scattered scales on the glaucous lower leaf surface. (Photo R.B.G., E.)
4. *R. brachyanthum* ssp. *hypolepidotum* 'Blue Light' A.M., Glendoick.

R. *charitopes*
Balf.f. & Farrer 1922. H3-4

Height to 1.5m, an often dense, small shrub which can get leggy with age or in shade. **Leaves** 2.6-7 x 1.3-3cm, *elliptic to obovate*; upper surface dark green, lower surface with very marked veins, very glaucous, with scales of varying density, *but not touching*. **Inflorescence** 2-6-flowered. **Corolla** broadly campanulate, c. 3cm across, clear apple-blossom pink or purplish, usually speckled with crimson; calyx usually relatively large, 5-9mm, *lobes rounded*; stamens 10, densely hairy at base; style bent.

The non-touching scales distinguish this species from *R. pruniflorum*, while the oval leaf shape differentiates it from *R. glaucophyllum* which has pointed leaves. It is easy to distinguish spp. *charitopes* from its relatives in the shape and colour of the flowers. We consider that ssp. *tsangpoense* is better associated with *R. pruniflorum*, as it used to be, than in the present arrangement.

Ssp. *charitopes* Cullen 1980

Corolla *pink* or pale pink fading to white; calyx *6-9mm*. **Distribution** Upper Burma-N.W. Yunnan, 3,200-4,300m (10,500-14,000ft), rocky slopes, cliffs, thickets and stream banks.

This is one of the most attractive of this subsection in flower, but the swelling buds are frost tender. It has a tendency to bloom in autumn. Introduced 1924. April-May.

Ssp. *tsangpoense* (Kingdon-Ward) Cullen 1978. H3-4

Corolla *purplish*; calyx *5-6mm*. **Distribution** S.E. Tibet, 2,400-4,100m (8,000-13,500ft).

This subspecies is not considered as attractive in flower as ssp. *charitopes*. It is common in the wild around the Tsangpo Gorge area, S.E. Tibet. This probably ought to be returned to specific status as there is no overlap in distribution with ssp. *charitopes* and their flowers are quite different. Introduced 1924, reintroduced 1995. April-May. So called var. *curvistylum* with smaller flowers than the type occurs on the Doshong La, S.E. Tibet. Cultivated plants under this name under Kingdon-Ward 5843 appear to be natural hybrids, crossed with *R. campylogynum*

Picture 4

4. *R. charitopes* ssp. *tsangpoense*, Glendoick, showing the calyx which is smaller than that of ssp. *charitopes*.

Picture 1

Picture 2

Picture 3

1. *R. charitopes* ssp. *charitopes* Forrest 25570, Glendoick.
2. *R. charitopes* ssp. *charitopes* Forrest 19872 showing the typical wide leaves and campanulate flowers. Glendoick.
3. *R. charitopes* ssp. *tsangpoense* on the south side of the Doshong La, Pemako, S.E. Tibet.

R. glaucophyllum
Rehder 1945 (*R. glaucum* Hook.f.) H(3-)4

Height 0.3-1.5m, a bushy and somewhat spreading shrub. Bark peeling. **Leaves** *very aromatic*, 3.5-9 x 1.3-2.5cm, *lanceolate to narrowly elliptic to elliptic*, tip pointed; upper surface dark brownish-green, lower surface *very glaucous and densely lepidote*, with small pale yellow and larger brown scales. **Inflorescence** 4-10-flowered. **Corolla** campanulate, 1.5-2.7cm long, pinkish-purple, rose to pale pink (white in var. *album*); calyx *large, leafy and hairy* with *pointed lobes*; style impressed, *sharply bent and shorter than the corolla* or long and straight (var. *tubiforme*).

R. glaucophyllum is distinguished by the combination of its peeling bark, very glaucous and densely lepidote lower leaf surface, its aromatic leaves, the colour of the flowers and large, hairy calyx with pointed lobes.

Var. *glaucophyllum* H4

Corolla *campanulate, pink*; style sharply deflexed. **Distribution** E. Nepal, India, Bhutan, S. Tibet, 2,700-3,700 (9,000-12,000ft), forest clearings and rocky slopes,

This variety has been long cultivated and is quite common in old collections. It is free-flowering, relatively late-flowering and makes a fine plant for the edge of woodland. Introduced 1850, reintroduced 1980s->. Late April-May.

Var. *album* Davidian 1981 H3

Corolla to 2cm long, *white*. Style sharply deflexed. **Distribution** This has a large, uniformly white-flowered population in E. Nepal.

We have found it a little less hardy than typical var. *glaucophyllum*. Introduced 1971->. Late April-May.

Var. *tubiforme* Cowan & Davidian 1948 H4

Corolla *tubular*, rose pink; style *long and straight*. **Distribution** E. Bhutan, W. Arunachal Pradesh, S.E. Tibet.

The flowers of this variety are vulnerable to being frosted in the swelling bud. Cullen considers that this may be a natural hybrid. Introduced 1936. April-May.

Picture 4

Picture 5

4. *R. glaucophyllum* var. *album*, Glendoick. This variety may only occur in E. Nepal.
5. *R. glaucophyllum* var. *tubiforme* showing the characteristic long, straight style. Glendoick.

Picture 1

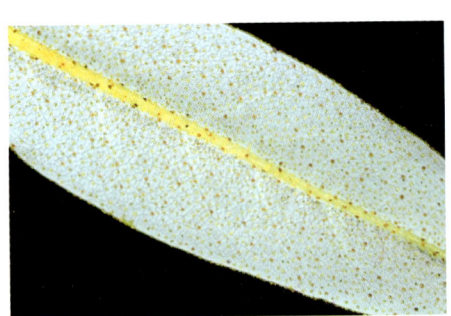

Picture 2

Picture 3

1. A clone from Branklyn growing at Glendoick of *R. glaucophyllum* var. *glaucophyllum* showing the relatively narrow leaves and campanulate flowers compared to those of *R. charitopes* ssp. *charitopes*.
2. Leaf lower surface of *R. glaucophyllum* var. *glaucophyllum* showing the very glaucous surface and the small pale and large brown scales. (Photo R.B.G., E.)
3. *R. glaucophyllum* var. *album*, Milke Danda, E. Nepal.

R. luteiflorum
(Davidian) Cullen 1978 H3

Height 0.9-1.5m, a compact to stiff, erect, shrub, tidier than *R. glaucophyllum*. Bark smooth, flaking. **Leaves** very aromatic, 3.5-9 x 1.3-2.5cm, elliptic; upper surface dark brownish-green, lower surface *very glaucous and sparsely lepidote*, with small pale yellow and larger brown scales. **Inflorescence** 3-6-flowered. **Corolla** campanulate, 2-2.2cm long, *bright clear yellow*; calyx lobes *rounded, not hairy*. **Distribution** Upper Burma, 2,700-3,000m (9,000-10,000ft), in thickets and exposed ridges.

This species differs from *R. glaucophyllum* in the yellow flowers and in the non-hairy calyx with rounded lobes. The leaves of *R. luteiflorum* are more sparsely lepidote on the lower surface and the leaf shape is different. Much earlier flowering than *R. brachyanthum*.

R. luteiflorum is a fine species for milder gardens but its swelling buds are tender and in colder gardens they often fail to open. It is not always very easy to please and seems to resent fertiliser. Introduced 1953 under two Kingdon-Ward numbers. April-May.

elepidote, lower surface *densely lepidote, the smaller scales almost touching,* pale yellow, clouded or milky, larger scales brown. **Inflorescence** 3-7-flowered. **Corolla** campanulate, 1-1.3cm long, *dull-crimson to plum-purple*; calyx lobes rounded; stamens pubescent for most of their length. **Distribution** Arunachal Pradesh, Upper Burma, S.E. Tibet, 2,400-4,000m (8,000-13,000ft), in forest, rhododendron thickets and rocky hillsides.

In flower this species is usually easily recognised, though some forms of *R. charitopes* ssp. *tsangpoense* are very similar. In leaf it may also closely resemble *R. charitopes* var. *tsangpoense* differing in the scale distribution on the lower leaf surface, more densely lepidote in *R. pruniflorum*. These two taxa probably merge in S.E. Tibet.

Picture 1

Picture 2

Picture 3

1. *R. luteiflorum* Kingdon-Ward 21040 from the Triangle, Upper Burma, at Glendoick, showing the campanulate flowers and the very glaucous leaf underside.
2. *R. luteiflorum*, a seedling of Kingdon-Ward 21040.
3. *R. luteiflorum* growing with *R. leucaspis* at Baravalla, W. Scotland.

Picture 1

Picture 2

1. *R. pruniflorum* at the Younger Botanic Garden, W. Scotland, showing the campanulate flower shape and the distinctive flower colour. (Photo R.B.G., E.)
2. *R. pruniflorum* at Glendoick.

R. pruniflorum

Hutch. 1930 (*R. tsangpoense* var. *pruniflorum* Cowan & Davidian, *R. sordidum* Hutch.) H3-4

Height to 1m, a bushy shrub. Bark brown, shredding. **Leaves** aromatic, 3-4.2 x 1.4-2.5cm; upper surface dark green, almost

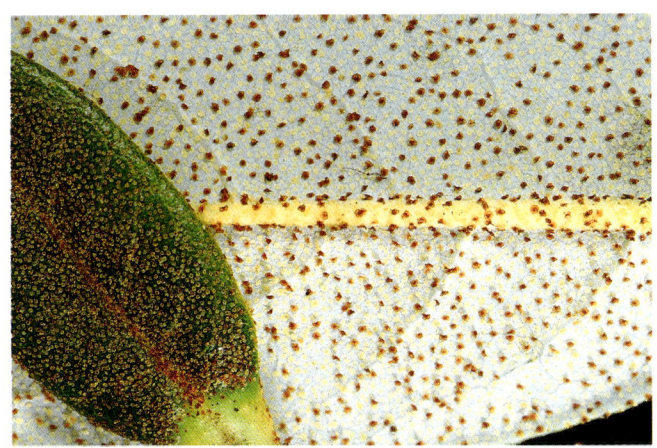

Picture 3

3. The lower leaf surface of *R. pruniflorum* showing the almost touching pale scales and more distant brown scales. These scales distinguish this species from *R. charitopes* var. *tsangpoense*. (Photo R.B.G., E.)

Picture 1

1. A clone, from the Royal Botanic Garden, Edinburgh, cultivated as *R. shweliense*. It does not match the one poor herbarium specimen of this species, in that the outside of the corolla is lepidote. Glendoick.

R. shweliense
Balf.f. & Forrest 1919. H3-4

Height 0.3-0.7m, a compact shrub (as cultivated). **Leaves** 3.2-4 x 1.5-1.6cm, narrowly elliptic to narrowly obovate; lower surface glaucous with mixed small yellow and larger brown scales of variable density. **Inflorescence** 2-4-flowered. **Corolla** campanulate, c.1.1cm long, *yellow, flushed pink*. **Distribution** Shweli-Salween divide, W. Yunnan, 3,000-3,400m (10,000-11,000ft), open cliffs and grassy slopes.

Cullen maintains that most if not all cultivated plants under *R. shweliense* are hybrids, as the corollas are lepidote outside. The original herbarium material is inadequate but the one poor flower does have an elepidote outside to the corolla. Davidian says that it usually has a lepidote corolla but this description is probably based on cultivated material. *R. shweliense* may be a natural hybrid but its distribution appears to be isolated from those of its relatives. It differs from other members of the subsection in its smaller stature, the less glaucous lower leaf surface and the smaller flowers.

Cultivated plants under the name *R. shweliense* are probably natural hybrids of *R. brachyanthum* possibly crossed with *R. lepidotum* and have small flowers of a rather unusual colour. Introduced 1924. May-June.

Subsection Heliolepida

Height 1-10m, medium shrubs to small trees. **Leaves** often *very aromatic*, evergreen, lower surface with *large conspicuous scales*. **Inflorescence** terminal, 5-10-flowered. **Corolla** funnel-shaped, *conspicuously lepidote outside*, white, pink, mauve, lavender to purple, often spotted; calyx disc-like, rarely somewhat lobed; stamens 10, pubescent towards base; ovary densely lepidote; style impressed. **Distribution** N. to S.W. Sichuan, N.E. to W. Yunnan, Upper Burma and S.E. Tibet.

This subsection is related to subsection Triflora, but is characterised by the purely terminal inflorescence, the less zygomorphic corolla and the usually greater density of scales on the lower leaf surface. *R. heliolepis* and *R. bracteatum* are further distinguished by their very aromatic leaves and late (June-August) flowering.

These are vigorous and generally easily-grown plants often succeeding in soils too heavy or too dry for most elepidote species. *R. rubiginosum* is very common in gardens, *R. heliolepis* less so and *R. bracteatum* is very rare.

R. bracteatum
Rehder & E.H. Wilson 1913 H4-5?

Height c.1.8m, an shrub erect. Branchlets *thin, lepidote, with persistent leaf-bud scales instead of leaves on lower part.* **Leaves** *very aromatic*, 3.8-5 x 2-2.5cm, *ovate to elliptic*, rounded at base; upper surface dark green, lepidote, lower surface sparsely lepidote with *large* (to 0.25 mm in diameter) golden scales. **Inflorescence** 4-6-flowered. **Corolla** openly funnel shaped, 1.5-2.5cm long, white to white-flushed-rose with or without a crimson blotch at the base; calyx weakly 5-lobed, margins ciliate; ovary lepidote and sparsely puberulent; style *shorter than the longest stamens*. **Distribution** Central Sichuan c.3,300m (10,800ft), in woodland and on cliffs.

This species differs from *R. heliolepis* in its smaller, rounded leaves and smaller flowers and the persistent leaf-bud *scales* and in addition it has an isolated distribution. *R. rubiginosum* has larger, less aromatic leaves, much more densely lepidote below, and is earlier flowering.

R. bracteatum is very rare in cultivation. It is quite attractive in forms with a large blotch but is rather slow to start blooming and seems to be susceptible to powdery mildew. Introduced 1910. June-July.

Picture 1

1. *R. bracteatum* Wilson 4253 collected near Wenchuan, Sichuan, China, in 1908. This form is from the Royal Botanic Garden, Edinburgh.

R. heliolepis Franch. 1887 H4

Height 2-4.5m, a usually erect shrub. Branchlets *lepidote*; bud scales deciduous. **Leaves** usually *aromatic*, (4-) 7.6-11 x (1.3-) 2.2-3.8cm, oblong, ovate-oblong to oblong-elliptic; upper surface lepidote with whitish, scurfy, deciduous scales, lower surface with close but not touching, large, glistening, golden or brownish scales. **Inflorescence** 4-10-flowered. **Corolla** funnel-shaped, 2-3.4cm long, white to pink, rarely purplish, usually with reddish, greenish or brownish spots or a bold purple blotch; calyx variable to 3mm, sparsely lepidote, margins ciliate; ovary densely lepidote; style *straight, shorter than the longest stamens*. **Distribution** Yunnan, Tibet, Sichuan, N.E. Burma, 2,400-3,800m (8,000-12,500ft), conifer forest and thickets.

R. heliolepis shares the aromatic foliage of *R. bracteatum* but has larger leaves and flowers and it has a more southerly distribution. The leaves of *R. rubiginosum* differ from those of *R. heliolepis* in usually being much less aromatic and more densely lepidote on the lower surface. *R. rubiginosum* is much more vigorous and also differs in its earlier flowering period and in its declinate style, longer than the stamens.

Useful for its late flowering, *R. heliolepis* is very variable in foliage and flower with forms with bold blotches often being the most showy. The pungently aromatic leaves are an interesting feature. Plants under the names var. *fumidum* and Pholidotum Group may be intermediate/hybrid forms with *R. rubiginosum*. Introduced 1912-> reintroduced 1981->. June-August.

Var. *heliolepis* (*R. oporinum* Balf.f. & Kingdon-Ward. *R. plebeium* Balf. & W.W. Sm.)

Leaves *rounded or truncate at the base*. **Inflorescence** *5-8-flowered*. **Distribution** More common in Yunnan than Sichuan.

Var. *brevistylum* Cullen 1978 (*R. brevistylum* Franch., *R. pholidotum* Balf.f. & W.W. Sm., *R. porrosquameum* Balf.f. & Forrest)

Leaf base *cuneate*. **Inflorescence** *6-10-flowered*. **Distribution** more common in Sichuan than in Yunnan.

Var. *fumidum* (Balf.f. & W.W. Sm.) R.C. Fang.

Branchlets sparsely lepidote. **Leaves** 4-7 x 1.5-3cm, base cuneate. **Corolla** purplish-red. **Distribution** N.E. Yunnan.

This taxon appears to be intermediate between *R. heliolepis* and *R. rubiginosum* and probably has less aromatic leaves than the other varieties. Introduced from Wumengshan, N.E. Yunnan, in 1994->.

HELIOLEPIDA

Picture 1

Picture 2

Picture 3

Picture 4

Picture 5

Picture 6

1. *R. heliolepis* var. *heliolepis* Forrest 6762, collected in the Cangshan region of C.W. Yunnan, China, in 1910.
2. *R. heliolepis* var. *heliolepis* Rock 129 collected in 1948/49 in Yunnan, growing at the Royal Botanic Garden, Edinburgh.
3. Leaf lower surface of *R. heliolepis* var. *heliolepis* showing the rounded leaf base. (Photo R.B.G., E.)
4. *R. heliolepis* var. *brevistylum,* a white-flowered clone of Kingdon-Ward 7108 collected in Burma in 1926 and growing at the Royal Botanic Garden, Edinburgh.
5. *R heliolepis* var. *brevistylum* Kingdon-Ward 7108, a clone with a deep purple blotch in the centre, growing at Glendoick.
6. *R heliolepis* var. *brevistylum* showing the cuneate leaf base characteristic of this variety. (Photo R.B.G., E.)

R. rubiginosum

Franch. (*R. leclerei* Lév, *R. catapastum* Balf.f. & Forrest, *R. desquamatum* Balf.f. & Forrest, *R. stenoplastum* Balf.f. & Forrest, *R. leprosum* Balf.f., *R. squarrosum* Balf.f.) H3-4

Height 6-9m, an erect to spreading, well-branched shrub. Branchlets lepidote, sometimes purplish. **Leaves** 6.3-11.3 x 2.5 4.5cm, oblong-clliptic to elliptic-lanceolate, lower surface *rusty to brownish to brownish-green, with flaky* scales, variable in size and shape, *dense and often overlapping*; petiole densely lepidote. **Inflorescence** to 10 (usually 4-8)-flowered. **Corolla** openly to widely funnel-shaped, 3-4cm long, pale or deep pink, lavender-pink, mauve-pink to almost white flushed pink or purple, with purple, brown or crimson blotch or spotting; calyx very small, lepidote; stamens 10; style glabrous, *declinate, longer than stamens*. **Distribution** S.E. Sichuan, Yunnan, S.E. Tibet, Upper Burma, 2,300-4,300m (7,500-14,000ft) common and widespread in thickets, bamboo and open forest etc.

R. rubiginosum differs from *R. heliolepis* and *R. bracteatum* in its usually much less aromatic leaves which are very densely lepidote on the lower surface, its earlier flowering period and the declinate style, longer than the stamens.

A widely grown, easy, vigorous species which can grow very large and is said to tolerate neutral or slightly alkaline soil. *R. rubiginosum* is variable in hardiness and flower size. Plants under Desquamatum Group have larger widely-funnel-shaped flowers and relatively large leaves and include some of the less hardy forms generally from the western end of its distribution. *R. rubiginosum* perhaps merges with *R. heliolepis* as var. *fumidum*. Introduced 1889->, reintroduced 1981->. March-April.

Picture 1

Picture 2

Picture 3

Picture 4

1. A fine dark-flowered form of *R. rubiginosum* near Zhuzipo, Salween-Mekong divide, N.W. Yunnan, China, with David Farnes standing underneath. It is rare to find a well-shaped specimen of this species in the wild.
2. *R. rubiginosum*, a dark-flowered form at Arduaine Gardens, W. Scotland.
3. *R. rubiginosum* Desquamatum Group with larger, more widely funnel-shaped flowers than typical *R. rubiginosum*.
4. *R. rubiginosum*, a pale-flowered clone at Glendoick.

Subsection Lapponica

Low to small shrubs. **Leaves** +/- evergreen, +/- aromatic, 0.2-7 x 0.1-2.6cm, densely lepidote with early deciduous leaf bud scales. **Inflorescence** terminal, occasionally axillary, 1-several-flowered; pedicel short. **Corolla** broadly funnel-shaped (except *R. intricatum*), predominantly in the lavender-mauve-violet range, also deep purple, pink, yellow and white, calyx usually conspicuous with scales and ciliate margins; stamens 5-10(-11); ovary 5-locular, lepidote; style straight or declinate; seeds unwinged and obscurely finned. **Distribution** Central, E. & S.E. Asia, N. Japan, N. Europe and N. North America.

Subsection Lapponica is the largest lepidote subsection, containing most of the smallest-leaved and many of the dwarfest species. These are essentially plants of moorland habitat on bare windswept mountainsides and peaks or in clearings among larger rhododendrons. *R. lapponicum* has a circum-polar distribution while the remainder are found in the Himalaya and especially in western China. Species from this subsection (and their numerous hybrids) can make excellent garden plants in cooler, relatively northern gardens where they can be planted to form undulating carpets. Although mostly very hardy, they dislike summer heat and are not easy to grow in California or E. U.S.A., Australia or most of New Zealand. Many species have flowers that are frost hardy, even when fully open.

The Lapponica species are among the most taxonomically complex as they are usually very variable, and most of the important characters used by botanists are not consistent even within one population of the 'same' species. The Philipsons in their revision of 1975 greatly reduced the number of species but, had they had access to field studies amongst the wild populations in China, they would probably have reduced the number still further. Frank Kingdon-Ward in *The Romance of Plant Hunting* pps 212-217 eloquently and amusingly discusses the virtual impossibility of identifying subsection Lapponica species in the wild and he challenges the botanists to come up with valid and consistent identifying characters.

We follow the Philipsons' revision, mentioning a few instances where we think things could be improved. Almost all the species hybridise/merge with their relations and so the boundary between one species and another can be rather arbitrary. The most problematic species in this respect seem to be *R. tapetiforme*, *R. telmateium*, *R. nivale* ssp. *boreale* and ssp. *australe*, *R. websterianum*, *R. nitidulum* and *R. yungningense*. The most common species in cultivation are *R. fastigiatum*, *R. impeditum*, *R. russatum* and *R. hippophaeoides*. Many subsection Lapponica species are rare in cultivation, being largely confined to specialist collections in Scotland, Scandinavia and W. North America.

R. bulu Hutch. 1929 H4?

Height 0.3-*1.6m, usually an erect, straggly* shrub. **Leaves** 0.9-2.3 x 0.3-0.8cm, elliptic to oblong-elliptic, upper surface dull green, lower surface pale with dense pale scales with some larger dark scales. **Inflorescence** 1-3-flowered. **Corolla** widely funnel-shaped, 1-1.6cm long, pinkish-purple, magenta, deep violet or occasionally white, *pale scales on outer surface*, pubescent in throat; calyx 1-3mm long, lobes trianglular or irregularly rounded; stamens 10; style longer than stamens. **Distribution** S.E. Tibet, 3,000-3,800m (10,000-12,500ft), in open woodland or scrubby hillsides.

R. bulu is one of the larger-growing Lapponica with an upright straggly habit. The key characters are the mostly light-coloured scales on the leaf lower surface with a scattering of tan-coloured scales, and the pale scales on the outside of the corolla. Both these characters may be present in the very closely related *R. nivale* which occurs at higher elevations, is smaller in stature and has darker, smaller leaves.

Unusually for a species in this subsection, it occurs at low altitudes in relatively dry scrub and it may be more heat- and drought-tolerant than its relatives. Not particularly striking in flower, and similar in appearance to many of its relatives, so its introduction will not cause a sensation! Introduced 1995->. April-May?

Picture 1

Picture 2

1. *R. bulu* at Tse near the Tsangpo River in S.E. Tibet showing the large-growing plant of lax habit.
2. *R. bulu* at Tse in S.E. Tibet on a dry, scrubby hillside amongst evergreen oak.

R. capitatum
Maxim. 1877. H4-5

Height to 1.5m, a *usually erect,* small shrub. **Leaves** 0.6-2.5 x 0.5-0.9cm, elliptic; upper surface rather pale and shiny, lower surface pale brown with darker speckling, *with dimorphic scales, pale with pale golden centres and dark with dark centres,* touching or non-touching. **Inflorescence** 3-5-flowered. **Corolla** broadly funnel-shaped, 1-1.5cm long, pale lavender to deep purple, elepidote; calyx very variable in size, lobes usually oblong, margins ciliate; stamens *10*; style declinate, longer than or equal to stamens. **Distribution** Gansu, E. Tibet, Sichuan, Shaanxi 3,000-4,300m (10,000-14,000ft), in forests, meadows and on mountainsides.

R. capitatum is usually identified by its fastigiate shoots giving a narrowly erect habit. It merges with *R. nitidulum* which has a fewer-flowered inflorescence and is generally later flowering.

Quite rare in cultivation, *R. capitatum* is very hardy and is one of the earliest of subsection Lapponica to flower. Introduced 1925 and reintroduced 1986. March-April.

3. Upper and lower leaf surfaces of *R. capitatum*. The scales in this example are less obviously of two colours than is usual. (Photo R.B.G., E.)

1. *R. capitatum* at Glendoick.
2. *R. capitatum* near Huanglongsi, N. Sichuan, China.

R. complexum
Balf.f. & W.W. Sm. 1916 H5

Height 0.08-0.6m, a low and compact to broadly upright shrub. **Leaves** 0.3-1.1 x 0.2-0.6cm, elliptic to ovate; lower surface *uniformly rust-coloured with contiguous scales*. **Inflorescence** 3-5-flowered. **Corolla** usually narrowly funnel-shaped, 0.9-1.3cm long, pale lilac to rosy purple; calyx very small, margins lepidote and/or ciliate; stamens 5-6; style *very short*. **Distribution** N.W. Yunnan 3,400-4,600m (11,000-15,000ft), moorland and stony slopes.

This species is quite closely related to *R. intricatum* with a similar short style and stamens (but only 5-6 stamens as opposed to 10), but without the tubular corolla and spreading corolla lobes. The 5-6 stamens distinguish this from most other Lapponica species.

1. *R. complexum* at Tian shi, near Zhongdian, N.W. Yunnan, China, showing the short style and typical flower shape.

LAPPONICA

R. complexum is a rarely cultivated species with small flowers and foliage and a neat habit. Introduced 1917, 1922, reintroduced 1992. April-May.

Picture 2

Picture 3

2. *R. complexum* Forrest 15392, the original 1917 introduction from Zhongdian, N.W. Yunnan, China, growing at Glendoick.
3. *R. complexum* foliage showing the uniformly rust-coloured scales. (Photo R.B.G., E.)

R. cuneatum

W.W. Sm. 1914 (*R. cheilanthum* Balf.f.& Forrest, *R. cinereum* Balf.f. & Forrest, *R. ravum* Balf.f. & W.W. Sm., *R. sclerocladum* Balf.f. & Forrest) H4-5

Height 1-2m or taller, an upright or spreading shrub. **Leaves** sometimes aromatic, *1.1-7cm x 0.5-2.6cm*, narrowly to broadly elliptic; lower surface with touching to overlapping, brown, fawn to creamy-yellow scales. **Inflorescence** 3-6-flowered. **Corolla** funnel-shaped, *1.6-3.4cm long*, rose-lavender to deep rose-purple; calyx 4-12mm long with ciliate margins; stamens 10; style longer or rarely equal in length to stamens. **Distribution** N. & W. Yunnan into S.W. Sichuan, 2,700-4,300m (9,000-14,000ft), in forests or above the treeline.

The giant of the subsection with the largest leaves, corolla and calyx, *R. cuneatum* is usually quite easily recognised. Taxonomically it appears to link subsection Lapponica to subsection Heliolepida. Natural hybrids with *R. yunnanense* and *R. rubiginosum* occur.

This variable species lacks the frost-hardy flowers of most of its relatives and is generally only of interest to keen collectors. Introduced 1913->, reintroduced 1980s->. March-May.

Picture 1

Picture 2

Picture 3

1. *R. cuneatum*, Gang-he-ba, Yulong Shan. N.W. Yunnan, China.
2. *R. cuneatum* Rock 11392 collected in N.W. Yunnan, China, in 1923-24 and growing at the Royal Botanic Garden, Edinburgh.
3. *R. cuneatum* Cox, Hutchison & Maxwell MacDonald 2636 with fairly deep-coloured flowers, from Yulong Shan, Yunnan, China, growing at Glendoick.

Picture 4

4. *R. cuneatum* showing the fawn to rust-coloured scales on the leaf lower surface. (Photo R.B.G., E.)

R. dasypetalum
Balf.f. & Forrest 1919. H4-5

Height 0.3-0.9m, a fairly compact, spreading shrub. **Leaves** 0.8-1.6 x 0.3-0.75cm, oblong-elliptic; upper surface often shiny, lower surface densely covered with uniform, contiguous, brown scales. **Inflorescence** 1-4-flowered. **Corolla** broadly funnel-shaped, 0.8-1.6cm long, bright rose-purple, *pubescent outside*; calyx 3mm long, *with long hairs on margins*; stamens 10; style longer than stamens. **Distribution** Litiping, N.W. Yunnan. 3,500m (11,500ft), open stony pasture.

As cultivated, this species has distinctive leaves, 'v'-shaped in cross section. It is characterised by the pubescent corolla and the relatively long hairs on the calyx.

R. dasypetalum is doubtfully deserving of specific status. Cultivated plants appear to be hybrids between *R. saluenense* and a subsection Lapponica species. Quite showy and distinctive. Introduced 1917. April-May.

Picture 1

1. *R. dasypetalum*, a probable natural hybrid of a member of subsection Lapponica with *R. saluenense*, from the Li Ti Ping, Yunnan, China, growing at Glendoick.

R. fastigiatum
Franch. 1886 (*R. capitatum* Franch. not Maxim., *R. nanum* Lév.) H4-5

Height usually under 0.6m, a compact, prostrate or mounding shrub. **Leaves** 0.5-1.6 x 0.3-0.6cm, oblong to ovate; upper surface *glaucous-grey or pale glaucous green*, lower surface fawn to greyish with *pale and opaque* scales. **Inflorescence** 1-5-flowered. **Corolla** funnel-shaped, 0.9-1.8cm long, light purple to deep blue-purple, occasionally pinkish; calyx 2.5-5.5mm long, margins ciliate; stamens usually 10; style exceeding stamens. **Distribution** N., N.E. & Central Yunnan, 3,200-4,900m (10,500-16,000ft), open stony pastures, screes, cliffs and in forests.

R. fastigiatum differs from the very closely related *R. impeditum* in its glaucous rather than green upper leaf surface and its opaque rather than brown scales on the lower surface. It is arguable that these two species should be merged as the differences between them are not very significant, although the distribution of *R. impeditum* is to the north-east of that of *R. fastigiatum* and they overlap in only a limited area.

This species is very common and rightly popular, although many plants are mistakenly sold under the name *R. impeditum*. In relatively cool climates, it is hardy and easy to please and makes a dense carpet with fine bluish-purple flowers. Introduced 1906->, reintroduced 1981->. April-May.

Picture 1

1. *R. fastigiatum* SBEC 0804 from the Cangshan, C.W. Yunnan, China, collected in 1981. This is a form with relatively dark-coloured flowers.

Picture 2

Picture 3

Picture 4

Picture 5

Picture 6

6. *R. fastigiatum* leaves showing the typical pale opaque scales of this species. (Photo R.B.G., E.)

R. flavidum
Franch. 1895 (*R. primulinum* Hemsl.) H4-5

Height to 1m, occasionally to 2.5m, habit erect. **Leaves** 0.7-1.5 x 0.3-0.7cm, broadly elliptic to oblong; upper surface *shiny*, lower surface with *prominent, widely-spaced, reddish scales*. **Inflorescence** 1-3-flowered. **Corolla** broadly funnel-shaped, 1.2-1.8cm long, pale yellow, pubescent inside and outside; calyx lobes 2-4mm, sparsely lepidote with ciliate margins; stamens (8-)9-10; style exceeding the stamens. **Distribution** N.W. Sichuan 3,000-4,000m (10,000-13,000ft), in a limited area.

This species differs from *R. rupicola* var. *chryseum* and var. *muliense* in its more erect habit, and its dark green, shiny leaves with widely-spaced, reddish scales on the lower surface. The yellow flowers separate it from other subsection Lapponica species.

Many cultivated plants labelled *R. flavidum* have white flowers are larger than the type in all parts and are probably hybrids. The true yellow-flowered *R. flavidum* is rare in cultivation, not easy to please, and prone to die-back; hopefully recent introductions will be more successful. Introduced 1905, 1908, reintroduced 1992. April-May.

There are several natural hybrids of this species. *R. flavidum* var. *psilostylum* which differs from the type in its bicoloured scales may be a hybrid with *R. rupicola* var. *chryseum*. *R. lysolepis* with tiny shiny leaves and pinkish-purple flowers is considered to be a hybrid between *R. flavidum* and *R.*

2. *R. fastigiatum* in the Cangshan, C.W. Yunnan, China.
3. A group of *R. fastigiatum* SBEC 0804 growing at Glendoick, showing colour variation within the species.
4. *R. fastigiatum* SBEC 0804, a pale blue-flowered selection.
5. *R. fastigiatum* pink-flowered SBEC 0753 growing in the Cangshan, Yunnan, China.

impeditum. *R. wongii* is almost certainly a natural hybrid between *R. flavidum* and *R. ambiguum*. This has yellow flowers and foliage mid-way in size between the putative parents.

Picture 1

Picture 2

Picture 3

Picture 4

4. *R. wongii* is almost certainly a natural hybrid between *R. flavidum* and *R. ambiguum*, Glendoick. Plants cultivated as *R. wongii* are very similar to ones observed in the wild alongside both putative parents.

1. *R. flavidum* var. *flavidum* at the Royal Botanic Garden, Edinburgh.
2. A plant often grown as *R. flavidum* white which is undoubtedly a hybrid. There are several similar clones.
3. *R. flavidum* var. *flavidum* leaves showing the shiny upper surface and the reddish scales on the lower surface. The scales are usually more widely spaced than in this example. (Photo R.B.G., E.)

R. hippophaeoides
Balf.f. & W.W. Sm. 1916 H5

Height to 1.25m, a compact to erect or sprawling shrub. **Leaves** 0.8-3.8 x 0.5-1cm, elliptic to oblong; upper surface *pale, slightly glaucous green*, lower surface densely lepidote with *overlapping*, yellowish buff, transparent scales. **Inflorescence** 4-8-flowered. **Corolla** broadly funnel-shaped, c.1.2cm long, nearly 2.5cm across, lavender-blue, bluish purple, to near rose, rarely white; calyx to 1.8mm; style 4-16mm; stamens *10*.

The glaucous foliage and the overlapping, creamy-yellow scales of this species distinguish it from most other members of subsection Lapponica. *R. intricatum* has similar greyish-green foliage but the shape of the corolla is very different.

Var. *hippophaeoides*
Style *4-10.5mm* **Distribution** N.W., W, S.W. & central Yunnan and S.W. Sichuan 2,400-4,800m open slopes, often marshy.

This is one of the finest and most widely-grown of the species in subsection Lapponica and the showy flowers are amongst the most frost-hardy of all rhododendrons, even when fully out. In the wild, it is often found in boggy ground. A white form has recently been introduced. Unfortunately this species is particularly prone to powdery mildew. Introduced 1913->, reintroduced 1990s. (March-) April-May.

LAPPONICA

Picture 1

Picture 2

Picture 3

1. Dr. David Chamberlain examining *R. hippophaeoides* var. *hippophaeoides* growing in a bog, Zhongdian Plateau, N.W. Yunnan, China.
2. A very rare white-flowered *R. hippophaeoides* var. *hippophaeoides* (now in cultivation) along with a typical lavender-blue one, Zhongdian Plateau, N.W. Yunnan, China.
3. The upright-growing *R. hippophaeoides* var. *hippophaeoides* 'Haba Shan', Glendoick.

Var. *occidentale* Philipson & Philipson 1975 (*R. fimbriatum* Hutch.)

Style *13-16mm*. **Distribution** N. & Central Yunnan, 3,500-4,250m, (11,000-14,000ft), open stony slopes.

This rare variety generally has a narrower leaf and a smaller inflorescence than in var. *hippophaeoides* and seems to be identical to what used to be known as *R. fimbriatum* (now Fimbriatum Group) which we have therefore moved as a synonym from var. *hippophaeoides*. As we have seen it, it tends to have purple rather than bluish flowers. April-May. Introduced 1993.

Picture 4

Picture 5

4. The low-growing *R. hippophaeoides* var. *hippophaeoides* Yu 13845, Glendoick.
5. *R. hippophaeoides* var. *occidentale* with its smaller flower truss and longer style than var. *hippophaeoides*. Do La Guo, Weixi, N.W. Yunnan, China.

R. impeditum
Balf.f. & W.W. Sm. 1916 (*R. litangense* Hutch., *R. semanteum* Balf.f.) H3-5.

Height to 0.9m+ in the wild but rarely over 0.3m in cultivation, a dense and mounded dwarf shrub. **Leaves** 0.4-1.5 x 0.3-

0.6cm, elliptic to ovate to oblong; upper surface dark green, *not glaucous*, lower surface pale, with contiguous or near-contiguous, large, brown-speckled or rust-coloured scales. **Inflorescence** to 4-flowered. **Corolla** broadly funnel-shaped, 0.7-1.6cm long, violet or purple to rose lavender, (white); calyx 2.5-4mm long, lobes strap-shaped, thin, margins ciliate; stamens variable in number, but usually 10; style length very variable. **Distribution** N. Yunnan, S.W. Sichuan 2,700-4,900m (9,000-16,000ft), open slopes, alpine meadows etc.

Many plants labelled *R. impeditum* in cultivation are in fact its closest relative *R. fastigiatum*. *R. fastigiatum* has glaucous foliage and opaque scales on the lower leaf surface while *R. impeditum* differs in its green foliage and dark brown scales.

Most forms of *R. impeditum* are hardy and easy to please wherever summers are not to hot. It is a parent of many very popular hybrids such as R. 'Blue Tit'. The Philipsons merged *R. litangense* into *R. impeditum* but it should be maintained as a Group as its upright/erect habit is very different from that of typical, cultivated *R. impeditum*. Introduced 1910->. April-May.

Picture 1

Picture 2

1. *R. impeditum* plus other species of subsection Lapponica, Lake Kong Wu, Muli, S.W. Sichuan, China. (Photo W.E. Berg)
2. *R. impeditum* from Gigha growing at Glendoick. This clone has slightly glaucous leaves and may be intermediate between this species and *R. fastigiatum*.

Picture 3

Picture 4

3. *R. impeditum* Forrest 20454, Glendoick. This form has dark green leaves and is very low-growing.
4. Foliage of *R. impeditum* with the dark upper leaf surface on left and the lighter lower surface with near-touching rusty-brown scales on the right. (Photo R.B.G., E.)

R. intricatum

Franch. 1895 (*R. blepharocalyx* Franch., *R. peramabile* Hutch.) H4-5

Height to 1.5m in the wild, to 0.9m in cultivation, usually less, a fairly compact, bushy shrub. **Leaves** 0.5-1.4 x 0.3-0.7cm, oblong to elliptic; upper surface *pale greyish-green* with overlapping gold, translucent scales, margins often hairy, lower surface uniformly buff to straw-coloured, scales contiguous to overlapping. **Inflorescence** 2-8-flowered, compact. **Corolla** *tubular with spreading lobes* (almost hypocrateriform), 0.8-1.4cm long, pale lavender to mauve and dark-blue, usually pale lavender-blue in cultivation; calyx 0.5-2mm; stamens usually 10, *very short*; style *very short*, shorter than stamens. **Distribution** N. Yunnan, S.W. and Central Sichuan, open moist meadows, hillsides, forest margins, 2,800-4,900m (9,300-16,000ft).

The distinctive flowers of this species have more in common with section Pogonanthum than subsection Lapponica. Its

LAPPONICA

closest relative is *R. complexum* which has 5-6 rather than 10 stamens, a less greyish-green upper leaf surface and a rust-coloured rather than pale lower leaf surface.

An attractive species with unusual and very frost-hardy flowers, *R. intricatum* is hard to propagate and remains rare in cultivation but should become more widely grown. Introduced 1904->, reintroduced 1990s. March-May.

Picture 1

Picture 2

Picture 3

Picture 4

4. Leaves of *R. intricatum*; the scales on the upper surface are overlapping while those on the lower surface are overlapping or touching. (Photo R.B.G., E.)

1. *R. intricatum* and other species of subsection Lapponica with *R. phaeochrysum* in the background, Muli, S.W. Sichuan, China. (Photo W.E. Berg)
2. *R. intricatum* Forrest 20, Royal Botanic Garden, Edinburgh, showing the tubular corolla with spreading lobes, unique in the subsection.
3. *R. intricatum* showing the short stamens, included within the corolla tube.

R. lapponicum

Wahlenberg 1812 (*R. confertissimum* Nakai, *A. ferruginosa* Pall., *R. palustre* Turcz., *R. parviflorum* F. Schmidt, *A. parvifolia* Adams) H5

Height to 1m, habit prostrate to erect. **Leaves** 0.4-2.5 x 0.2-7cm, oblong-elliptic to elliptic-ovate, retained for only one year (or almost deciduous with the few remaining leaves hanging); upper surface greyish to dark green and densely lepidote, usually bronze in winter, lower surface densely lepidote, with bicoloured, straw to fawn or rust-coloured, contiguous scales. **Inflorescence** 3-6-flowered. **Corolla** broadly funnel-shaped, 0.7-1.4cm long, pale magenta-rose to purple (very rarely white); calyx 1-2mm long, lobes triangular; stamens 5-10; style longer than stamens. **Distribution** Circum-polar distribution, including Scandinavia, Russia, Siberia, Korea, N. Japan, Alaska, Canada and Greenland, and relict populations on mountains in northern U.S.A., in a large variety of habitats.

The distinguishing features of this species (especially in cultivation) are its elliptic, densely lepidote leaves with bicoloured scales, its early flowering, and its sparse, hanging leaves and bronze winter colouring which can make the plant look as if it is dead.

R. lapponicum has a wide circum-polar distribution but only plants known as Parvifolium Group, from relatively near the Pacific coast in Japan, Korea and Siberia, are easy to establish and grow in cultivation. The best Parvifolium selections make very useful garden plants as their frost-hardy flowers can open as early as January when very little else is in flower. The really dwarf forms from the Arctic have proven almost impossible to cultivate; they often grow over permafrost in the wild and cannot tolerate warmer conditions. Introduced 1825->. January-March in cultivation, April-June in the wild).

4. Shoot of *R. lapponicum* showing the contiguous scales. (Photo R.B.G., E.)

R. nitidulum
Rehder & E.H. Wilson 1913 H4-5

Height to 1.3m, habit broadly upright to spreading. **Leaves** 0.5-1.1 x 0.3-0.7cm, ovate to elliptic; lower surface with contiguous or near contiguous scales. **Inflorescence** 1-2-flowered. **Corolla** funnel-shaped, 1.2-1.5cm long, rosy-lilac to violet purple, pubescent inside; calyx 1-3mm long, variable in shape; stamens (8)-10; style exceeding stamens. **Distribution** N.W. & Central Sichuan (var. *omeiense* only recorded from Emei Shan), 3,000-5,000m (10,000-16,000ft).

It is hard to distinguish this species from some forms of *R. nivale* ssp. *boreale*, especially those formerly under *R. violaceum* which is supposed to have a smaller corolla and darker scales. *R. nitidulum* may merge with *R. capitatum* in the wild. Very variable in flower and leaf even on the same plant. The two varieties of *R. nitidulum* are just botanical hair-splitting; the differences are not significant and the name var. *omeiense* is misleading in that both varieties can be found on Emei Shan.

R. nitidulum is a neat, dense grower with tiny leaves and flowers of good colour, although not always very free-flowering. First introduced 1908 but was lost to cultivation. Reintroduced 1980->. April-May.

Var. *nitidulum* (var. *nubigerum* Rehder & E.H. Wilson)

Leaves with uniform pale scales on the lower surface.

1. *R. lapponicum* growing north of the Arctic Circle, near Tromso, N. Norway.
2. *R. lapponicum* Parvifolium Group from Hokkaido, Japan, growing at Glendoick. This form is much easier to grow than those from the Arctic.
3. *R. lapponicum* Parvifolium Group, probably from Siberia, growing at Glendoick. A very early-flowering form.

Var. *omeiense* Philipson & Philipson 1975

Leaves with pale scales and a few scattered darker scales on the lower surface.

Picture 1

Picture 2

1. *R. nitidulum* var. *omeiense* K. Rushforth 185, Glendoick.
2. Leaves of *R. nitidulum* var. *omeiense* showing the few scattered dark scales amongst the paler, nearly touching ones. (Photo R.B.G., E.)

R. nivale Hook.f. 1849 H4-5

Height to 1m, a prostrate, low, compact or erect shrub. **Leaves** 0.5-0.9 x 0.2-0.5cm, variable in shape but +/- elliptic or ovate; lower surface yellowish to fawn with +/- contiguous, *dimorphic scales, the majority pale gold, with few to many darker*. **Inflorescence** 1-2(-3)-flowered. **Corolla** broadly funnel-shaped, 0.9-1.3cm long, varying from rich purple through magenta to lilac and pink; calyx variable; stamens usually 10; style variable in length, usually longer than stamens.

This species is characterised by the contrasting pale and dark scales but is very variable and is very hard to separate from its nearest relatives such as *R. telmateium* and *R. nitidulum*. In all its varying forms, this species covers vast areas of Chinese and Himalayan hillsides, looking like heather moorland from a distance when in flower.

Ssp. *nivale*
(*R. paludosum* (part) Hutch. & Kingdon-Ward)

Height to 0.45m rarely to 1.5m in the wild, habit prostrate to low and compact to upright. **Leaves** 0.2-1cm long on the typical high elevation plant, elliptic to broadly elliptic. **Corolla** widely funnel-shaped, 0.8-1.1cm long, rich purple to magenta; calyx *2-4mm long, usually coloured with a band of dark scales on the margin*; style longer than stamens. **Distribution** Nepal, Sikkim, Bhutan, S.E. & S. Tibet, 3,000-5,800m (10,000-19,000ft), open mountainsides and screes.

The scales on the calyx are an important feature of ssp. *nivale*. The significant rather than +/- obsolete calyx lobes separates it from ssp. *boreale*. *R. telmateium* has a pointed leaf while ssp. *nivale* usually has a rounded one. This is the only subsection Lapponica species in the Himalaya apart from the aberrant *R. setosum*.

The high elevation forms of ssp. *nivale* have amongst the smallest leaves and shortest annual growth of any rhododendron. This subspecies is the rhododendron recorded from the highest ever altitude: said to occur at 5,800m in the Himalaya. The high altitude forms are shy-flowering and hard to grow in cultivation, only successful in cool northern gardens. Plants under the name Paludosum Group from S.E. Tibet have relatively large leaves, a more upright habit and are easier to please than the type. Introduced 1915->, reintroduced 1970s->. April-May.

Ssp. *boreale* Philipson & Philipson 1975
(*R. nigropunctatum* Franch., *R. ramosissimum* Franch., *R. alpicola* Rehder & E.H. Wilson, *R. violaceum* Rehder & E.H. Wilson, *R. oresbium* Balf.f. & Kingdon-Ward, *R. stictophyllum* Balf.f., *R. vicarium* Balf.f., *R. batangense* Balf.f., *R. oreinum* Balf.f., *R. yaragongense* Balf.f.)

Generally taller than ssp. *nivale* particularly in cultivation. The contrast between dark and light scales on the leaf lower surface is less pronounced than in ssp. *nivale*. Calyx lobes *+/- obsolete*. **Distribution** N.W. & Central Yunnan, Tibet and Sichuan 3,100-4,300m (10,250-14,000ft), in open moorland, dry rocky slopes and swampy grassland.

The Philipsons claim the style is predominantly shorter than the stamens which we cannot accept, as the style is longer in many of the former species now made synonyms of ssp. *boreale*. It hybridises and intergrades with many other Subsection Lapponica species in the wild making firm identification very difficult.

Ssp. *boreale*, described by the Philipsons, is a great amalgamation of former species which, despite considerable variation, do seem to merge into one cline. Some former species are given Group status: Stictophyllum with tiny leaves, Ramosissimum with fine blue flowers (as cultivated) and Nigropunctatum with an upright habit and tiny leaves. The Philipsons consider that *R. edgarianum*, which is very similar to Ramosissimum Group as we grow it, is a hybrid of *R. nivale* ssp. *boreale*. Some cultivated clones are close to *R. nitidulum*. Introduced c.1905->. April-May.

Ssp. *australe* Philipson & Philipson 1975

Habit usually relatively upright compared to the other subspecies. **Leaf** apex +/- acute, scale colour contrast less pronounced than in ssp. *nivale*. Calyx lobes *ciliate*.

This subspecies is rather similar to *R. telmateium* and is probably not in cultivation. Its distribution is to the south of and is isolated from that of ssp. *boreale*.

Picture 1

Picture 2

Picture 3

Picture 4

Picture 5

Picture 6

1. Tufts of *R. nivale* ssp. *nivale* on the top of the Nyima La, S.E. Tibet.
2. *R. nivale* ssp. *nivale* on the Nyima La, S.E. Tibet.
3. Tall-growing low elevation *R. nivale* ssp. *nivale* in the Rong Chu Valley, S.E. Tibet, probably referable to Paludosum Group.
4. White *R. nivale* ssp. *nivale* on the Temo La, S.E. Tibet.
5. *R. nivale* ssp. *nivale* on Kanchenjunga, E. Nepal.
   (Photo R. McBeath)
6. *R. nivale* ssp. *boreale*, Bei-ma Shan, N.W. Yunnan.
   (Photo R. McBeath)

LAPPONICA

Picture 7

Picture 8

Picture 9

7. *R. nivale* ssp. *boreale*, Balang, N.W. Sichuan, China.
8. *R. nivale* ssp. *boreale* Stictophyllum Group Rock 24385, showing its small, shiny, dark green leaves. Royal Botanic Garden, Edinburgh.
9. *R. nivale* ssp. *boreale* Ramosissimum Group with its dense habit and small leaves. Glendoick.

R. orthocladum
Balf.f. & Forrest H4-5

Height to 1.3m, a compact to erect, much-branched, low shrub. **Leaves** 0.8-1.6 x 0.25-0.5cm, narrowly elliptic to *lanceolate*; upper surface slightly glaucous, lower surface yellowish-brown to fawn, with contiguous, golden to fulvous scales with a few to many darker scales intermixed. **Inflorescence** 1-5-flowered. **Corolla** funnel-shaped, 0.7-1.3cm long, pale to deep-lavender-blue to purple (white in var. *microleucum*); calyx 0.5-1.5mm long; stamens 8-10; style equal to or shorter than the stamens. **Distribution** N. Yunnan, S.W. Sichuan 3,440-4,300m (11,000-14,000ft), forest margins, cliffs, thickets.

R. polycladum is similar to *R. orthocladum* but has longer stamens and style and a more rangy, open growth habit. *R. orthocladum* is also closely related to *R. thymifolium* which has uniformly straw-coloured scales on the usually narrower leaves, and a more erect habit.

Var. *orthocladum*

Corolla in shades of lavender and purple.

Var. *microleucum*
(Hutch.) Philipson and Philipson 1975

Corolla white.

This species is rarely grown in its blue-purple forms but is quite commonly cultivated under the name var. *microleucum* which is an albino form from wild seed and which therefore should not strictly have varietal status. It has masses of tiny white, frost-hardy flowers and it provides a useful contrast to all the blue-purple Lapponica species. Introduced 1913->. April-May.

Var. *longistylum* (probably not in cultivation) seems to be a hybrid/intermediate with *R. polycladum* with a longer style. It would perhaps be better treated as a form of *R. polycladum*.

Picture 1

1. *R. orthocladum* var. *orthocladum* Forrest 20488, Royal Botanic Garden, Edinburgh showing the short style.

Picture 2

Picture 3

Picture 4

2. Leaves of *R. orthocladum* var. *orthocladum* showing the mixed, dense, dark and light-coloured scales on the lower surface. (Photo R.B.G., E.)
3. Beds of dwarf rhododendrons at the Royal Botanic Garden, Edinburgh. The white-flowered plants in the middle are *R. orthocladum* var. *microleucum*.
4. *R. orthocladum* var. *microleucum* Forrest 22108, Royal Botanic Garden, Edinburgh. (Photo R.B.G., E.)

R. polycladum

Franch. 1886 (*R. compactum* Hutch., *R. scintillans* Balf.f. & W.W. Sm.) H4-5

Height to 1.2m, an upright and spreading shrub often with some vigorous, straggly shoots. **Leaves** 0.8-1.8 x 0.3-0.6cm, *narrowly elliptic* to elliptic; upper surface dark green, lower surface greyish with brown stippling or more uniformly reddish-brown, the scales contiguous in groups or not contiguous. **Inflorescence** to 5-flowered. **Corolla** broadly funnel-shaped, 0.7-1.3cm long, lavender to rich purple-blue, (white); calyx variable, lobes shorter than 2.5mm; stamens 10; style *longer* than stamens. **Distribution** N. N.W. & Central Yunnan, 3,000-4,400m (10,000-14,500ft), in dry and wet sites in alpine meadows amongst scrub and at forest margins.

This species is characterised by its very narrow leaves on rangy, slender, arching shoots. It differs from *R. orthocladum* in its ranging habit and in the longer style.

This species is usually cultivated as the clone *R. polycladum* Scintillans Group F.C.C. which has very fine blue-purple flowers, perhaps the finest shade of blue among the subsection Lapponica species. This species benefits from pruning to improve its shape. It seems to be rather prone to powdery mildew. Introduced 1913->. April-May.

Picture 1

Picture 2

1. The F.C.C. clone of *R. polycladum* with its rich purple-blue flowers. Glendoick.
2. A pale-flowered clone of *R. polycladum*, Royal Botanic Garden, Edinburgh.

LAPPONICA

Picture 3

3. Foliage of *R. polycladum* showing the narrow typical leaves and touching scales on the lower surface. (Photo R.B.G., E.)

Picture 1

Picture 2

R. rupicola
W.W. Sm. 1914 H4-5

Height to 0.6m or more, a compact to straggly shrub. **Leaves** 0.6-2.1 x *0.3-1.2cm*, broadly elliptic to elliptic; lower surface with prominent scales, *dark and pale mixed*, overlapping to near contiguous. **Inflorescence** to 6-flowered. **Corolla** broadly funnel-shaped, 1-1.7cm long, *intense purple, reddish purple, occasionally crimson to magenta, yellow* in var. *chryseum* and var. *muliense*, lobes lepidote outside; calyx 4-5mm long, with a central band of pale scales, margins rarely ciliate; stamens 5-10; style usually slightly longer than stamens.

Var. rupicola
(*R. achroanthum* Balf.f. & W.W. Sm., *R. propinquum* Tagg)

Corolla purple to crimson (rarely white). **Distribution** N. Burma. N. to Central Yunnan, S.W. Sichuan, S.E. Tibet, 3,000-4,600m (10,000-15,000ft), mountainsides and meadows.

The very deep purple flowers distinguish this from all species in subsection Lapponica except its closest relative *R. russatum*. *R. rupicola* has pale scales on the calyx lobes and the outside of the corolla, while *R. russatum* usually has a heavily pubescent style and hairs on the calyx lobes. As cultivated *R. rupicola* generally has narrower leaves than *R. russatum*. These two species merge in the wild and out of flower they can be hard to distinguish.

The flowers of the best forms of this variety are a very striking colour, especially with the sun shining through them. Forms under the name Luridum Group are a very dark reddish-purple. Introduced 1910->, reintroduced 1980s->. April-May.

Picture 3

1. *R. rupicola* var. *rupicola* Forrest 16579, Royal Botanic Garden, Edinburgh.
2. *R. rupicola* var. *rupicola* Luridum Group, Kingdon-Ward 7048, a particularly dark flowered selection. Glendoick.
3. *R. rupicola* var. *chryseum*, Zhongdian Plateau, N.W. Yunnan, China. (Photo R. McBeath)

LAPPONICA

Var. *chryseum* (Balf.f. & Kingdon-Ward) Philipson & Philipson 1975.

Corolla *yellow*; calyx lobes margined with hairs only. Stamens 5-10. N.E. Burma, N.W. Yunnan, S.E. Tibet, 3,300-4,700m (11,000-15,500ft), forest clearings and moorland.

R. flavidum differs in its more erect habit and one-coloured scales.

The attractive yellow flowers provide a fine contrast with the other Lapponica species. April-May.

Var. *muliense*
(Balf.f. & Forrest) Philipson & Philipson 1975.

Leaves tend to be more oblong than elliptic. **Corolla** yellow; calyx lobes margined with scales and hairs. Stamens usually 10. S.W. Sichuan, 3,000-4,900m (10,000-16,000ft), woodland clearings and alpine meadows. April-May.

Horticulturally this is identical to var. *chryseum* and, despite the geographical isolation, we are not convinced that the two varieties are usefully distinguished. In Muli, S. Sichuan, this variety merges/crosses with other Lapponica species producing hybrids in a range of colours.

Picture 4

Picture 5

4. *R. rupicola* var. *chryseum*, Bei-ma Shan, N.W. Yunnan, China.
5. *R. rupicola* var. *chryseum*, Zhongdian Plateau, N.W. Yunnan, China. (Photo R. McBeath)

Picture 6

Picture 7

Picture 8

6. A white-flowered clone of *R. rupicola* var. *chryseum*, Glendoick.
7. Leaf lower surface of *R. rupicola* var. *chryseum* showing the prominent mixed dark and light scales. (Photo R.B.G., E.)
8. *R. rupicola* var. *muliense* Yu 14641, Royal Botanic Garden, Edinburgh.

R. russatum
Balf.f. & Forrest 1919 (*R. cantabile* Hutch., *R. osmerum* Balf.f. & Forrest) H4-5

Height 0.3-1.8m, a variable shrub, low and compact to tall and straggly. **Leaves** +/- aromatic, 1.6-4 x *0.6-1.7cm*, narrowly to broadly elliptic or oblong, obtuse or rounded, lower surface heavily speckled with +/- contiguous, bicolor, pale and dark brown/rust, or uniformly red-brown scales. **Inflorescence** 4-6-flowered. **Corolla** broadly funnel-shaped, 1-2cm long, deep to medium indigo-blue, deep to medium reddish-purple, pinkish lavender (rose, white); calyx to 6mm long, *with long hairs on margins*; stamens 10, pubescent towards base; style, *usually heavily pubescent,* usually longer than stamens. **Distribution** N, N.W. Yunnan and S.W. Sichuan 3,400-4,300m (11,000-14,000ft), in alpine pasture and forest margins.

The relatively large leaves and deep-coloured flowers distinguish this species from most of subsection Lapponica. Its closest relation is *R. rupicola* but *R. russatum* can be distinguished by its wider leaves, pubescent style, the non-lepidote corolla and the long hairs on the calyx margins.

Its striking, deep purple-blue flowers have made *R. russatum* one of the most widely-grown of the Lapponica species and it has been much used in hybridising. The two deepest forms we have seen are known as the Collingwood Ingram form and 'Blue Black'. Selected forms are recommended as many wild forms are mediocre. Introduced 1917->, reintroduced 1980s->. April-May.

Picture 1

Picture 2

Picture 3

Picture 4

1. *R. russatum*, on open moorland, Duo La Guo, Weixi, N.W. Yunnan, China. (Photo F. Hunt)
2. *R. russatum*, a fine selection from Collingwood Ingram growing at Glendoick.
3. *R. russatum*, a compact clone from Hill of Tarvit growing at Glendoick.
4. Leaf upper and lower surfaces of *R. russatum* showing the nearly touching pale and dark scales on the lower surface. (Photo R.B.G., E.)

R. setosum
D. Don 1821 H4-5

Height to 1.2m but usually less than 0.45m, a low but often open shrub. Branchlets *bristly*. **Leaves** 1-1.9 x 0.6-0.8cm, elliptic, oblong to obovate, margins *hairy*; upper surface with golden scales, lower surface with dimorphic scales, vesicular and golden, and flat and pale brown; petiole *bristly*. **Inflorescence** 1-6-flowered. **Corolla** openly funnel-shaped, 1.5-1.8cm long, *purple to pinkish to wine red;* calyx 5-8mm *reddish or crimson-purple*; stamens 10; style longer than stamens. **Distribution** Nepal, Sikkim, W. Bengal, Bhutan, S. Tibet, 2,700-4,900m (9,000-16,000ft), open hillsides and slopes.

This species was removed from the Lapponica subsection by the Philipsons but returned there by Cullen. In our opinion, the Philipsons have the stronger case as the bristles on the leaves, branches and leaf underside are atypical of other species in subsection Lapponica. We would suggest that *R. setosum* be moved to a subsection Setosa with *R. fragariflorum* which is an essentially eastern variation of this species, differing in its smaller leaves and lower habit.

A distinctive species both in flower and foliage and worth growing in cooler gardens, *R. setosum* is fastidious as to drainage and is sometimes liable to partial die-back. Introduced 1825-1949, reintroduced 1970s->. May.

Picture 1

Picture 2

Picture 3

4. Foliage of *R. setosum* showing the characteristic hairy leaf margins, the golden scales on the upper surface and the pale scales on the lower surface. (Photo R.B.G., E.)

Picture 4

1. *R. setosum* Cox, Hutchison & Maxwell MacDonald 2049, Jaljale Himal, E. Nepal.
2. *R. setosum* Schilling 2260, Glendoick, showing the large coloured calyces.
3. *R. setosum*, Glendoick.

R. tapetiforme
Balf.f. & Kingdon-Ward 1916 H4-5

Height to 0.9m, usually much lower, a *matted or rounded shrub*. **Leaves** 0.4-1.5 x 0.3-0.8cm, broadly elliptic or rounded; upper surface dark green, covered with translucent scales, lower surface rufous, with uniform, contiguous, rust-coloured scales. **Inflorescence** 1-4-flowered. **Corolla** broadly funnel-shaped, 0.9-1.6cm long, purplish to purplish-blue, (violet, rose, yellow); calyx variable to 2mm; stamens 10 or fewer (5-6); style usually *longer than stamens*. **Distribution** Tibet/Burma/Yunnan border regions, 3,500-4,600m (11,500-15,000ft), open moorland, covering large areas.

R. tapetiforme is characterised by its low growth habit, its rounded leaves and conspicuous scales on the lower leaf surface. The closely related *R. yungningense* has a more upright habit and usually a larger flower and calyx.

Picture 1

1. *R. tapetiforme*, Big Snow Mountain, N. Yunnan, China.

LAPPONICA

Quite variable in the wild, especially in size and shape of leaf, and rare in gardens; most plants labelled *R. tapetiforme* in cultivation do not entirely agree with the published description. Introduced 1917-21, 1992->. April-May.

Picture 2

Picture 3

Picture 4

2. Hillside of *R. tapetiforme* on Bei-ma Shan, N.W. Yunnan, China.
3. A form of *R. tapetiforme* with pale flowers on the Salween-Mekong divide near Londre.
4. Foliage of *R. tapetiforme* with translucent scales on upper leaf surface and large, touching, uniform, rust-coloured scales on the lower surface.
   (Photo R.B.G., E.)

R. telmateium

Balf.f & W.W. Sm. 1916 (*R. diacritum* Balf.f. & W.W. Sm., *R. drumonium* Balf.f. & W.W. Sm. *R. idoneum* Balf.f. & W.W. Sm., *R. pycnocladum* Balf.f. & W.W. Sm.) H4-5

Height to 0.9m, a prostrate to erect small shrub. Branchlets often very slender. **Leaves** 0.3-1.2 x 0.15-0.5cm long, lanceolate, elliptic to oval, *with a distinct point at the tip*; upper surface dull grey-green, covered in pale gold scales, lower surface *pale,* overlaid with *dense overlapping* scales, dimorphic, pale gold to rufous and dark (dark occasionally absent). **Inflorescence** 1-2(-3)-flowered. **Corolla** broadly funnel-shaped, 0.7-1.2cm long, lavender or rose-pink to purple; calyx small (1-2.5mm), margins with scales and or hairs; stamens 10 (8-11); style shorter, equalling or longer than stamens. **Distribution** N, N.W. & Central Yunnan and S.W. Sichuan 2,900-5,000m (9,500-16,000ft), on open moorland, oak forest and boggy ground.

This species is hard to identify conclusively, as few characters are constant and the former species placed in synonymy here are very varied. The Philipsons say that the defining characters are the small calyx and dimorphic scales over a pale background, but admit that 'individual plants' may not have the dimorphic scales. The very small leaves are another useful character.

Scarce in cultivation under the name *R. telmateium* and more commonly seen under the former specific names. Diacritum Group is usually low and compact and has +/- oval or elliptic leaves, Pycnocladum Group has relatively wide leaves, Drummonium Group has a shorter style and is very dwarf while Idoneum Group has creamy-yellow to fawn scales and long stamens. Introduced 1910->, reintroduced 1980s->. April-May.

Picture 1

1. *R. telmateium* Diacritum Group, Yulong Shan, N.W. Yunnan, China, growing on limestone.
   (Photo A.D. Schilling).

LAPPONICA

rim-like, lobes often coloured rose, sometimes lepidote and sometimes ciliate; stamens 10 (8-12); style short or long. **Distribution** E. Tibet, N. Sichuan, Qinghai and Gansu 2,600-4,600m (8,500-15,000ft).

This species is characterised by its usually upright habit, its narrow leaves with fawn scales and its usually solitary flowers. Its closest relations are *R. polycladum* and *R. orthocladum* which both have much deeper-coloured scales and usually more flowers to the truss. The latter has a more compact rounded growth habit.

R. thymifolium is quite a pretty plant in flower and foliage and is fairly distinctive. Introduced 1915-32. April-May.

Picture 2

Picture 3

Picture 4

Picture 4

1. *R. thymifolium* showing the erect habit and narrow leaves. Glendoick.
2. *R. telmateium* Diacritum Group, Yulong Shan, N.W. Yunnan. This form is usually low and compact.
3. *R. telmateium*, a tall-growing form with slender branchlets. Royal Botanic Garden, Edinburgh.
4. Foliage of *R. telmateium* with pale gold scales on upper leaf surface and dense pale gold scales with occasional rufous scales on the lower surface. (Photo R.B.G., E).

R. thymifolium

Maxim. 1877 (*R. polifolium* Franch., *R. spilanthum* Hutch.) H4-5

Height to 1.2m, an *erect*, low shrub. **Leaves** 0.5-1.2 x 0.2-0.5cm, usually *narrowly obovate to oblanceolate*, lower surface straw-coloured, densely covered with contiguous to overlapping, pale fawn scales (rarely with a few darker scales). **Inflorescence** *1-(2)-flowered*. **Corolla** broadly funnel-shaped, 0.7-1.1cm long, lavender or purplish; calyx 0.5-1.2mm, or

R. tsaii W.P. Fang 1939. H5

Shrub to 1.3m. **Leaves** 0.6-1.2 x 0.25-0.5cm, narrowly elliptic or oblong-lanceolate; lower surface buff, densely covered in overlapping pale scales. **Inflorescence** *3-7-flowered*. **Corolla** pale purplish, broadly funnel-shaped, 0.45-0.7cm long, elepidote outside; calyx to 1mm, densely lepidote; stamens *4-7*, shorter than the corolla; ovary densely pale lepidote; style c. 2mm, slightly shorter than the stamens, glabrous. **Distribution** E. Yunnan, 2,900m (9,500ft).

This species is closely related to *R. hippophaeoides*, differing in its smaller leaves and only 4-7 stamens which lack basal hairs.

R. tsaii is recorded only in N.E. Yunnan, overlooking the Yangtze river, and has not yet been introduced. Across the

Yangtze in Sichuan there are populations from subsection Lapponica that lie botanically between *R. hippophaeoides* var. *hippophaeoides* and *R. tsaii*. The Philipsons regard these as being hybrids between these two taxa but they are better considered intermediate forms, forming a cline from one species to the other. These intermediate forms were introduced in 1990 and 1995 (from different localities) as *R. tsaii* aff.

Picture 2

2. *R. websterianum* C. 5080, collected at Mujizo Lake, W. Sichuan, in 1990, at first time of flowering in 1994 at Glendoick.

R. websterianum
Rehder & E.H. Wilson 1913 H4-5

Height to 1.5m, a fastigiate, much-branched shrub. New growth *coincides with flowering.* **Leaves** 0.6-1.5 x 0.3-0.9cm, ovate to ovate-lanceolate; upper surface *pale greyish-green,* lower surface straw-coloured, densely covered with contiguous, buff-coloured scales. **Inflorescence** *single* occasionally 2-flowered. **Corolla** funnel-shaped, 1.3-1.9cm long, light purple (yellow?); calyx lobes *2.8-5mm*, usually coloured and lepidote; stamens 10; style reddish-purple, longer than stamens. **Distribution** Sichuan around Kangding, 4,100-4,900m (13,500-16,000ft), moorland and bog.

R. websterianum is closely related to *R. nitidulum* and *R. minyaense* but is more erect in habit, with buff rather than golden scales on the lower leaf surface, and is characterised by its new growth established at time of flowering. It also somewhat resembles a smaller-leaved *R. hippophaeoides*.

Although plants have been cultivated as *R. websterianum*, it is thought that all were incorrectly labelled (usually turning out to be *R. hippophaeoides*) and that the true species was not successfully introduced until 1990 >. April May.

Picture 1

1. *R. websterianum*, Mujizo Lake, W. Sichuan, China, showing the typical erect habit with pale greyish-green leaves.

R. yungningense
Balf.f. 1930 (*R. glomerulatum* Hutch.) H4-5

Height to 1m (1.3m), an erect, rounded, low shrub. **Leaves** 0.8-2 x 0.4-0.8cm, elliptic to oblong; upper surface greyish-green, mat, lower surface with fawn to rust-coloured scales with darker centres, contiguous to near contiguous, sometimes also with scattered darker scales. **Inflorescence** 1-4-flowered. **Corolla** broadly funnel-shaped, 1.1-1.4cm long, deep purplish-blue, rose-lavender (white); calyx *very variable* and often irregular, usually 2-3mm, margins usually with long hairs and some scales; stamens 10 (8-12); style length variable. **Distribution** N., N.W. Yunnan, S.W. Sichuan, 3,200m-4,300m (10,500-14,000ft), alpine slopes.

R. yungningense is characterised by its dull green leaves, densely scaly on the lower surface, and its very variable calyx. It often grows with *R. fastigiatum* in the wild but is usually

Picture 1

1. *R. yungningense* Forrest 16282, Royal Botanic Garden, Edinburgh.

taller and more upright, and is closely related to *R. tapetiforme* which has a more rounded leaf and usually a smaller calyx.

Plants of this species are most often cultivated as *R. glomerulatum* which is synonymous. Not one of the most showy or interesting Lapponica species and likely to remain a collector's plant. Introduced 1922->. April-May.

Picture 2

Picture 3

2. *R. yungningense* Forrest 29260, Royal Botanic Garden, Edinburgh. This example has a short style but it can also be longer. (Photo R.B.G., E.)
3. Leaves of *R. yungningense* showing the touching scales on the upper surface and touching scales with darker centres on the lower surface. (Photo R.B.G., E.)

Subsection Ledum

Height to 2m, small shrubs; young shoots lepidote and *covered with rufous tomentum*, puberulous hairs or glands. **Leaves** evergreen, 0.6-8 x 0.2-2cm, *linear to +/- elliptic, very aromatic*, margins usually strongly recurved, upper surface dark green, lower surface *white papillate, often white setulose* and sometimes with +/- rufous indumentum; scales rimless, golden. **Inflorescence** terminal, many-flowered, +/- compact. **Corolla** rotate, 0.4-1cm long, *white*; calyx small or obsolete; stamens 7-12; style straight.

A very distinct subsection, characterised by the aromatic leaves, the presence of tomentum and indumentum on the young shoots and leaves and the many flowered inflorescence. The closest relative is probably *R. micranthum* which has similar flowers but lacks tomentum or indumentum on the branches and leaves. The only other species in subgenus Rhododendron with indumentum on the leaves are the members of subsection Edgeworthia which are very different in many other characters.

Until recently *Ledum* was considered a separate genus but recent published work by Kron & Judd 1990 and Harmaja 1991 has concluded that *Ledum* should be included within subgenus Rhododendron of the genus *Rhododendron*. Subsection Ledum contains a number of very hardy species which make fairly accommodating garden plants. All have freely-produced, small white flowers in many-flowered, roundish inflorescences. They appreciate similar growing conditions to those enjoyed by most other dwarf rhododendrons.

R. groenlandicum

(Oeder) Kron & Judd 1990 (*Ledum groenlandicum* Oeder, *L. latifolium* Jacq., *L. pacificum* Small) Labrador Tea H5

Height 0.5-2m, an erect shrub, young shoots with rufous tomentum. **Leaves** 1.2-6 x 0.5-1.5cm, linear-elliptic, margins recurved, lower surface with thick rufous indumentum, usually hiding the midrib, epidermis papillose with white setulose hairs, scales dense, mixed with red-brown glands; petiole 1-5mm. **Inflorescence** flowers numerous. **Corolla** 4-8mm long, white; calyx minute, stamens 7-10; ovary glandular, style glabrous. **Distribution** Greenland, Canada, N. U.S.A., sea level to 1,800m (0-6,000ft),

This species differs from *R. tomentosum* ssp. *tomentosum* in its wider leaf. *R. hypoleucum* differs in the presence of indumentum on the midrib of the leaf lower surface only.

R. groenlandicum is the commonest of the subsection in cultivation and is often regarded as the best garden plant, although some forms of *R. hypoleucum* have larger inflorescences. The common name Labrador Tea refers to the use of the leaves for making an aromatic hot drink. Plants under the name Compactum or Compactus (probably best considered at Group status) are dwarfer than the type with short, very woolly shoots, short but broad leaves and small inflorescences. Introduced 1763. April-June

Picture 1

Picture 2

1. *R. groenlandicum* at Askival, W. Scotland. (Photo P. Stone)
2. The distinctive many-flowered inflorescence of small white flowers of *R. groenlandicum*. (Photo P. Stone)

R. hypoleucum

(Kom.) Harmaja 1990 (*Ledum hypoleucum* Kom., *L. palustre* var. *diversipilosum* Nakai, *L. nipponicum* (Nakai) Tolm.) H5

Height 0.5-1m, an erect shrub, young shoots with rufous tomentum. **Leaves** 1.7-8 x 0.5-2cm, oblong-elliptic, margins recurved, ciliate with long, brown, crisped hairs, upper surface with rufous hairs, lower surface glaucous, +/- papillate, *densely white-pubescent,* scales 1-3 x their own diameter apart, with *long, crisped, rufous hairs confined to the midrib*; petiole 2-7mm. **Inflorescence** flowers numerous. **Corolla** 0.5-0.7mm long, white; calyx lobes 1-2mm, stamens 9-12; ovary densely pubescent and scaly, style glabrous. **Distribution** N.E. Russia, Japan.

This species differs from its relatives in its conspicuous white pubescence on the leaf lower surface, with hairs only on the midrib.

The best selections of this species have the most showy flowers in this subsection. Introduced in 1893 to U.S.A.

R. neoglandulosum

Harmaja 1990 (*Ledum glandulosum* Nutt., *L. californicum* Kellogg) Trapper's Tea, Glandular Labrador Tea H4

Height 0.5-2m, an erect shrub, young shoots puberulent, dotted with glands. **Leaves** 1.5-4 x 0.5-2cm, broadly elliptic-oval, *margins flat* or slightly recurved, lower surface *light green, papillate, glabrous or +/- pubescent*, scales 1-2 x their own diameter apart; petiole 4-10mm. **Inflorescence** flowers numerous. **Corolla** c.6mm, white; calyx small, lobes rounded, margins ciliate; stamens 8-12; ovary densely glandular and scaly, style sparsely glandular. **Distribution** N.W. U.S.A.

This species differs from other members of the subsection in its usually flat leaves, without hairs or with few hairs on the lower surface.

Long grown under the name *Ledum glandulosum*, this required to have its name changed as there is already a *Rhododendron glandulosum*. Sometimes rather straggly in cultivation. Introduced 1894. June-August.

Picture 1

1. *R. neoglandulosum* on block scree, Lake Chelaw, Wa, U.S.A. (Photo P. Stone)

Picture 2

2. Several plants of *R. neoglandulosum* on block scree, Lake Chelaw, Wa, U.S.A. (Photo P. Stone)

R. x columbianum (Piper) Harmaja 1990 is considered to be a natural hybrid between *L. neoglandulosum* and *L. groenlandicum*. Leaves relatively narrower than *R. neoglandulosum*, 4-6 x 1-2cm, margins slightly reflexed, lower surface sometimes with a few rufous hairs. Oregon to California, North America, in swamps and bogs.

R. tolmachevii

Harmaja 1990 (*Ledum macrophyllum* Tolm.)

We have no record of this species from E. Russia being cultivated. It is closely related to *R. hypoleucum*, differing in the presence of hairs on the whole lower leaf surface.

R. tomentosum Harmaja 1990 H5

Height 0.3-1.2m, an erect or decumbent shrub, young shoots rufous-lanate, glandular. **Leaves** *0.6-5 x 0.1-1.2cm*, linear to narrowly elliptic-oblong, margin strongly recurved, lower surface with dense rufous-lanate indumentum, with or without short, setulose hairs underneath and sometimes with reddish glands. **Inflorescence** many-flowered. **Corolla** 4-8mm long, white; calyx minute; stamens 7-10; ovary glandular, style glabrous.

Ssp. *tomentosum* (*Ledum palustre* L., *L. palustre* var. *dilatatum* Wahlenb.) Wild Rosemary, Crystal Tea, Marsh Ledum

Habit *upright, bushy*. **Leaves** *1.2-5 x 0.2-1.5cm*, lower surface with *short setulose hairs* below indumentum. **Distribution** N. and Central Europe, European Russia to S. Siberia, sea level to 2,000m (0-6,500ft), moors, pine forests and bogs.

The leaves of this subspecies are half as wide as the closely related *R. groenlandicum*.

Ssp. *tomentosum* was used by the Norwegians to make beer more intoxicating. May-June

Ssp. *subarcticum* (Harmaja) G. Wallace 1992 (*Ledum minus* Hort., *L. palustre* var. *decumbens* Aiton, *L. subarcticum* Harmaja)

Habit prostrate or decumbent. **Leaves** *0.6-2 x 0.1-0.3cm*, lower surface with *few or no setulose hairs*. **Distribution** Arctic regions of Europe, Asia including Hokkaido (N. Japan) and Korea and North America.

This subspecies is becoming popular amongst alpine gardeners for its neat prostrate growth and very narrow leaves.

Picture 1

Picture 2

1. *R. tomentosum* ssp. *tomentosum* collected by F. Doleshy, at Glendoick.
2. A showy form of *R. tomentosum* ssp. *subarcticum* (as var. *decumbens*) at the garden of I. & M. Young, Aberdeen.

LEPIDOTA

Subsection Lepidota

Height to 2.4m, creeping, mounding or upright shrubs with thin branchlets. **Leaves** evergreen, semi-deciduous or deciduous. **Inflorescence** 1-5-flowered. **Corolla** *rotate-campanulate*, white, yellow, pink, reddish or purple; calyx well-developed; stamens 10; style *very short and sharply deflexed*. **Distribution** Kashmir, Himalayas, S.E. Tibet & N.W. Yunnan.

Subsection Lepidota species are characterised by the rotate-campanulate corolla and short, deflexed style. Closely related to subsection Baileya and less so to subsections Trichoclada and Lapponica.

R. lepidotum has been introduced many times but is not all that commonly cultivated. *R. cowanianum* and *R. lowndesii* are rare.

R. cowanianum
Davidian 1952 H4

Height to 2.4m, an upright or spreading shrub, sparse with age. Foliage often colours well in autumn. **Leaves** deciduous, *thin*, 2.5-6.5 x 2.2-3cm, broadly elliptic to obovate, margins hairy; upper and lower surfaces with yellowish-pale brown scales. **Inflorescence** 2-5-flowered. **Corolla** shortly campanulate, 1.4-2cm long, pink, purplish-magenta to deep wine; calyx large, reddish; style short, sharply bent. The flowers of each inflorescence *open separately and the first drops before the last opens*. **Distribution** Nepal, 3,000-4,000m (10,000-13,000ft), clearings, forest margins, gorges and river beds.

This species is taller and more sparse in habit than the other members of the subsection. It resembles *R. baileyi* but is deciduous with fewer flowers to the inflorescence.

R. cowanianum is one of the least spectacular rhododendrons, and is therefore only of interest to the keen collector. It was very rare in cultivation until its recent reintroduction. Introduced 1954, reintroduced 1990s. May.

Picture 1

1. *R. cowanianum* Stainton, Sykes and Williams 9097 from a 1954 expedition to Nepal growing at Glendoick, showing the completely deciduous branchlets with new leaves emerging with the flowers.

Picture 2

2. *R. cowanianum* Stainton, Sykes & Williams 9097 showing the typical, rather sparse, upright habit.

R. lepidotum
(Wallich ex) G. Don 1834. (*R. cremnastes* Balf.f. & Farrer, *R. elaeagnoides* Hook.f., *R. obovatum* Hook.f., *R. salignum* Hook.f., *R. sinolepidotum* Balf.f.) H3-5

Height 0.05-2m, a low and compact to upright and leggy shrub, often stoloniferous. **Leaves** evergreen to semi-deciduous (rarely deciduous), 0.4-3.8 x 0.4-1.2cm, narrowly elliptic, obovate or rarely lanceolate; upper surface dark green and usually *densely lepidote*, lower surface pale greyish-green, densely lepidote, with distant to overlapping, large, *brownish scales with translucent rims*. **Inflorescence** 1-3 (rarely-4)-flowered. **Corolla** rotate-campanulate, 1.2-2.4cm across, *white, yellow, pink, red or various shades of purple*, often spotted darker; pedicels lepidote, 12-25cm; calyx variable in shape and colour, lepidote; style very short, deflexed. **Distribution** N. India, Nepal, Sikkim, Bhutan, N.E. Burma, N.W. Yunnan and S.E. Tibet, 2,400-4,900m (8,000-16,000ft), in a great variety of habitats.

The closely related *R. lowndesii* differs from *R. lepidotum* in its creeping habit and its hairy branchlets, leaves and petioles. *R. baileyi* is usually taller, with more flowers to each inflorescence, and has larger leaves with crenulate scales on the lower surface.

R. lepidotum grows best in a sunny position and is useful for its relatively late flowering. Its distribution covers a huge area of China and the Himalaya and is very variable; botanists named many species and varieties which have now been reduced in synonymy with *R. lepidotum*. These are best treated at group status, as the variability shows no geographical pattern and numerous intermediate forms occur. Plants under Elaeagnoides Group are completely deciduous and usually have yellow flowers, those under Obovatum Group have flowers more rounded and less star-shaped than the type while Salignum Group have star-shaped yellow flowers. Introduced 1850-> and many times recently. May-June.

Picture 1

Picture 2

Picture 3

Picture 4

Picture 5

3. *R. lepidotum* growing in the wild in Nepal. (Photo R. McBeath)
4. A pale pink to white-flowered form of *R. lepidotum* at Glendoick.
5. Leaves of *R. lepidotum* showing the scales on both leaf surfaces. (Photo R.B.G., E.)

1. *R. lepidotum* Cox, Hutchison & Maxwell MacDonald 2029, a purple-flowered form collected in Nepal, showing the rotate-campanulate flower shape and the short deflexed style.
2. *R. lepidotum* Gould 18, a fine yellow-flowered form at the Royal Botanic Garden, Edinburgh.

R. lowndesii
Davidian 1952 H3-4

Height 0.05-0.3m, c. 0.2m in cultivation, a low shrub, forming a dense, creeping mat, stoloniferous; stems hairy. **Leaves** *deciduous, thin*, 1.5-2.5 x 0.6-1.1cm, narrowly-elliptic to oblanceolate, margins *hairy*, lower surface pale green with distant yellow scales with translucent margins. **Inflorescence** 1-2-flowered. **Corolla** rotate-campanulate, 1.5-2cm long, 1.2-2.4cm across, *pale yellow*, usually lightly spotted or flushed red; pedicel erect, holding corolla above the foliage; calyx sparsely lepidote, fringed with hairs, greenish or reddish; style short, bent, glabrous; ovary lepidote. **Distribution** Central &

MADDENIA

W. Nepal, in drier areas only, 3,000-4,600m (10,000-15,000ft) in rock crevices, cliff ledges and peaty banks.

This species differs from *R. lepidotum* in its creeping habit and in the hairy and completely deciduous leaves with widely-spaced yellow scales on the lower surface.

R. lowndesii is a plant for the alpine enthusiast as it is tricky to grow, especially as a young plant. Like many high-altitude alpine plants, it should be covered with a cloche in winter. Natural hybrids with *R. lepidotum* are quite common in the wild and the cultivar 'Pipit' is a named selection of this cross. Introduced 1952->. May-June.

Picture 1

Picture 2

Picture 3

1. *R. lowndesii* at the Royal Botanic Garden, Edinburgh. (Photo R.B.G., E.)
2. *R. lowndesii* from the Stainton, Sykes & Williams expedition to Nepal in 1954, growing at Glendoick.
3. *R. lowndesii* growing on a rock in Nepal. (Photo G.F.Smith)

Subsection Maddenia

Height to 15m, often *epiphytic*, compact to very straggly shrubs. **Leaves** evergreen; lower surface whitish or greyish with (usually) unequal scales in various densities. **Inflorescence** 1-12-flowered. **Corolla** funnel-campanulate, rarely +/- campanulate or very narrowly funnel-campanulate, *white, pink or yellow, often fragrant;* calyx variable; stamens 8-27, filaments usually pilose towards the base; ovary tapering into style or impressed, seeds *winged and finned.* **Distribution** E. Himalaya, W. & S. Yunnan, S.W. China into Indo-China.

Most of the largest-leaved and largest-flowered lepidote species (excluding Section Vireya) are in subsection Maddenia. Most are rather straggly. The flowers are white and/or pale pink or yellow; most of the white species are +/- scented, the yellow ones have little or no scent.

Coming from relatively low altitudes, these are among the least hardy of the non-vireya species. They make excellent garden plants in mild areas such as western coastal British Isles and France, California, W. Vancouver Island (Canada), S.E. Australia, New Zealand and similar climates. In colder climates where they are not reliably hardy outdoors, they can also make fine conservatory or greenhouse plants, either planted in beds or grown in pots which can be brought inside the house when in flower. Many species benefit from pruning every few years after flowering. The epiphytic species require especially good drainage.

Some species such as *R. lindleyi* and *R. ciliatum* are fairly common in cultivation while others such as *R. levinei* and *R. excellens* have only recently been introduced. New species may be named from recent expeditions to Vietnam. Taxonomically, the species in subsection Maddenia are amongst the most confusing in the genus. Using the Balfourian system, much of the work was done by Hutchinson 1919 – 1930 at the R.B.G., Kew. His work was examined in detail by H. Sleumer in 1949 and 1958 and found to be have several flaws. Many of the the characteristics which Hutchinson's classification uses, such as density of scales on the corolla and leaves, as well as leaf shapes, leaf bases etc. are not consistent when specimens other than the type specimen are examined. Much cultivated material does not match the wild material and many so-called species were named from cultivated material alone. There is a great scarcity of herbarium material of many so-called species; many only exist as the type specimen. The picture is particularly complex in the Ciliicalyx aggregate which Hutchinson rather absurdly managed to increase to 30 species. Cullen 1980 quite rightly considerably reduces this number to 12, and it is likely, after further field work, that this whole group will be reduced to only 2-3 taxa. (See H. Sleumer in *Blumea* Suppl. IV: 40-47 (1958) for a lucid and convincing discussion of the taxonomy of subsection Maddenia). Cullen renamed many herbarium specimens he considered incorrectly identified and all this unheaval has meant that plant material often circulates under more than one specific name .

To aid in understanding this complex and confusing subsection, we use the following subdivisions which correspond roughly to the old subseries:

Group 1. Maddenii Alliance
Group 2. Dalhousiae Alliance
Group 3. Megacalyx
Group 4. Ciliicalyx Alliance:
    4(a) RR. ciliatum, valentinianum, formosum etc.
    4(b) Johnstoneanum Aggregate
    4(c) Ciliicalyx Aggregate

1. Maddenii Alliance

Now considered a single, variable species characterised by the (15-) 17-25 stamens.

R. maddenii
Hook.f. 1849 H2-3

Height 1-6m, a compact, sprawling or straggly shrub, sometimes trailing on cliffs and steep slopes. Bark rough and flaking. **Leaves** 5-20 x 2-8cm, lanceolate, oblong-lanceolate, obovate or elliptic; lower surface densely lepidote, scales almost contiguous to overlapping; very variable in size and shape. **Inflorescence** 2-11-flowered. **Corolla** tubular-funnel-shaped, 4.5-12.5cm long and to 12.5cm across, white, white flushed or lined pink, orange, rose or purple with or without greenish-yellow base, (yellow), sweetly scented, partly or fully lepidote on the outside; calyx 1-20mm, not hairy; stamens *(15-)17-25*; style lepidote half to whole of its length; ovary *(8-)10-12-celled*. **Distribution** 1,500-3,700m (5,000-12,000ft) in a variety of habitats, in dense forest, on open ridges and slopes, rocks and cliffs.

R. maddenii differs from other species in the subsection in the greater number of stamens: (15-)17-25 rather than 10-12 (except *R. excellens* which has 15.) In addition the ovary of *R. maddenii* has 8-12 chambers while all the other species in this subsection have only 5-7.

Quite rightly, all the many former species in this group have been reduced to synonymy under the name *R. maddenii*. Davidian makes a very unconvincing case for the maintenance of any of the former species but we have given Group status for those which have horticulturally significant features. This species is useful for its scent, its relatively late flowering and its heat tolerance. Unlike most other members of this subsection, *R. maddenii* is more commonly terrestrial than epiphytic in the wild and is therefore often less fastidious as to soil conditions than many of its relatives.

Ssp. maddenii (*R. brevitubum* Balf.f. & R.E. Cooper, *R. brachysiphon* Balf.f., *R. calophyllum* Nutt., *R. jenkinsii* Nutt., *R. polyandrum* Hutch., *R. maddenii* var. *longiflora* Watson)

Leaves *less than 6cm broad*, usually 6-11 x 2.8-6cm, often obovate; stamens often glabrous; capsule ovoid-globose (looking as if it has been cut off at the top). **Distribution** W. Arunachal Pradesh, Bhutan, Sikkim.

This is the more westerly form of the species and is usually less hardy than ssp. *crassum*. Plants under Brachysiphon Group often have pink flowers while certain forms of Polyandrum Group have the largest flowers: those of a form known as the Gigha form in the UK and as the Bodnant form in New Zealand are magnificent. Introduced 1849->, reintroduced 1965->.

Ssp. crassum (Franch.) Cullen 1978 (*R. chapaense* P. Dop, *R. manipuriense* Balf.f. & Watt, *R. odoriferum* Hutch.)

Leaves *more than 5.5cm broad*, usually 9-18 x 5.5-8cm, usually +/- elliptic; stamens usually pubescent; capsule oblong cylindrical (rounded at the top, not looking as if it has been cut

MADDENIA

off). **Distribution** Yunnan, S.E. Tibet, Upper Burma, Indo-China, N.E. India.

This is the more easterly occurring and hardier subspecies. Ssp. *crassum* includes selections which can be grown outdoors in a sheltered site in relatively cold gardens. Plants under Manipurense Group have larger than average leaves, while Odoriferum Group have small leaves and tubular flowers. Introduced 1906->, reintroduced 1981->.

Picture 1

Picture 2

Picture 3

Picture 4

Picture 5

1. *R. maddenii* ssp. *maddenii* at Arduaine showing the large number of stamens which distinguishes this from other subsection Maddenia species.
2. *R. maddenii* ssp. *maddenii* Cox and Hutchison 438 from Subansiri, India, collected in 1965; the pink-purple outside of the corolla is unusual.
3. *R. maddenii* ssp. *maddenii* Polyandrum Group. A form from Gigha growing at Brodick Castle, W. Scotland. (Photo J. Basford)
4. *R. maddenii* ssp. *maddenii* Polyandrum Group 'Bodnant form', a magnificent clone, popular in New Zealand where it was photographed.
5. *R. maddenii* ssp. *crassum* Ludlow and Sherriff 12248 which seems to be intermediate between the two subspecies.

Picture 6

Picture 7

6. *R. maddenii* ssp. *crassum* a fairly hardy form from Maryborough, S. Ireland, growing at Baravalla, W. Scotland
7. *R. maddenii* ssp. *crassum* showing the typical foliage, growing at Fernhill, near Dublin, Ireland.

2. Dalhousiae Alliance

Characteristics:
1. Main vein raised on upper leaf surface.
2. Corolla *large* (except *R. levinei*).
3. Pedicels and calyx not covered with waxy, whitish 'bloom'.
4. Calyx *large*, conspicuous, deeply lobed.
5. Stamens 10-15.
6. Ovary with 5 chambers.
7. Capsule longer than persistent calyx.
8. Ovary *tapered* into the style.

R. dalhousiae Hook.f. 1849 H2

Height to 3m, a usually leggy and untidy, erect or sprawling, often epiphytic shrub. Young shoots *covered with small bristles*; bark *brown, cinnamon, peeling*. **Leaves** only at ends of branches, 6.3-20 x 2.5-7.5cm, obovate to oblanceolate; lower surface greyish- or brownish-green with small reddish scales more than their own diameter apart. Leaf margins +/- bristly. **Inflorescence** 2-6-flowered; pedicel *pubescent*. **Corolla** tubular-campanulate, c.10cm long; +/-cream, calyx glabrous.

This species is closely related to *R. lindleyi* and *R. taggianum* but is distinguished by the bristly new growth and pubescent pedicel and usually by the wider leaf and the flower colour which is more yellow in *R. dalhousiae*.

Var. *dalhousiae*

Corolla *usually greeny-yellow in bud, fading to cream when open*, or rarely white, slightly scented (occasionally strongly). **Distribution** E. Nepal, Sikkim, Bhutan, S.E. Tibet 1,800-2,700m (6,000-9,000ft), usually epiphytic on trees, stumps or rocks.

This variety has fine flowers in its best forms but is hard to keep tidy indoors. It is only hardy outdoors in mildest gardens and, like most epiphytes, it needs perfect drainage. This variety shows some indication of heat and drought-resistance. Introduced 1850->, reintroduced 1970s->. April-July.

Var. *rhabdotum* (Balf.f. & R.E. Cooper) Cullen 1978

Corolla creamy white to pale yellow *with 5 bold red stripes down the lobes*. Flowers usually open several weeks later than in var. *dalhousiae*. **Distribution** Bhutan, Arunachal Pradesh 1,500-2,700m (5,000-9,000ft), in rain forest, cliffs, rocks, epiphytic or terrestrial.

This variety is perhaps easier to grow indoors than var. *dalhousiae* and is useful for its later flowering. Its extraordinary flowers never fail to attract comment. Introduced 1925->. Late June-August.

Picture 1

1. *R. dalhousiae* var. *dalhousiae* at Glendoick, grown indoors, showing the distinctive creamy-yellow, tubular-campanulate flowers.

MADDENIA

Picture 2

Picture 3

Picture 4

Picture 5

5. Foliage of *R. dalhousiae* var. *rhabdotum*. The hairs on the leaf margins in this form are sometimes absent.

R. excellens
Hemsl. & E.H. Wilson 1910 H1-2?

Height to 3m, probably an upright and straggly shrub. **Leaves** 15-20 x 3.8-13.3cm, oblong-elliptic; upper surface bullate, lower surface with slightly unequal scales c. their own diameter apart. **Inflorescence** 3-5-flowered. **Corolla** funnel-campanulate, 10-11cm long, white, scented, lepidote outside; calyx conspicuous, deeply 5-lobed, lobes +/- ovate, *glabrous*; stamens *(12)-15* very short; ovary densely lepidote, tapered into the style. **Distribution** S.W. Yunnan, N. Vietnam.

As we have seen it, this species looks like a smaller-leaved and flowered version of *R. nuttallii*. The leaves have prominent brown scales on the upper surface of the new growth which persists on the petiole and midrib. The lower surface is covered with rusty brown scales with occasional scattered larger ones. The effect of all the scales gives a greyish-green appearance.

Recently introduced from China and N. Vietnam, this appears to be a distinct species, although what the Chinese call *R. excellens*, which was the form first introduced, does not entirely agree with the published description. Davidian claims that *R.*

2. A form of *R. dalhousiae* var. *dalhousiae* in Sikkim. (Photo W.E. Berg)
3. A huge clump of *R. dalhousiae* var. *dalhousiae* growing outdoors at Brodick Castle, W. Scotland. (Photo J. Basford)
4. *R. dalhousiae* var. *rhabdotum* showing the easily-recognised red-striped flowers which usually open later than those of var. *dalhousiae*.

Picture 1

1. *R. excellens* in N. Vietnam. (Photo K. Rushforth).

excellens is related to *R. maddenii* due to the number of stamens but in all other characters it is obviously far more closely related to other members of the Dalhousiae alliance. Introduced c.1980->. May-June-July?

The recently introduced *R. levinei*, although not very spectacular, develops into quite a neat tidy plant with distinctive hairy foliage. It requires good drainage. Introduced 1981. April-May.

Picture 2

Picture 3

Picture 1

Picture 2

2. *R. excellens* at the Kunming Botanic Garden.
3. *R. excellens* showing the bullate leaves and pointed flower buds.

1. *R. levinei* showing the characteristic flower shape.
2. *R. levinei* showing the very hairy leaf margins and dark leaves which characterise this species.

R. levinei
Merrill 1918 H1?

Height 3-4m, less in cultivation, a loose, arching shrub. Branchlets *sometimes bristly*; bark mahogany, peeling or smooth. **Leaves** 5-10 x 2-5cm, obovate to lanceolate, *margins hairy*; upper surface *covered in long hairs at first*, lower surface brownish with unequal golden scales, 1-2 x their diameter apart. **Inflorescence** 1-3-flowered. **Corolla** funnel-campanulate, 4.5-5cm long, white, scented; calyx 7-10mm long, deeply 5 lobed, sparsely lepidote outside; stamens 10; ovary tapering into style. **Distribution** Guangdong, hillsides and rock crevices, c.950m.

Characterised by its small, very hairy leaves and small flowers *R. levinei* has little in common with the other members of this alliance and perhaps belongs on its own.

R. liliiflorum
H. Lév. 1913 H1-2?

Height to 3m+, an upright and straggly shrub. Branchlets red to green, lepidote; bark reddish, peeling. **Leaves** 10-13 x 3-5cm, oblong-elliptic to narrowly oblong-elliptic; lower surface brownish or greyish-green with unequal brownish scales 1-3 x their own diameter apart. **Inflorescence** 2-3-flowered. **Corolla** narrowly funnel-campanulate, 5.5-8cm long, white, densely lepidote outside, scented; calyx lobed more than half its length; stamens 10; style tapered into ovary. **Distribution** Guizhou, Guangxi 600-1,400m (1,950-4,600ft).

This species is a Chinese relative of *R. lindleyi* and *R. taggianum* and as cultivated differs from these two species in its more lepidote corolla and later flowering time.

R. liliiflorum was introduced in 1985 under Guiz 163. We are now growing a second generation of very uniform seedlings. Its distribution is isolated from that of its relatives. June.

Picture 1

Picture 2

1. *R. liliiflorum* Guiz 163, the original introduction from Guizhou province, China, flowering at Glendoick.
2. *R. liliiflorum* Guiz 163 showing the funnel-campanulate flowers.

R. lindleyi

T. Moore 1864 *(R. bhotanicum* Clarke, *R. grothausii* Davidian, *R. basfordii* Davidian) H2-3

Height to 4.5m, a straggly shrub, usually epiphytic in the wild. Young shoots sometimes tinged brown or red; bark reddish-brown peeling. **Leaves** 5.5-12(-14) x 1.5-5.5cm, narrowly elliptic to oblong-elliptic, rarely somewhat obovate, lower surface greyish-green with distant, reddish-brown scales. **Inflorescence** 2-12, commonly 3-7-flowered, loose; flower buds *large, rounded*. **Corolla** open funnel-campanulate, 7-11.6cm long, white, white with yellow/orange base, sometimes tinged pink, sweetly scented; calyx large, pinkish or green, *covered in small hairs but elepidote*; style lepidote only at base; capsule *large*. **Distribution** E. Nepal east to Arunachal Pradesh, S.E. Tibet and Manipur, 1,800-3,400m (6,000-11,000ft), usually epiphytic in trees and on cliffs.

R. lindleyi differs from *R. dalhousiae* in the non-bristly new growth, its more rounded flower buds, its white rather than cream corolla and its hairy calyx. *R. taggianum* is so closely related to *R. lindleyi* that it is doubtful whether it should be maintained as a separate species. Only the presence of scales and absence of hairs on the calyx separate it from *R. lindleyi*.

R. lindleyi is one of the most spectacular Maddenia species and it is widely grown in milder gardens. It can be grown indoors but is very straggly and hard to keep in bounds even with pruning. The scent was originally described as 'like lemon and nutmeg' but is sweeter than this sounds. Davidian's *R. grothausii* covers the white to pink-tinged forms from Ludlow and Sherriff 6562, including the award clone 'Geordie Sherriff'. Davidian states that *R. grothausii* is distinguished from *R. lindleyi* by a smaller, more bullate leaf, reddish-brown flower buds turning crimson-purple and a smaller corolla. We consider these distinctions worthy of no more than varietal or group status. *R. lindleyi* is quite variable in hardiness, the Ludlow and Sherriff forms being amongst the hardiest; these can be successful in a sheltered site even in moderately severe climates such as at Glendoick. *R. lindleyi* is best planted in clumps or groups, at the foot of a wall, a raised bed or in a tree-stump where the long shoots can trail over the edge. This species tends to have a small root-system and so often requires staking. Introduced c.1850->, reintroduced 1965->. Late April-May.

Picture 1

1. *R. lindleyi* Cox, Hutchison & Maxwell MacDonald 2011, below the Milke Danda ridge, E. Nepal. This form is similar to the earliest introductions of this species with few large flowers to the inflorescence and a broadly tubular-campanulate corolla.

conspicuous network of veins. **Inflorescence** 2-7(-12)-flowered. **Corolla** funnel-campanulate, margins sometimes wavy or reflexed, *very large, up to 12.5cm long and across*, white to tinged pink with a yellow blotch, slightly to strongly fragrant; calyx conspicuous, to 2.6cm long, deeply 5-lobed, sometimes with a few small hairs; stamens 10; ovary tapered into the style. **Distribution** Bhutan, Arunachal Pradesh, N.W. Yunnan, S.E. Tibet, 1,100-3,700m (3,500-12,000ft), epiphytic or in thickets, forest margins, rocky slopes and cliffs.

This species is easily recognised by its huge wrinkled oblong/oval leaves, which are the biggest of any lepidote species (excluding Section Vireya) and by its massive flowers. The reddish-purple new growth is characteristic of many but not all forms of the species.

Although a very spectacular species in flower, often with a strong scent, *R. nuttallii* is seldom seen to perfection due to its tenderness, requiring a very favourable climate or a large greenhouse. Plants under Sinonuttallii Group from China may be slightly hardier. So called *R. goreri* is said to have less bullate leaves, but is just a Tibetan form of a varied and widely-distributed species. Introduced 1852->. April-May.

Picture 2

Picture 3

Picture 4

2. *R. lindleyi* showing the typical straggly habit and plentiful flowers. Glendoick. 2nd generation from Ludlow and Sherriff 6562, (*R. grothausii* Davidian.)
3. *R. lindleyi* with fine pink-flushed flowers, at Inverewe, N.W. Scotland.
4. *R. lindleyi* showing its typically rounded flower buds.

R. nuttallii

Booth 1853 (*R. sinonuttallii* Balf.f. & Forrest *R. goreri* Davidian) H1(-2)

Height to 10m, a large, sometimes epiphytic shrub or small tree, often leggy in cultivation. New growth sometimes *reddish, brown or purple*. **Leaves** *large*, 13-26 x 5.5-12cm, elliptic, rounded at both ends; upper surface *heavily bullate*, lower surface densely covered with unequal scales and with a

Picture 1

Picture 2

1. *R. nuttallii*, Subansiri division of Arunachal Pradesh, N.E. India.
2. *R. nuttallii* showing its characteristic conspicuous calyces. Glendoick.

Picture 3

3. *R. nuttallii* showing the often reflexed corolla lobes. Glendoick.

R. taggianum

Hutch. 1931 (*R. headfortianum* Hutch.) H2

Height 2-3m, a leggy shrub. Bark dark reddish-brown and grey, peeling. **Leaves** to 12.5 x 6.3cm, oblong-lanceolate to oval; lower surface with small, glandular scales, 2-3 x their own diameter apart. **Inflorescence** (1-)3-5-flowered. **Corolla** broadly tubular funnel-shaped, 5.5-9.5cm long, light amber to cream in bud, opening pure white with a yellow throat, usually very fragrant; calyx to 2cm long, *not hairy, often margined with deciduous scales*. **Distribution** S.E. Tibet, Arunachal Pradesh, Upper Burma and W. Yunnan, 1,800-3,700m (6,000-12,000ft), conifer forest margins, rocky slopes, thickets or epiphytic.

This species is very similar to *R. lindleyi* and perhaps should be made a variety of it. Unlike *R. lindleyi*, it is seldom epiphytic and it differs in its non-hairy but often lepidote margins to the calyx lobes. It typically has a more elongated flower bud. wider corolla and wider leaves than *R. lindleyi*.

R. taggianum is equal in hardiness to the more tender forms of *R. lindleyi* and is spectacular in flower and usually very fragrant. Rather large for indoor culture and it is reluctant to respond to pruning, sending out only single shoots. It usually produces a better root system than *R. lindleyi*. Plants under the name Headfortianum Group, from S.E. Tibet, are lower-growing than the type, with narrower leaves, a 1-3-flowered inflorescence and smaller flowers. Introduced 1924->. April-May.

Picture 1

Picture 2

Picture 3

1. *R. taggianum* at Glendoick.
2. *R. taggianum* showing the broadly tubular funnel-shaped corolla. Glendoick.
3. *R. taggianum* showing the elongated flower bud and wide leaves. Glendoick.

3. Megacalyx
(a single species)

R. megacalyx
Balf.f. & Kingdon-Ward 1916 H2-3

Height 2-7.6m, a rarely epiphytic, loose, floppy shrub to small open tree. Bark light brown, peeling, fairly smooth. **Leaves** to 19 x 7.6cm, elliptic to obovate-elliptic; upper surface +/- bullate, lower surface with *conspicuous, sunken veins*, glaucous and densely lepidote with two different sizes of golden or brownish scales, the smaller ones *sunk* in pits; petiole *with a shallow groove* on upper surface. **Inflorescence** 2-6-flowered. **Corolla** tubular funnel-shaped with *larger lower lobes*, 7.5-11cm long, often flushed purple in bud, opening pure white, yellowish or greenish at the base with potter's thumb marks, with a nutmeg-like scent; calyx *consistently large,* green or pinkish green, *elepidote*; stamens 10; ovary densely lepidote, tapering into the style; capsule *very short*. **Distribution** N.W. Yunnan, E. & S.E. Tibet and Upper Burma 1,800-4,000m (6,000-13,000ft), usually on steep cliffs and banks, also forests and thickets.

An easily distinguished species, characterised by its sunken leaf veins and scales, the grooved petiole, the large, elepidote calyx, the large lower corolla lobes and the small capsule.

R. megacalyx is a fine plant with showy, sweetly-scented flowers, making a neater plant than *R. lindleyi* and *R. nuttallii*. It is only successful outside in relatively mild gardens but can make a good, large conservatory plant. Introduced 1917->. May.

Picture 1

1. *R. megacalyx* showing the large calyces indicated in its name. Glendoick.

Picture 2

Picture 3

2. *R. megacalyx* showing the potter's thumb marks on the corolla. Glendoick.
3. Foliage of *R. megacalyx* showing the bullate upper surface with grooved leaf midrib and petiole.

4. Ciliicalyx/ Johnstoneanum Alliance

Characteristics:
1. Main vein totally impressed above.
2. Pedicels and calyx not covered with waxy, whitish 'bloom'.
3. Calyx often small, usually fringed with hairs.
4. Stamens (8)-10.
5. Ovary with 5 chambers.

This alliance, based on the old Ciliicalyx subseries, causes the most taxonomic problems. Sleumer (in our view, correctly) reduced the number of species in this group considerably further than Cullen, but more field work will need to be done before a more or less definitive reclassification can take place. Cullen makes three sub-groups within Alliance 4: a large group including the yellow-flowered species and *R. formosum* and its relatives, another group centred around *R. johnstoneanum* and a third centred around *R. ciliicalyx*. Although in cultivation identifiable clones of each species exist, with many of the 'species' only the flowers can be used for positive identification.

4A.

These are grouped together because they do not fit into either of the aggregates 4B and 4C.

Smallish shrubs, usually under 2m. Style impressed.

R. burmanicum
Hutch. 1914 H2-3

Height to 2m, a usually compact shrub, straggly in shade; young shoots with bristles. **Leaves** 4-8 x 1.6-4cm, obovate, margins with bristles when young; upper surface dark green, moderately to densely lepidote, lower surface densely lepidote with contiguous to overlapping, brown scales; petiole densely lepidote and sparsely bristly. **Inflorescence** 4-6(-10)-flowered. **Corolla** funnel-campanulate, densely lepidote, 2.8-5cm long, yellow, greenish-yellow to cream, sometimes scented; calyx disc-like, lepidote and ciliate; stamens 10; style impressed. **Distribution** Central Burma, (Mt Victoria) 2,700-3,000m (9,000-10,000ft), fringes of forest.

This species is larger-leaved and less bristly/hairy than R. valentinianum without the hairs on the upper-surface midrib. R. valentinianum is never scented.

True R. burmanicum is very rare in cultivation; the plant usually seen, sometimes known as R. burmanicum Cox form, is a hybrid, perhaps with R. valentinianum, which makes a very fine, tidy, free-flowering, unscented yellow for milder areas. True R. burmanicum is less hardy, less compact and larger in all parts, usually with paler flowers. Introduced 1910 >, reintroduced 1956. April-May.

Picture 1

1. R. burmanicum, probably Kingdon-Ward 21921 collected on Mt Victoria, Burma, in 1956 and grown at Brodick Castle, W. Scotland.

Picture 2

2. R. burmanicum hybrid. This is the plant most commonly grown under the name R. burmanicum (sometimes as the 'Cox' form) but is almost certainly a hybrid. It is nevertheless an excellent plant for milder gardens. This fine specimen is at Glenarn, W. Scotland.

R. ciliatum
Hook.f. 1849 (R. modestum Hook.f.) H3-4

Height to 2m, usually less, a fairly compact shrub which can be straggly in shade. Branchlets hairy; bark roughish, reddish-brown, peeling. **Leaves** 3.8-9 x 2.1-3.4cm, elliptic to narrowly-elliptic, margins *hairy*; upper surface dark green, lower surface with scattered, brown, unequal scales, 2-3 x their own diameter apart. **Inflorescence** 2-5-flowered. **Corolla** campanulate to funnel-campanulate, 2.6-5cm long, white, white flushed pink, pale pink; calyx conspicuous, lobes ovate, unequal, margins *bristly/hairy;* stamens 10; style impressed. **Distribution** E. Nepal, Sikkim, Bhutan, S. Tibet 2,400-4,000m (8,000-13,000ft), conifer and rhododendron forest, rocky hillsides and beside water.

R. ciliatum is identified by the hairs on the leaves, leaf-margins, petiole, and calyx, coupled with the white-pale pink flowers. Its closest relatives R. valentinianum and R. fletcherianum have yellow flowers, and bristles rather than hairs on the branchlets.

An often-introduced Himalayan species which has been a successful parent of hybrids such as R. 'Praecox' and R. 'Cilpinense'. It is very variable in flower quality and hardiness; some recent introductions have been good on both counts. Introduced 1850->, reintroduced 1971->. March to May.

Picture 1

Picture 2

Picture 3

Picture 4

Picture 5

5. *R. ciliatum* showing upper leaf surface (above) and lower leaf surface (below) with hairy margins and brown scales, 2-3 x their own diameter apart. (Photo R.B.G., E.)

R. coxianum
Davidian 1972 H2

Height to 3m, an often epiphytic, usually straggly shrub. Branchlets moderately to sparsely bristly. **Leaves** 5-11 x 1.5-3cm, *oblanceolate*, lower surface with unequal, translucent to brown scales, 2-5 x their own diameter apart; petiole with or without wings at margins. **Inflorescence** 2-4-flowered. **Corolla** tubular funnel-campanulate, c.7cm long, creamy-white, greenish-yellow blotch at base, lepidote outside, strongly scented; pedicel lepidote, not bristly; calyx unequal, +/- hairy; stamens 10; ovary 7-celled, densely lepidote; style impressed, long and slender. **Distribution** Arunachal Pradesh, India, 1,700m (5,500ft), in marshy ground or epiphytic.

This species may be an extreme form of *R. formosum*, closest to var. *inaequale* with a similar scent, but differing in the stiff leathery leaves and the non-frilled corolla lobes. The lower two corolla lobes are sometimes split.

Picture 1

1. A pink-flowered selection of *R. ciliatum* Beer, Lancaster and Morris 324 collected in E. Nepal in 1971.
2. *R. ciliatum* in C. Bhutan with *R. thomsonii* behind. (Photo A.D. Schilling)
3. A large bush of *R. ciliatum* at Stonefield Castle, W. Scotland.
4. A pink to white-flowered form of *R. ciliatum*.

1. *R. coxianum* Cox & Hutchison 475B from the Subansiri division of Arunachal Pradesh, N.E. India, collected in 1965, showing the oblanceolate leaves and characteristic flower shape of this species.

R. coxianum was discovered and collected by Cox and Hutchison in 1965 and remains rare in cultivation, partly because it is not always easy to cultivate. Introduced 1965. April-May.

Picture 2

2. A large plant of *R. coxianum* at Glendoick.

Picture 1

Picture 2

1. *R. fletcherianum* 'Yellow Bunting' A.M. (colour slightly deeper than it really is) which is a selection from Rock 22302.
2. *R. fletcherianum* 'Yellow Bunting' at Glendoick.

R. fletcherianum
Davidian 1961 H4

Height to 1.20m, a small shrub, compact when young, leggy with age. Bark roughish, reddish-brown, peeling. **Leaves** 2.3-5.6 x 1.1-2.8cm, oblong-lanceolate to oblong-elliptic, margins *crenate and bristly*, lower surface with pale brown scales 3-6 x their own diameter apart; petiole *flat on upper surface and winged*. **Inflorescence** 2-5-flowered, compact, pedicel pilose. **Corolla** widely funnel-shaped, nearly 5cm long, *pale yellow*; calyx conspicuous, pilose; stamens 10; style impressed. **Distribution** Found only twice by Rock in S.E. Tibet, 4,000-4,300m (13,000-14,000ft), in alpine regions and forests.

Taxonomically, this species falls somewhere between *R. valentinianum* and *R. ciliatum* but its distribution is isolated from both. It is distinguished by its pale yellow flowers, crenate leaf margins and the flat winged petiole. It tends to have a pointed leaf while that of *R. valentinianum* is elliptic. This species was known as *R. valentinianum* aff./Rock for many years.

Hardier than *R. valentinianum* and more free-flowering, *R. fletcherianum* is quite a distinctive species which requires good drainage to avoid leaf spot and die back. It has proven to be a good parent, of R. 'Curlew' and R. 'Patty Bee' for example. Introduced 1932. March-May.

R. formosum
Wallich 1832 H2-3

Height to 3m, a fairly compact to straggly and open shrub. Young growth bristly. **Leaves** 3-12 x 1-5cm, narrowly-elliptic to linear-elliptic, margins bristly, especially when young; upper surface dark green, lower surface with unequal scales 1-3 x their own diameter apart; petiole *often with bristles*. **Inflorescence** 2-6-flowered. **Corolla** openly-funnel-campanulate, 4-7.5cm long, white to white flushed pink, often with a yellow blotch and pink markings, usually pilose at the base, weakly to strongly scented; calyx *disc-like, lepidote with few hairs;* stamens 10, pubescent towards the base; style impressed; ovary and lower part of style lepidote. **Distribution** N.E. India 600-1,800m (2,000-6,000ft), in open, forest or river banks.

This species differs from *R. coxianum* in its less rigid, non-leathery leaves. The somewhat bristly petiole is a diagnostic feature as is the disc-like, lepidote calyx with a few hairs.

Var. *formosum* (*R. assamicum* hort., *R. gibsonii* Paxton, *R. iteophyllum* Hutch.) H2-3

Leaves *narrow, to 1.5cm wide*. **Corolla** often flushed pink, usually with some scent, though not as strong as in var. *inaequale*. Plants under Iteophyllum Group have narrower leaves.

This variety is usually hardier and easier to grow than var. *inaequale*. One clone has proved quite hardy, growing outside on a wall for us at Glendoick. Introduced 1845->, reintroduced 1965. May-June.

Var. *inaequale* (Hutch.) Cullen 1978. H2

Leaves *1.5-5cm wide*. **Corolla** usually larger than in var. *formosum*, white with yellow blotch, scent stronger than in var. *formosum*.

Var. *inaequale* is closely related to *R. horlickianum* which differs in the distinctive leaf shape, the absence of scent and in the minute calyx fringed with hairs.

Var. *inaequale* is more tender than var. *formosum* but has a much stronger scent, one of the finest in the subsection. It needs to be hard pruned if grown indoors and requires excellent drainage for best results. Introduced 1927, reintroduced 1965. April-June.

Picture 1

Picture 2

Picture 3

Picture 4

1. *R. formosum* var. *formosum* Cox & Hutchison 302, Baravalla, W. Scotland, showing the small, rather narrow leaves and white-flushed pink flowers.
2. *R. formosum* var. *formosum* Iteophyllum Group, Brodick Castle, W. Scotland, showing the very narrow leaves typical of this group.
3. *R. formosum* var. *formosum* 'Khasia' A.M., Cox and Hutchison 320. This clone is intermediate between var. *formosum* and var. *inaequale* in both foliage and flower. Glendoick.
4. *R. formosum* var. *inaequale* Cox & Hutchison 301, showing the foliage and flowers larger than in var. *formosum* var. *formosum* and usually with a stronger scent. Glendoick.

R. cuffeanum Craib. ex. Hutch. is only known from cultivated material. Similar to var. *inaequale*.

R. scopulorum Hutch. 1930. H1-2

Height to 4.5m, an upright or bushy shrub. **Leaves** 5-7.5 x 2-3cm, rigid in texture, *elliptic to obovate-elliptic;* upper surface pale green, lower surface with *distant, unequal, golden* scales. **Inflorescence** 2-4-flowered. **Corolla** funnel to funnel-campanulate, to 6.2cm long, white or white flushed pink with a golden or yellow blotch inside, *lepidote and pubescent all over outside*, sometimes fragrant; calyx 5-lobed, usually not ciliate, lobes broadly triangular; stamens 10, densely pubescent towards the base; style impressed, lepidote at base. **Distribution** S.E. Tibet, 1,800-2,400m (6,000-8,000ft), on steep slopes, cliffs, rocks and screes and in forests.

This species is characterised by its lepidote and pubescent corolla and by the sparse scales on the leaf underside.

Picture 1

Picture 2

1. *R. scopulorum* Kingdon-Ward 6354 collected in the Tsangpo Gorge, Tibet in 1924, growing at the Royal Botanic Garden, Edinburgh.
2. *R. scopulorum* Kingdon-Ward 6354 at the Royal Botanic Garden, Edinburgh.

The fairly rare *R. scopulorum* is quite free flowering but rather tender and variable in the degree of scent. Introduced 1924, 1947, 1996. April-May

Picture 3

3. *R. scopulorum* by the Po Tsangpo river in S.E. Tibe

R. valentinianum
(Forrest ex.) Hutch. 1919 H2-3

Height to 1.3m, usually less, a rounded to very compact shrub. Young branchlets very hairy/bristly; bark *peeling*. **Leaves** 2.6-4.6 x 1.6-3.1cm, *elliptic, rounded at both ends*; upper surface *dark* green, hairy at the margins and sometimes on part of midrib, lower surface *densely lepidote*; scales brown, overlapping; petiole very hairy/bristly. **Inflorescence** 1-4(-6)-flowered. **Corolla** funnel-campanulate, 2.6-3.5cm long, *bright yellow*; calyx deeply 5-lobed with bristly margins; stamens 10; ovary densely lepidote; style impressed, long and thin. **Distribution** N.E. Burma-Yunnan frontier 2,700-3,700m (9,000-12,000ft), in open scrub, rocky slopes and cliffs.

R. valentinianum is closely related to *R. fletcherianum* which has a more upright habit, larger, less rounded leaves with crenate margins and with the scales on the lower surface of the leaf much more scattered. In appearance, the closest relatives are *R. megeratum* and *R. leucaspis*. The former has generally smaller leaves (in cultivation) which are glaucous on the lower surface and the style is sharply deflexed. *R. leucaspis* has almost rotate, pure white flowers.

At its best *R. valentinianum* is a handsome foliage plant with deep green hairy leaves, but it tends to be rather shy-flowering and needs good drainage, enjoying a mossy rock, tree-stump or raised bed. It survives most winters at Glendoick. Introduced 1917->. March-April.

R. valentinianum aff. (var. *olongifolium*? Fang)

Height to 2m+, upright to rounded shrub. **Leaves** rugose, medium-green, larger than the type, margins bristly, lower surface somewhat glaucous, larger **Inflorescence** 2-4-flowered, buds red. **Corolla** unseen, said to be yellow; capsules cylindrical, style impressed. **Distribution** S.E. Yunnan, 2,750-2,850m, on steep banks and epiphytic on broad-leaved trees.

While probably related to *R. valentinianum*, it can hardy be considered a variety of it as it is so much larger in all parts, with more rugose leaves. Perhaps related to *R. burmanicum*. Probably deserving of specific status. Introduced 1995.

Picture 1

Picture 2

Picture 3

Picture 4

3. Young growth of *R. valentinianum* showing the elliptic leaves with hairy margins.
4. *R. valentinianum* aff. Cox & Hutchison 7186. A new species from S.E. Yunnan, a larger plant with larger, stiffer leaves than *R. valentinianum*. (Possibly described as var. *oblongifolium* Fang.)

1. *R. valentinianum* growing at Glendoick, showing the dark green leaves and the yellow flowers with a long, thin style.
2. *R. valentinianum* at Glenarn, W. Scotland, showing its typical habit of opening its flowers gradually over an extended period.

4B. (R. Johnstoneanum aggregate)

Largish shrubs. **Corolla** large; style impressed.

R. dendricola

Hutch. 1919 (*R. atentsiense* Hand.-Mazz, *R. notatum* Hutch., *R. taronense* Hutch.) H1

Height to 4.5m, a usually straggly shrub. New growth rarely bristly; bark smooth, dark purple to mahogany-red, peeling. **Leaves** 7-12 x 3-5cm, narrowly-elliptic to narrowly obovate, lower surface with scales of variable density. **Inflorescence** 2-5-flowered. **Corolla** funnel-shaped, to 10cm long, white, often with a greenish, yellow or orange blotch/flare, and/or flushed pink, usually fragrant; calyx disc-like or with tiny lobes, *not hairy*; stamens 10; ovary lepidote; style impressed. **Distribution** N. Burma, Arunachal Pradesh, S.E. Tibet, N.W.

Yunnan, 900-3,000m (3,000-10,000ft), epiphytic or terrestrial on rocks, cliffs, forest margins and rocky slopes.

This is a variable species, differing from *R. johnstoneanum* in the non-bristly foliage, from *R. walongense* in the absence of calyx hairs, and from *R. veitchianum* in the impressed rather than tapering style.

R. dendricola is one of the most tender species of subsection Maddenia but it makes a showy conservatory shrub in its best forms. Introduced 1918->. May.

Picture 1

Picture 2

Picture 3

1. A fine *R. dendricola* from Kingdon-Ward collection. It has the number 280 at the Royal Botanic Garden, Edinburgh, but it should probably be 281.
2. *R. dendricola* Kingdon-Ward 21512 from the Triangle, N. Burma. (formerly labelled *R. supranubium*) at Glenarn, W. Scotland.
3. Close up of the style and ovary of *R. dendricola* showing the non-hairy calyx and the impressed style. (Photo R.B.G., E.)

R. johnstoneanum

(Watt ex) Hutch. 1919 (*R. parryae* Hutch. type only) H2-3

Height to 4.6m, usually less, a usually rather untidy shrub, growing broader than high with age. Branchlets *bristly*; young growth aromatic. **Leaves** 5-10 x 2.4-3cm, elliptic to broadly elliptic, margins *usually bristly*, lower surface with dense and nearly contiguous or overlapping brown scales; petiole hairy and lepidote. **Inflorescence** 2-5-flowered. **Corolla** funnel-campanulate, 4.6-6cm long, white, cream, pale greenish-yellow, often with a yellowish blotch and pink/purplish flush, sometimes fragrant, (sometimes double or semi-double), pedicel densely lepidote; calyx *very short with long hairs on margins*; style impressed. **Distribution** Manipur, Mizoram, N.E. India 1,800-3,000m (6,000-10,000ft), in open scrub or grass, forest margins or epiphytic.

This is quite a variable species, characterised by its bristly/hairy branchlets, petiole and young leaves, the dense scales on the lower leaf-surface and the often creamy-yellow flowers. It differs from *R. lyi* in its impressed rather than tapered style, more bristly leaves and petioles and wider leaves.

A popular and widely grown species for milder areas, especially good in western Scotland and Cornwall and similar climates. One of the hardiest of the subsection Maddenia species and worth attempting in a sheltered site in climates as severe as Glendoick. The double forms are generally more tender than the type. Introduced 1882 and 1927. May

Picture 1

1. *R. johnstoneanum* showing the typical creamy-yellow flowers with deeper flushing.

Picture 2

Picture 3

Picture 4

2. A fine bush of *R. johnstoneanum* at Glenarn, W. Scotland.
3. One of the double forms, *R. johnstoneanum* 'Double Diamond' from Windsor Great Park, S. England, growing in the garden of Mrs J. Sinclair, Washington, U.S.A.
4. Leaf bases of *R. johnstoneanum* showing the hairy margins and the hairy and densely lepidote petioles. (Photo R.B.G. E.)

R. walongense Kingdon-Ward 1953 H2?

Height 2-3m, a sometimes epiphytic shrub. **Leaves** c. 10cm long, elliptic, slightly pointed at the tip, lower surface brownish with lax, unequal scales. **Inflorescence** 3-4-flowered. **Corolla** funnel-shaped. 6-7cm long, creamy-white with a greenish blotch, densely pilose/pubescent all-over, scented; calyx small, *hairy*. **Distribution** Arunachal Pradesh, India, 1,500-2,100m (5,000-7,000-ft), in thin mixed forest, ravines, rocks or epiphytic.

The identifying features of *R. walongense* are the pubescent/pilose corolla and the persistent calyx hairs.

This species may only be in cultivation under *R. walongense* Affinity Cox & Hutchison 373 (which was previously identified as *R. parryae*). This form, which was epiphytic in the wild, has peeling, deep, mahogany-coloured bark and the flowers have a spicy scent. Introduced 1928/1965. April-May.

Picture 1

Picture 2

1. *R. walongense* affinity Cox & Hutchison 373 collected in the Subansiri division of Arunachal Pradesh, N.E. India, in 1965.
2. *R. walongense* affinity Cox & Hutchison 373 showing the characteristic mahogany-coloured, peeling bark.

4C. (R. Ciliicalyx aggregate)

This aggregate is the most confusing; although their distribution covers an area from India to Burma, Thailand, Laos and China, there is every indication that all these so-called species are simply variations of only two or three variable taxa.

This aggregate is characterised by:
1. Style tapering into ovary.
2. Young growth, leaf margins and calyx usually with hairs/bristles.

R. carneum
Hutch. H2

Height to 1-1.8m, a rather open shrub. Young growth *not* hairy. Bark rough, flaking but becoming smooth with age. **Leaves** 5-11 x 3-4cm, narrowly elliptic; upper surface lepidote, lower surface with scales their own diameter apart. **Inflorescence** *3-7-flowered*. **Corolla** funnel-shaped, c.7cm long, *flesh pink* fading to near white, moderately lepidote all over, lightly scented; calyx lepidote and fringed with hairs; stamens *10-12*; ovary tapering into the style; style densely lepidote, pink. **Distribution** Burma, open, grassy hillsides, c.2,300m (7,500ft).

This species is only known from cultivated plants and is probably an extreme form of *R. ciliicalyx*/*R. pachypodum*. It is distinguished by its pink flowers, the lepidote leaf upper surface, lack of bristles/hairs on the new growth and the moderately lepidote corolla. It differs from *R. veitchianum* in the hairs on the calyx lobes as well as in the deeper flower colour. *R. horlickianum* has flowers of a similar pink colour but with fewer flowers to the truss, and also differs in the less densely lepidote upper leaf surface.

This species is worth growing for its pink flowers which are deeper-coloured if the plant can be grown outdoors. Introduced 1912. April-May

Picture 2

2. *R. carneum* showing the foliage.

R. horlickianum
Davidian 1972 H1-2

Height 1.5-3m, a straggly shrub. Young growth setose at first; bark greenish-brown, roughish, ultimately cinnamon-brown, peeling. **Leaves** 5-11 x 2-4.5cm, elliptic-lanceolate to obovate-lanceolate, apex *acuminate*, lower surface with unequal scales, 1-1½ x their own diameter apart. **Inflorescence** *2-4-flowered*. **Corolla** widely funnel-shaped, 6.5-7cm long, white to creamy white, *strongly flushed rose*, orange flare, *lepidote and pubescent* outside; calyx *minute, fringed with long bristles*; stamens 10-11; ovary tapering into style. **Distribution** N. Burma, rocks or epiphytic on trees. 1,200-2100m (4,000-7,000ft).

R. horlickianum is a recently described and fairly distinct species, perhaps most closely related to *R. carneum*, differing in its hairy young leaves which are partly elepidote above and with fewer flowers to the truss.

Picture 1

1. *R. carneum* grown indoors at Glendoick, lacking the characteristic pink colour which would be apparent on a plant grown outdoors.

Picture 1

1. *R. horlickianum*. All cultivated plants are seedlings of the original introduction of Kingdon-Ward 9403 collected in 1931 in Northern Burma.

The strong pink flushing in the flower colour makes this an attractive but unfortunately +/- unscented species. Introduced 1931. March-April.

Picture 1

2. *R. horlickianum* showing the pink flushing on the corolla which is characteristic of this species.

R. ludwigianum Hosseus 1911. H1

Height to 1.5m, a straggly shrub. Bark peeling with age, purplish-pink. **Leaves** to 10 x 5cm, obovate, *thick and leathery*, margins fringed with tiny hairs on the new growth; lower surface with scales c. half their own diameter apart.

Picture 1

1. *R. ludwigianum* Smitinand 7819 from Thailand, grown at the Royal Botanic Garden, Edinburgh. (Photo J. Cubey)

Inflorescence 2-4-flowered. **Corolla** funnel-shaped, up to 12cm across, white sometimes flushed rose, rose, with or without a yellow flare, pubescent inside, *not scented*; pedicel *hairy*; calyx minute, margins hairy; ovary tapered into style which is lepidote in lower half. **Distribution** Thailand 1,600-2,200m (5,250-7,250ft),

R. ludwigianum is characterised by the hairy pedicel, small calyx, lack of scent, and according to Davidian, the branchlets are commonly warted with leaf scars. The species was almost lost to cultivation and what is grown under this name is very similar to *R. pachypodum* and also to *R. dendricola*.

This species has a spectacular flower which unfortunately is not scented. It is very rare in cultivation. Introduced 1905->. April-May.

R. lyi
Lév. 1914 (*R. leptocladon* Dop, *R. saravanense* Dop) H1-2

Height to 1.8m or more, a *very straggly* shrub. Branchlets trailing, lepidote and bristly. **Leaves** 7-8 x 2.5-3cm, narrowly obovate, lower surface with dense but non-contiguous brown scales. **Inflorescence** 2-6-flowered. **Corolla** funnel-campanulate, c.5.5cm, white, *with the whole corolla surface sparingly lepidote*, sometimes fragrant; calyx minute, *densely lepidote with a few long hairs*; style long, lepidote in the lower two-thirds. **Distribution** Guizhou, Laos, Vietnam, Thailand 1,200-2,800m (4,000-9,000ft), dense woods and dry limestone plateau.

The distinguishing features of this species are the lepidote corolla, the long style, the minute but densely lepidote calyx and the non-touching scales on the lower leaf surface. Sleumer merges this with *R. johnstoneanum* although these two species are quite distinct as cultivated. It is more likely to be confused with *R. pachypodum* and its allies.

Until recently *R. lyi* has been rare in cultivation, mainly seen in forms introduced by P. Valder from Thailand which are straggly and shy-flowering. Hopefully recent introductions from Vietnam will make better garden plants. Introduced c.1912, reintroduced 1974->. April-June.

MADDENIA

Picture 1

Picture 2

1. *R. lyi* in New Zealand. (Photo J. Oldham)
2. *R. lyi* in N. Vietnam. (Photo K. Rushforth)

R. pachypodum
Balf.f. & W.W. Sm. 1916 (*R. pilicalyx* Hutch,
R. scottianum Hutch., *R. supranubium* Hutch). H2-3

Height 1.2-7.6m, a fairly compact to very straggly shrub. Branchlets *lepidote, not hairy*; bark light reddish-brown, peeling. **Leaves** 3-10 x 1.3-3cm, obovate, elliptic-obovate to oblanceolate, rarely elliptic, lower surface densely lepidote, scales of variable density, almost contiguous to 4 x their own diameter apart. **Inflorescence** 1-4-flowered. **Corolla** widely funnel-shaped, 3.8-10cm long, white, white flushed pink, rose, sometimes with a brown to yellow blotch, corolla *lepidote outside on both tube and lobes, pubescent within*, sometimes scented; pedicel glabrous; calyx usually small but sometimes with one long lobe; stamens *10 (11-13)*; style glabrous. **Distribution** W. Yunnan, Guangdong and Upper Burma 1,800-3,700m (6,000-12,000ft), forest margins, scrub, slopes and cliffs.

This is a very variable species into which Cullen has placed many specimens formerly labelled *R. ciliicalyx*. There seems little reason not to merge these two species which occur in the same mountains and which are both variable and appear to completely overlap. These two species should probably also be merged with the very similar *R. roseatum* which usually has wider leaves and occurs further west. *R. ludwigianum* differs in its hairy pedicel and style.

R. pachypodum is much more common in cultivation than is realised, but is still often grown under other specific names, especially *R. ciliicalyx*. Scottianum Group have larger flowers than those under Supranubium Group. Introduced 1910->, reintroduced 1981. March-June.

Picture 1

Picture 2

Picture 3

1. A selection of *R. pachypodum* SBEC 0115 with pink flushed flowers, from the Cangshan, C.W. Yunnan, China, growing at Glendoick.
2. *R. pachypodum* growing on a cliff face at Qingbixu, Cangshan, C.W. Yunnan, China.
3. *R. pachypodum* on the lower slopes of the Cangshan, C.W. Yunnan, China.

R. ciliicalyx Franch. 1886 (*R. missionarum* Lév., *R. pseudociliicalyx* Hutch.) H2 Only differs from *R. pachypodum* in that the scales on the corolla are sparse and confined to the lobes.

Most plants labelled *R. ciliicalyx* in cultivation have been re-identified as *R. dendricola* or *R. pachypodum* and the species as described is probably not in cultivation.

R. roseatum

Hutch. 1919 (*R. lasiopodum* Hutch., *R. parryae* Hutch. A.M. form) H1-2

Height to 5m, habit usually upright. Bark red and peeling. **Leaves** 5.5-10.5 x 3.5.5cm, +/- *obovate*, with a sharply pointed tip, the lower surface with lax to dense brownish scales. **Inflorescence** 2-4-flowered. **Corolla,** narrowly funnel-shaped, 7-8cm long, white, white-flushed pink, with a yellow blotch, sometimes fragrant, lepidote outside; calyx *shortly-lobed with long hairs on the margins*. **Distribution** W. & S.W. Yunnan, forests, hillsides, scrub or epiphytic, 1,500-2,700m (5,000-9,000ft).

There seems little point in separating this species from *R. pachypodum* which occurs further east and has narrower leaves but is otherwise indistinguishable. Also very similar to *R. dendricola* which has a different leaf shape, less dense scales on the lower leaf surface and a smaller calyx.

R. roseatum is rare in cultivation. The plant we grow under this name (formerly *R. parryae* A.M.) is useful for its relatively late flowering. Introduced 1918->. April-May.

Picture 1

1. *R. roseatum* A.M. clone from the Royal Botanic Garden, Edinburgh (formerly *R. parryae* A.M.) at Glendoick.

Picture 2

2. *R. roseatum* A.M. at Glendoick, showing the pointed leaves and the yellow blotch on the corolla.

R. veitchianum

Hook.f. 1857 (*R. smilesii* Hutch, *R. cubittii* Hutch. (type specimen)) H1-2

Height to 3.7m, a compact shrub to small tree. Bark brown, flaking. **Leaves** 5-10 x 2-4cm, obovate to elliptic-obovate, lower surface *pale green with distant, brown, unequal scales*. **Inflorescence** 2-5-flowered. **Corolla** openly funnel-campanulate, 5-7cm long, up to 12cm across, white, often blotched yellow, *margins often frilled or wavy*, strongly to slightly scented, (said to be like verbena); calyx disc-like, lepidote, margins ciliate; stamens 10; ovary tapering into style. **Distribution** Burma, Thailand, Laos, 900-2,400m (3,000-8,000ft), usually epiphytic, or terrestrial on disturbed land.

This species is characterised by its relatively large flowers with frilled or wavy edges and by the almost glabrous leaf underside with scattered scales. The leaf shape is quite distinctive.

R. veitchianum is amongst the most spectacular in this alliance, in terms of flower-size, but is very tender, barely being hardy outside anywhere in the U.K. Introduced c. 1850 and several times recently. February to July.

R. cubittii (now Cubittii Group) has been sunk into *R. veitchianum* by Cullen. It has white flowers, flushed pink outside with a bold orange or yellow flare. Described from inadequate material, it was only found once in the wild in 1909 in N. Burma. Hardy in mildest U.K. gardens. March-April.

MICRANTHA

Picture 1

Picture 2

Picture 3

1. *R. veitchianum* at Glendoick, grown from a P. Valder collection from Thailand in 1974.
2. *R. veitchianum* P. Valder 29 or 30, a very frilled form, at the St Andrews Botanic Garden.
3. Leaf lower surfaces of *R. veitchianum* (older leaf at top) showing the distant brown scales (at maturity). (Photo R.B.G. E.)

Picture 4

4. *R. veitchianum* Cubittii Group 'Ashcombe' F.C.C. showing the strong orange flare in the corolla throat. Arduaine Gardens, W. Scotland.

R. fleuryi Dop 1929 H1? Seed reputed to be from this species has recently been introduced from North Vietnam. The description of the corolla being white with yellow lines on the tube makes it sound interesting. Introduced 1994.

Subsection Micrantha

A monotypic subsection.

R. micranthum

Turcz. 1848 (*R. pritzelianum*, Diels *R. rosthornii* Diels) H4-5?

Height to 1.8m or more in cultivation; a shrub, inclined to be straggly when young, more bushy with age. Branchlets slender; new growth lightly pubescent. **Leaves** 3.2-4.4 x 0.6-1.3cm, oblanceolate, gradually narrowed to the base, lower surface with prominent light brown scales. **Inflorescence** terminal or sometimes axillary in upper 3 leaf axils, usually more than 20-flowered; rachis conspicuous. **Corolla** campanulate, *0.6-0.8cm long*, white, *densely lepidote on the outside*; pedicel puberulent and sparsely lepidote; calyx 1-2mm, lepidote and fringed with hairs; stamens 10; style *short straight and slender*. **Distribution** Shandong to Gansu, also W. Hubei, W. Sichuan, and rarely N.E. Yunnan, 1,600-3,200m (5,200-10,500ft), thickets and scrub.

In flower this distinctive species most resembles the members of subsection Ledum (formerly the genus *Ledum*) from which it differs in the absense of shoot and leaf hairs (apart from a little pubescence). Out of flower it resembles some subsection Lapponica species such as *R. polycladum* but its long shoots and small narrow leaves are fairly distinctive.

Although the flowers are small, *R. micranthum* is quite a useful garden plant as it is late flowering. It is rather prone to bark-split as a young plant but most forms are very hardy at maturity. It is distributed over a wide area in the wild, but apparently shows little variation. Introduced 1901-10. (May)-June-July.

Picture 1

Picture 2

1. *R. micranthum* with its distinctive white flowers.
2. Close-up of *R. micranthum* showing the many flowered trusses from multiple (terminal and axillary) buds. (Photo R.B.G., E.)

Subsection Moupinensia

Height to to 1.5m, compact to rather straggly shrubs, often epiphytic. Branchlets lepidote and bristly. **Leaves** evergreen, dark green, thick and rigid, margins ciliate. **Inflorescence** 1-2-flowered. **Corolla** openly funnel-campanulate, white to pink; calyx conspicuous; stamens 10; ovary lepidote, tapering into the style. **Distribution** Sichuan, N.E. Yunnan, Guizhou.

Subsection Moupinensia is related to subsection Maddenia, differing in its generally smaller leaves and characterised by the openly funnel-campanulate corolla.

R. moupinense is relatively common in cultivation; *R. dendrocharis* has only recently been introduced but should become popular.

R. dendrocharis
Franch. 1886 (*R. petrocharis* Diels) H4

Height to 0.7m, a *compact*, usually epiphytic shrub. Branchlets lepidote and bristly. **Leaves** *1.3-2.5 x 0.4-1.5cm*, elliptic to oval; upper surface dark green and shiny, elepidote, lower surface scales touching to 1½ x their own diameter apart. **Inflorescence** 1-2-flowered. **Corolla** widely funnel-shaped to almost flat, 3.5-5.5cm across, rose-pink (usually) to white, often heavily spotted; calyx 1-3mm; style *shorter than or equal in length to stamens, deflexed.* **Distribution** W. Central Sichuan, 1,800-3,200m (6,000-10,500ft), epiphytic on conifers and broad-leaved trees and on cliffs and rocks.

This species is closely related to *R. moupinense* but is smaller in all parts except the corolla, more compact growing and (as cultivated) later flowering. At present, *R. petrocharis* with white flowers, densely pubescent inside the tube, is not considered sufficiently distinct to merit specific status, though may be given subspecific or varietal status after further fieldwork.

Picture 1

Picture 2

Picture 3

1. *R. dendrocharis* growing on a mossy tree at Wolong, W. Central Sichuan, China. (Photo P.C. Hutchison)
2. *R. dendrocharis* Chamberlain, Cox & Hutchison 4012, growing under glass at Glendoick. The flowers would be deeper-coloured if grown outside.
3. *R. dendrocharis* Cox 5016 at Glendoick showing the characteristic large flower/leaf size ratio which will attract breeders to use it for hybridising.

R. dendrocharis promises to be one of the best dwarf rhododendron introductions in the combination of its small dark leaves and large rose-pink flowers, and it may have breeding potential. It seems relatively easy to cultivate so far, but as it is usually epiphytic in the wild, it requires good drainage. Some introductions appear to be freer-flowering than others. Introduced 1980s-> from several locations. March-April.

R. moupinense
Franch. 1886 H3-4

Height to 1.5m, a fairly compact to lax, somewhat straggly shrub. Branchlets bristly at first; bark cinnamon-brown, peeling; new growth bronzy-green. **Leaves** thick and stiff, *2-4.6 x 1.1-2.3cm*, narrowly ovate to elliptic or oval; upper surface dark shiny green, elepidote, lower surface paler, with scales their own diameter apart; petiole bristly. **Inflorescence** 1-2-flowered. **Corolla** openly funnel-campanulate, to 5cm across, white, white tinged pink to deep rose-red, with or without spots; calyx 1-4mm; style *straight, longer than stamens;* capsule large, cylindrical. **Distribution** Central Sichuan, N.E. Yunnan and Guizhou, 2,000-3,300m (6,500-10,800ft), epiphytic on broad leaved trees and on rocks and cliffs.

This species is closely related to *R. dendrocharis* but is larger in all parts except the corolla, earlier flowering, less compact and the straight rather than deflexed style is longer than the stamens.

R. moupinense is a beautiful, early-flowering species, very free flowering and surprisingly frost-hardy in bud and flower and with attractive young growth. It is a little tender and prone to bark-split when young; the white forms of Wilson 879 are less hardy than the pink forms. It dislikes excessive heat in summer, has rather brittle shoots and tends to have rather a small root system. It appreciates good drainage and is quite drought resistant, doing well even near moisture-greedy tree roots. Introduced 1909, reintroduced 1991. January-April.

A recent introduction from Guizhou has rounder smaller leaves than the type with bristly margins and later, pale pink to white flowers. This introduction seems to lie taxonomically between *R. moupinense* and *R. dendrocharis,* with the style typically longer than the stamens. The pure white forms are very promising but the hardiness is as yet untested. Introduced in 1985 as Guiz. 120.

RHODODENDRON

Subsection Rhododendron

Height to 1.5m, usually compact shrubs, growing wider than high. Branchlets densely lepidote. **Leaves** evergreen, densely lepidote, sometimes hairy. **Inflorescence** terminal, 3-16-flowered. **Corolla** tubular with spreading lobes, rosy-crimson to pink or rarely white; calyx small to medium, lobed; stamens 10; ovary densely lepidote; style straight. **Distribution** Europe.

A distinct subsection which shows little relationship with other subsections and which has an isolated distribution in Europe's high mountain ranges. This subsection is characterised by the densely lepidote branchlets and lower leaf surface, the pink (white) flowers and the elongated rachis.

The species in this subsection have long been cultivated and are particularly useful for their hardiness and lateness of flowering. *R. ferrugineum* is the commonest in cultivation and *R. myrtifolium* is fairly rare. All are susceptible to gall fungus.

R. ferrugineum L. 1753 (Alpenrose) H5

Height to 1.5m, a usually fairly compact shrub with spreading branches, growing wider than tall. Branchlets densely lepidote. **Leaves** 1.6-5 x 0.6-1.6cm, elliptic, lanceolate to oblanceolate; upper surface dark shiny green, lower surface *reddish-brown* with overlapping or contiguous scales. **Inflorescence** 5-16-flowered, rachis *tall*. **Corolla** tubular with spreading lobes, 1.2-1.7cm long, rosy-crimson to rose, rarely white; calyx *minute, to 1.5mm,* with scales and long *hairs;* stamens *as long as corolla tube*; ovary *densely lepidote*; style *glabrous, to 2 x as long as ovary*. **Distribution** Common in Pyrenees and Alps as far east as Austria, 900-2,100m (3,000-7,000ft), forest floor, pine scrub to open moorland on acid soil.

This species differs from *R. hirsutum* in its glabrous, darker leaves, with overlapping or touching scales, and its usually darker flowers with a glabrous style. It is usually larger in stature than *R. myrtifolium* with larger, more densely lepidote leaves, darker flowers and with the style longer than the ovary.

Picture 1

Picture 2

Picture 3

Picture 4

1. *R. moupinense*, a fine pink clone from the Royal Botanic Garden, Edinburgh, growing at Glendoick.
2. *R. moupinense* Wilson 879 collected in Sichuan in 1909.
3. *R. moupinense*, the typical reddish-bronzy new growth.
4. An attractive pure white-flowered clone of *R. moupinense* aff. Guiz. 120, quite different from Wilson 879. This may be referable to *R. petrocharis*.

311

RHODODENDRON

R. ferrugineum is extensively cultivated in Europe but is rare in North America. It will not tolerate high temperatures or drought and while useful for its hardiness and late flowering, the variable *R. ferrugineum* is not the most free-flowering species. Introduced to Britain pre 1740. June-July.

R. x intermedium Tausch 1839 is a natural hybrid occurring where *R. ferrugineum* and *R. hirsutum* meet, often forming a whole population of crosses and back-crosses showing variation from the one species to the other. White-flowered forms also occur. This hybrid is quite rare in cultivation.

Picture 1

Picture 2

Picture 3

1. *R. ferrugineum* at Glendoick, showing the typical tubular flowers with spreading lobes.
2. *R. ferrugineum* in the Dolomites, N.E. Italy.
3. *R. ferrugineum,* a pale-flowered form at the Royal Botanic Garden, Edinburgh.

Picture 4

Picture 5

4. *R. ferrugineum,* a deep-coloured form from Hill of Tarvit, E. Scotland, growing at Glendoick.
5. *R. ferrugineum,* a rare white-flowered form which is less easily grown than the pink-flowered forms.

R. hirsutum
L. 1753 H5

Height to 1m, a compact to spreading shrub. Branchlets lepidote, often pubescent and bristly. **Leaves** 0.8-3 x 0.4-1.4cm, elliptic, obovate to oblanceolate, *margins hairy*; upper surface glabrous, (*usually paler* than *R. ferrugineum*), lower surface *pale, scales 2-4 x their own diameter apart.* **Inflorescence** 4-12-flowered, rachis *usually shortish.* **Corolla** tubular with spreading lobes, 1-2cm long, *pale to deep pink,* sometimes to crimson, rarely white; calyx *2-5mm, bristly*; style *pubescent at base, as long or a little longer than ovary.* **Distribution** European Alps, 360-1,800m (1,200-6,000ft), on or adjacent to limestone, often on drier soil than *R. ferrugineum,* in woodland, screes, dwarf pine and moorland.

This species differs from *R. ferrugineum* in its hairy leaves with a paler lower surface and more scattered scales, its usually paler flowers and its larger, bristly calyx. From *R. myrtifolium, R. hirsutum* differs in its hairy leaves with more distantly spaced scales on the lower surface, and in the longer style.

R. hirsutum is useful for its ability to grow on near-alkaline soils. It is not perhaps as vigorous or easy to grow as *R.*

ferrugineum and is liable to hide its flowers in the young foliage. White flowered forms are in cultivation. There is a pretty but often short-lived double-flowered form known as 'Flore Pleno' with just a few hairs on the leaves. Introduced to Britain in 1656->. June-July.

Picture 1

Picture 2

Picture 3

1. *R. hirsutum* growing on limestone in the Dolomites, N.E. Italy
2. *R. hirsutum* in the Dolomites, N.E. Italy.
3. *R. hirsutum* showing the characteristic hairy branchlets and leaves. (Photo H. Gunn)

Picture 4

4. *R. hirsutum*, a rare white-flowered form growing at Glendoick.

R. myrtifolium
Schott & Kotschy 1851 (*R. kotschyi* Simonkai) H4-5

Height rarely over 0.6m, a compact to creeping shrub. Branchlets not bristly. **Leaves** *0.9-2.3 x 0.5-0.8cm*, elliptic to oblanceolate; upper surface dark green, shining, elepidote, lower surface dark, lepidote, scales contiguous to ¹/₂ x their own diameter apart. **Inflorescence** 3-7-flowered. **Corolla** *narrowly* tubular with spreading lobes, 1.3-2cm long, rosy to mauvy-pink, (white forms exist but are probably not in cultivation), corolla *densely pubescent* outside; calyx minute, c. 1mm; stamens *shorter* than corolla tube; style *shorter or about as long* as ovary. **Distribution** Bulgaria, Slovenia, Romania, S.W. Russia, 1,500-2,400m (5,000-8,000ft), habitat as *R. ferrugineum* but sometimes on limestone.

This species differs from *R. ferrugineum* in its smaller stature, smaller leaves, less densely lepidote on the lower surface, the more pubescent and narrower corolla tube and the shorter style. From *R. hirsutum* it differs in its non-bristly branchlets and darker leaves, with denser scales on the leaf lower surface, the narrower corolla tube and the shorter style.

Picture 1

1. *R. myrtifolium,* a form from Branklyn, E. Scotland, growing at Glendoick.

Dwarfer and earlier flowering than *R. ferrugineum*, *R. myrtifolium* is not as easy to please and it occasionally suffers bark-split and fungal problems. Introduced to Britain 1846->. May or later.

Picture 2

2. *R. myrtifolium* in the Retezat Mountains, Romania. (Photo J.E. Barrett)

Subsection Rhodorastra

Height 0.3-4m, dwarf, spreading to upright shrubs. Branchlets thin and covered with scales; young growth comes early from below the flowers. **Leaves** *thinly-textured, deciduous to semi-evergreen*, lepidote on both surfaces. **Inflorescence** terminal and sometimes also axillary. **Corolla** purple, reddish-purple, mauve, pink or white; calyx minute; stamens 10, pubescent; ovary lepidote; style glabrous, straight, usually longer than corolla and stamens. **Distribution** E. Asia and Japan.

A very distinctive subsection of 2 (possibly 4) very hardy species which are invaluable for their early flowering. Some forms have frost-hardy flowers and, though the early growth is often frosted, it seldom results in permanent damage. Both species are widely cultivated and much used in breeding, particularly in E. North America.

R. dauricum

L. 1753. (*R. ledebourii* Pojark, *R. sichotense* Pojark) H5.

Height to 2.8m or more, a usually erect shrub. Branchlets lepidote and pubescent; bark light grey. **Leaves** *semi-deciduous, elliptic, rounded at each end;* lower surface densely lepidote, scales translucent, turning brown, *overlapping to $1^1/_2$ x their own diameter apart*. Remaining leaves usually turn purplish-brown in winter. **Inflorescence** 1-3, terminal or terminal and axillary. **Corolla** widely funnel-shaped, c.1cm long, 2-5cm across, pale to deep rosy-purple, purple, almost pink or white; calyx minute, lepidote; style glabrous. **Distribution** E. Russia, Siberia, Mongolia, N. China, Japan, altitudes and habitat very varied.

This species is distinguished from *R. mucronulatum* in that the leaves, rounded at each end, are partly retained in winter and are smaller and of thicker substance with a denser covering of scales on the lower surface. Intermediate forms between the species do occur.

R. dauricum is a very useful species for its hardiness and early flowering. Many fine selected clones are now available with white or double flowers although some of these are rather subject to root problems if drainage is not perfect. Two Russian-named species *R. ledebourii* and *R. sichotense* are probably regional variations of *R. dauricum*. Both have more evergreen leaves than the type and the latter has particularly large flowers. Plants under var. *sempervirens* also have better-than-average winter leaf-retention and may be referable to *R. ledebourii*. Dwarf forms, selected in Japan, are now well-distributed in cultivation but we have found that the flowers of these are easily frosted. Introduced 1780->. December-April (cultivation), later in the wild.

Picture 1

1. *R. dauricum* at Glendoick, showing the clusters of deep-green leaves at the end of the branchlets which usually persist through the winter.

R. mucronulatum Turcz. 1837 H5.

Height to 2m, an open, erect and sometimes sprawling shrub; compact in var. *chejuense*. Bark light grey to light brown, flaking. **Leaves** *deciduous*, 3-7.5 x 1.5-3cm, *elliptic-lanceolate to lanceolate*, of thin texture, usually *acute at both ends*, upper surface of young leaves with a few hairs, lower surface with scales *2-3 x their own diameter apart*. **Inflorescence** single-flowered, often from multiple buds, terminal or terminal and axillary. **Corolla** widely funnel-shaped, 2.1-3.3cm long, 3-5cm across, deep rosy-purple, mauve-pink, rose-pink or white; calyx minute, lepidote; style glabrous.

R. mucronulatum is fully deciduous while *R. dauricum* usually retains at least a few leaves at the branch tips. *R. mucronulatum* has leaves which are thinner in texture, more lanceolate in shape and have more scattered scales on the lower surface and the corolla is usually larger than in *R. daucicum*. In cultivation, *R. mucronulatum* is usually seen in its pink forms while pink forms of *R. dauricum* are very rare. Intergrading forms between the two species do exist.

Like *R. dauricum*, *R. mucronulatum* is very hardy and suitable for the coldest growing areas. It can, however, be severely damaged by late spring frosts when growth is well advanced, even in relatively mild climates. Tall forms benefit from pruning after flowering, especially when young. Introduced 1882->. February-April.

Var. *mucronulatum*

(*R. dauricum* var. *mucronulatum* (Turcz.) Maxim., *R. taquettii* Lév.)

Height over 1m. Habit usually upright. **Distribution** E. Siberia, China (Hubei, Shandong), Mongolia, Korea, Japan, 300-1,700m (1,000-5,500ft), dry slopes and larch forest.

This is the fairly common, tall form of the species which has been much used in hybridising in E. North America.

Var. *chejuense* Davidian 1989.

Height to 0.5-1m. Habit usually mounding or spreading. **Distribution** Cheju Island, Korea.

This variety has recently been described to distinguish a stable, isolated, low-growing population from Cheju Island.

The flowers range from pink to purple and the autumn colour is often good. Introduced 1976, 1982. March-April.

Picture 2

Picture 3

Picture 4

Picture 5

2. *R. dauricum*, a fine, free-flowering form at Glendoick.
3. *R. dauricum* 'Midwinter' F.C.C., the earliest clone to flower that we have seen, usually in December or January at Glendoick.
4. *R. dauricum* 'Hokkaido' A.M., a fine white-flowered clone which is unfortunately hard to propagate and is particularly susceptible to root rot.
5. *R. dauricum*, a dwarf form obtained from Japan which is later flowering and less hardy in the swelling bud than most forms.

SALUENENSIA

Picture 1

Picture 2

Picture 3

Picture 4

Picture 5

5. *R. mucronulatum* var. *chejuense* a pink-flowered selection from Cheju Island, Korea, collected by Warren and Pat Berg and H. Suzuki. Glendoick.

1. *R. mucronulatum* var. *mucronulatum* 'Cornell Pink' A.M., one of the finest taller pink-flowered clones.
2. *R. mucronulatum* var. *mucronulatum*, a white-flowered clone at Glendoick.
3. *R. mucronulatum* var. *mucronulatum*, a purple-flowered form at the Royal Botanic Garden, Edinburgh.
4. *R. mucronulatum* var. *chejuense*, a selection with deep-coloured flowers which is very low growing. Collected on Cheju Island, Korea, by Warren and Pat Berg and Hideo Suzuki. Glendoick.

Subsection Saluenensia

Height to 1.5m, prostrate, bushy to erect shrubs. Young growth usually lepidote and sometimes with small hairs/bristles. **Leaves** evergreen, densely lepidote on lower surface with *overlapping scales*. **Inflorescence** 1-3 (occasionally 7)-flowered. **Corolla** *openly*-funnel campanulate, pink, reddish-crimson, magenta, purple, purplish-crimson, pilose on the outside; pedicels long; calyx usually large, coloured and hairy; stamens 10; style *always red*. **Distribution** Arunachal Pradesh, N.E. Burma, S. & S.E. Tibet, N. & N.W. Yunnan into Sichuan.

This subsection has been reduced from eight to two species by Cullen and could even be reduced to a single species as there are intergrades between *R. saluenense* and *R. calostrotum*. Most of the former species do have horticultural distinctions which are maintained under subspecific, varietal or group status. Species in this subsection are chracterised by the +/– open funnel-campanulate corrolla and the overlapping scales.

Both species and most of their subspecies and varieties are quite widely grown in cultivation in the U.K. and similar climates, especially *R. calostrotum* ssp. *calostrotum* and ssp. *keleticum*. They are valuable, free-flowering dwarfs with high ornamental value and are relatively easily-grown.

R. *calostrotum*
Balf.f. & Kingdon-Ward 1920 H4-5

Height to 1.5m, much less in cultivation in most forms, an erect, mounded, creeping or prostrate shrub. Young growth lepidote but *not hairy or bristly* or the hairs quickly deciduous. **Leaves** 0.7-3.5 x 0.4-2cm, sub-orbicular to oblong-ovate, rarely oblong-ovate, margins sometimes sparsely hairy/bristly; upper surface sometimes glaucous, lower surface with dense, overlapping scales, usually *in 3-4 distinct tiers*. **Inflorescence** 1-5-flowered. **Corolla** widely funnel-shaped or rotate, 1.8-2.8cm long, magenta, pink, purple, rose-crimson, often spotted darker on the upper lobes; pedicel lepidote; calyx fringed with hairs; ovary lepidote, *without hairs*.

This species differs from *R. saluenense* in the non-hairy ovary and in the absence of hairs and bristles on the branchlets, petioles, and midrib, though they are sometimes present on the leaf margins.

Most forms of *R. calostrotum* are free-flowering and form neat, compact plants and there are many fine selections. Some forms are particularly susceptible to powdery mildew. Cullen has divided this species into 4 subspecies:

Ssp. *calostrotum*

Leaves usually *glaucous* on upper surface. **Inflorescence** usually 1-2-flowered. Pedicels 1.6-2.7cm. **Distribution** N. Burma, W. Yunnan, 3,300-4,300m (10,500-14,000ft), alpine meadows.

The most popular clone is 'Gigha' F.C.C. with a particularly compact habit and lovely rose-crimson flowers; others are of various shades of purple. May. Introduced 1919.

Ssp. *riparium* (Kingdon-Ward) Cullen 1978
(*R. rivulare* Kingdon-Ward, *R. riparium* Kingdon-Ward, *R. calciphilum* Hutch. & Kingdon-Ward, *R. nitens* Hutch., *R. kingdonii* Merrill). H4-5.

Leaves rarely glaucous on upper surface. **Inflorescence** 2-5-flowered. **Corolla** pink-purplish; pedicels 1-(1.5)cm.

This subspecies is an assemblage of former species which as cultivated appear to be quite distinct from one another, but which seem to merge into a cline in the wild. Two former species currently have group status but some may merit varietal status when wild populations have been studied. **Distribution** Arunachal Pradesh, India, N.E. Burma, S. & S.E. Tibet, N.W. Yunnan, 3,050-4,550m (10,000-14,500ft), alpine meadows, rocky slopes and in damp ground.

Calciphilum Group. H4. **Leaves** to 1.3cm long, elliptic, greyish/bluish-green. **Corolla** pink. **Distribution** N.E. Burma, 4,000-4,300m (13,000-14,000ft). Flowers open later than average for *R. calostrotum* in late May.

Nitens Group. H4. Height 0.3-0.45m, compact to erect. **Leaves** 0.7-2.5cm long, oblong-obovate to oblong-elliptic; upper surface *shiny*, young growth green. **Corolla** deep purplish-pink to magenta-pink. **Distribution** N.E. Burma, 3,700m (12,000ft). This is useful for its late flowering, but is not as easy to please as its relatives. Introduction under Kingdon-Ward 5482 only, in 1922. June-July.

Ssp. *riparioides* Cullen 1978

Height 0.3-1.5m. **Leaves** 2.2-3.3cm long; upper surface usually *glaucous*, lower surface smooth, scales *not tiered*. **Distribution** Weixi in N.W. Yunnan only, 3,700-4,400m (12,000-13,500ft) alpine slopes and cliffs.

Usually cultivated as the form *R. calostrotum* Rock which is a handsome plant with deep purple flowers, half of which may open in autumn. Susceptible to powdery mildew. May.

Picture 1

1. *R. calostrotum* ssp. *calostrotum* 'Gigha' F.C.C., a very fine selection with bright rosy-red flowers and greyish foliage.

SALUENENSIA

Picture 2

Picture 3

Picture 4

Picture 5

Picture 6

Picture 7

Picture 8

2. *R. calostrotum* ssp. *calostrotum* 'Gigha' on a rock at the Rhododendron Species Botanical Garden, Tacoma, Washington State, U.S.A.
3. *R. calostrotum* ssp. *riparium* on the Dokar (Showa) La, Pome, S.E. Tibet, with *R. wardii* behind.
4. *R. calostrotum* ssp. *riparium* Kingdon-Ward 8229, collected in Arunachal Pradesh, N.E. India, in 1927-28 and growing at the Royal Botanic Garden, Edinburgh.
5. *R. calostrotum* ssp. *riparium* Calciphilum Group with greyish green foliage and pink flowers. (Photo H. Gunn)
6. *R. calostrotum* ssp. *riparium* Nitens Group Kingdon-Ward 5482, collected in N.E. Burma in 1922 and characterised by its late flowering.
7. *R. calostrotum* ssp. *riparioides* Rock which has fine flowers and attractive blue-grey foliage.
8. Leaf lower surfaces of *R. calostrotum* showing the dense, overlapping scales. (Photo R.B.G., E.)

Ssp. *keleticum*
(Balf.f. & Forrest) Cullen 1978 (*R. radicans* Balf.f. & Forrest) H4-(5)

Height 0.05-0.4m, a prostrate to mounding shrub. Branchlets *non-bristly*; young growth sometimes lepidote. **Leaves** 0.7-2.1 x 0.2-0.7cm, oblong to elliptic to lanceolate, *usually with an acute apex*; upper surface *elepidote,* lower surface densely lepidote with overlapping, brown to fawn scales; petiole *rarely bristly.* **Inflorescence** 1-3-flowered. **Corolla** widely funnel shaped or rotate, to 3cm long and 3.8cm across, pale to deep purplish-crimson with crimson spots. **Distribution** S.E. Tibet, S.E. Burma, N.W. Yunnan, 3,400-4,600m (11,000-15,000ft) on moorland, scree, rocks and cliffs.

The +/- pointed leaf tip, the non-bristly branchlets and (usually) petiole, the low habit of growth, the green rather than bronzy new leaves and the relatively late flowering all distinguish this subspecies from its relatives. It merges with ssp. *riparium* at lower altitudes; Cullen says that the Calciphilum and Nitens Groups of *R. calostrotum* may be intermediates between the two subspecies.

In its best forms this is a fine, showy subspecies which forms neat mounds. Particularly good forms have come from Rock 58, introduced in 1949 from N.W. Yunnan. Plants under Radicans Group are the most dwarf, with a creeping or mounding habit, tiny leaves and relatively large, almost flat-faced flowers held on upright pedicels above the foliage. Introduced 1919->, reintroduced 1992. May-June.

Picture 9

9. *R. calostrotum* ssp. *keleticum* Rock 58 from N.W. Yunnan, China, the best form we have seen of this subspecies.

Picture 10

10. *R. calostrotum* ssp. *keleticum* Radicans at Glendoick showing the characteristic single flowers held up above the foliage.

R. saluenense Franch. 1898. H4-5

Height 0.05-1.5m, habit prostrate to upright. Young growth, branchlets and petioles *covered in persistent hairs or bristles*. **Leaves** 0.8-3.6 x 0.5-1.5cm, oblong-orbicular to oblong-elliptic, rarely oblong-obovate, *margins bristly or hairy*, lower surface with dense, overlapping scales *borne in several tiers*, midrib usually with a few hairs/bristles. **Inflorescence** 1-3-flowered. **Corolla** widely funnel-shaped, 1.7-3.1cm long, magenta to shades of purple and reddish-purple; pedicel lepidote and hairy/bristly; ovary *lepidote*.

This differs from *R. calostrotum* in its hairy new growth, branchlets, petiole and pedicel, the lepidote ovary and the normally paler lower leaf surface.

Picture 1

1. *R. saluenense* ssp. *saluenense* above Londre, Salween-Mekong divide, N.W. Yunnan, China.

SALUENENSIA

Ssp. *saluenense* (*R. amaurophyllum* Balf.f. & Forrest)

Erect in habit, to 1.5m. **Leaves**, upper surface persistently lepidote and usually bristly. **Distribution** N.E. Burma, N.W. Yunnan, S.E. Tibet 3,400-4,300m (11,000-14,000ft), forest margins to stony hillsides.

This is less common in the wild than ssp. *chameunum* and may be a stable intermediate between the latter and *R. calostrotum* ssp. *riparium*. Some forms have very hairy leaves and branchlets. Introduced 1914. April-May.

Picture 2

Picture 3

Picture 4

2. *R. saluenense* ssp. *saluenense* Forrest 21760 collected in N.W. Yunnan, China, in 1921-22 and growing at the Royal Botanic Garden, Edinburgh.
3. *R. saluenense* ssp. *saluenense* Forrest 21772 collected in N.W. Yunnan in 1921-22 and growing at the Royal Botanic Garden, Edinburgh.
4. The hairy foliage typical of ssp. *saluenense* Kingdon-Ward 9633 from the Burma-Tibet frontier region, collected in 1931.

Picture 5

5. The tiered or overlapping scales on the lower leaf surface separate *R. saluenense* from *R. calostrotum*. (Photo R.B.G., E.)

Ssp. *chameunum* (Balf.f. & Forrest) Cullen 1978
(*R. colobodes* Balf.f., *R. cosmetum* Balf.f. & Forrest, *R. charidotes* Balf.f. & Forrest, *R. pamprotum* Balf.f. & Forrest, *R. prostratum* W.W. Sm., *R. sericocalyx* Balf.f. & Forrest)

Height to 0.6m, a prostrate to fairly compact shrub; new growth sometimes coloured. **Leaves** often reddish, coppery or purple in winter, upper surface usually *glossy and elepidote*, without bristles. **Distribution** Common and widespread, S.E. Tibet, N.E. Burma, N.W. Yunnan, S.W. Sichuan, 3,500-5,200m (11,500-17,000ft), usually above the tree-line.

This is a more common and widespread plant in the wild than ssp. *saluenense*. Best selected forms have most striking reddish-purple flowers of a colour rarely seen in other species. Prostratum Group are the lowest growing forms from the highest altitudes and are not as easy to grow as typical ssp. *chameunum*. Charidotes Group has green new growth and a bristly calyx, ovary and capsule and is intermediate between the 2 subspecies. Introduced 1914-> and reintroduced c. 1990->. Late April-May/June.

Picture 6

6. *R. saluenense* ssp. *chameunum* at Li ti ping, N.W. Yunnan, China. (Photo R. McBeath)

SALUENENSIA

Picture 7

Picture 8

Picture 9

Picture 10

10. *R. saluenense* ssp. *chameunum* Prostratum Group. This group includes the lowest-growing forms which are harder to please in cultivation than typical ssp. *chameunum*. Glendoick.

7. *R. saluenense* ssp. *chameunum* near the east side of the Doker La, Salween-Mekong divide, N.W. Yunnan, China. This pale-flowered form seems typical for this area.
8. *R. saluenense* ssp. *chameunum* Forrest 12968 (the original introduction) from the Zhongdian Plateau, N.W. Yunnan, China, collected in 1914.
9. *R. saluenense* ssp. *chameunum* Charidotes Group probably introduced by Farrer and Cox in 1919 from N.E. Burma and growing at Glendoick.

Subsection Scabrifolia

Height 0.15-3m, low and compact to upright and spreading, often straggly shrubs. Young growth lepidote and usually covered with fine hairs and bristles. **Leaves** evergreen; lower surface covered with almost rimless scales, ½ to 4 x their own diameter apart. **Inflorescence** axillary or terminal and axillary (Both Davidian and Cullen mistakenly assert that in subsection Scabrifolia all flowers are borne in axillary inflorescences which is not true in the case of *R. spinuliferum*). **Corolla** openly-campanulate, funnel-campanulate to tubular, to 3cm long, white, pink, orange-red to red; stamens (8)-10; style impressed, usually declinate. **Distribution** S.W. Sichuan, Yunnan and Guizhou.

This is a fairly uniform group of species related to subsections Triflora and Virgata but distinguished by the flowers from axillary buds (except *R. spinuliferum*). Only *R. racemosum* is widely grown; the others are rare and/or rather tender. They are often found in relatively hot, dry conditions in the wild and their heat and drought tolerance could be exploited by breeders. The species *R. hemitrichotum* and *R. mollicomum* are not very distinct from *R. racemosum* and arguably should be considered extreme forms of this species. Most forms of the species in this subsection tend towards straggliness and benefit from pruning after flowering.

R. hemitrichotum
Balf.f. & Forrest 1918 H4

Height 0.3-2m+, a fairly compact to leggy shrub. Branchlets pubescent. **Leaves** 1.2-4.5 x 0.7-1.3cm, *narrowly elliptic*; upper surface with *downy or pubescent hairs,* lower surface *glabrous* (except for a few hairs on midrib), *glaucous*, with +/- rimless scales; petiole with pubescent hairs. **Inflorescence** 2-3 flowers per leaf axil, in uppermost leaves. **Corolla** widely funnel-shaped, 0.9-1.4cm long, pink or white-edged-pink, with or without purple spots; calyx lepidote and fringed with small hairs. **Distribution** N. Yunnan, S.W. Sichuan, 2,900-4,300m (9,500-14,000ft), dry rocky pastures, scrub, rocky slopes.

The downy or pubescent hairs on the upper surface of the leaves, branchlets and petioles and the narrower leaf distinguish *R. hemitrichotum* from *R. racemosum*. The glabrous and glaucous leaf underside separate this species from *R. mollicomum* and *R. pubescens*.

R. hemitrichotum is rarely as good as the better forms of the closely related *R. racemosum* and so it remains rare in cultivation. Introduced 1918->. April.

Picture 3

3. The narrowly elliptic leaves of *R. hemitrichotum* showing the pubescent upper surfaces and petioles and the glabrous and glaucous lower surfaces. (Photo R.B.G., E.)

Picture 1

Picture 2

1. *R. hemitrichotum*, Glendoick.
2. *R. hemitrichotum*, Savill Gardens, Windsor Great Park, S. England.

R. mollicomum
Balf.f. & W.W. Sm. 1921 (var. *rockii* Tagg) H3-4.

Height 0.6-3.6m, a compact to erect and straggly shrub. Branchlets *densely pubescent*. **Leaves** 1.2-3.6 x 0.7-1.3cm, lanceolate (-oblong), pubescent on upper and lower surfaces, lower surface lepidote, *pale green, not glaucous*. **Inflorescence** 1-3 flowers per leaf axil in uppermost leaves. **Corolla** *narrowly* funnel-shaped, *1.7 to 3cm* long, pale to deep pink; calyx 0.5-1mm, variably lepidote and pubescent; style long, *straight*. **Distribution** N. Yunnan, S.W. Sichuan, 2,400-3,800m (8,000-12,500ft), dry bouldery hillsides, thickets or forest margins.

This species has larger flowers than *R. hemitrichotum* and differs in the pale green rather than glaucous leaf underside. It differs from *R. pubescens* and *R. scabrifolium* var. *spiciferum* in its pubescent but not bristly leaves. Its pubescent leaves and green leaf lower-surface distinguish it from *R. racemosum*. The specimens of this species in the herbarium of the R.B.G. Edinburgh appear to cover a range from very close to *R. racemosum* to very close to *R. scabrifolium* var. *spiciferum* suggesting that *R. mollicomum* may be a hybrid between these two species.

R. mollicomum is quite pretty but not as hardy as the average *R. racemosum* and is therefore likely to remain rare. Introduced 1913->. April.

SCABRIFOLIA

Picture 1

Picture 2

1. *R. mollicomum*, in June Sinclair's garden, Hood Canal, Washington State, W. U.S.A.
2. *R. mollicomum* showing the typical lanceolate leaves, Rhododendron Species Botanical Garden, W. U.S.A.

scabrifolium var. *spiciferum*.

R. pubescens is not of much merit with small leaves and flowers. The best, deep coloured forms of *R. scabrifolium* var. *spiciferum* give a better show but are usually less hardy. Introduced 1918. April.

Picture 1

Picture 2

Picture 3

1. *R. pubescens* 'Fine Bristles' A.M., Valley Gardens, Windsor Great Park, S. England.
2. *R. pubescens* Kingdon-Ward 3953, Royal Botanic Garden, Edinburgh.
3. The narrowly lanceolate leaves of *R. pubescens* showing the strongly recurved margins. The bristles separate this species from *R. mollicomum*. (Photo R.B.G., E.)

R. pubescens
Balf.f. & Forrest 1920 H3-4

Height to 1.3m, a compact to sparse and leggy shrub. New growth covered in *hairs and bristles*. **Leaves** *very narrowly elliptic to very narrowly lanceolate*, 1.8-2.4 x 0.6-0.6cm, margins strongly recurved; upper and lower surfaces covered with hairs and bristles. **Inflorescence** 2-3-flowered from each axillary bud. **Corolla** funnel-shaped, 0.6-1cm long, rose pink; calyx fringed with long hairs; stamens 8-10; style *usually glabrous*. **Distribution** N. Yunnan, S.W. Sichuan, 2,400-3,800m (8,000-12,500ft), dry bouldery hillsides, thickets or forest margins.

Davidian placed this species in synonymy under *R. spiciferum* while Cullen retains it at specific status. It differs from *R. scabrifolium* var. *spiciferum* in its very narrow, recurved leaves, its greater hardiness and its isolated, more northerly distribution. It would probably be best treated as a variety of *R. scabrifolium*. Some plants cultivated as *R. pubescens* are *R.*

SCABRIFOLIA

R. racemosum

Franch. 1886, (*R. motsouense* Lév., *R. iochanense* Lév.), H3-4(-5).

Height. 0.15-4.6m, a low and compact to tall and leggy shrub. Branchlets usually glabrous and frequently red. **Leaves** 1.5-5 x 0.7-3cm, broadly obovate to oblong-elliptic; upper surface *glabrous* except for a few hairs on the midrib, lower surface glabrous and *glaucous*, densely lepidote with almost rimless scales. **Inflorescence** 1-4 per axil in uppermost leaves. **Corolla** widely funnel-shaped, 0.8-2.3cm long, white tinged pink, pale pink, pink to purplish-pink to pure white; style glabrous. **Distribution** N. half of Yunnan (very common), S.W. Sichuan and W. Guizhou, 800-4,300m (2,500-14,000ft) on plains, in scrub, forest margins and rocky slopes in dry and occasionally wet conditions.

The glabrous upper leaf-surface (with only a few hairs on the midrib), and the glaucous lower leaf surface distinguish *R. racemosum* from its relatives. *R. hemitrichotum* differs only in its hairy upper leaf surface and narrower leaves.

This is a widely grown and very variable species which has been much used in hybridising. There have been many recent introductions of this species some of which should provide hardier and more varied colour forms than have been previously cultivated. Taller forms need pruning when young for best results. *R. racemosum* is sometimes found in very dry situations in China and may be useful for similar conditions in cultivation. Introduced 1889-> and reintroduced 1981->. March-May.

Picture 1

Picture 2

Picture 3

Picture 4

1. *R. racemosum*, Zhongdian Plateau, N.W. Yunnan, China.
2. A very fine rich pink-flowered selection of *R. racemosum* SSNY 47, Zhongdian Plateau, N.W. Yunnan, China.
3. Picked shoots of *R. racemosum*, Zhongdian Plateau, N.W. Yunnan, China, showing the variations in flower colour, size and shape.
4. *R. racemosum* 'Rock Rose' A.M., a good medium-pink selection. Glendoick.

Picture 5

Picture 6

5. *R. racemosum* 'Glendoick' a particularly deep pink-flowered selection. Glendoick.
6. Leaf lower surface of *R. racemosum* showing the glaucous, glabrous surface with dense, almost rimless scales. (Photo R.B.G., E.)

R. scabrifolium
Franch. 1886 H2-3.

Height 0.3-3m, a compact and bushy to tall and straggly shrub. **Leaves** 1.5-9 x 0.4-1.5cm, narrowly elliptic to oblanceolate; upper surface covered in *hairs and bristles*, lower surface lepidote and covered with *hairs*. **Inflorescence** 2-3(-5) per axillary bud. **Corolla** variable in size and shape, white to deep pink; calyx rim-like or 5-lobed, fringed with hairs; style sparsely pubescent at base.

The non-glaucous lower leaf distinguishes this species from *R. racemosum* and *R. hemitrichotum* and the hairs and bristles on the leaves separate it from *R. mollicomum*. Out of flower, it can look close to *R. spinuliferum* which has a more rigid habit and lacks the hairs on the leaf underside.

Var. *scabrifolium*

Leaves *broad and large, 2.3-9.5 x 0.6-2.8cm.* **Corolla** openly funnel-shaped, 0.9-1.7cm long. **Distribution** N. Yunnan. 1,500-3,400m (5,000-11,000ft).

This variety is taller growing and with larger leaves than var. *spiciferum* and with a more northerly distribution. Most introductions are rather tender and are not very showy with small flowers for the size of the leaves. Introduced 1885, reintroduced 1980s->. March-April.

Var. *spiciferum* (Franch.) Cullen 1978.

Leaves *narrow and small, 1.2-3.5 x 0.2-1.3cm.* **Corolla** narrowly funnel-shaped, 1-1.5cm long, tube 0.6-1.7cm. **Distribution** Central & S. Yunnan, c. 2,400m (8,000ft), common on dry rocky slopes.

Var. *spiciferum* is a more showy plant than var. *scabrifolium*; some forms with rich pink flowers can make fine plants for milder areas. This has a more south-easterly distribution than var. *scabrifolium* and is very closely related to *R. pubescens* which differs in its more northern distribution and its very narrow, recurved leaves. Introduced 1921, reintroduced 1980->. March-May

Picture 1

1. *R. scabrifolium* var. *scabrifolium* SBEC K160 showing the relatively large hairy leaves. Glendoick.

SCABRIFOLIA

Var. *pauciflorum* Franch. 1898 (*R. dielsianum* Hand.-Mazz.; *R. x duclouxii* Lév. is the name used by the Chinese)

Leaves 2.5-9 x 0.8-2.5cm. **Corolla** 1.6-2.3cm long.

This variety is a natural hybrid or intergrading form which commonly occurs between var. *spiciferum* and *R. spinuliferum*. It is variable, closer to one species or the other, though most forms we have seen are close to var. *spiciferum* but with larger flowers and leaves.

Picture 2

Picture 3

Picture 4

Picture 5

5. *R. scabrifolium* var. *pauciflorum* McLaren S 33, Royal Botanic Garden, Edinburgh. This form is closer to var. *spiciferum* than *R. spinuliferum*.

R. spinuliferum

Franch. (*R. duclouxii?* Lév., *R. fuchsiiflorum* Lév.) H2-3.

Height 2.4m+, an erect and often rather straggly shrub. Branchlets softly pubescent; bark fairly smooth, dark purple-brown. **Leaves** 3-9 x 1.5-4cm, oblanceolate to obovate, margins hairy; upper surface *bullate*, lower surface with corresponding conspicuous veining, hairs persistent only on midrib. **Inflorescence** 1-4 per bud, *usually terminal*, occasionally axillary. **Corolla** *tubular, filled with watery nectar, erect, contracted at each end*, c.2.5cm long, pink, orange, peach-red, brick-red to crimson; stamens and style protruding from corolla, **Distribution** Central & S. and N.E. Yunnan, S. Sichuan, 1,700-2,600m (5,500-8,500ft), in thickets and pine forest.

This species is easy to identify in flower with the clusters of upright tubular flowers. Out of flower it closely resembles *R. scabrifolium*, differing in the near absence of bristles on the leaf underside. These two species can be separated from others in the subsection by their bullate leaves. It crosses or merges with *R. scabrifolium* as var. *pauciflorum*. The Chinese use the name *R. x duclouxii* to describe such plants.

R. spinuliferum has long-lasting, curious flowers which always attract attention. It can be grown successfully on a wall even in gardens as cold as Glendoick. Introduced 1907-> reintroduced 1981->. April-May.

2. A compact selection of *R. scabrifolium* var. *spiciferum* at J. & P. Warren's garden, Timaru, New Zealand.
3. A fine rich pink selection of *R. scabrifolium* var. *spiciferum* from near Kunming, Yunnan, growing at Glendoick.
4. The hairy leaf lower surface of *R. scabrifolium* var. *spiciferum* showing the bristly leaf margins. (Photo R.B.G., E.)

Subsection Tephropepla

Height 0.3-2m+, dwarf and compact to upright and spreading shrubs. Bark sometimes peeling. **Leaves** evergreen, usually relatively narrow. **Inflorescence** mainly terminal. **Corolla** campanulate to tubular-campanulate, pink, purple, white, cream or yellow; calyx 5-lobed, conspicuous; stamens 10; *style long, slender and straight*, tapering or impressed. **Distribution** E. and S.E. Tibet, N.W. and N. Yunnan, N.E. Upper Burma, Central and W. Sichuan and N.E. India.

This subsection is related to subsections Cinnabarina, Virgata, Triflora and Boothia. *R. hanceanum* and *R. longistylum* are somewhat aberrant here but are closer to subsection Tephropepla than to subsection Triflora where they were placed prior to Cullen's revision.

R. hanceanum is generally hardier than the other species in this subsection and is commonly cultivated. *R. tephropeplum*, *R. auritum* and *R. xanthostephanum* are found in good collections in milder areas while *R. longistylum* is rare.

Picture 1

Picture 2

Picture 3

1. *R. spinuliferum* 'Blackwater' A.M., Brodick Castle, W. Scotland, showing the characteristic, upright tubular flowers with protruding stamens and style. (Photo J. Basford)
2. *R. spinuliferum* deep red-flowered selection, P. Grigg's garden, near Timaru, South Island, New Zealand.
3. *R. spinuliferum*, a deep pink-flowered form from SBEC K58, St. Andrews Botanic Garden, E. Scotland.

R. auritum Tagg 1931 H2-3

Height to 3m, habit erect and sometimes untidy. Bark *peeling, coppery-red*. **Leaves** 2.5-6.6 x 1-2.7cm, narrowly-elliptic to elliptic; lower surface brown with unequal, contiguous or overlapping brown scales, *only the smaller scales slightly sunken in pits*. **Inflorescence** 3-7-flowered, sometimes axillary. **Corolla** tubular-campanulate, 2-2.5cm long, *very pale yellow to cream*, often tinged pink on the lobes; pedicel lepidote; calyx *lobes reflexed*; style lepidote at the base. **Distribution** Only recorded from the Tsangpo Gorge, S.E. Tibet, 2,100-2,600m (7,000-8,500ft), on sheltered cliffs and open stony banks.

This species is similar to *R. xanthostephanum* but is distinguished by the paler flowers, the reflexed calyx and the less sunken scales on the lower leaf surface.

R. auritum is only reliably hardy in relatively mild gardens but can be quite showy and attractive at its best. Introduced 1924, 1947. April.

R. hanceanum
Hemsl. 1889 H3-5.

Height 0.15-2m (cultivated plants mostly the dwarfer forms), a low and very compact to tall and sprawly shrub. Young growth *bronzy-brown*. **Leaves** 2.5-12.8cm x 1.6-5.5cm, narrowly-obovate to oblong-elliptic, *thick and rigid*, tip sometimes acuminate, scales on lower surface golden brown, 1-5 x their own diameter apart. **Inflorescence** 5-15-flowered, rachis *elongated, to 12mm*. **Corolla** funnel-campanulate, 1.3-2.1cm long, white to yellow; calyx lobes to 5mm, sparsely fringed with scales; ovary lepidote; style impressed. **Distribution** Central Sichuan, 1,500-3,000m (5,000-10,000ft), in thickets and on cliffs.

A distinctive species, most closely related to *R. longistylum* which is less hardy, has narrower leaves, a fewer-flowered inflorescence, a longer style and lacks the long rachis. The bronzy new growth and the compact habit of cultivated plants of *R. hanceanum* are easily recognised features. According to the original description, this species is characterised by an acuminate leaf apex with a conspicuous drip tip. This character is not apparent in the older cultivated forms of the species but is evident on the recently introduced taller forms.

This species has long been cultivated in two forms, cream and yellow. The cream-coloured ones include the clone 'Canton Consul' A.M. while the rarer yellow form is more dwarf and later flowering and is grown as Nanum Group. The name Nanum is often incorrectly applied to cream forms in North America and elsewhere. One of the hardiest dwarf species outside subsection Lapponica, the long-cultivated dwarf forms of *R. hanceanum* perform well in quite severe climates but taller introductions with larger leaves are less hardy. Introduced 1909-10, reintroduced 1989-> May-June.

Picture 1

Picture 2

Picture 3

1. *R. auritum* Kingdon-Ward 6278, Royal Botanic Garden, Edinburgh, showing the typical, pale yellow-tinged pink flowers.
2. *R. auritum*, Arduaine Gardens, W. Scotland, showing the typically erect growth habit.
3. *R. auritum* showing the characteristic peeling, coppery-red stems.

Picture 1

1. *R. hanceanum* 'Canton Consul' A.M., sometimes wrongly labelled 'Nanum' showing the bronzy young growth. Glendoick.

A dainty rather than showy species which is not very hardy and which is mainly of interest to collectors. There may only be one white-flowered clone of this species currently in cultivation. Introduced 1908-10. May

Picture 2

Picture 1

Picture 3

Picture 2

2. The true *R. hanceanum* Nanum Group, very compact with clear yellow flowers. Glendoick.
3. *R. hanceanum* Nanum Group, showing the dense trusses of yellow flowers. Glendoick.

1. *R. longistylum*, Arduaine Gardens, W. Scotland.
2. *R. longistylum* showing the characteristic long stamens and style. (Photo G.F. Smith)

R. longistylum
Rehder & E.H. Wilson 1913. H3

Height 0.5-2m, a dwarf shrub, erect but inclined to sprawl. Young shoots +/- lepidote, usually puberulent. **Leaves** 1.6-6 x 0.6-1.5cm, oblanceolate, lanceolate to oblong-lanceolate; lower surface with unequal scales, 2-4 x their own diameter apart. **Inflorescence** terminal or terminal and axillary, 1-3 flowers per bud. **Corolla** narrowly funnel-shaped or funnel-campanulate, 1.3-1.8cm long, white, white tinged pink; calyx lobes narrowly triangular; style *impressed and long*; stamens and style *much exserted from the corolla*. **Distribution** W. Sichuan 900-2,300m (3,000-7,500ft), scrub-clad rocky slopes, in thickets and on cliffs.

R. longistylum is easily identified in flower by the long stamens and style protruding far out from the corolla. It is less hardy and with smaller leaves than *R. hanceanum* and without the long rachis of that species. It was originally confused with *R. micranthum* which has narrower leaves and much smaller, later flowers.

R. tephropeplum
Balf.f. & Farrer 1922 (*R. spodopeplum* Balf.f. & Forrest, *R. deleiense* Hutch. & Kingdon-Ward) H3(-4)

Height 0.5-1.3m(-1.8m), a bushy to upright and sprawly shrub. Bark brownish, peeling. **Leaves** very variable, 3-13 x 0.8-4cm, narrowly oblanceolate to narrowly elliptic to oblanceolate; upper surface usually dark and shiny, lower surface *densely lepidote* with unequal dark scales slightly sunk in pits, contiguous to their own diameter apart; petiole moderately or densely lepidote. **Inflorescence** 3-9-flowered. **Corolla** campanulate to tubular-campanulate, 1.7-3.2cm long, *purplish through carmine rose to pink, pale pink (rarely white)*, usually not or only sparsely lepidote; pedicels densely lepidote; calyx *large, leafy and lepidote 5-7mm*, with a few hairs; style longer than the corolla, impressed. **Distribution** E. Arunachal Pradesh, S.E. Tibet, N.E. Burma, N.W. Yunnan, 2,400-4,300m

TEPHROPEPLA

(8,000-14,000ft), on rocks/cliffs/scree and meadows.

The flower colour separates this species from the others in this subsection but out of flower some forms resemble *R. auritum* and *R. xanthostephanum*. The former has overlapping or touching scales on the lower leaf surface; the latter has a silvery brown leaf lower surface and non-touching scales and both have a more peeling bark.

R. tephropeplum is a very variable species in habit, leaf shape, flower colour and hardiness. It can be very showy and should be more widely grown in selected forms. Forms under Deleiense Group have the largest, widest leaves and often have a larger-than-average inflorescence. Introduced 1921->. Late April-May.

Picture 1

Picture 2

1. *R. tephropeplum* Forrest 26457, Nymans, S. England.
2. *R. tephropeplum* Deleiense Group Kingdon-Ward 8165 with the large leaves and flowers typical of this group. Royal Botanic Garden, Edinburgh.

Picture 3

3. A pale-flowered *R. tephropeplum*, Guavis, New Zealand.

R. xanthostephanum
Merrill 1941 (*R. aureum* Franch. non Georgi) H2-3

Height 0.3-3m, a loose, upright or spreading shrub. Bark smooth, reddish-brown. **Leaves** 5-10 x 1-3.4cm, oblong, narrowly-elliptic to elliptic; lower surface glaucous, scales unequal; the smaller and more numerous ones deeply sunk in pits, *1/2 to their own diameter apart*. **Inflorescence** 3-5-flowered, terminal and sometimes axillary. **Corolla** campanulate, 1.8-2.8cm long, *bright to creamy yellow*; calyx lobes *erect*, not reflexed; style long, slender and straight, lepidote at the base; ovary tapering into style. **Distribution** N.E. Burma, S.E. Tibet, N.W. & Central Yunnan, N. Vietnam, 1,500-4,000m (5,000-13,000ft), forests and their margins, scrub, pastures and cliffs.

The yellow flowers, the non-reflexed calyx, the sunken scales and the silvery brown leaf underside distinguish *R. xanthostephanum* from *R. tephropeplum* and *R. auritum*. It also resembles *R. sulfureum* which has a deflexed style and leaves with a glaucous lower surface.

Picture 1

1. *R. xanthostephanum*, Felix Dury's garden, near New Plymouth, New Zealand.

R. xanthostephanum is a showy plant with bright yellow flowers, suitable for warmer and milder areas. Relatively hardy forms from high altitudes seem to have been lost to cultivation. Introduced 1906->. April-May.

Picture 2

2. *R. xanthostephanum*, Brodick Castle, W. Scotland. (Photo N. Price)

Subsection Trichoclada

Height 0.3-2.5m, low and compact to upright and straggly shrubs. Branchlets *usually bristly*. **Leaves** *deciduous, semi-evergreen to evergreen*, often hairy, scales *vesicular*. **Inflorescence** 1-6-flowered, terminal and sometimes terminal and axillary. **Corolla** usually campanulate to funnel-campanulate, *yellow, creamy-yellow*, sometimes tinged red and spotted; calyx minute to large; stamens (8-)10; ovary lepidote; style long, straight or bent. **Distribution** W. to N.W. Yunnan, Upper Burma and S.E. Tibet.

This subsection, characterised by the yellow flowers and the hairy or bristly leaves, contains closely related species and shows some relationship with *R. megeratum* of subsection Boothia.

The species in this subsection have pretty but rarely spectacular yellow flowers and some are valued more for their fine glaucous foliage. They are generally fairly easy to grow and are quite commonly cultivated with the exception of the rare *R. caesium*.

R. caesium
Hutch. 1933 H3

Height to 1.5m, a fairly compact, upright and spreading shrub. Branchlets usually *not bristly*. **Leaves** semi-evergreen, slightly bristly, 2.3-5 x 1.2-1.6cm, oblong to oval; upper surface +/- glaucous, lower surface *glaucous*, scales unequal, brown 2-5 x their own diameter apart. **Inflorescence** (1-)2-3-flowered. **Corolla** funnel-campanulate, 1.5-2.5cm long, greenish to pale yellow, greenish spots; calyx 1-2mm, sparsely lepidote with a few hairs; style long, straight. **Distribution** S.W. & Central Yunnan, 2,400-3,000m (8,000-10,000ft), rocky slopes among scrub.

R. caesium resembles *R. viridescens* which is a stronger-growing, hardier, more bristly plant, not glaucous on the leaf lower surface. *R. lepidostylum* differs in its more glaucous leaf upper surface and its more bristly branchlets and petiole.

R. caesium is not as hardy as its relatives and the foliage is not as striking. The flowers are a good colour but tend to get lost in the foliage. Rare in cultivation. Introduced 1925. May.

Picture 1

1. *R. caesium* Forrest 26798,

R. lepidostylum
Balf.f. & Forrest 1920 H4-5

Height 0.5 rarely to 1.2m, a compact, flat-topped shrub, growing wider than high. Branchlets *densely bristly*. **Leaves** evergreen, margins bristly, 2-4.3 x 1-2cm, oblong-obovate to oval; upper surface *highly glaucous, especially when young*, lower surface moderately glaucous, bristly, with large scales,

TRICHOCLADA

1-4 x their own diameter apart. **Inflorescence** 1-3-flowered. **Corolla** funnel-campanulate, 2-3.3cm long, yellow, with or without spots; pedicel bristly; calyx 1-7mm, bristly; style long and straight. **Distribution** S.W. Yunnan only, 3,000-3,700m (10,000-12,000ft), on boulders, cliffs and crevices.

This species differs from *R. caesium* in its more glaucous, more bristly foliage and later flowers and from *R. viridescens* in its spreading habit and more glaucous, more bristly young leaves.

R. lepidostylum is a popular garden plant on account of its excellent, glaucous-blue young foliage. The flowers are often hidden in the new growth. It appears to grow best in a light, well-drained soil. Introduced 1924. June.

Picture 3

Picture 4

3. *R. lepidostylum* showing the typical character of the flowers becoming hidden in the developing young growth. Glendoick.
4. Leaf lower surface of *R. lepidostylum* with the characteristic large scales and bristles on margin and blade. (Photo R.B.G., E.)

Picture 1

Picture 2

1. *R. lepidostylum*, Baravalla, W. Scotland, showing the flat-topped habit and the bristles which in this example have caught water droplets on them.
2. *R. lepidostylum* showing its typically very glaucous leaves with bristly margins. Glendoick.

R. mekongense
Franch. 1898 H4-5

Height to 2.5m, a broadly upright shrub, leggy in shade. Branchlets variably bristly. **Leaves** *deciduous*, 2.5-5.5 x 1.4-2.1cm, usually obovate but sometimes obovate-elliptic; upper surface usually glabrous, lower surface variably bristly, *scales very unequal, dense*. **Inflorescence** 2-5-flowered. **Corolla** funnel-campanulate, 1.4-2.3cm long, yellow to greenish-yellow, sometimes flushed red; pedicel *usually bristly*; calyx to 7mm, lobes of variable length, lepidote, sometimes hairy.

This species has proven to be taxonomically very problematic. Cullen has quite rightly made a large reduction in the many former species in this subsection. He maintains *R. mekongense* and *R. trichocladum* at specific rank and distinguishes them by the scales: unequal in *R. mekongense*, equal in *R. trichocladum*. Using this distinction, almost all cultivated plants under either name are referable to *R. mekongense*. Chamberlain points out in the report of the 1981 Cangshan expedition that Cullen's description of *R. mekongense* seems to encompass the type specimen of *R. trichocladum*. Previous to the Edinburgh revisions and according to Davidian *R. trichocladum* differs in its more bristly/hairy branchlets, leaf margins, petiole, pedicel and in the presence of long hairs on the calyx margins. We have reverted to the original description, moving Cullen's *R. mekongense* var. *longipilosum* to *R. trichocladum* as it appears to more or less match the type description of *R. trichocladum*. After examining these two species in the field over a large range of their distribution, it has become quite clear to us that there is virtually no basis for maintaining the two species at all, as their distributions overlap and there are so many intermediate forms.

R. mekongense can can be quite showy, especially in the deepest yellow-flowered forms, but is often best planted in clumps for effect.

Var. *mekongense* (var. *melinanthum* Balf.f. & Kingdon-Ward, *R. chloranthum* Balf.f. & Forrest, *R. semilunatum* Balf.f. & Forrest)

Corolla yellow, sometimes spotted, not marked or flushed with red or pink. **Distribution** E. Nepal, S. & S.E. Tibet, N..E. Burma, N.W. & Central Yunnan, 2,700-4,500m (9,000-14,500ft), forest margins, scrub and moorland.

Var. *melinanthum* has quite rightly been placed in synonymy here in the latest revision of this subsection. Introduced 1913-1948, 1981->. April-July.

Var. *rubrolineatum* (Balf.f. & Forrest) Cullen 1978

Leaves *almost completely lacking bristles.* **Corolla** yellow *flushed red,* creamy-yellow *flushed rose,* occasionally cream. **Distribution** Arunachal Pradesh, N.W. Yunnan, S. & S.E. Tibet, 3,400-4,300m (11,000-14,000ft), rocky slopes and forest margins.

There are two distinct taxa grown under this name. One taxon is completely deciduous and has yellow flowers with red or pink markings. This is simply a colour variation of var. *mekongense* and should not have any more than group status. The second taxon is a group of natural hybrids. These are semi-deciduous plants with creamy or pink flowers derived from *R. racemosum* crossed with *R. mekongense* or *R. trichocladum*. We have observed these hybrids in the wild on the Cangshan and near Weixhi, Yunnan, and such hybrids have also appeared as rogues in seed of *R. racemosum*. Introduced 1917, reintroduced 1981 (hybrid). April-June.

Picture 1

Picture 2

Picture 3

1. A hillside of *R. mekongense* var. *mekongense* above Londre, Salween-Mekong divide, N.W. Yunnan, China. This hillside is heavily grazed, preventing the re-establishment of forest.
2. *R. mekongense* var. *mekongense* near Zhuzipo, Salween-Mekong divide, N.W. Yunnan, China. Deep yellow-flowered forms are uniform throughout this area.
3. *R. mekongense* (formerly var. *melinanthum*) Forrest 13900, Royal Botanic Garden, Edinburgh.

TRICHOCLADA

R. trichocladum is quite showy if planted in a group, and it contrasts well with purple-blue lepidotes. Of moderate garden value, horticulturally it is more or less identical to *R. mekongense* var. *mekongense*. Introduced 1924. April-May.

Picture 4

Picture 5

4. Deciduous *R. mekongense* var. *rubrolineatum* on the south side of the Doshong La, S.E. Tibet. Most plants seen in this area had some degree of red-flushing on the corolla but grew amongst pure yellow forms referable to var. *mekongense*.
5. *R. mekongense* var. *rubrolineatum*, Royal Botanic Garden, Edinburgh. This is a natural hybrid of *R. mekongense/R. trichocladum* x *R. racemosum*.

Picture 1

Picture 2

R. trichocladum

Franch. 1886 (*R. lithophilum* Balf.f. & Kingdon-Ward, *R. lophogynum* Balf.f. & Forrest, *R. oulotrichum* Balf.f. & Forrest, *R. xanthinum* Balf.f. & W.W. Sm. *R. mekongense* var. *longipilosum* Cullen.) H(3-)4-5

Height to 1.8m, habit loose, upright. Branchlets usually bristly. **Leaves** deciduous, obovate to obovate-elliptic; upper surface with *bristles and/or soft hairs*, lower surface with +/- *dense, twisted bristles,* scales +/- *equal-sized, distant*. **Inflorescence** 1-5-flowered. Flowers before leaves. **Corolla** funnel-campanulate, 1.5-2.3cm long, yellow to greenish-yellow; calyx 1-5mm, *lepidote and bristly*. **Distribution** Upper Burma, Central & W. Yunnan, 2,000-4,300m (6,500-14,000ft), rocks, scrub and forest margins.

See under *R. mekongense* for a detailed discussion of the differences between it and *R. trichocladum*.

Picture 3

1. *R. trichocladum*, on the east flank of the Cangshan, C.W. Yunnan, China, with the Dali plain in the background. There is a *Piptanthus* sp. on the left with bright yellow flowers.
2. A fine specimen of *R. trichocladum* SBEC 0750, Huadianba, Cangshan, C.W. Yunnan, China.
3. *R. trichocladum* SBEC 0504, Cangshan, C.W. Yunnan, China.

TRICHOCLADA

Picture 4

4. The late-flowering *R. trichocladum* Kingdon-Ward 21079 with glaucous leaves. Glendoick.

Picture 1

Picture 2

1. *R. viridescens* with evergreen leaves and yellow flowers with greenish spots
2. *R. viridescens* Rubroluteum Group with evergreen leaves and flowers tinged red with reddish spots.

R. viridescens

Hutch. 1933 (*R. rubroluteum* Davidian) H4-5

Height to 1.5m, a broadly upright, compact shrub. Branchlets bristly. **Leaves** *evergreen*. 2.3-6.7 x 1.3-3cm, obovate; upper surface glaucous when young, lower surface pale green, *usually not bristly,* scales unequal. **Inflorescence** 3-6-flowered. **Corolla** funnel-campanulate, 1.6-2.4cm long, pale yellow with greenish spots. (Rubroluteum Group creamy-yellow, tinged or flashed with red or pink and reddish spots); pedicel usually bristly; calyx 1-2mm. **Distribution** S.E. Tibet, 3,000-3,400m (10,000-11,000ft), in boggy areas and alpine slopes.

R. viridescens has been returned to specific status as extensive wild populations of it have been observed in S.E. Tibet. Although *R. mekongense* occurs in the same area, the two species are always easily distinguishable as *R. viridescens* is evergreen rather than deciduous, much later flowering and lower growing. *R. viridescens* differs from *R. lepidostylum* in its more upright habit with less glaucous, less hairy leaves, and from *R. caesium* in its less glaucous leaf underside, its more bristly branchlets and its greater hardiness. Rubroluteum Group covers forms with cream-coloured flowers with red or pink markings on the corolla.

This species is well worth growing for its late flowers and attractive glaucous young foliage rivalling *R. lepidostylum*. The flowers tend to get lost in the emerging new growth. Introduced 1924, reintroduced 1996. June-August.

Subsection Triflora

Height *0.2-10m+,* low, mound-forming, spreading to upright or umbrella-shaped, sometimes straggly shrubs. Branchlets usually lepidote. **Leaves** usually evergreen, sometimes semi-deciduous, occasionally +/- deciduous, to 12(-15) x 5cm, lanceolate through oblong to almost orbicular; lower surface lepidote, scales entire, overlapping to 8 x their diameter apart. **Inflorescence** terminal or terminal and axillary (in upper few leaf axils), 1-15-flowered. **Corolla** *usually widely/openly funnel-shaped, zygomorphic*, white, pink, yellow, purple, lavender-blue, violet and reddish-purple, spotted or unspotted; calyx usually minute; stamens 10, long and protruding; ovary densely lepidote; style elepidote. **Distribution** Nepal eastwards through Bhutan, N.E. India, S.E. Tibet, through Yunnan and Burma to Sichuan, Hubei, Shaanxi and Guizhou, Japan (one species).

TRIFLORA

Apart from the Japanese *R. keiskei*, these are relatively tall-growing. The widely funnel-shaped, zygomorphic corolla is the key identifying feature of this subsection. Most species are plentiful in the wild, and they often come from relatively dry areas, which allows them to tolerate poor, dry or heavy soils in cultivation better than most larger-leaved species. Very variable in hardiness, those from Sichuan tend to be the hardiest but even these can suffer bark-split when early growth is frosted. The more straggly species can be pruned quite hard to maintain a compact shape. The best forms of some of the species in this subsection are amongst the most free-flowering and spectacular of the larger growing species.

The species commonest in cultivation are *R. ambiguum, R. augustinii, R. concinnum, R. davidsonianum, R. keiskei, R. lutescens, R. oreotrephes, R. rigidum* and *R. yunnanense*. Less common are *R. polylepis, R. searsiae, R. siderophyllum, R. trichanthum, R. triflorum* and *R. zaleucum*.

Subsection Triflora can be divided into two fairly distinct alliances: the purple and yellow species such as *R. augustinii, R. concinnum,* and *R. ambiguum* form one group while the the pink and white ones such as *R. yunnanense* and *R. davidsonianum* form the other. *R. oreotrephes* is considered on its own, as it is somwhat aberrant in this subsection and seems to be related to subsection Cinnabarina.

1. Augustinii/Triflorum Alliance

Corolla white, pink, blue, purple, lavender, yellow.

R. ambiguum

Hemsl. 1911 (*R. chiengshienianum* Fang.) H5

Height 0.5-5.7m, *a usually dense and fairly compact* shrub. Bark *rather rough*. **Leaves** 2.2.3-8 x 1.2-3.2cm, lanceolate, oblong-lanceolate, ovate-lanceolate to elliptic; upper surface lepidote, usually darkish green, midrib *usually pubescent for part of its length*, lower surface glaucous, with dark brown, unequal scales, contiguous to their own diameter apart. **Inflorescence** usually terminal, 2-7-flowered. **Corolla** widely funnel-shaped, 2-2.6cm long, *greenish to pale yellow*, lepidote outside; ovary lepidote; style usually glabrous. **Distribution** W. Sichuan, 2,600-4,500m (8,500-14,000ft), thickets, rocks and forest.

This species differs from *R. triflorum* in its more compact and dense habit, the pubescent midrib on the upper leaf surface and the dense, unequal, brown scales on the leaf lower surface. *R. keiskei* is usually lower growing than *R. ambiguum* and has a different leaf shape. *R. ambiguum* merges with the purple *R. concinnum* and is often indistinguishable from it out of flower. Some seedlings from recent wild seed are a muddy pinkish-yellow intermediate/hybrid between the two species.

R. ambiguum is one of the hardiest subsection Triflora species and makes a neater, more compact bush than most of its relatives. It is not usually as showy as *R. lutescens* but is later flowering and considerably hardier. Introduced 1904-10, reintroduced 1980->. April-May

Picture 1

Picture 2

1. A selected clone of *R. ambiguum* K. Rushforth 195 from Emei Shan, W. central Sichuan, China.
2. *R. ambiguum*, a so-called dwarf form which is only marginally slower growing than average for the species. Glendoick.

R. wongii Hemsl. & E.H. Wilson 1910 is almost certainly a hybrid of *R. ambiguum* x *R. flavidum*. We have seen this 'species' growing amongst the putative parents in Sichuan, China. It is similar to *R. ambiguum* with smaller leaves and is lower growing.

Picture 3

Picture 4

3. *R. ambiguum* K. Rushforth 139 a hybrid/intermediate form with *R. concinnum*.
4. The glaucous leaf lower surface of *R. ambiguum* with characteristic, unequal scales, touching to their own diameter apart. (Photo R.B.G., E.)

R. augustinii
Hemsl. 1889 H3-4

Height 1-7m, a usually erect shrub. Branchlets often hairy, young foliage *usually covered in downy hairs*, often tinted red. **Leaves** 3.3-12 x 1.1-4.5cm, lanceolate to oblong and somewhat obovate; lower surface with unequal, distant, golden or brown scales, *midrib usually hairy for a long part of its length*; petiole sometimes fringed with long hairs. **Inflorescence** 2-6-flowered. **Corolla** widely funnel-shaped, 2-4.3cm long, pale or blue, deep or pale lavender-blue to mauve or violet-mauve, purplish-red, pink or white, blotched or spotted olive green, purple or ochre; calyx usually fringed with long hairs; stamens of various colours, pubescent at base.

The most useful diagnostic feature of *R. augustinii* and all its subspecies is the hairs on the leaf-underside midrib. This distinguishes it from *R. concinnum* and other subsection Triflora species with the exception of *R. trichanthum* which is hairy over the whole leaf surface and in most other parts.

R. augustinii is one of the most popular species for gardens in areas with a moderate climate. The flowers in various shades and depths of blue to lavender-blue with green and ochre 'eyes' and different colours of stamens can give a range of effects. Many popular clones have been bred by the selective crossing of wild forms and we find that paler forms tend to be hardier. The foliage sometimes suffers from leaf spot.

Ssp. *augustinii* (*R. vilmorinianum* Balf.f. type only)

Leaf upper surface with hairs on most of the veins; petiole *fringed with hairs*. **Corolla** usually blue to purple, also pink, lavender and white, tube *lepidote*. Usually earlier flowering than ssp. *chasmanthum*. **Distribution** Hubei and Central, N.W., S. Central and E. Sichuan (not Yunnan) 1,200-3,400m (4,000-11,000ft), forest margins and especially in open, rocky situations.

Picture 1

Picture 2

1. A floriferous specimen of *R. augustinii* ssp. *augustinii* in June Sinclair's garden, Hood Canal, Washington State, W. U.S.A.
2. *R. augustinii* ssp. *augustinii*, Glendoick.

This subspecies, ssp. *chasmanthum* and crosses between them such as 'Electra' have provided many fine forms with very showy blue-purple flowers. Plants under the name *R. vilmorinianum* are garden hybrids of *R. augustinii* x *R. yunnanense* and have white to pink flowers. Introduced 1890-> and reintroduced 1990s. April-May.

Picture 3

Picture 4

Picture 5

3. *R. augustinii* ssp. *augustinii* in Wolong Panda Reserve, W. Central Sichuan, China, where the quality of the flowers was as good as any cultivated selections.
4. Leaf lower surface of *R. augustinii* ssp. *augustinii* showing the characteristic hairy midrib and the unequal, distant, golden-brown scales. (Photo R.B.G., E.)
5. The popular *R. augustinii* 'Electra', Glendoick. This is a cross between ssp. *augustinii* and ssp. *chasmanthum*.

Ssp. *chasmanthum* (Diels 1912) Cullen 1978
(*R. chasmanthoides* Balf.f. & Forrest, *R. hirsuticostatum*. Hand.-Mazz.)

Leaves often wider than in ssp. *augustinii*, sometimes almost deciduous, petiole *glabrous or minutely hairy* (c.f. ssp. *augustinii*). **Corolla** very widely funnel-shaped, sometimes partly reflexed, giving a tighter truss than in ssp. *augustinii*, purple, blue, pink or white. **Distribution** Yunnan, S.E. Tibet & W. Sichuan to the south and west of locations for ssp. *augustinii*.

The flowers of ssp. *chasmanthum* are often just as fine as those of ssp. *augustinii* and it is useful for its typically later flowering. Pure white and pale pink forms with yellowish-green spotting are common around Weixi; these may be referable to ssp. *hardyi*. Introduced 1919->, reintroduced 1988->.

Picture 6

Picture 7

Picture 8

6. *R. augustinii* ssp. *chasmanthum* showing the characteristic very widely funnel-shaped corolla with the lobes tending to reflex. Glendoick.
7. Three forms of *R. augustinii* ssp. *chasmanthum* showing the variation in colour from near white to lavender-blue. Near Londre, Salween-Mekong divide, N.W. Yunnan, China.
8. *R. augustinii* ssp. *chasmanthum* with pale pink, heavily blotched flowers, Ta Pao Shan, N.W. Yunnan, China.

TRIFLORA

Ssp. *rubrum* (Davidian 1963) Cullen 1978 (*R. bergii* Davidian 1976) H3-4

A low, compact shrub. **Leaves** evergreen, *dark green*; petiole with loriform (strap-like) and filiform (threadlike) hairs. **Corolla** *purplish-red, earlier* than its relatives. **Distribution** N.W. Yunnan 4,000m (13,000 ft).

This is a distinct plant, perhaps deserving specific status as *R. bergii*, although we suspect that it is a natural hybrid of *R. augustinii* ssp. *chasmanthum* x *R. rubiginosum*. Horticulturally, of limited interest and not easy to propagate. Introduced 1924. March-April.

Picture 9

Picture 10

9. A view of white-flowered *R. augustinii* ssp. *chasmanthum*, Weixi, N.W. Yunnan, China. This may be better considered a form of ssp. *hardyi*.
10. *R. augustinii* ssp. *chasmanthum* white form, Weixi, N.W. Yunnan, China.

Ssp. *hardyi* (Davidian 1974) Cullen 1978 H3-4

Leaves *deciduous* or almost deciduous (depending on clone and climate). Young foliage bronzy. **Corolla** *white or greenish white*, faintly tinged lavender, spotted yellowish-green. **Distribution** E. Tibet, N.W. Yunnan, 3,350-3,660m (11,000-12,000 ft).

Quite a showy plant but hard to propagate and so remains rare. Introduced 1949. May. Recent introductions from Weixi, collected under ssp. *chasmanthum* may be referable to this subspecies.

Picture 11

11. *R. augustinii* ssp. *hardyi*, Royal Botanic Garden, Edinburgh. The flowers of this clone have a particularly bold blotch.

Picture 12

12. *R. augustinii* ssp. *rubrum*, Glendoick, with the typical purplish-red flowers.

R. concinnum
Hemsl. 1910 (*R. apiculatum* Rehder & E.H. Wilson, *R. atroviride* Dunn, *R. benthamianum* Hemsl., *R. coombense* Hemsl., *R. hutchinsonianum* W.P. Fang, *R. laetevirens* (Balf.f. ex) Hutch., *R. pseudoyanthinum* Balf.f. ex Hutch., *R. subcoombense* Balf.f., *R. yanthinum* Bur. & Franch.) H4-5

Height 0.6-4.5m, a fairly compact shrub. **Leaves** 2.5-8.5 x 1.2-3.5cm, oblong-lanceolate, oblong-elliptic, ovate to ovate-lanceolate; upper surface lepidote, midrib puberulent, lower surface *densely lepidote*, grey or brownish with *large, unequal scales*, yellowish-brown, half their own diameter apart to contiguous (rarely wider apart); petiole densely lepidote. **Inflorescence** terminal or terminal and axillary, 2-5-flowered. **Corolla** widely funnel-shaped, 2-3cm long, *deep to reddish purple to near ruby red* (clones usually seen in cultivation), also lavender pink, mauve-pink, white spotted brown, with or without green or crimson spots, lepidote outside; style glabrous or with downy hairs. **Distribution** W. and Central Sichuan, Hubei, Shaanxi, Gansu, Henan, 1,500-4,400m (5,000-14,500 ft), in woodland/forest and cliffs.

This species is characterised by the near ovate or ovate-lanceolate leaf with a densely lepidote lower surface, and by the usually purple or reddish-purple flowers. The absence of hairs on the leaf and petiole separates this species from *R. augustinii* and *R. trichanthum*. Only the flower colour separates it from the yellow *R. ambiguum* with which it crosses/merges.

R. concinnum has a wide distribution and is very variable which accounts for all the synonyms. Horticulturally some of the former species are retained as groups. Pseudoyanthinum Group contains many of the deepest-coloured, red and reddish-purple forms. Benthamianum Group tend to be paler, of a more violet/lavender colour, usually with pale leaves. *R. concinnum* is one of the hardiest of the subsection, so is useful in severe climates. Introduced 1904-7->, reintroduced 1990->. April-May

R. amesiae Rehder & E.H. Wilson 1913 H4-5? Differing from *R. concinnum* in its bristly petiole, *R. amesiae* is either an extreme form of *R. concinuum* or a natural hybrid with *R. trichanthum*. It is not of any particular horticultural merit. **Distribution** W. Sichuan 2,300-3,000m (7,500-10,000 ft).

Picture 1

Picture 2

Picture 3

Picture 4

1. *R. concinnum* at Rilong, N.W. Sichuan, China.
2. A hillside covered with *R. concinnum* growing with *R. vernicosum* at Rilong, N.W. Sichuan, China.
3. *R. concinnum* Benthamianum Group, a deep purple form; most forms under this name have paler-coloured flowers. Glendoick.
4. A reddish-flowered clone of *R. concinnum* Pseudoyanthinum Group raised at Tower Court, S. England, growing at Glendoick.

Picture 5

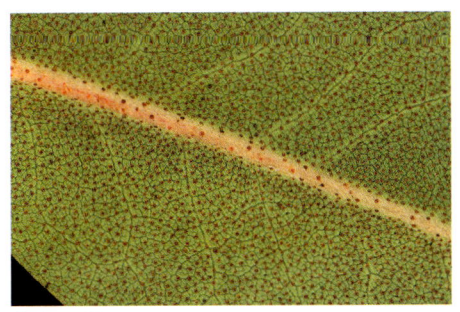

Picture 6

Picture 7

5. *R. concinnum* Pseudoyanthinum 'Chief Paulina' raised by Del James in Eugene, Oregon, U.S.A.
6. Leaf lower surface of *R. concinnum* showing the large scales which are usually more densely distributed than in this example. The densely lepidote lower leaf surface is an identifying feature of this species. (Photo R.B.G. E.)
7. *R. amesiae*, essentially just a form or hybrid of *R. concinnum* with bristly petioles. (Photo R.B.G., E.)

R. keiskei Miquel 1866, (*R. laticostum* Ingram, *R. trichocalyx* Ingram) H4-5.

Height 0.05-3m, a very variable shrub, prostrate and creeping to tall and leggy. New growth often reddish. **Leaves** 2.5-7.5 x 0.8-2.8cm, lanceolate to oblong-lanceolate to narrowly elliptic; upper surface olive to medium to dark green, *partly puberulent*, lower surface with *medium to large, brown scales*. **Inflorescence** 2-6-flowered. **Corolla** widely funnel-shaped, 1.8-2.8cm long, *pale to lemon yellow*, unspotted; calyx usually small; style slender. **Distribution** Japan, from Yakushima northwards to Central Japan, 600-1,850m (2,000-6,000ft), on hills and rocks, occasionally as an epiphyte.

This is the only Japanese species in subsection Triflora. In cultivation (apart from the rare tall form), *R. keiskei* is by far the dwarfest member of the subsection. *R. keiskei* also differs from the other yellow members of subsection Triflora in the pubescence on the upper leaf surface and in the relatively large, brown scales on the lower leaf surface.

Fairly common in cultivation, especially in E. North America, *R. keiskei* is probably the hardiest member of the subsection. The so-called 'dwarf' forms grow to 0.5-1m with often rather thinly-textured flowers. The clone 'Ebino' is one of the best with particularly red new growth. Cordifolia Group (var. *prostratum*), often grown as the clone 'Yaku Fairy' is the most dwarf form, forming a dense mound, seldom attaining 30cm, with fine flowers. The leaves occasionally turn black in cold wet winters. The tall forms, although common in the wild, are rare in cultivation and are less hardy than the dwarf and alpine forms. We also have an upright-growing, epiphytic form which grows to c.1m. Introduced 1908->, Cordifolia Group 1960s. March-May.

Picture 1

Picture 2

1. *R. keiskei* a pale-flowered form of medium height at the Royal Botanic Garden, Edinburgh.
2. The striking reddish-purple new growth of the 'Ebino' form of *R. keiskei*; most forms have less spectacular reddish or bronzy colouring on the young leaves.

TRIFLORA

Picture 3

Picture 4

3. An epiphytic form of *R. keiskei* collected on Yakushima Island, Japan by Warren Berg, and growing at Glendoick.
4. *R. keiskei* 'Yaku Fairy' A.M. collected on Yakushima Island, Japan. There may be two very similar clones in commerce under this name.

R. lutescens

Franch. 1886 (*R. blinii* Lév., *R. costulatum* Franch., *R. lemeei* Lév.) H3-4

Height 0.9-6m, habit a usually upright and willowy shrub. Young growth usually *bronzy-red*. **Leaves** 4.8-9.3 x 1.3-3.7cm, lanceolate, oblong-lanceolate to ovate lanceolate, apex *acutely acuminate*; upper surface +/- lepidote, lower surface with broadly-rimmed scales, yellowish or brown, 1/2-5 x their own diameter apart. **Inflorescence** mostly axillary, 1-3-flowered. **Corolla** widely funnel-shaped, 1.3-2.6cm long, 2-4.5cm across, *pale to deeper primrose-yellow* with green spots, lepidote and *pubescent* outside. **Distribution** W. Sichuan, N.E. Yunnan, also Hubei and Guizhou, 550-3,200m (1,750-10,500ft), in thickets in the open or amongst trees.

The long, reddish/bronze-tinged leaves usually make this one of the easiest subsection Triflora species to recognise. Its flowers are characterised by the pubescent corolla. It is taller and less dense-growing than *R. ambiguum*, lacks the glaucous leaf underside of *R. triflorum* and is taller growing than *R. keiskei* which can also have reddish-bronze leaf colouring.

R. lutescens is a showy and distinctive species, quite variable in hardiness. The selected clones have much larger-than-average flowers but are relatively tender. We find the F.C.C. clone more tender than 'Bagshot Sands'. Smaller-flowered forms from higher altitudes should prove to be hardier. Introduced 1904->, reintroduced 1990->. February-April.

Picture 1

Picture 2

1. *R. lutescens* 'Bagshot Sands' A.M. from Tower Court, S. England, a form with particularly large flowers. Glendoick.
2. *R. lutescens* 'Bagshot Sands' A.M. at Glendoick.

Picture 3

Picture 4

3. The large-flowered but rather tender *R. lutescens* F.C.C. clone from Exbury, S. England, growing at Glendoick.
4. *R. lutescens* at Glenarn, W. Scotland. The flowers of this form are small compared to those of the two award clones. This is the typical flower size of this species in the wild.

R. polylepis
Franch. 1886 (*R. harrovianum* Hemsl.) H3-4

Height 0.9-5m, a loose, broadly upright shrub. Branchlets with dense flaky scales. **Leaves** 4.5-10.2 x 1.2-3.7cm, oblong-lanceolate, lanceolate to oblanceolate; upper surface dark green, lower surface with *large, unequal, flaky, brown* scales, overlapping to contiguous, rarely wider apart. **Inflorescence** terminal or rarely axillary and terminal, 3-5-flowered. **Corolla** widely funnel-shaped, 2.1-3.5cm long, *usually pale to deep purple* to purplish-violet or rosy-mauve, +/- yellow spots; calyx minute; ovary lepidote; style glabrous. **Distribution** W. Sichuan, 2,000-3,000m (6,500-10,000ft), common in woodland, thickets and cliffs.

The dry and flaky scales on the branchlets and leaves characterise this species. The related *R. concinnum* generally has a more ovate or elliptic leaf than *R. polylepis*, while *R. searsiae* has a bluish-green leaf underside, densely covered with three different types of scales.

R. polylepis is rare in cultivation and only of limited merit horticulturally, although selected forms in purple colours can be quite striking. Introduced 1904->, reintroduced 1986->. April-May.

Picture 1

Picture 2

1. *R. polylepis*, Wolong Panda Reserve, Central W. Sichuan, China.
2. *R. polylepis* Cox, Hutchison & Maxwell MacDonald 2500 from Wolong, at Glendoick.

R. searsiae
Rehder & E.H. Wilson 1913 H4-5

Height 2.5-5m, a somewhat compact to upright and spreading shrub. **Leaves** 2.5-8 x 1-2.6cm, *lanceolate, oblanceolate* to rarely oblong-elliptic, apex acuminate; upper surface lepidote, lower surface *bluish-glaucous* with *three sizes* of scales: a. small, milky and golden, b. medium, milky and golden or brown, c. large and golden or brown, $1/2$ to rarely their own diameter apart. **Inflorescence** terminal or terminal and axillary, 3-8-flowered. **Corolla** widely funnel-shaped, 2-

3.4cm long, white to pale rose-purple, spotted light green, *rarely lepidote* outside. **Distribution** W. Sichuan (Wa Shan) 2,300-3,000m (7,500-10,000ft), thickets and woods.

R. searsiae is quite a distinct species, characterised by the elepidote corolla, the pointed leaves, longer and narrower than most in the subsection, and the conspicuous scales in three sizes on the glaucous lower leaf surface.

R. searsiae is a free-flowering and easily-grown species, hardier than most of its relatives but not particularly showy or distinctive. Introduced 1908. April-May.

Picture 1

Picture 2

Picture 3

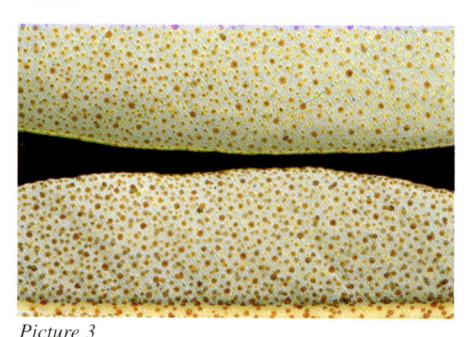

1. *R. searsiae* at the Asian Garden, University of British Columbia, Vancouver, Canada.
2. *R. searsiae* from Corrour, C. Scotland, growing at Glendoick, showing the narrow, often recurved leaves.
3. The leaf of *R. searsiae* showing the glaucous lower surface with the different types of yellowish and brown scales. (Photo R.B.G., E.)

R. trichanthum

Rehder 1945 (*R. villosum* Hemsl. & E.H. Wilson not Roth) H4

Height 1-6m, a broadly upright shrub. Branchlets *bristly*. Young growth with dense bristles. **Leaves** 4-11 x 1.5-3.7cm, oblong-lanceolate, lanceolate to ovate-lanceolate, apex acuminate; upper surface lepidote, bristly, lower surface often bristly, with unequal scales, 1-4 x their own diameter apart; petiole *bristly* and often pubescent. **Inflorescence** terminal, 3-5-flowered. **Corolla** widely funnel-shaped, 2.8-3.8cm long, variable in colour; selected cultivated forms dark plum-, bluish- or reddish-purple, also pale purple to bluish-mauve, sometimes with a green or brown flare, lepidote outside, tube *bristly*; pedicel *bristly*; calyx and ovary *bristly*. **Distribution** W. Sichuan 1,600-3,150m (5,250-10,350ft), common in woodlands, often forming dense thickets.

The key feature of this species is the dense covering of hairs/bristles on the leaf, petiole, pedicel, flower etc. which separates it from all other species in this subsection.

Selection of the best forms is desirable in this very variable species and both dark and paler forms can be attractive. One of its values is that it is amongst the latest to flower of subsection Triflora. The foliage and new growth are amongst

Picture 1

Picture 2

1. *R. trichanthum*, a dark-flowered form at the Royal Botanic Garden, Edinburgh.
2. *R. trichanthum* Wilson 1342, dark-flowered form at the Royal Botanic Garden, Edinburgh, showing the characteristic, bristly branchlets.

the best of the subsection and the habit is quite dense and compact but it needs good drainage to avoid leaf spot. Introduced 1904-10, reintroduced 1990->. May-June.

Picture 3

Picture 5

3. A pale-flowered *R. trichanthum* at Glendoick.
4. Flowers of *R. trichanthum* showing the bristly pedicels, corolla tubes and calyces. (Photo R.B.G., E.)
5. Leaf bases of *R. trichanthum* showing the bristly margins, very bristly petioles and the hairy midrib on leaf lower surface. (Photo R.B.G., E.)

R. triflorum
Hook.f. 1849. H3-4

Height to 5(-7)m, an upright shrub, often thinly-clothed with leaves. Bark *smooth, peeling, fawn, pink to reddish-brown, sometimes non-peeling* (Tibetan Mahogani Group). **Leaves** *slightly to moderately aromatic*, 3-7.2 x 1.3-2cm, oblong-lanceolate to elliptic; upper surface *glabrous;* lower surface *glaucous to pale green*, with brown, *very small* scales, ½ to their own diameter apart. **Inflorescence** terminal, 2-4-flowered. **Corolla** almost flat to openly or widely funnel-shaped, 2-3.3cm long, 2-4.3cm across, cream to pale, greenish, or bright yellow, spotted greenish yellow or suffused reddish, pink, or bronzy (Mahogani Group), lepidote and +/- pubescent outside; calyx usually minute; ovary lepidote; style usually glabrous.

A very variable species, usually characterised by the distinctive peeling bark and slightly to moderately aromatic leaves, glabrous on the upper surface, glaucous on the lower surface, with minute scales. Plants in S.E. Tibet (including Mahogani Group) have neither peeling bark nor glaucous leaves and are more compact and hardier. These populations should be given varietal status to distinguish them from the eastern Himalayan forms.

Var. *triflorum*
(*R. triflorum* var. *mahogani* Hutch (*R. deflexum* Griff.) H3-4

Corolla *widely funnel-shaped 2.3-3.6cm across*. **Distribution** E. Nepal, Sikkim, Bhutan, S.E. Tibet, Arunachal Pradesh and Burma-Tibet frontier, 2,100-4,000m (7,000-13,000 ft), in a wide assortment of habitats.

This variety is variable in flower, bark and hardiness; it is hard to find clones which have the best in all three. For milder gardens this is an attractive species, later flowering than most of its relatives and with fine, peeling bark in the western forms. Populations in S.E. Tibet have a non-peeling bark and the flowers are usually of very mixed colours, varying from pure and unmarked yellow to cream or yellow overlaid and spotted with red; such plants can be striking but the flowers of some forms look brownish from a distance. The reddish forms were described as var. *mahogani* (now Mahogani Group) which is botanical nonsense, as the different colours grow mixed together. Introduced 1850->, reintroduced 1970s->. May-June.

Var. *bauhiniiflorum*
(Watt ex Hutch.) Cullen 1978 H3

Corolla *saucer-shaped, 3.8-4.3cm across*, greenish to clear yellow. **Distribution** Nagaland & Manipur, India, 2,400-2,900m (8,000-9,500ft).

This variety differs from var. *triflorum* in its larger, more saucer-shaped corolla. The flowers are often more showy than in var. *triflorum* but the peeling bark is not as good and it is generally less hardy. Introduced 1928. Apparently the name *bauhiniiflorum* is invalid so this variety will have to be reduced to group status.

TRIFLORA

Picture 1

Picture 2

Picture 3

Picture 4

Picture 4

Picture 6

5. *R. triflorum* var. *triflorum* Mahogani Group, in the Rong Chu Valley, S.E. Tibet. The corolla colour and markings can vary considerably.
6. *R. triflorum* var. *bauhiniiflorum* Kingdon-Ward 7731, Royal Botanic Garden, Edinburgh.

1. *R. triflorum* var. *triflorum* below Darjeeling, W. Bengal, India.
2. *R. triflorum* var. *triflorum* Kingdon-Ward 6409, Royal Botanic Garden, Edinburgh.
3. *R. triflorum* var. *triflorum* Mahogani Group at Glenarn, W. Scotland.
4. *R. triflorum* var. *triflorum* Mahogani Group in the Rong Chu Valley, S.E. Tibet.

R. zaleucum
Balf.f. & W.W. Sm. 1917 H3(-4)

Height 0.6-10.6m, a bushy to broadly upright shrub. Young growth often reddish or bronzy. **Leaves** *aromatic*, 3.2-8.8 x 1-3cm, usually lanceolate to oblong-lanceolate, rarely oblong, elliptic to obovate; upper surface dark green, margins +/- bristly, lower surface *very glaucous or nearly white*, especially on young leaves. Scales on lower leaf surface *unequal, large, flat, 1½-4 x their own diameter apart*. **Inflorescence** terminal or terminal and axillary, 3-5-flowered. **Corolla** widely funnel-shaped, *2.6-4.8cm long*, very variable: white, white flushed lilac, purple or rose, pale lilac, pale yellow, pink to almost salmon, with +/- crimson spotting, sometimes fragrant, *lepidote outside*. **Distribution** W. Yunnan-Burma border area, 1,800-4,000m (6,000-13,000ft), on dry, rocky slopes, in scrub, thickets or damp, coniferous or rhododendron forest.

This distinctive species is characterised by the glaucous or whitish lower leaf surface with conspicuous scales, and by the lepidote corolla.

One of the least hardy of Subsection Triflora, R. zaleucum has large and attractive flowers in many different shades and often has fine, reddish young growth. Its early growth makes it vulnerable to spring frosts. Introduced 1912->. March-May.

Var. *zaleucum* (R. erileucum Balf.f. & Forrest)

Leaves to 8.8cm long. **Corolla** *not pale yellow*.

Var. *flaviflorum* Davidian 1984

Leaves to 10cm long. Corolla *pale yellow*. **Distribution** N. Triangle, Upper Burma, 2,700m (9,000ft).

Introduced in 1953 under Kingdon-Ward 20837. There are also pink-flowered forms from this number which may be hybrids with *R. oreotrephes*.

Picture 1

Picture 2

Picture 3

Picture 4

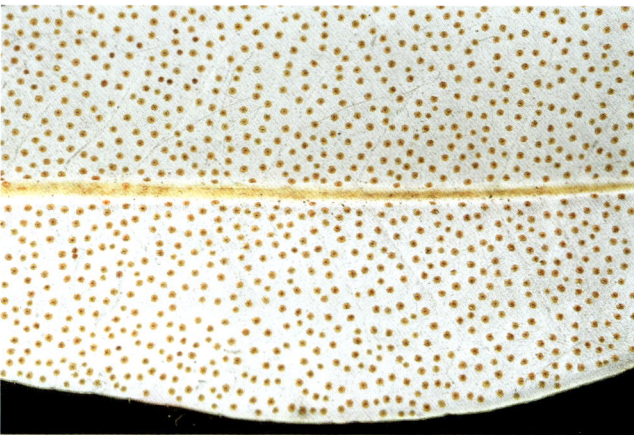

Picture 5

1. A pale pink to white-flowered clone of *R. zaleucum* var. *zaleucum* at Inverewe Gardens, N.W. Scotland.
2. A pink-flowered clone of *R. zaleucum* var. *zaleucum* at Inverewe, N.W. Scotland, showing the characteristic white, glaucous lower surface of the leaf.
3. A pink and yellow-flowered clone of *R. zaleucum* var. *zaleucum* probably Kingdon-Ward 20837 from Brodick Castle, W. Scotland, at a rhododendron show.
4. *R. zaleucum* var. *flaviflorum* Kingdon-Ward 20837 at Glenarn, W. Scotland. This variety differs from the type in its yellow flowers.
5. *R. zaleucum* showing the distinctive white glaucous leaf lower surface and the large, brown scales 1½-4 x their own diameter apart. (Photo R.B.G., E.)

TRIFLORA

2. YUNNANENSE ALLIANCE

Corolla white, pink or lavender.

These 6 species (including *R. pleistanthum*) are very closely related. Cullen refers to them as 'intergrading microspecies' as they are very variable and seem to overlap with one another both in taxonomic characters and in distribution.

R. davidsonianum
Rehder & E.H. Wilson 1913 (*R. charianthum* Hutch.) H(3-)4

Height 0.9-5m, an erect or angular, spreading shrub. Branchlets lepidote. **Leaves** evergreen, 2.3-7.8 x 0.8-2.6cm, lanceolate to oblong, *often 'V'-shaped in section,* lower surface *densely* lepidote, scales *small and brown with narrow rims and darker centres, contiguous to their own diameter apart.* **Inflorescence** terminal or terminal and axillary, 3-6(-10)-flowered. **Corolla** widely funnel-shaped, 1.9-3.3cm long, mauve to pale lavender, white tinged pink to pink, rarely white, with or without purple to red spots and/or blotch. Most selected clones are pink. **Distribution** Central & S.W. Sichuan 1,800-3,500m (6,000-11,500ft), common in open situations, conifer forest, on cliffs and stream banks.

R. davidsonianum is very closely related to several other members of the alliance but has a more northerly distribution. The leaves are more evergreen and more lepidote than *R. yunnanense* and are usually characterised by the bending up of the leaf blades, forming a 'v' in cross-section. *R. tatsienense* differs in its crimson branchlets and smaller flowers.

Very free-flowering from a young age, in its best forms *R. davidsonianum* rivals *R. yunnanense* as the showiest species in subsection Triflora. Variable in hardiness; bushes often sprout again from the base after being severely cut back. Introduced 1904->, reintroduced 1990->. April-May.

Picture 1

Picture 2

Picture 3

Picture 4

1. *R. davidsonianum*, a form with heavily-blotched flowers from Caerhays, S.W. England, at Baravalla, W. Scotland.
2. *R. davidsonianum*, another form with a heavily-blotched corolla from Caerhays, S.W. England, at Glendoick.
3. *R. davidsonianum* F.C.C. clone from Bodnant, N. Wales, at Glendoick, with only faint, dark markings.
4. *R. davidsonianum* 'Ruth Lyons' with unmarked corollas, raised by James Barto in Eugene, Oregon. We find this clone rather lacking in vigour, but the flowers are very fine.

Picture 5

5. The lower leaf surface of *R. davidsonianum* showing the unequal brown scales with darker centres. The scales are often more widely spaced than in this example. (Photo R.B.G., E.)

Picture 1

Picture 2

Picture 3

R. rigidum

Franch. 1886 (*R. caeruleum* Lév., *R. eriandrum* (Lév. ex) Hutch., *R. hesperium* Balf.f. & Forrest, *R. rarosquameum* Balf.f., *R. sycnanthum* Balf.f. & W.W. Sm.) H3-4

Height 0.6-3m in cultivation, to 10m in the wild, habit erect, compact to loose. **Leaves** evergreen, *leathery*, 2.5-6.8 x 1-3.2 cm, elliptic to oblanceolate, upper and lower surfaces *usually glaucous*, the lower surface with golden or brown, unequal scales, 2-8 x their own diameter apart. **Inflorescence** terminal or terminal and axillary, 2-6-flowered. **Corolla** widely funnel-shaped, 1.8-3.1cm long, white, through lilac-rose to to deep rose lavender, +/- spotted or marked gold, olive brown, to reddish brown, (as cultivated) often with a curious fragrance which is very attractive to insects; calyx minute, usually glabrous. **Distribution** N. Yunnan & S. Sichuan (800) -2,000-3,400m ((2,500-) 6,500-11,000 ft), in the open, forest margins, cliffs and rocks and in coniferous forest.

This species is characterised by its leathery, usually glaucous, evergreen leaves and its rigid, relatively compact habit. After fieldwork in Yunnan, it is hard to accept that this species is any more than a variety of the very variable *R. yunnanense*. *R. tatsienense* has red stems in contrast to *R. rigidum*'s green-blue stems, while *R. oreotrephes* shares the glaucous foliage but has a wider leaf and different flower shape.

Although reported in many different colours in the wild, *R. rigidum* is almost exclusively cultivated in its very showy, pure white forms. It is often a neat grower, tidier than its relatives. Introduced 1910->. March-May.

1. A pinkish-flowered *R. rigidum* Rock 11288, Royal Botanic Garden, Edinburgh.
2. *R. rigidum* Rock 59207 at Nymans, S. England, with a compact inflorescence.
3. *R. rigidum* with a loose inflorescence, growing at Glendoick.

R. siderophyllum

Franch. 1898 (*R. ioanthum* Balf.f., *R. jahandiezii* Lév., *R. leucandrum* Lév., *R. obscurum* (Franch. ex) Balf.f., *R. rubro-punctatum* Lév. & Vant., *R. sequini* Lév. (part)) H3(-4)

Height 1-3(-7)m, a rounded to upright shrub. Young growth brownish. **Leaves** 3-9 x 1.5-4cm, oblong-lanceolate, elliptic to lanceolate; upper surface usually slightly lepidote, lower surface densely lepidote, *with unequal, brown scales with dark*

TRIFLORA

centres, 1-2 x their own diameter apart; petiole densely lepidote. **Inflorescence**, terminal and axillary, *very densely crowded* 3-6-flowered. **Corolla** widely funnel-shaped, 1.5-3cm long, white to pink, rose, pale lavender-blue to purple with +/- yellow to rose spots. **Distribution** S.W. Sichuan, Guizhou, Central, N.E. & S. Yunnan, 1,800-3,400m (6,000-11,000ft) in dry woods, scrub-covered hills or thickets.

This species is closely related to *R. tatsienense, R. yunnanense* and *R. davidsonianum* but is usually distinguished by its dense inflorescence of small flowers and by the distinctive scales on the lower leaf surface. Davidian but not Cullen draws attention to its usually pale, matt, greyish-green leaves.

R. siderophyllum is generally not as showy as its relatives, and tends to be rather tender, but as it comes from relatively dry and hot areas, it could be useful for heat tolerance. Most forms in cultivation have white flowers. Introduced 1918->, reintroduced 1981->. May.

Picture 1

Picture 2

1. *R. siderophyllum*, Younger Botanic Garden, Benmore, W. Scotland.
2. *R. siderophyllum* grown from seed from near Kunming, C. Yunnan, China, showing the typical fairly dense inflorescence of small flowers. Glendoick.

Picture 3

3. *R. siderophyllum* flowering in autumn on Lao jing shan, S.E. Yunnan, China.

R. tatsienense

Franch. 1898 (*R. heishuense* Fang, *R. hypophaeum* Balf.f. & Forrest, *R. leilungense* Balf. f. & Forrest, *R. stereophyllum* Balf.f. & W.W. Sm., *R. tapelouense* Lév.) H3-4

Height 0.3-2.7m, a spreading to erect, often rather untidy shrub. Branchlets *deep red to crimson*. **Leaves** rigid and leathery, 1.6-6 x 1-3.1cm, elliptic, obovate, oval, oblong to oblong-lanceolate, apex usually rounded; upper surface usually lepidote, lower surface pale to glaucous-green, with unequal, brown scales, $1/2$ to their own diameter apart, rarely wider; petiole lepidote. **Inflorescence** terminal or terminal and axillary, 1-6-flowered. **Corolla** widely funnel-shaped, *1.4-2.3cm* long, blush-white, to rose pink to rose-lavender or purple, +/- red spots; calyx disc-like or undulate, often ciliate. **Distribution** N.W. Yunnan and W. & S.W. Sichuan, 2,100-3,700m (7,000-12,000 ft) in open areas.

This species is characterised by its small leathery leaves and red stems. It is very variable, sometimes resembling *R. racemosum* and at the other extreme *R. yunnanense* and *R. davidsonianum* and it may merge completely with these species.

Smaller flowered than its closest relatives and rarely as showy. This species may be derived from hybrids of *R. yunnanense* x *R. racemosum*. Introduced 1917->, reintroduced 1981->. March-May.

usually lepidote and ciliate. **Distribution** N.W. & N.E. Yunnan, S.W. Sichuan, Burma, S.E. Tibet, Guizhou, 900-4,300m (3,000-14,000 ft), in dry and sometimes moist situations in a huge variety of habits, where it is often the only rhododendron species.

This is a very variable species which is closely related to and/or merges with several others. It is taller growing with thinner, less glaucous, more deciduous leaves than *R. rigidum*. *R. davidsonianum* has leaves which form a 'V' shape in cross section and has more lepidote but not bristly leaves. *R. siderophyllum* has a more densely lepidote leaf underside and a more tightly packed inflorescence of small flowers. *R. yunnanense* is larger in all parts than *R. tatsienense*.

This species is one of the great sights of Yunnan province, China, where it is extremely plentiful over a wide area and great altitudinal range. One of the most free-flowering of all species, from a very young age, but it is vulnerable to bark split from late spring frosts and needs protection when young. It often benefits from pruning as it can get very straggly. *R. yunnanense* will tolerate drier sites than most rhododendrons and will flower freely even in considerable shade. Very variable in hardiness; for severer climates it is worth trying to obtain forms collected at higher altitudes, while for hot, dry climates, low-altitude forms should be sought. *R. yunnanense* now includes several former species: Hormophorum Group covers completely deciduous forms while Suberosum Group has relatively bristly leaves and usually white flowers. Introduced 1889->, reintroduced 1981->. May.

Picture 1

Picture 2

1. *R. tatsienense* Forrest 16249, Castle Howard, N.E. England, showing the small flowers resembling those of *R. racemosum*.
2. *R. tatsienense* Lancaster 951 from C.W. Sichuan, a selected clone, at Glendoick.

R. yunnanense

Franch. 1886 (*R. aechmophyllum* Balf.f. & Forrest, *R. bodinieri* Franch., *R. chartophyllum* Franch., *R. hormophorum* Balf.f. & Forrest, *R. pleistanthum* Balf. f. ex Wilding, *R. seguini* Lév. (part), *R. strictum* Lév., *R. suberosum* Balf.f. & Forrest, *R. hesperium* Balf.f. & Forrest (part)) H3-4

Height 0.9-3.6m, often taller in cultivation, an erect, often sparse shrub. **Leaves** usually rather thin, not rigid, evergreen, *semi deciduous to deciduous*, 2.5-10.4 x 0.8-2.8cm, oblanceolate to narrowly elliptic or obovate, often with variation on one plant, margins sometimes with hairs/bristles; upper surface +/- lepidote, lower surface +/- bristly, scales unequal, brown, 2-6 x their own diameter apart; petiole +/- bristly and puberulous. **Inflorescence** terminal or terminal and axillary, 1-6-flowered. **Corolla** widely funnel-shaped, 1.8-3.4cm long, white, white flushed rose-purple, rose, pinkish purple to blue-lavender, *usually marked or blotched* olive, deep green, deep rose to brownish crimson; calyx minute,

Picture 1

1. *R. yunnanense* with pink, blotched flowers at Glendoick.

TRIFLORA

Picture 2

Picture 5

Picture 6

Picture 3

Picture 4

5. *R. yunnanense* probably Forrest 10056, with white, flowers blotched yellow, collected in Yunnan from 1912-14 and growing at Glendoick.
6. A white-flowered *R. yunnanense* with only faint markings photographed on the SSNY expedition to N.W. Yunnan in 1992.

R. pleistanthum (Balf.f. ex) Wilding 1923. H3-4 This species is said to differ from *R. yunnanense* in its non-bristly leaf margins and petiole. It was made synonymous with *R. yunnanense* in 1963 but was rather inexplicably resurrected by Cullen. Recent field work has indicated that the presence and absence of bristles can even be found on different branches on the same plant and that there is very little basis for maintaining *R. pleistantum* at specific status, although it is said to have a more northern and western distribution than typical *R. yunnanense*.

2. A hillside of *R. yunnanense* near Deqin, N.W. Yunnan, China. (Photo I. Sinclair)
3. A bank of *R. yunnanense* at Tian Bao Shan, N.W. Yunnan, China.
4. A fine, pink-flowered *R. yunnanense* at Xiao Huadianba, Cangshan, Central W. Yunnan, China.

Picture 7

7. *R. pleistanthum* Forrest 15002 at Royal Botanic Garden, Edinburgh. This so-called species is simply a non-bristly-leaved form of the very variable *R. yunnanense*.

The foliage of *R. oreotrephes* looks close to that of some forms of *R. cinnabarinum*, and recent fieldwork in S.E. Tibet has revealed what appear to be intermediate forms between these two species.

This species appears to bridge subsections Triflora and Cinnabarina in its glaucous, +/- elliptic leaves and funnel-campanulate flower. Very variable and common in the wild; a wide range of colours from dark to pale can sometimes be found within one population. For foliage and habit, this is perhaps the finest species of the subsection and some forms have very showy flowers. Introduced 1910->, reintroduced 1980s->. April-May.

3. Oreotrephes Alliance
(now reduced to a single species).

R. oreotrephes

W.W. Sm. 1914 (*artosquamatum* Balf.f. & Forrest, *R. cardoeoides* Balf.f. & Forrest, *R. depile* Balf.f. & Forrest, *R. exquisetum* Hutch., *R. hypotrichum* Balf.f. & Forrest, *R. oreotrephoides* Balf.f., *R. phaeochlorum* Balf.f. & Forrest, *R. pubigerum* Balf.f. & Forrest, *R. siderophylloides* Hutch., *R. timeteum* Balf.f. & Forrest, *R. trichopodum* Balf.f. & Forrest) H4.

Height 0.6-7.6m, a rounded to broadly upright shrub. **Leaves** usually evergreen, sometimes semi-deciduous, 1.8-8.9 x 1.2-4.3cm, oblong-elliptic, elliptic, oblong, ovate to almost orbicular; upper surface usually medium to palish green young growth, often *glaucous*, lower surface *glaucous to pale glaucous green*, with +/-equal sized scales, near touching to 3-4 x their own diameter apart, *purplish, reddish-brown or greyish, opaque, narrowly rimmed*. **Inflorescence** terminal or terminal and axillary, 1-3(-10)-flowered. **Corolla** widely funnel-shaped to sometimes funnel-campanulate, 1.8-4cm long. whitish-pink, light purple through pink to rose and bright purple (yellowish and apricot reported in wild), spotted or unspotted; calyx reduced to a rim, sparsely lepidote, sometimes ciliate. **Distribution** S.E. Tibet, Sichuan, N. Yunnan and Burma, 2,700-4,300m (9,000-14,000 ft), in many different habitats.

The glaucous, +/- elliptic leaves and the distinctive scales distinguish this species from the others in subsection Triflora.

Picture 1

Picture 2

1. *R. oreotrephes* in the Rong Chu Valley, S.E. Tibet. This population has uniformly dark-coloured flowers with more funnel-shaped corollas than those found in Yunnan, and is thus more closely related to *R. cinnabarinum*. There were up to 10 flowers in each truss, more than is usual for *R. oreotrephes*.

2. *R. oreotrephes* at Glendoick showing some of the possible variation in flower colour. The dark one on the right is 'Pentland' A.M., from the Royal Botanic Garden, Edinburgh.

TRIFLORA

Picture 3

Picture 4

Picture 5

3. *R. oreotrephes* at Glendoick, a moderately deep-coloured form.
4. A compact, pale-flowered clone of *R. oreotrephes* growing in J. Ramsay's garden in Washington State, W. U.S.A.
5. Glaucous foliage of *R oreotrephes* at Glendoick.

Subsection Uniflora

Height 0.07-0.9m, *prostrate to mound-forming*, sometimes suckering low shrubs. **Leaves** +/- evergreen, to 3cm long; lower surface with scales ½-6 x their own diameter apart. **Inflorescence** 1-3-flowered, terminal. **Corolla** funnel-shaped to campanulate, *densely pilose outside*, pink, purple or yellow; calyx 1-7mm; stamens 10; ovary lepidote; style slender and straight. **Distribution** Upper Burma, N.W. Yunnan, S. & S.E. Tibet, N.E. India, Bhutan and E. Nepal.

The species in this subsection are characterised by their low habit and the densely pilose corolla. Relationships have been suggested with subsections Saluenensia, Tephropepla, Campylogyna, Lepidota and even Cinnabarina. Cullen remarks that apart from the widespread *R. pumilum*, the other species were only known from a total of 7 wild collections, and the lack of herbarium material meant that his treatment of the subsection could only be considered provisional. We have since found large populations of *R. imperator* and *R. pemakoense* in S.E. Tibet, suggesting that the earlier collectors only reached the fringes of the distributions of some of the species. We have made some alterations to Cullen's work based on recent field observation.

The species in this subsection are free-flowering but not always easy to grow. *R. pemakoense* is quite common in cultivation, the others are uncommon or very rare.

R. imperator

Kingdon-Ward 1930 (*R. patulum* Kingdon-Ward) H3-4?

Height to 0.3m, a dwarf or near prostrate shrub. **Leaves** semi-deciduous, 1.8-3.8 x 0.4-1cm, lanceolate or oblanceolate, *very pointed at both ends*; upper surface *dark and shiny*, lower surface pale green with brown scales 1-6 x their own diameter apart. **Inflorescence** 1-2-flowered. **Corolla** widely funnel-shaped, 2.3-3.2cm long, bright to dark pinkish-purple, with or without deep crimson spots, outside +/- pubescent; style long, slender and straight. **Distribution** S.E. Tibet, Arunachal Pradesh (as *R. patulum*), Upper Burma, 3,000-3,400m (10,000-11,000ft), forming mats in a shady gully where snow lingers late (Burma), on rocks and steep slopes (S.E. Tibet).

Previous to recent fieldwork in S.E. Tibet, the accepted wisdom was that *R. uniflorum* occured in S.E. Tibet and *R. imperator* in Burma. Two expeditions have revealed large populations of typical *R. imperator* on the southern slopes of the Doshong La while there was no sign of *R. uniflorum*. It is obvious that *R. imperator* is the 'true' species with quite a wide distribution, and we have therefore returned it to specific status. The narrow leaves, pointed at both ends and the large flowers make this species very distinctive. The type specimen of *R. patulum* from Arunachal Pradesh is referable to *R. imperator*, although cultivated plants under this name are usually *R. pemakoense*.

This species is very pretty in full flower. We find the original introduction a little tender; covering it with a cloche from September to May helps to stop die-back from wet and frost. It is not very vigorous and rarely forms a good specimen. Recent introductions from S.E. Tibet may be easier to cultivate. Introduced 1926, 1995, 1996. April-May.

Picture 1

Picture 2

Picture 3

1. *R. imperator* showing the deep coloured flowers and tiny pointed leaves, growing at Glendoick.
2. *R. imperator* on the south side of the Doshong La, S.E. Tibet.
3. Close-up of *R. imperator* on the southern side of the Doshong La, S.E. Tibet. The corolla of this population is noticeably larger and flatter than the original introduction from Upper Burma.

R. ludlowii
Cowan 1937 H4

Height to 0.3m, usually less, a spreading or creeping but *rarely dense* shrub. Branchlets lepidote. **Leaves** 1.2-1.6 x 0.8-1cm, broadly obovate to oblong-obovate, *margins notched*; upper surface dark green, lower surface pale green with large brown scales, 2-3 x own diameter apart. **Inflorescence** 1-2-flowered. **Corolla** *broadly campanulate*, 1.5-2.5cm long, *yellow* with reddish-brown spots; calyx *large, 5-7mm*, margin ciliate; style longer than stamens, shorter than corolla. **Distribution** S. Tibet, 4,000-4,300m (13,000-14,000ft), creeping amongst moss on open rocky hillsides amongst other Ericaceae.

This is a distinct species differing from other members of the subsection in its toothed leaves, yellow flowers and large calyx.

In cultivation, the rare *R. ludlowii* is a very slow growing and difficult species, preferring soil rich in organic matter and some shade, but even then it is not always successful. The bowl-shaped flowers are very large for the size of leaf and this trait has been passed on to several fine hybrids such as 'Curlew'. Introduced 1938. May.

Picture 1

Picture 2

1. *R. ludlowii* at Glendoick. This species often has a weak root system and dislikes fertiliser.
2. *R. ludlowii*, Ludlow, Sherriff and Taylor 6600 at R.B. Cooke's garden, N.E. England, one of the best specimens of this species ever grown in cultivation. (Photo R.B.G., Kew)

UNIFLORA

R. pemakoense
Kingdon-Ward 1930 H4

Height 0.3-0.6m, a dense, mound-forming, often *stoloniferous* shrub. Branchlets lepidote and pubescent. **Leaves** 1-3 x 0.5-1.5cm, *oblanceolate*; upper surface dark green and lepidote, lower surface glaucous, scales unequal, ¹/₂-1¹/₂ x their own diameter apart. **Inflorescence** 1-2-flowered. **Corolla** tubular-funnel-shaped, 2.4-3.5cm long, pale pinkish-purple to near pink; calyx 1-4mm, coloured; style longer than stamens, as long or longer than corolla. **Distribution** S.E. Tibet, 3,000-3,700m (10,000-12,000ft), carpeting steep moss-covered rocks and slopes.

This species is closely related to *R. Imperator*, differing in its more evergreen foliage, the rounded rather than pointed leaf apex and the tubular-funnel-shaped rather than broadly-funnel-shaped corolla.

R. pemakoense is a popular, easily grown, very free-flowering species with large flowers but the swelling buds are very prone to frost damage and it often fails to open its flowers. Davidian quite rightly states that the type specimen of *R. patulum* is almost identical to that of *R. imperator*. Cultivated plants labelled *R. patulum* are *R. pemakoense*, slightly less vigorous than the type, with a deeper-than-normal flower colour. Introduced 1924. March-May.

Picture 1

Picture 2

Picture 3

Picture 4

1. *R. pemakoense* on the Dokar (Showa) La, Pome, S.E. Tibet.
2. *R. pemakoense* at Trehane, S.W. England. It is sad that the opening buds are easily frosted and such a fine show of flower is seldom seen.
3. *R. pemakoense* showing the broadly funnel-campanulate flowers which differentiate this species from *R. uniflorum* and *R. imperator*.
4. *R. pemakoense* so-called 'Patulum', at the Royal Botanic Garden, Edinburgh. Plants cultivated under the name 'Patulum' are referable to *R. pemakoense* while the type specimen is a form of *R. imperator*.

R. pumilum
Hook.f. 1849 H3-4

Height 0.05-0.6m, a prostrate to erect, rather open shrub. Young branchlets lepidote and pubescent. **Leaves** evergreen, sometimes semi-evergreen, 0.8-1.9 x 0.4-1.1cm, elliptic to obovate to oval; upper surface dark green, rather shiny, lower surface slightly glaucous, scales 1-3 x their own diameter apart. **Inflorescence** 1-3-flowered. **Corolla** *campanulate*, 0.8-2.1cm long, pale pink through rose to purple; calyx 1-3mm, coloured; style *longer or equalling stamens, about ¹/₂ as long as corolla*. **Distribution** E. Nepal, through Sikkim, N.E.

Bhutan to Arunachal Pradesh, S.E. Tibet and Upper Burma, 3,400-4,300m (11,000-14,000ft), open hillsides, mossy rocks, screes and grass.

This species differs from its relatives in its sparser habit, small leaves, small pink, campanulate corolla and in the style half as long as the corolla. In cultivation it tends to over-flower and it can often be identified by the masses of seed capsules held on long pedicels which seem to inhibit the new growth. It is usually self-pollinating.

Picture 3

3. *R. pumilum* McBeath 1120 collected in E. Nepal and growing at Glendoick.

Picture 1

Picture 2

1. *R. pumilum* on the south side of the Doshong La, S.E. Tibet.
2. *R. pumilum* in E. Nepal. (Photo R. McBeath)

R. pumilum has pretty little thimble-like flowers on a rather sparse plant. Not the easiest species to satisfy, being liable to suffer winter damage and die-back and, like *R. ludlowii*, it requires plentiful organic matter. Plants labelled *R. pumilum* are often wrongly named. Probably introduced 1850 but the oldest plants in cultivation are likely to be from 1924->. Several new introductions since 1950. April-May.

R. uniflorum
Kingdon-Ward 1930 H(3-)4

Height to 1m. **Leaves** often semi-deciduous, 1.3-2.4 x 0.6-1.2cm, *oblong-oval, oval, obovate or oblong-obovate*, underside glaucous, scales 3-6 x their own diameter apart. **Inflorescence** 1-3-flowered. **Corolla** *widely funnel-shaped*, 2.2-2.8cm long, mauve pink to bright to pinkish-purple; calyx 1-2mm; style about 2/3 length of corolla. **Distribution** Only found on Doshong La, S.E. Tibet, 3,400-3,700m (11,000-12,000ft), moraines and gravel.

Picture 1

1. *R. uniflorum* Kingdon-Ward 5876 from the Doshong La, S.E. Tibet, growing at Glendoick.

R. uniflorum appears to be simply an extreme form or hybrid of *R. imperator* and should be reduced to synonymy with it. The cultivated material does not match the one poor herbarium specimen. *R. uniflorum* differs from *R. imperator* in the wider leaves, less pointed at both ends. It differs from the closely related *R. pemakoense* in its more pointed leaves, its broadly funnel-shaped corolla and its style only 2/3 the length of the corolla.

UNIFLORA

While it is as free-flowering as *R. pemakoense*, *R. uniflorum* often has inferior foliage. Introduced 1924. April-May.

Picture 2

2. The leaves of *R. uniflorum* as cultivated, wider and less pointed than those of *R. imperator*

Subsection Virgata

A monotypic subsection.

R. virgatum
Hook.f. 1849 H2-3

Height 0.3-2.4m, habit usually ungainly and sprawling with long arching branches. Terminal bud *always vegetative*. **Leaves** 1.8-8 x 0.5-2cm, recurved, lanceolate to oblong-oval; upper surface dark green, lower surface pale with unequal, brown scales, ½ (-2-4) x their own diameter apart. **Inflorescence** 1(-2)-flowered. Flowers *always axillary in upper 1-12 leaves*. **Corolla** funnel- to tubular-funnel-shaped, 2.5-3.9cm long, pale to deep pink to mauve, rarely white; calyx 0.5-3mm with 5 rounded lobes; stamens 10; ovary densely lepidote, glabrous; style straight, longer than stamens. Seeds *tailed*.

This distinct but variable species has flowers from axillary inflorescences only. It is also characterised by the funnel-shaped corolla, pubescent outside, and the lobed calyx. The subsection Scabrifolia species have more flowers to the inflorescence, the style and stamens are exserted from the corolla and the upper leaf surface is pubescent and ciliate.

R. virgatum is variable in hardiness and even hardier forms are too tender to succeed permanently outdoors in most of the British Isles and equivalent climates. Easily grown and floriferous where successful, enjoying fairly dry conditions. The flowers are often spaced out along the branchlets. Introduced 1850->. March-May.

Ssp. *virgatum*

Corolla *2.5-3.9cm long*, often deeper-coloured than in ssp. *oleifolium*. **Distribution** E. Nepal through Bhutan, Arunachal Pradesh to S.E. Tibet, 2,400-3,800m (8,000-12,500ft), forest margins, scrub and dry banks.

The white forms of this subspecies are particularly fine.

Ssp. *oleifolium*
(Franch.) Cullen 1978 (*R. sinovirgatum* Hort.)

Corolla *1.5-2.5cm* long, often paler-coloured than in ssp. *virgatum*. **Distribution** Widely distributed in N.W. to S.W. Yunnan and S.E. Tibet, 2,000-4,000m (6,500-13,000ft) (highest altitude probably exaggerated), in warm, dry sites.

Picture 1

Picture 2

1. *R. virgatum* ssp. *virgatum*, a form from Reuben Hatch growing in the garden of R. Shumm. (Photo R. Shumm)
2. *R. virgatum* ssp. *virgatum* in C. Bhutan. (Photo A.D. Schilling)

VIREYA

Section Vireya

Subsection Pseudovireya

Pseudovireya Alliance
(formerly Vaccinioides series)

Height to 1.5m, compact to spreading, usually epiphytic shrubs. Branchlets *usually scabrid* (warty). **Leaves** evergreen, rather thick and fleshy, to 5 x 2.4cm, usually in whorls at ends of shoots; lower surface lepidote. **Inflorescence** 1-7-flowered. **Corolla** campanulate, lepidote outside, small, yellow, lilac-pink to white, waxy; calyx 0.5-4mm; stamens 10; ovary lepidote; style short, consistently not tapered into ovary. Capsule opens from top downwards, seeds with *long tails at each end.* **Distribution** Nepal, Bhutan, Sikkim, Arunachal Pradesh to Upper Burma, N.W. to S.W. Yunnan, S.E. Tibet, Guizhou, Taiwan and N. Vietnam.

Section Vireya contains around 300 species which grow in the tropical regions of S.E. Asia such as New Guinea and Borneo. The species in this section are characterised by their seeds which have tails or fins on them. The species in the former Vaccinioides series are characterised by the small flowers of thick, fleshy texture, the small leaves in whorls at the ends of the warty branchlets and the long-tailed seeds.

These species have recently been reclassified in a subsection of section Vireya so strictly are outwith the confines of this book. We have included the species that were formerly in the Vaccinioides Series because, unlike the rest of section Vireya, their distribution extends well into the Sino-Himalayan region and some are hardy enough to grow outdoors in milder parts of the British Isles and equivalent climates.

These species vary in hardiness and garden value and, as they are mainly epiphytic, are best grown outdoors on mossy rocks and logs and in very well-drained planting medium indoors. All are rare in cultivation.

Picture 3

Picture 4

Picture 5

3. *R. virgatum* ssp. *virgatum,* a fine white-flowered form from Gigha, W. Scotland, only suitable for the mildest British gardens.
4. *R. virgatum* ssp. *oleifolium* SBEC 0621 on the Cangshan, Yunnan, China, growing on a fairly dry roadside bank.
5. *R. virgatum* ssp. *oleifolium* SBEC 1227 on the Cangshan, Yunnan, China.

R. euonymifolium
Lév. 1913, (*R. poilanei* Dop) H1-2

Height to 0.6m, a small shrub with spreading branchlets. **Leaves** 3-4 x 1.5-2cm, notched, obovate to oblong-obovate; lower surface with small, unequal, brown scales, 2-3 x their own diameter apart. **Inflorescence** *1-2*-flowered. **Corolla** campanulate, c.1.3cm long, yellow; calyx *minute*; stamens shorter than corolla; ovary *glabrous;* style longer than stamens, shorter than corolla. **Distribution** S.W. Yunnan, N. Vietnam, 2,000m (6,500ft).

This species differs from *R. kawakamii* in its 1-2-flowered rather than 3-7-flowered inflorescence, its smaller calyx and the glabrous ovary. Chinese botanists have named several other species which are very similar to this one.

This species has been recently introduced from N. Vietnam. It may be hardy enough for cultivation outside in the U.K. or similar climates. Introduced 1991-2. The very closely related *R. emarginatum* may also be in cultivation.

Picture 1

Picture 2

1. *R. euonymifolium* in N. Vietnam. (Photo K. Rushforth).
2. *R. euonymifolium* in N. Vietnam. (Photo K. Rushforth).

R. kawakamii
Hayata 1911, (*R. taiwanianum* S.S. Ying, *R. kawakamii* var. *flaviflorum* Liu & Chuang) H2-3

Height 1-1.5m, a bushy or straggly shrub. Branchlets *not scabrid.* **Leaves** 2-5 x 0.9-2.4cm, thick and leathery, oblong-obovate to obovate; upper surface bright green, lower surface yellowish-green with small, unequal, brown scales, 2-3 x their own diameter apart. **Inflorescence** *3-7*-flowered, loose, pedicel long, to 2.5cm. **Corolla** campanulate, 1-1.2cm long, bright yellow (white), lepidote outside; calyx 1-2mm; stamens as long or shorter than corolla; ovary *densely hairy;* style straight or bent, as long as stamens or shorter. **Distribution** 1,800-2,600m (6,000-8,500ft), epiphyte in rich rain forest.

R. kawakamii differs from *R. euonymifolium* in the absence of warts on the branchlets, the greater number of flowers to the inflorescence and the hairy ovary. As far as we know, the white-flowered form has not been introduced into cultivation.

Quite a pretty species in a modest way and useful for its very late flowering. It has survived out of doors in our Argyll garden on a mossy rock for many years and also for a few years at Glendoick in a raised bed near a wall. It requires very good drainage but is otherwise not difficult to grow. Introduced 1969. July-August.

Picture 1

Picture 2

1. *R. kawakamii* showing the bright green leaves and yellow flowers on long pedicels, Glendoick.
2. *R. kawakamii* at Glendoick.

R. rushforthii
Argent & D.F. Chamb. 1996. H2?

Height to 1.5m, an erect shrub. **Leaves** elliptic, 1.8-11 x 1.7-4.2cm, apex pointed, upper surface *bluish or silvery-green*, lower surface moderately to finely lepidote, scales variable. **Inflorescence** 3-8-flowered, flower bud scales *fringed with two types of hairs and scales*. **Corolla** yellow, 1-1.5cm long, outside lepidote, inside of tube pubescent; calyx 5-lobed, lepidote and sparsely hairy; ovary hairy and lepidote, style sparsely hairy. **Distribution** Laokai province Vietnam c. 1,800m (6,000ft).

With a longer and more glaucous leaf than its relatives and with more flowers in the inflorescence than all but *R. kawakamii* from Taiwan, this species seems to be fairly distinctive. The two types of hairs on the lepidote bud scales are another diagnostic character.

This newly introduced and newly-named species is said to have withstood -8°c. *R. rushforthii* may be as hardy as *R. kawakamii* and should be tough enough to grow out of doors in milder gardens. Introduced 1992.

R. santapauii
Sastry, Kataki, P.A. & E.P. Cox, P.C. Hutchison H2-3?

Height to 0.6m, a fairly compact, dwarf shrub in cultivation. Branchlets scabrid. **Leaves** 2.5-4 x 1.1-1.5cm, thick and fleshy, not notched, obovate-lanceolate; upper surface dark green, lower surface pale, scales unequal, brown, 2-3 x their own diameter apart. **Inflorescence** 2-4-flowered. **Corolla** campanulate, 1.2-1.3cm long, *creamy-white, waxy with red scales outside*; calyx 1mm; stamens shorter than corolla; style bent, *shorter* than corolla, stamens and ovary.

Arunachal Pradesh, c.1,600m (5,400ft), epiphytic on large trees.

This is a distinct species, differing from *R. euonymifolium* and *R. kawakamii* in its darker leaves, white flowers and the style shorter than the ovary.

The rare *R. santapauii*, only discovered in 1965, is a pretty little plant and is free-flowering from a small size. It needs excellent drainage. Hardiness unknown. Introduced 1965 only. July-August.

Picture 1

Picture 2

1. *R. santapauii*, Royal Botanic Garden, Edinburgh. (Photo J. Cubey)
2. *R. santapauii* showing the thick fleshy leaves and waxy flowers.

R. vaccinioides
Hook.f. 1851 H2-3 (-4?)

Height to 1m, a compact to straggly, epiphytic shrub. Branchlets densely scabrid. **Leaves** 0.6-2.5 x 0.4-1.1cm, spathulate-obovate to obovate; upper surface shiny, lower surface with unequal scales, 2-5 x their own diameter apart. **Inflorescence** 1-2(-3)-flowered. **Corolla** campanulate, *0.6-1.1cm long,* white to lilac-pink, lepidote outside; calyx 1-2mm; stamens as long as corolla; style straight or bent, shorter than corolla and stamens. **Distribution** Nepal, Sikkim, N. India, Bhutan to S.E. Tibet, upper Burma and N.W. & W. Yunnan, 1,800-4,300m (6,000-14,000ft), epiphytic, usually in shade and on mossy rocks and cliffs.

THERORHODION

This is a distinct species which superficially resembles a dwarf *Vaccinium*. It is easily distinguished from its relatives in its small leaves, pale flowers and small stature.

The high elevation forms, which should prove hardy, have never been introduced, although with such small flowers, they will not have much garden value. Introduced 1850->. July-September.

Picture 1

Picture 2

1. *R. vacciniodes* Bowes Lyon 33 at the Royal Botanic Garden, Edinburgh, showing the small, shiny leaves and the white-tinged-lilac-pink, waxy flowers. (Photo R.B.G., E.)
2. *R. vacciniodes* Bowes Lyon 33, Royal Botanic Garden, Edinburgh. (Photo R.B.G., E.)

SUBGENUS THERORHODION

Height 10-30cm, dwarf to prostrate shrubs. **Leaves** usually *deciduous*, 0.5-5 x 0.3-2.5cm, obovate to spathulate, margins bristly. **Inflorescence** 1-3-flowered, pedicel a *leaf-like bract*. **Corolla** +/- rotate, 1.5-2.5cm long, purple to rose-purple, pink or white; calyx 5-18mm long, bristly; stamens 10; style pubescent towards base, ovary pubescent.

Some authorities have placed this very distinct subgenus in a separate genus *Therorhodion*. The current consensus seems to be for it to remain within the genus *Rhododendron* as it has been shown that the flowers are in fact borne on pedicels with leaf-like bracts rather than on leaf-bearing shoots as previously thought.

R. camtschaticum
Pall. 1784. II5

Height to 0.3m, a prostrate to bushy, +/- deciduous shrub, which roots and suckers as it spreads. Branchlets with few or no glandular hairs. *Leafy shoots from separate buds below the inflorescence.* **Leaves** *deciduous,* (occasionally evergreen?), thin, 1.6-5 x 0.8-2.5cm, obovate to spathulate, margins bristly; lower surface with soft hairs on the veins. **Inflorescence** 1-3-flowered. **Corolla** +/- rotate, c.2 cm long, 4cm across, purple to rose-purple, also red, pink or white, *pubescent outside, margins ciliate*; calyx large, 8-18mm long, bristly; stamens 10, unequal; style curved, pubescent towards base; ovary pubescent.

Ssp. camtschaticum
(*Rhodothamnus camtschaticus* (Pall.) Lindl., *Therorhodion camtschaticum* (Pall.) Small.

Corolla lobes pubescent outside, margins ciliate; leaves of vegetative shoots *without* or with *sparse* glandular hairs. **Distribution** Kamtschatica, N. Japan, Sakhalin through Kuriles, Aleutians to S. Alaska, in gravelly loam and crevices, often on hilltops.

Almost all cultivated *R. camtschaticum* is this subspecies rather than ssp. *glandulosum*. An excellent, very hardy plant for cool, northern climes such as N. Germany where it flourishes and is commercially produced in considerable numbers. Although somewhat prone to spring frost damage, it grows well in E. Scotland but further south and west it becomes harder to please. The Philipsons surprisingly state that this species can be evergreen while Davidian takes the more accepted line that it is always deciduous. Introduced to Britain 1799. May-June and also often in autumn.

THERORHODION

Ssp. *glandulosum*

(Small) Hultén 1930 (*Therorhodion glandulosum*, *Rhododendron glandulosum* (Small) Hutch.

Corolla lobes glabrous outside, margins not ciliate; leaves of vegetative shoots glandular, *hairy*. **Distribution** Siberia, W. Alaska.

Lower-growing than ssp. *camtschaticum*, this subspecies is very rarely cultivated and is likely to be harder to please than ssp. *camtschaticum*.

Picture 1

Picture 2

Picture 3

Picture 4

Picture 5

Picture 6

1. *R. camtschaticum* ssp. *camtschaticum* at Bremen Rhododendron Park, Germany showing its low and spreading habit.
2. A typical purple-flowered *R. camtschaticum* ssp. *camtschaticum* at Glendoick showing the almost rotate flowers on long pedicels.
3. Selected red-flowered *R. camtschaticum* ssp. *camtschaticum* at Glendoick.
4. The rare white-flowered *R. camtschaticum* ssp. *camtschaticum* which is becoming more readily available. Glendoick.
5. Autumn colour on *R. camtschaticum*.
6. Flowers of *R. camtschaticum* showing the pubescence on the +/- rotate corolla with ciliate margins; also the hairy leaves, pedicels and calyces. (Photo R.B.G., E.)

R. redowskianum
Maxim. 1970, probably not in cultivation

Differs from *R. camtschaticum* in its smaller flowers c.1.5cm long and shorter calyx c.5mm long. **Distribution** E. Siberia, Manchuria and Korea.

Seed of this little known species has been introduced at least twice in recent years but we do not know of anyone who has successfully raised seedlings beyond the first winter. Horticulturally it is probably best considered a difficult-to please, dwarf version of *R. camschaticum*.

SUBGENUS TSUTSUSI

Height 0.5-4.5m, creeping to upright shrubs. Vegetative buds and inflorescence *from within the same bud scales*. **Leaves** persistent and/or deciduous, in pseudowhorls (section Brachycalyx) or alternate (section Tsutsusi), covered in either simple, ribbon-like and flattened bristles or stiff, glandular hairs. **Inflorescence** 1-15-flowered. **Corolla** rotate to tubular-campanulate; stamens (4-)5-10(-12); ovary bristly to glandular; style usually glabrous; seeds without wings or tails. **Distribution** Japan, E. Asia, S.E. Asia, Taiwan, China, Burma.

The two sections which form subgenus Tsutsusi are rather distantly related but are classified together due to the characteristic that the vegetative buds and inflorescence are enclosed within the same bud scales. This subgenus differs from subgenus Pentanthera mainly in cotyledon and seedling characteristics which of course are not visible on mature plants. There are no yellow-flowered species in this subgenus. All species in subgenus Pentanthera have either 5 or 10 stamens (section Pentanthera have 5, the other sections have 10) while most of subgenus Tsutsusi have 6-10.

Section Brachycalyx

Height to 5m, small to large, deciduous to semi-evergreen shrubs. Branchlets hairy at first, becoming glabrous. Flowers and growth develop from within the same bud scales. **Leaves** *almost always in pseudowhorls* near the ends of branchlets, monomorphic, mostly in 3s, usually *rhombic*, with some long, sometimes adpressed hairs, denser on the midrib. **Inflorescence** 1-4-flowered; flowers appear before or with the leaves. **Corolla** funnel-shaped to funnel campanulate, purple, mauve, lavender, magenta, red, rose, pink, orange to white; calyx minute; stamens usually 10; style glabrous or with hairs near the base. **Distribution** Japan and S.E. Asia.

This section contains a group of fairly closely related species characterised by their 3-leaved, whorled, deciduous, +/- rhombic leaves. The species in section Tsutsusi differ in their mostly evergreen, often dimorphic, linear to broadly ovate leaves. Section Sciadorhodion (of subgenus Pentanthera) species differ in their non-rhombic leaf shape, and their more rotate or rotate-campanulate corolla shape.

We have followed the recent classification of this section by D.F. Chamberlain and S.J. Rae (*Edinburgh Journal of Botany Vol. 47, No. 2*). This revision was unfortunately completed with inadequate access to herbaria and wild material in Asia, prompting the authors to state:

'Among the more recently described Japanese Taxa, particularly in section Brachycalyx, there are several for which we have seen no material, and from the descriptions it appears that very narrow species, subspecies and varietal limits have been used. Where we have seen no material, we have accepted the original taxonomic circumscriptions and this should be borne in mind when reading this account'.

Essentially this is an admission that the rigorous standards of taxonomy which were used in the Edinburgh revisions of Sections Hymenanthes and Rhododendron were not applied here. The revision can therefore be considered only provisional and of limited value. One cannot help questioning the purpose of publishing such an incomplete treatment without study in Japan and China when access to both wild material and herbaria in both countries is now easily gained. We have anticipated the inevitable future revision of this section by dividing it into 3 alliances: Farrerae, Reticulatum and Weyrichii, reflecting our opinion that it will probably be reduced to 3-5 species. Of the current species, a few will probably retain subspecific and varietal status, while many of the others will be reduced to synonymy.

The Japanese species of this subsection perform well in areas of high or moderately high summer temperatures which ripen the wood and set flower buds. In areas with cool summers they are rather shy flowering and lacking in vigour. The Chinese species are more tender and of limited garden merit.

Farrerae Alliance

R. farrerae
Tate ex Sweet 1831 (*R. cinereoserratum* P.C. Tam, *Azalea squamata* Lindl.) H1-2?

Height to c. *0.5m*, a low, densely-branched shrub. Branchlets becoming glabrous. **Leaves** deciduous or semi-evergreen, *1.5-3 x 1-2cm*, ovate, upper and lower surface with long, brown hairs, becoming glabrous; petiole *very short, 1-3mm, hairy*. **Inflorescence** 1-2-flowered. **Corolla** open-campanulate, 3cm or more across, lavender to purplish-pink, spotted; stamens 10; ovary densely hairy; style glabrous; capsule *ovoid*. **Distribution** China, Hong Kong and adjacent territory to Hunan to Fujian, mixed woodland, c. 600m (2,000 ft).

R. farrerae is closely related to *R. mariesii*, differing in its much smaller stature, smaller leaves, shorter, hairy petiole, paler purplish flowers and ovoid capsule.

This rarely cultivated species is rather tender and of limited garden value. Introduced 1829. December onwards in Hong Kong.

R. mariesii
Hemsl. & E.H. Wilson 1907 (*R. gnaphalocarpum* Hayata, *R. shojoense* Hayata, *R. umbelliferum* Lév.). H3?

Height *1-3m*. Branchlets with adpressed, yellowish hairs at first, becoming glabrous and grey. **Leaves** deciduous, *3-10 x 2-4.5cm*, ovate-lanceolate to broadly ovate; upper surface dark green, lower surface pale; petiole 5-20mm, glabrous. **Inflorescence** 1-2-flowered; flowers appear before the leaves. **Corolla** funnel-shaped, c.5cm across, rose to purple, spotted; calyx minute; stamens 10; ovary yellowish-grey, hairy; style glabrous; capsule *cylindrical*. **Distribution** Widespread in S. Central and S.E. China and Taiwan, 350-1,200m (1,200-4,000ft) in China, 1,600-1,900m (5,300-6,300ft) in Taiwan, cliffs and thickets.

R. mariesii is closely related to *R. farrerae*, differing in its much larger stature, larger leaves, longer, glabrous petiole, deeper-coloured flowers and cylindrical capsule.

This rarely-cultivated species is of limited ornamental value. It is not very hardy and was killed at Glendoick by a cold winter. Introduced 1886->. April.

Reticulatum Alliance

The so-called species in this alliance are essentially members of a single variable taxon *R. reticulatum*. Contrary to his usual position as a botanical 'splitter', this treatment has been followed by H.H. Davidian 1995. We reluctantly retain the specific names as described in *Edinburgh Journal of Botany*, Vol. 47, No. 2.

The species in this alliance can be treated as one entity from a horticultural point of view. Although all the species appreciate a +/- continental climate, they can be relatively successful in a warm, open situation in cooler regions with a maritime climate. Forms with larger, rich reddish-purple or pure white flowers can be very showy, and the autumn colour is often good. Rabbits seem to be particularly fond of the foliage of these species. Introduced from 1834 onwards. April-June.

BRACHYCALYX

R. reticulatum

D. Don ex G. Don 1834 (*R. rhombicum* Miq., *R. sakawanum* Makino). Rose Azalea. H4-5

Height 1-8m (usually 1.5-2.5m in cultivation), a broadly upright or tiered shrub or tree. Young shoots soon becoming glabrous. **Leaves** 3.8-6.3 x 2.5-5.5cm, rhombic-ovate to lanceolate, apex acute; upper surface with short hairs at first, soon glabrous, lower surface with short brown hairs, mainly on midrib and veins; petiole *pilose*. **Inflorescence** *1-2(-3)-flowered*; flowers appear before the leaves. **Corolla** funnel-campanulate, 2.5-5cm across, lavender to deep purple to magenta, rarely white, usually unspotted; stamens 10; ovary hairy; style usually glabrous. **Distribution** Japan, S. Honshu, Shikoku, Kyushyu, 15-1,800m (50-6,000ft), often in thickets, dry rocky slopes to stream sides. April-June.

Picture 1

1. *R. reticulatum*, Royal Botanic Garden, Edinburgh, showing the flowers opening before the leaves expand.

R. decandrum

Makino 1917 (*R. inobeanum* Honda)

Young branchlets soon become glabrous. **Leaves** broadly rhombic, apex acuminate; petiole *sparsely glandular* and hairy. **Corolla** magenta, spotted; stamens 10; ovary *glandular* with a few hairs. **Distribution** S. Honshu, Shikoku.

Picture 2

Picture 3

2. A dark-flowered *R. decandrum*, Royal Botanic Garden, Edinburgh, showing the ten stamens.
3. A pale-flowered *R. decandrum*, Royal Botanic Garden, Edinburgh.

R. dilatatum Miq. 1863

Branchlets glabrous. **Leaves** rhombic, apex acuminate, with long weak hairs on both surfaces at first; petiole papillate. **Corolla** rose-purple or white; stamens 5; ovary *glandular*. **Distribution** S. Honshu.

Picture 4

4. *R. dilatatum* in Japan. (Photo H. Suzuki).

Picture 5

5. The lovely white-flowered form of *R. dilatatum* (also known as var. *leucanthum* or *R. reticulatum* var. *albiflorum*) at Glendoick.

R. hidakanum

Hara 1974 (*R. dilatatum* var. *boreale* Sugimoto)

Branchlets at first +/- glandular, greyish. **Leaves** broadly rhombic-ovate, apex cuspidate, petiole glandular. **Corolla** magenta; calyx c. 3mm, *purple*; stamens 10; ovary *shortly stalked-glandular*. **Distribution** S. Hokkaido.

R. kiyosumense

Makino 1931 (*R. shimidzuanum* (Honda ex) Makino)

Branchlets glabrous. **Leaves** rhombic, apex acuminate. **Corolla** purple; stamens 10; ovary *densely bristly*. **Distribution** E. Honshu.

Picture 6

6. *R. kiyosumense* in Japan. (Photo H. Suzuki).

R. mayebarae Nakai & Hara 1935

Branchlets glabrous. **Leaves** ovate-rhombic, apex acute, glabrous lower surface except for midrib. **Corolla** deep magenta; pedicels *with dense brown hairs*; stamens 10; ovary *brown, bristly*. **Distribution** Kyushu.

Picture 7

7. Foliage of *R. mayebarae* showing the typical acute leaf apex. Polly Hill's garden, Martha's Vineyard, E. U.S.A.

R. nudipes Nakai 1926

Branchlets glabrous. **Leaves** broadly rhombic, apex acute, tip blunt, upper and lower surfaces with long brown hairs at first, becoming glabrous on lower surface. **Corolla** rose-purple; stamens *(?5-)8-10*; ovary with dense brown villose hairs. **Distribution** Honshu, Kyushu.

This species has been recently further subdivided into subspecies and varieties but Chamberlain states that the type descriptions are inadequate and non-comparative. The only significant difference between the two subspecies, as outlined by Chamberlain, is in the size of the capsule.

Ssp. *nudipes* (*R. nagasakianum* Nakai)

Capsule 2-3mm long.

Ssp. *niphophilum* Yamazaki 1981

Capsule 8-13mm long.

Var. *lagopus* (Nakai) Yamazaki 1984

Leaves rhombic, both surfaces sparsely pilose, petiole *densely woolly*. **Corolla** rose-purple; calyx c.2mm; stamens 10; ovary with villose hairs; capsule 10-13mm long. **Distribution** S.W. Honshu, Shikoku.

This variety has been classified as a variety of *R. reticulatum* and *R. wadanum* by other authorities.

BRACHYCALYX

Picture 8

Picture 9

8. *R. nudipes* ssp. *nudipes*, Japan. (Photo H. Suzuki).
9. *R. nudipes* ssp. *niphophilum* var. *lagopus*, Glendoick.

R. viscistylum Nakai 1935

Young shoots soon glabrescent. **Leaves** rhombic, apex *acute*, upper surface pilose, brown at first, lower surface pale, *sometimes viscous*. **Corolla** reddish-purple, spotted; stamens 10; ovary *glandular with brown hairs at apex;* style *viscous*. **Distribution** Kyushu.

Picture 10

10. *R. viscistylum* in Japan. This species is characterised by the sticky glands on the style and sometimes on the leaf under-surface. (Photo H. Suzuki)

R. wadanum
Makino 1917 (*R. glandulistylum* Komatsu)

Branchlets villose. **Leaves** rhombic, upper surface sparsely hairy when young, lower surface sparsely hairy; petiole densely hairy. **Corolla** rich rose-pink; stamens 10; ovary densely hairy; style with *stalked glands* on lower half. **Distribution** S.E. Honshu.

Picture 11

Picture 12

Picture 13

11. *R. wadanum*, Royal Botanic Garden, Edinburgh.
12. *R. wadanum* showing the rhombic leaves, Royal Botanic Garden, Edinburgh.
13. A white-flowered form of *R. wadanum* in Japan. (Photo T. Takeuchi).

Weyrichii Alliance

The Japanese Botanist Hatusima reduced the following three species to one (*R. weyrichii*) in 1969. This seems to be the most valid treatment as the characters which separate them are taxonomically not very significant.

R. amagianum
Makino 1930 (*R. weyrichii* var. *amagianum* Hatusima).
Mt. Amagi Azalea. H4

Height to 5m, a usually erect shrub with spreading branches. Branchlets white pubescent at first, becoming glabrous. **Leaves** in 3s, *4.8-11.5 x 3-9cm*, broadly ovate-rhombic, upper surface with long, scattered, brown hairs, lower surface with adpressed, pubescent hairs. **Inflorescence** 1-3-flowered; flowers appear *with or after* the leaves. **Corolla** open-campanulate, 4.5-6cm across, *reddish-orange* with brown spots; pedicel *0.6-1.3cm* long; stamens 10; ovary brown, pubescent; style crimson, white pubescent at base. **Distribution** Only from Mts Amagi and Higane in Idzu Province, S. Honshu, Japan.

R. amagianum differs from its close relatives, *R. weyrichii* and *R. sanctum,* in its larger, broader leaves, longer pedicels and later, reddish-orange flowers.

This handsome species has quite impressive foliage and flowers and it has considerable untapped garden potential for warmer areas. It requires more heat than can be provided in Scotland to perform at its best. Introduced in late 1930s and 1960s->. June-July.

Picture 1

Picture 2

3. *R. amagianum* in Japan. (Photo H. Suzuki).

1. *R. amagianum* cultivated in Japan, showing the large, ovate-rhombic leaves, Japan. (Photo H. Suzuki).
2. *R. amagianum* at Baravalla, W. Scotland.

R. sanctum
Nakai 1932 (*R. weyrichii* var. *sanctum* Hatusima).
Shrine Azalea. H4-5

Height to 5m but often much less in cultivation, a broadly upright shrub. Branchlets pubescent at first, soon becoming glabrous, rusty-brown. **Leaves** 3-8 x 2.5-6cm, broadly rhombic, thick, upper surface hairy, becoming glabrous, *glossy*, lower surface with scattered hairs on the midrib; petiole densely covered with red-brown, pilose hairs. **Inflorescence** 3-4-flowered; flowers appear *after* the leaves. **Corolla** funnel-campanulate, c.3.8cm across, *deep aniline-rose to rosy-purple,* spotted, occasionally white; stamens 10; ovary densely pilose; style pilose in lower half. **Distribution** S. Honshu, Tokai and Ise provinces, mountain sides.

This species is closely related to *R. amagianum* and *R. weyrichii*, differing from the former in its rose-coloured flowers and from *R. weyrichii* in its flower colour, larger glossy leaves and later flowers.

Picture 1

1. *R. sanctum* 'Zeke', a clone with rich-coloured flowers and typical glossy leaves, at Polly Hill's garden, Martha's Vineyard, E. U.S.A.

BRACHYCALYX

Like its relatives, *R. sanctum* prefers warm summers and is of limited value in Scotland and similar cool climates. It is a handsome plant with bold foliage which often colours well in autumn. May-June.

Picture 2

2. *R. sanctum* at Glendoick showing the relatively small size of the flowers compared to the leaf size.

R. weyrichii

Maxim. 1970 (*R. shikokianum* Makino) H 3-4

Height 1-4.6m, a shrub or small tree (rarely tree-like in cultivation) with short ascending branches. Branchlets reddish, glabrous, grey, becoming dark purple. **Leaves** 3.8-8 x 1.5-6cm, *not* shiny, broadly rhombic, both surfaces with brown, pilose hairs at first, becoming almost glabrous. **Inflorescence** 2-4-flowered; flowers appear before, with or after the leaves. **Corolla** open funnel-campanulate, 3.8-6.3cm across, rich salmon-pink to brick-red with darker spots, occasionally white; stamens usually 10; ovary densely pilose; style glabrous or pubescent at base. **Distribution** S.E. Honshu, Shikoku, Kyushu, Japan and Cheju Island, Korea.

Picture 1

1. A free-flowering example of *R. weyrichii* in Japan. (Photo H. Suzuki).

This species differs from *R. amagianum* in its smaller, non-shiny leaves and from *R. sanctum* in its reddish flowers.

R. weyrichii is not as impressive a plant as its close relatives, and like them it requires warm summers to grow well. Introduced 1914. April-May-June.

Picture 2

Picture 3

2. *R. weyrichii* in Japan. (Photo T. Takeuchi).
3. Leaf of *R. weyrichii* showing the partially hairy upper surface and the hairy petiole. (Photo R.B.G., E.).

Section Tsutsusi

Dwarf to medium to occasionally tall shrubs, much branched. Shoots emerge from the axils of lower terminal bud scales. Leaves almost deciduous to partly evergreen, monomorphic (of one type), or *dimorphic (of two types)*: the spring leaves, largely deciduous, and the summer leaves persistent or leaving only the smaller leaves near the terminal bud. If there is one type of leaf only, these are scattered along the stems and are usually persistent. **Leaves** linear to broadly ovate with hairs present, even at maturity; hairs usually adpressed but can be stiff and glandular or weak and soft. **Inflorescence** 1-15-flowered. **Corolla** *rotate to tubular-campanulate,* purple, red, near orange, pink to white, none with yellow tones; calyx minute to large, often variable; stamens usually 5 or 10, occasionally 4-10; ovary hairy, sometimes glandular; style usually glabrous, sometimes hairy or glandular near base; seeds not winged. **Distribution** Upper Burma, W., S. and E. China, Taiwan, Japan, with outliers in several locations in S.E. Asia. While most come from relatively moist mountains, some are found in low elevation dryish scrub away from other rhododendrons.

Species in section Brachycalyx differ in the usually deciduous, rhombic to rhombic-ovate leaves in pseudowhorls of usually 3.

These species were recently revised by D.F. Chamberlain and S.J. Rae in *Edinburgh Journal of Botany*, Vol. 47, No.2, 1990. The hardy species in this large section and their many hybrids are generally known as 'Evergreen', 'Japanese' or 'Obtusum' azaleas. Those grown for the pot plant industry and outdoors in warm climates are known as 'Indian' or 'Indica' azaleas. Huge numbers of the latter are grown in specialised nurseries for year-round decoration in the home and are perhaps the most widely grown part of the genus Rhododendron world-wide. The majority of section Tsutsusi are not suitable for Scotland and parts of northern Europe where the wood is inadequately ripened leading to winter damage. The evergreen azaleas and their hybrids flourish in light shade in southern Britain and similar climates but further north more sun is required and they should be planted in the open. Petal blight fungus affects the flowers of these azaleas in many warmer areas.

Many new species in section Tsutsusi have recently been described in China but few are likely to prove hardy except in very mild areas. These are not included in this book as they have not yet been introduced to cultivation outside Asia.

hairy, style glabrous. **Distribution** S. Sichuan, N.E. Yunnan, 750-1,800m (2,250-6000ft) in thickets.

Apparently related to *R. simsii* and *R. microphyton* though seemingly larger growing than these two. The long leaf apex and the very hairy leaves, petioles etc are two diagnostic characteristics. Its distribution seems to be isolated from that of its relatives. Likely to be tender. Introduced 1996->.

R. breviperulatum
Hayata 1913 H3-4?

Height low, a much-branched shrub. Branchlets slender, covered with shiny, brown, adpressed hairs. **Leaves** persistent, 1-3 x 0.6-1.7cm, ovate-elliptic, both surfaces with scattered shiny, brown hairs. **Inflorescence** 2 or more-flowered. **Corolla** funnel-campanulate, to 2.5cm across, rose-pink, spotted crimson to reddish; calyx 2-4mm long, white, hairy; stamens 5-6; style pilose at base. **Distribution** Mountain forest of Taiwan, c.2,550m (8,500ft).

R. breviperulatum was introduced by J.L. Creech but is still little known. As cultivated at the Royal Botanic Garden, Edinburgh, it has a low spreading habit. March-April in wild.

R. eriocarpum
(Hayata) Nakai 1922 (*R. tamurae* Makino, *R. tawadae* Ohwi). H2-3?

Height to 0.4m+, a compact to occasionally upright shrub. Branchlets with adpressed, brown hairs. **Leaves** monomorphic, persistent, 1.7-2.5 x 1-1.5cm, obovate to elliptic, both surfaces with long-hairs. **Inflorescence** 1-2-flowered. **Corolla** funnel-

R. atrovirens
Franch 1886. H2?

A large shrub or small tree, young shoots covered with adpressed, brown hairs. **Leaves** *monomorphic*, persistent, elliptic 2-8 x 1-3cm, apex *long, acuminate*, margins entire, both surfaces *covered with shining, brown hairs*, glabrescent above, except on midrib; petiole strigose. **Inflorescence** 2-4-flowered. **Corolla** funnel-campanulate, 1.5-3cm long, glabrous, red with darker flecks at base of upper lobes; stamens 10, ovary densely

Picture 1

1. *R. eriocarpum* in Japan showing its +/- compact, spreading habit. (Photo H. Suzuki).

campanulate, 3-4cm across, red through purplish-pink to white; calyx 2-3mm long with long, adpressed hairs; stamens *8-10*; style glabrous. **Distribution** Kyushu, Yakushima. Ryukyu Is., S. Japan, open woodland, c.300m (1,000ft).

This species differs from *R. indicum* in its broader leaves and has 8-10 rather than 5 stamens.

The rarely cultivated *R. eriocarpum* has been successfully grown in E. U.S.A where it is useful for its late flowering, but its requirement of considerable summer heat to ripen the wood makes it a poor performer in N. Europe. The well-known Gumpo azaleas are thought to be hybrids of *R. eriocarpum*. July.

Picture 2

2. *R. eriocarpum* in Japan showing the typical obovate to elliptic, fairly broad leaves. (Photo H. Suzuki).

R. indicum

(L.) Sweet 1833 (*R. balsaminiflorum* Carrière, *R. breynii*, Planch., *R. crispiflorum* Hook.f., *A. danielsiana* Paxton, *R. decumbens* (D. Don ex) G. Don, *R. hannoense* Nakai, var. *lateritium* Lind., *A. macrantha* Bunge, *R. sieboldii* var. *serrulatum* Miquel.). H3-4

Height 0.60-2m, a much-branched and usually compact shrub. Branchlets with adpressed, bristly, chestnut-brown hairs. **Leaves** dimorphic, persistent, often *dark green*, spring leaves 2-3 x 0.8-1cm, *narrowly lanceolate to oblanceolate*, upper surface with scattered strigose hairs, lower surface with strigose hairs on the midrib, summer leaves smaller, 1-1.8 x 0.3-0.5cm. **Inflorescence** 1-2-flowered. **Corolla** broadly funnel-shaped, 5-6.3cm across, rose-red to bright scarlet; calyx c.1mm long, with brown strigose hairs; stamens *5*; style glabrous. **Distribution** S. & W. Honshu, Shikoku, Kyushu and Yakushima, ravines, stream sides and open country, at generally higher elevations than *R. eriocarpum*.

This species is characterised by its dark, narrow leaves, and its late flowers with 5 stamens. Its closest relative *R. eriocarpum* differs in its wider leaves and flowers with c. 9 stamens.

R. indicum, one of the parents of the Satsuki hybrids, exists in many named forms, some of which perform well in southern England but are not hardy enough for N. Germany or Scotland. This species should not be confused with the 'Indica' azaleas commonly sold as houseplants. Introduced 1833. June-July.

Picture 1

Picture 2

1. *R. indicum* on Yakushima, Japan. (Photo T. Takeuchi).
2. A double-flowered *R. indicum* cultivar.

R. kaempferi

Planch. 1854 (*R. sieboldii* Miquel, *R. macrogemmum* Nakai, *A. kaempferi* (Planchon) André). H3-5

Height 1-3m, an upright shrub. Branchlets with short, adpressed, red-brown, strigose hairs. Bark cinnamon-brown with age. **Leaves** dimorphic, persistent to semi-persistent, spring leaves 2-5 x 1-2.5cm, summer leaves 1-2 x 0-.5-1cm, lanceolate to elliptic, both surfaces strigose. **Inflorescence** 2-3-flowered. **Corolla** funnel-shaped, 4.4-6.3cm long, salmon-red through rosy-red, pink to white; calyx 3-5mm long with strigose hairs; stamens *5(-6)*; style glabrous. **Distribution** widespread in Japan, sea-level to 1,200m (4,000ft), sunny hillsides, volcanoes, also forests.

This species differs from *R. indicum* and *R. kiusianum* var. *sataense* in its upright, tiered, taller growth habit and its larger, broader leaves.

R. kaempferi is generally hardy and is widely cultivated in its numerous forms and in multitudes of hybrids. Naturally occurring curiosities, such as forms with petalloid stamens and pistil or with no corolla, are prized in Japan. Latisepalum Group (formerly var. *latisepalum*) is a northern form with pale salmon flowers. *R. transiens* Nakai is thought to be a natural hybrid of *R. kaempferi* x *R. macrosepalum*. Introduced 1892. May-June.

Picture 1

1. *R. kaempferi*, Royal Botanic Garden, Edinburgh, from seed collected by H. Suzuki in Japan.

Picture 2

Picture 3

Picture 4

Picture 3

2. *R. kaempferi* at Glendoick.
3. A bed of *R. kaempferi* forms in Japan. (Photo T. Takeuchi).
4. *R. kaempferi* Latisepalum Group collected in Hokkaido, Japan, growing at Glendoick.
5. *R. kaempferi* 'Polypetalum', one of the many extreme forms selected by the Japanese.

6. A double-flowered cultivar of *R. kaempferi* in Japan. (Photo T. Takeuchi)

2. A plant cultivated as *R. kanehirai* with very narrow leaves. This example has 5 rather than 10 stamens and is perhaps referable to *R. indicum* (Photo S. Ishida).

R. kanehirai
E.H. Wilson 1921 H2-3?

Height 1-2.5m, a much-branched shrub. Branchlets with broad adpressed, stiff, chestnut-brown hairs. **Leaves** dimorphic, persistent, oblanceolate to narrowly obovate, spring leaves 2-5 x 0.5-1.5cm, summer leaves 1.5-3 x 0.2-0.6cm, both surfaces sparsely strigose. **Inflorescence** 1-2-flowered. **Corolla** funnel-campanulate, 2.5-4.3cm long, pink to scarlet; calyx c.1mm long, with short, chestnut hairs; stamens 10; style glabrous or with few hairs at base. **Distribution** N. Taiwan, sea level to 400m (1,200ft).

R. kanehirai differs from *R. indicum* in its 10 rather than 5 stamens, and from *R. tashiroi* in its fewer-flowered inflorescence and its deeper, less spotted flowers.

R. kanehirae was introduced in 1969, and may no longer be in cultivation. We grew it for a while but it did not prove hardy. Taxonomically, this species could be treated as a Taiwanese variety of *R. tashiroi*.

1. *R. kanehirai* cultivated in Japan. (Photo S. Ishida)

R. kiusianum
Makino 1914. H4-5

Height 0.15-1m, *usually under 0.6m, a compact, much-branched* shrub, wider than high. Branchlets with adpressed, red-brown hairs. **Leaves** monomorphic, *almost deciduous*, or evergreen, 0.5-3 x 0.2-1.5cm, oval, ovate-elliptic to obovate, both surfaces with short, red-brown hairs. **Inflorescence** 2-3-flowered. **Corolla** funnel-shaped, 1.8-2.5cm across, salmon-red, orange-salmon, rose, pink to purple and white; calyx 2-3mm long with red-brown, short hairs; stamens 5; style glabrous. **Distribution** Kyushu, Japan, 1,200-1,700m (4,000-5,600m), confined to (often volcanic) high peaks, meadows, pumice flats and among dwarf pines.

R. kiusianum is distinct among Japanese species for its compact, dwarf habit and small, almost deciduous leaves.

Var. *kiusianum*
(*R. indicum* var. *amoenum* forma *japonicum* Maxim, *R. kaempferi* var. *japonicum* Rehder, *R. indicum* var. *japonicum* Makino, *R. obtusum* var. *japonicum* Kitamara.)

Leaves *usually deciduous or almost deciduous*, 0.5-2 x 0.2-1cm, oval to obovate.

R. kiusianum var. *kiusianum* is very hardy and is ideal for small gardens as it is slow-growing and retains a neat, compact habit. Many different forms are cultivated in a wide range of colours. It is one of the best azaleas for northern Europe, especially Scotland and Germany and it has been much used in breeding to produce hardy, low-growing hybrids. Introduced 1918. May.

Var. *sataense* (Nakai) D.F. Chamb. & S.J. Rae 1990

Leaves *evergreen*, ovate-elliptic, 1-3 x 0.5-1.5cm.

This variety is a hybrid/intermediate between *R. kaempferi* and *R. kiusianum* which in some places has formed quite large stable populations. It is rarely cultivated outside Japan. May.

Picture 1

Picture 2

Picture 3

Picture 4

Picture 5

4. White-flowered *R. kiusianum* var. *kiusianum* at Glendoick.
5. *R. kiusianum* var. *sataense*, Royal Botanic Garden, Edinburgh, showing the more elliptic leaves and larger flowers than those of var. *kiusianum*.

1. *R. kiusianum* var. *kiusianum* at Glendoick showing its compact growth habit. This is the most common flower colour in the wild.
2. Rosy-red-flowered *R. kiusianum* var. *kiusianum* at Glendoick.
3. Pink-flowered *R. kiusianum* var. *kiusianum* at Glendoick.

R. *macrosepalum*

Maxim. 1870 (*R. linearifolium* var. *macrosepalum* Makino, *R. hortense* Nakai). H2-3

Height 0.3-2m, a low shrub, growing wider than high. Branchlets with greyish, spreading, pilose hairs, sometimes glandular. **Leaves** dimorphic; spring leaves 2.5-7 x 1.5-2.5cm, oblanceolate, summer leaves 1.2-2 x 0.3-0.6cm, ovate-elliptic, semi-persistent, rugulose, both surfaces pilose, with a few bristles. **Inflorescence** 2-10-flowered. **Corolla** broadly funnel-shaped, 3.8-5cm across, *lilac-pink to rose-purple, often heavily spotted, sometimes scented*; calyx *1.5-3cm long*, glandular-pilose; stamens 5(-7); ovary setose-glandular; style glabrous. **Distribution** S. Honshu, Shikoku, Japan, in pine woods and thickets.

The fairly large, pilose leaves, often scented flowers and the large calyx make this species fairly distinctive. Its closest relative is probably *R. mucronatum* which has adpressed hairs on the branchlets, a smaller leaf and calyx, 10 stamens and a non-glandular ovary.

In its best forms, *R. macrosepalum* is a pretty species, but it remains rare in cultivation due to its tenderness. It is rarely

TSUTSUSI

hardy enough for N. Europe and the colder parts of N.E. North America. The clone 'Linearifolium' with very narrow leaves and narrow corolla lobes is hardier than the type and is quite commonly cultivated in Europe and North America. Introduced 1914. April-May.

Picture 1

Picture 2

Picture 3

1. *R. macrosepalum* in Japan. (Photo H. Suzuki).
2. A form of *R. macrosepalum* with heavily blotched flowers, in Japan. (Photo H. Suzuki).
3. *R. macrosepalum* 'Linearifolium' with strap-like leaves and flowers in E. Somers's garden, North Canterbury, New Zealand.

R. *microphyton*
Franch. 1886. H2-3

Height 0.3-2m, an upright to bushy and spreading shrub, branchlets with adpressed, brown hairs. **Leaves** persistent, monomorphic, 1-4 x 0.5-1.5cm, elliptic to lanceolate, both surfaces with adpressed, brown hairs, *denser on the lower surface*. **Inflorescence** 3-6-flowered. **Corolla** funnel-campanulate with a *long, cylindrical tube*, 1.3-1.9cm across, rose-purple through pink to almost white, spotted; calyx 1-5mm long, brown, strigose; stamens 5; style glabrous. **Distribution** E. Burma, Yunnan, S.W. Sichuan, Thailand?, said to be 1,800-3,000m (6,000-10,000ft) but we have not seen it above 2,500m (8,250ft), dryish slopes and gullies in foothills.

This species is distinct from other members of the section in its dense brown hairs on the leaf lower surface and in its long corolla tube.

R. microphyton is quite pretty in a modest way but it is too tender for much of Britain and N. Europe. Introduced 1913 and reintroduced in 1980s but rare in cultivation. April-May.

Picture 1

Picture 2

1. Forms of *R. microphyton* of various colours, Cangshan, C.W. Yunnan, China, showing the long corolla tube which characterises this species.
2. *R. microphyton*, Kunming-Dali road, C.W. Yunnan, China.

R. mucronatum
(Blume) G. Don 1834. H3-4

Height 1-3m, an upright, spreading shrub. Branchlets with *loose, adpressed, strigose hairs and softer, pilose, grey-brown sometimes glandular, hairs*. **Leaves** semi-persistent, dimorphic, spring leaves 3.5-5 x 1.5-2cm, ovate-lanceolate, summer leaves 1.5-3 x 0.5-1cm, upper and lower surfaces with *reddish-grey, pilose hairs*. **Inflorescence** 1-3-flowered. **Corolla** widely-funnel-shaped, c.3.8cm across, white or rose pink; calyx to 15mm, glandular-pubescent; stamens *10*; ovary *setose, not glandular;* style glabrous.

R. mucronatum differs from *R. macrosepalum* in its smaller leaves and calyx, 10 stamens and non-glandular ovary.

Var. *ripense* (Makino) E.H. Wilson 1921 H3

Corolla *rose-pink*. **Distribution** S.W. Honshu, Shikoku, N.E. Kyushu, Japan, on river banks.

This variety is hardy only in milder areas. April-May.

Picture 1

Picture 2

1. *R. mucronatum* var. *ripense* in Japan. (Photo H. Suzuki).
2. *R. mucronatum* var. *ripense* in Japan. (Photo H. Suzuki).

Var. *mucronatum* (*R. argyi* Lév., *R. burmannii* G. Don, *A. ledifolia* Hook.f., *R. leucanthum* Bunge, *A. liliiflora* Poit., *R. rosmarinifolium* (Burmann) Dippel). H4

This widely cultivated and easily grown variety differs from var. *ripense* in its pure white flowers. This variety is of cultivated origin, either an albino form of var. *ripense* or a hybrid. Introduced to Britain 1819.

Picture 3

3. *R. mucronatum* var. *mucronatum* at Glendoick

R. nakaharai
Hayata 1908. H3-4

Height prostrate to 0.3m, usually creeping and *very compact*. Branchlets with adpressed, shining, brown hairs. **Leaves** persistent, monomorphic, 0.5-1.2(-2.5)cm, elliptic to elliptic-obovate, upper surface with scattered pilose hairs borne on raised pustules, lower surface paler, with scattered, adpressed, shining, brown hairs. **Inflorescence** 2-3-flowered. **Corolla** funnel campanulate, to 3.8cm across, *brick-red to rose-red*, calyx c.2mm long, setose; stamens 10, pilose on lower half; ovary setose, style glabrous. **Distribution** N. Taiwan, 300-800m (1,000-2,500ft), on grassland.

This is a distinct azalea, easily recognised by its prostrate, creeping, very compact habit and its late, usually red flowers.

R. nakaharai is an excellent garden plant and is surprisingly hardy considering its low elevation in the wild. It has often been used as a parent in hybridising, giving rise to a range of low, late-flowering cultivars. Introduced to U.K. in 1957, U.S.A. 1960. June-August. (This species is spelt *R. nakaharae* by some authorities.)

Picture 1

Picture 2

1. *R. nakaharai* 'Mt. Seven Star' from Taiwan and named in U.S.A. A relatively strong grower with lighter-coloured leaves than 'Mariko' and bright red flowers. Glendoick.
2. *R. nakaharai* 'Mariko' A.M., A tight grower with rose-red flowers which we introduced from Japan in 1957. Glendoick.

R. noriakianum
Suzuki 1935. H3?

Height to 1m, a much-branched, *rather open* shrub. Branchlets with adpressed, short strigose hairs at first, soon *becoming almost glabrous*. **Leaves** deciduous, monomorphic, 0.7-1.5 x 0.4-0.6cm, ovate to ovate-oblong; upper surface almost glabrous. **Inflorescence** 3-4-flowered; **Corolla** funnel-shaped, to *2.5cm across*, red; calyx densely pilose; stamens 7-10; style glabrous. **Distribution** N. Taiwan, 2,000-3,000m (6,500-10,000ft), on open grassland.

This species is related to *R. nakaharai* but has a more erect habit and much smaller flowers.

R. noriakianum may no longer be in cultivation and the plants we grew under this name had purple flowers and were probably incorrectly identified. It is unlikely to prove such a good garden plant as *R. nakaharai*. May?

R. oldhamii
Maxim. 1870 (*R. ovatosepalum* Yamamoto) H2-3

Height 1.2-3m, a spreading shrub. Branchlets with *dense, red-brown, glandular* hairs and scattered, adpressed hairs. **Leaves** persistent, dimorphic, spring leaves *3.5-8.8 x 1.8-2.5cm*, summer leaves 1.5-2 x 0.8-1cm, ovate-lanceolate, upper and lower surfaces covered with long, pilose, light brown hairs. **Inflorescence** 1-4-flowered. **Corolla** funnel-shaped, 2.5-3.5cm long, 3.8-5cm across, brick-red to coral-pink, spotted; calyx c.2mm long, glandular and pilose; stamens (8-)10; style glabrous. **Distribution** Common in Taiwan, sea-level to 2,700m (9,000ft), on cliffs, grassy slopes and around water.

This is a distinct species, with very hairy and glandular stems, pedicels and calyx and comparatively large leaves and flowers.

Forms of *R. oldhamii* currently cultivated are only suitable for outdoors in mild areas. It might well be worth collecting hardier forms from the highest elevations. *R. oldhamii* is quite a pretty plant with a long flowering season and it has proven quite a successful indoor azalea. Introduced 1878 but probably lost, reintroduced 1918 and recently. April-May.

Picture 1

Picture 2

1. *R. oldhamii* in Japan. (Photo T. Takeuchi).
2. A bush of *R. oldhamii* in Japan. (Photo T. Takeuchi).

Picture 3

3. Leaf bases of *R. oldhamii* showing the long pilose hairs and the petioles with spreading pilose hairs. (Photo R.B.G., E.).

R. rubropilosum

Hayata 1911 (*R. caryophyllum* Hayata, *R. randainse* Hayata). H2-4?

Height to 3m, an upright to spreading shrub. Branchlets with dense, grey to reddish-brown, adpressed hairs. **Leaves** persistent, monomorphic, 1-3 x 0.5-1cm, oblong-lanceolate to elliptic, upper surface *dark green with pale grey, adpressed* hairs, lower surface light green with red-brown, adpressed hairs. **Inflorescence** 2-4-flowered. **Corolla** funnel-shaped, 1-1.5(-2.5)cm long, pink to lavender, spotted; calyx minute with red-brown bristles; stamens 7-10; style *with a few adpressed hairs at base*. **Distribution** Widespread in Central Taiwan, 2,000-3,200m (6,500-10,500ft), on sunny hillsides.

This is a fairly distinct species with dark, hairy leaves, smallish pink to lavender flowers and the style with a few hairs at the base. It differs from *R. subsessile* in its pink rather than lilac-purple flowers and in the denser covering of hairs on the leaves.

Picture 1

1. *R. rubropilosum* at Glendoick, Rhododendron Venture 74/005.

R. rubropilosum is probably quite variable in hardiness, but the forms we have grown have proven perfectly hardy. Although quite distinctive, the flowers are not particularly showy and this species is unlikely to become popular. Introduced 1918, reintroduced 1968-74->. May-June.

Picture 2

Picture 3

2. *R. rubropilosum* Page 10095, Royal Botanic Garden, Edinburgh.
3. Leaf upper surface of *R. rubropilosum* showing the dark green leaf covered with pale grey, adpressed hairs. (Photo R.B.G., E.).

R. saxicolum

Sleumer 1958. H1 2?

Height 3-6m, a shrub, with young shoots at first covered with adpressed, red-brown hairs, soon glabrescent. **Leaves** dimorphic, persistent; spring leaves 4-7.5 x 2-3.5cm, ovate to ovate-oblong, apex acuminate, gland-tipped, margin entire, upper surface glabrescent with a few persistent hairs on the midrib, lower surface with scattered, adpressed, strigose hairs; summer leaves 1.5-2 x 0.5-1cm, otherwise as for spring leaves; petiole strigose. **Inflorescence** 3-4(-5)-flowered, pedicels densely strigose; **Corolla** funnel-shaped, 1.5-2cm long, *white tinged with rose*, outer surface glabrous, inner surface *papillate*; calyx strigose; stamens 5, hairy in lower half, ovary densely strigose, style hairy at base. **Distribution** Vietnam 400-1,800m (2,500-6,000ft), rocky soil in forest.

Characterised by the white to pink corolla, papillate within. The flowers are rather small for the leaf size, so this species is likely to be of limited garden merit.

Introduced from Vietnam in 1992. Likely to be tender.

R. scabrum

G. Don 1834 (*R. liukiuense* Komatzu, *R. sublanceolatum*, Miq. *R. sublateritium* Komatsu, ?*R. yakuinsulare* Masamune). H2

Height 1-2m, a stiff to loosely-branched shrub. Branchlets with adpressed, grey-brown hairs *which gradually disappear*. **Leaves** dimorphic, persistent, spring leaves 3-9 x 2-3.5cm, summer leaves 3-4 x 1-1.5cm, elliptic to lanceolate; upper and lower surfaces with scattered, adpressed, pilose hairs. **Inflorescence** 2-6-flowered. **Corolla** broadly funnel-shaped, *4.5-6cm* long, *5-10cm* across, rose-red to brilliant scarlet; calyx c.5mm with adpressed pilose or glandular-pilose hairs; stamens 10; style *glabrous*. **Distribution** Ryukyu Archipelago, S. Japan, on scrubby hillsides and cliffs.

This species is the largest-flowered in section Tsutsusi. It resembles the equally tender *R. simsii* but differs in the generally larger leaves and flowers, the non-sticky bud scales, the less hairy leaves and branchlets and the glabrous rather than bristly/strigose style.

Picture 1

1. *R. scabrum* in Japan with its characteristically large flowers. (Photo H. Suzuki).

R. scabrum is a fine species for milder areas, quite commonly cultivated in S. Japan but rarely elsewhere. There are several named selections and many hybrids of this species. Introduced to Britain in 1909, U.S.A. in 1919. May.

Picture 2

2. *R. scabrum* in Japan. (Photo T. Takeuchi).

R. serpyllifolium

(A. Gray) Miq. 1865-6. H3-4

Height to 1.2m, a much-branched shrub. Branchlets *very thin*, covered in adpressed, chestnut-brown hairs. **Leaves** *largely deciduous*, monomorphic, *0.3-1 x 0.3-0.5cm*, obovate-oblong to elliptic, upper surface with scattered, short hairs with pustules, lower surface with hairs mainly on the midrib. **Inflorescence** usually 1-flowered. **Corolla** funnel-shaped, c.1.7cm long, pale to rose-pink or white; calyx small; stamens 5; style glabrous. **Distribution** Central and S. Japan on volcanic soil and boulders.

Picture 1

1. *R. serpyllifolium* is easily identified by its thin twiggy habit and mass of tiny leaves and flowers. Warren Berg's garden, Hood Canal, Washington State, W. U.S.A.

This is a distinct species, characterised by its tiny leaves, very thin shoots and small flowers. The white forms could be confused with *R. tschonoskii* which has larger leaves and smaller, much less showy flowers.

R. serpyllifolium is a neat plant, showy as a mature bush, and perhaps best in its white-flowered forms which are sometimes known as var. *albiflorum*. It prefers hot summers to ripen its growth. Introduced 1882. April-May.

Picture 2

2. The white-flowered form of *R. serpyllifolium* in Japan. (Photo H. Suzuki).

R. simsii

Planchon 1854 (*R. annamense* Rehder, *R. bicolor* P.C. Tam, *R. indicum* var. *formosanum* Hayata, *R. indicum* var. *ignescens* Sweet, *R. viburnifolium* W.P. Fang). H1-2

Height 1-3m, a much-branched, sometimes straggly shrub. Branchlets with dense, adpressed, short, shining, brown hairs, *bud scales sticky inside*. **Leaves** usually persistent, dimorphic, spring leaves 3-7 x 1-2cm, summer leaves 1-2 x 0.5-1cm, ovate-lanceolate to linear-elliptic, upper surface with sparse, short, adpressed hairs, lower surface paler with denser, short hairs. **Inflorescence** 2-6-flowered. **Corolla** broadly funnel-shaped, *2.5-6cm long*, various shades of red, spotted (white to rose pink in var. *mesembrinum*); calyx 3-7mm, with short strigose hairs; stamens usually 10; style *strigose at base*. **Distribution** Wide distribution from N.E. Upper Burma, Thailand, Indo-China, W. to E. China, S. Taiwan and extreme S. Japan, 300-2,400m (1,000-8,000ft), dry slopes, cliffs to river banks.

This is a widespread, variable species. Its closest relative is probably *R. scabrum* which generally has larger leaves and flowers, is less hairy and has a glabrous style and non-sticky bud scales.

R. simsii is best known as the chief parent of the common indoor or 'Indica' (a misnomer) azaleas. Its hybrids and selections are widely cultivated in many warm countries, in streets, parks and gardens. Introduced pre. 1812. May (in cultivation).

Picture 1

Picture 2

Picture 3

1. *R. simsii* on the Hunchbacks, Hong Kong New Territories, in April.
2. *R. simsii*, Cangshan, W.C. Yunnan, China, in May.
3. *R. simsii* Guiz 233, Royal Botanic Garden, Edinburgh, from Guizhou, S.C. China.

R. subsessile Rendle 1896. H2?

A much-branched, low shrub. Branchlets with adpressed, brown hairs. **Leaves** persistent, dimorphic, spring leaves 2.5-4 x 0.9-1.2cm, summer leaves 1.5 x 0.7cm, elliptic-lanceolate, midrib prominent, both surfaces *at first clothed with silky rufous-grey hairs,* at maturity with white adpressed hairs only on the upper surface only. **Inflorescence** 2-4-flowered. **Corolla** funnel-campanulate, 1.5-2cm long, c. 2.5cm across, lilac to violet-purple; calyx small with adpressed, brown hairs; stamens 6-10; style with a few adpressed, brown hairs at base. **Distribution** Luzon, Philippines, on mountains in pine, oak and mossy forest.

This species appears to be closely related to the Taiwanese *R. rubropilosum*, differing in its less hairy leaves and in the lilac-purple rather than pink flowers.

We grew *R. subsessile* indoors for a number of years but it is not very showy and seems unlikely to have much future as a garden plant. May.

Picture 1

Picture 2

Picture 3

R. tashiroi
Maxim. 1887. H3-4?

Height 2-6m in wild, usually much less outside Japan, a much-branched shrub. Branchlets *with weak brown hairs at first, becoming glabrous*. **Leaves** *persistent,* probably monomorphic, 4.5-7 x 1.5-2.5cm, *2-3 at ends of branchlets*, elliptic-obovate, both surfaces at first with adpressed, grey-brown hairs, soon becoming glabrous except on the midrib on the lower surface. **Inflorescence** 2-5-flowered. **Corolla** broadly funnel-campanulate, 2-4cm long, ivory pink, pink, to pale rose-purple, spotted; calyx minute, with dense brown hairs; stamens 10(-12); style glabrous. **Distribution** S. Kyushu, Japan to (probably) Taiwan, sea level to 500m (1,600ft), in woods and thickets.

This distinct species provides a possible link between sections Tsutsusi and Brachycalyx as it hybridises quite readily with some species in the latter section. Its closest relative is *R. kanehirai* which has deeper-coloured flowers, with a fewer-flowered inflorescence.

At its best in a suitable climate with summer heat, *R. tashiroi* is a fine plant which resembles *R. yunnanense* when in full flower. It is capable of withstanding drought and high temperatures and it and its hybrids have potential in warm parts of the world. We have attempted to grow *R. tashiroi* at Glendoick but it has always been short-lived, and it remains rare in Europe and North America. April-June.

1. *R. tashiroi* in Japan showing its resemblance to some members of subsection Triflora. (Photo H. Suzuki).
2. *R. tashiroi* in Japan. (Photo H. Suzuki).
3. A fine clone of *R. tashiroi*, selected by K. Wada, growing in Warren Berg's garden, Hood Canal, Washington State, W. U.S.A.

R. tosaense
Makino 1892 (*R. komiyamae* Makino, *R. miyazawae* Nakai & Hara, *R. surugaense* Kurata). H3-4

Height 1.5-2m, a much-branched, *often erect* shrub. Branchlets slender with adpressed, grey-brown hairs. **Leaves** *deciduous or partly persistent*, dimorphic, spring leaves 0.7-4 x 0.2-1cm, summer leaves 0.3-0.7cm long, oblanceolate to oblanceolate-spathulate, both surfaces with scattered, adpressed grey hairs. **Inflorescence** 1-6-flowered. **Corolla** funnel-shaped, 1.8-2.5cm long, pink to lilac-purple or white, lightly flushed pink; calyx minute with adpressed strigose hairs; stamens 5-6(-10?); style glabrous. **Distribution** Shikoku, Kyushu and W. of Mt Fuji, Honshu (*R. komiyamae*), Japan, 0-300m (0-1,000ft), on exposed slopes or in forest.

This species is quite distinct with its almost deciduous leaves, generally upright habit and pink flowers.

Uncommon in cultivation, *R. tosaense* is pretty in its better forms. It generally performs best in climates with a hot summer but has proved quite successful in S. England. Plants under the name *R. komiyamae* are said to have 10 rather than 5-6 stamens. Introduced 1914 and also 1940s and 50s. April-May.

Picture 1

Picture 2

Picture 3

1. *R. tosaense* Komiyamae Group with deep-coloured flowers, Glendoick.
2. A pale pink-flowered *R. tosaense* in Japan. (Photo H. Suzuki)
3. A pretty white-flowered *R. tosaense* in Warren Berg's garden, Hood Canal, Washington State, W. U.S.A. This shows the almost deciduous leaves.

R. tschonoskii Maxim. 1870. H5

Height 0.3-1.5m, a spreading, erect to mat-forming shrub. Branchlets with *dense, adpressed, rufous hairs*. **Leaves** *largely deciduous*, monomorphic, 1-2 x 0.3-1cm, lanceolate to elliptic, both surfaces with scattered, adpressed, long, weak hairs. **Inflorescence** 3-6-flowered. **Corolla** funnel-shaped, *0.7-0.9cm* long, white; pedicels *c.3mm* long; calyx minute with adpressed pale brown strigose hairs; stamens 4-5; style glabrous. **Distribution** Japan, S. Korea, Sakhalin, Kamchatka, on mountain tops, rocks, cliffs and in moist forest.

This is a distinct, almost deciduous species with leaves which turn orange-red to reddish-brown in autumn. The flowers are very small on short pedicels, almost hidden in the leaves.

R. tschonoskii is very hardy but is of little ornamental value apart from its autumn colour. It is uncommon in cultivation. Introduced 1878, June.

Picture 1

Picture 2

1. *R. tschonoskii* showing the tiny white flowers which are small in comparison with the leaf size. Royal Botanic Garden, Edinburgh.
2. *R. tschonoskii* at Glendoick.

R. tsusiophyllum

Sugimoto 1956 (*R. tanakae* Ohwi, *Tsusiophyllum tanakae* Maxim.). H4-5

Height to 0.45m, a mounding to prostrate shrub. Branchlets covered with adpressed strigose hairs. **Leaves** semi-persistent, monomorphic, 1-1.2 x 0.5-0.7cm, obovate, margins ciliate; upper surface +/- glabrous when mature, lower surface glabrous except for the slightly strigose midrib. **Inflorescence** 1-4-flowered. **Corolla** *tubular-campanulate*, c. 0.9cm long, pubescent outside, tube 0.6cm long, pink buds open to white; calyx minute, lobes ciliate; stamens (4-)5; ovary *3-celled*; style glabrous. **Distribution** rare in the wild, confined to a small part of Central Honshu on open mountain slopes.

This is a very distinct species which was formerly classified in a separate genus as *Tsusiophyllum tanakae*. Its chief distinctions are its tubular flowers and its 3-celled ovary (most other species in this section have a 5-celled ovary).

R. tsusiophyllum is a dainty little creeping or mound-forming shrub with small but unusual tubular, white flowers. It is rare in cultivation. Introduced 1915. June.

Picture 1

Picture 2

1. *R. tsusiophyllum* showing the spreading habit. Glendoick.
2. *R. tsusiophyllum* showing the distinctive tubular-campanulate flowers. Glendoick.

R. yedoense var. poukhanense (Lév.)

Nakai 1920 (*R. coreanum* Rehder, *R. hallaisanense* Lév.). H4-5

Height to 2m, usually much less, usually a semi-upright shrub in cultivation though sometimes prostrate in the wild. Branchlets with adpressed strigose hairs, almost glabrous in second year. **Leaves** +/- *deciduous*, dimorphic, 3-8 x 1-2.5cm, elliptic-lanceolate to oblanceolate, upper and lower surfaces with *scattered* adpressed shining brown strigose hairs, summer leaves thicker, upper surface almost glabrous at maturity. **Inflorescence** 2-3-flowered. **Corolla** broadly funnel-shaped, 3.5-4cm long, to 5cm across, *rose to pale lilac-purple*, slightly fragrant; calyx 5-8mm long, with adpressed, strigose hairs; stamens 10; style glabrous or pilose towards base. **Distribution** Central and S. Korea and islands including Tsushima, Japan, c. 1,100m (3,700ft), among rocks and scrub.

This is a fairly distinct species, differing from *R. mucronatum* in its extreme hardiness and its largely deciduous leaves. It has less hairy leaves and branchlets than the other species in the section with a similar flower size.

R. yedoense var. *poukhanense* is one of the toughest azalea species and has been used in hybridisation to impart hardiness. It is quite variable in flower size and colour, but the best forms are quite showy. Due to botanical rules, this taxon has to be classed as a variety of its own cultivated offspring which was described many years earlier as *R. yedoense*. *R. yedoense* var. *yedoense* has double flowers. The wild var. *poukhanense* was introduced to the U.S.A. in 1905, U.K. in 1913->. April-May.

Picture 1

Picture 2

1. Flowers of a deep-coloured selection of *R. yedoese* var. *poukhanense* at Glendoick.
2. A pale-flowered *R. yedoense* var. *poukhanense* at the Royal Botanic Garden, Edinburgh.

Glossary

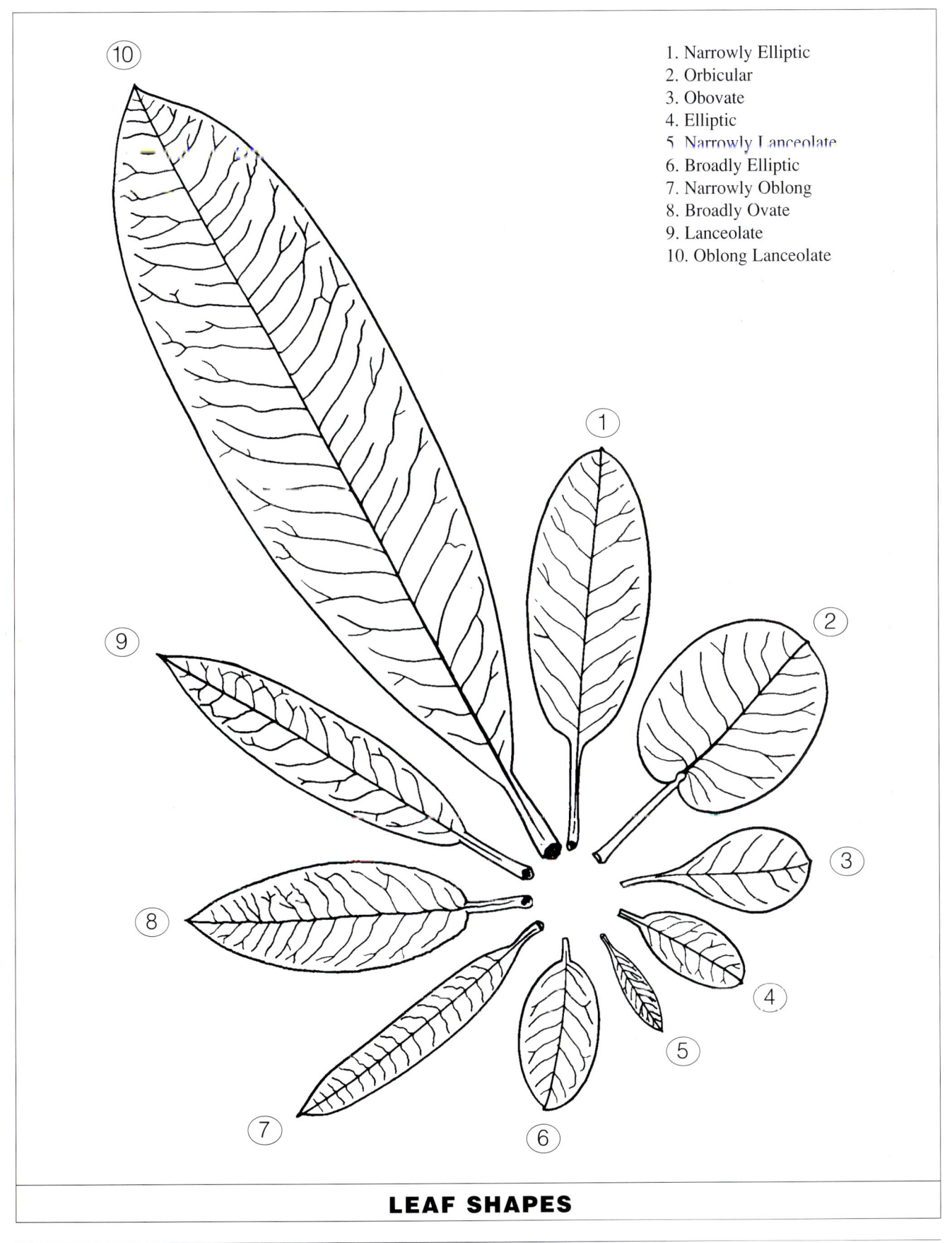

1. Narrowly Elliptic
2. Orbicular
3. Obovate
4. Elliptic
5. Narrowly Lanceolate
6. Broadly Elliptic
7. Narrowly Oblong
8. Broadly Ovate
9. Lanceolate
10. Oblong Lanceolate

LEAF SHAPES

Widely funnel-campanulate

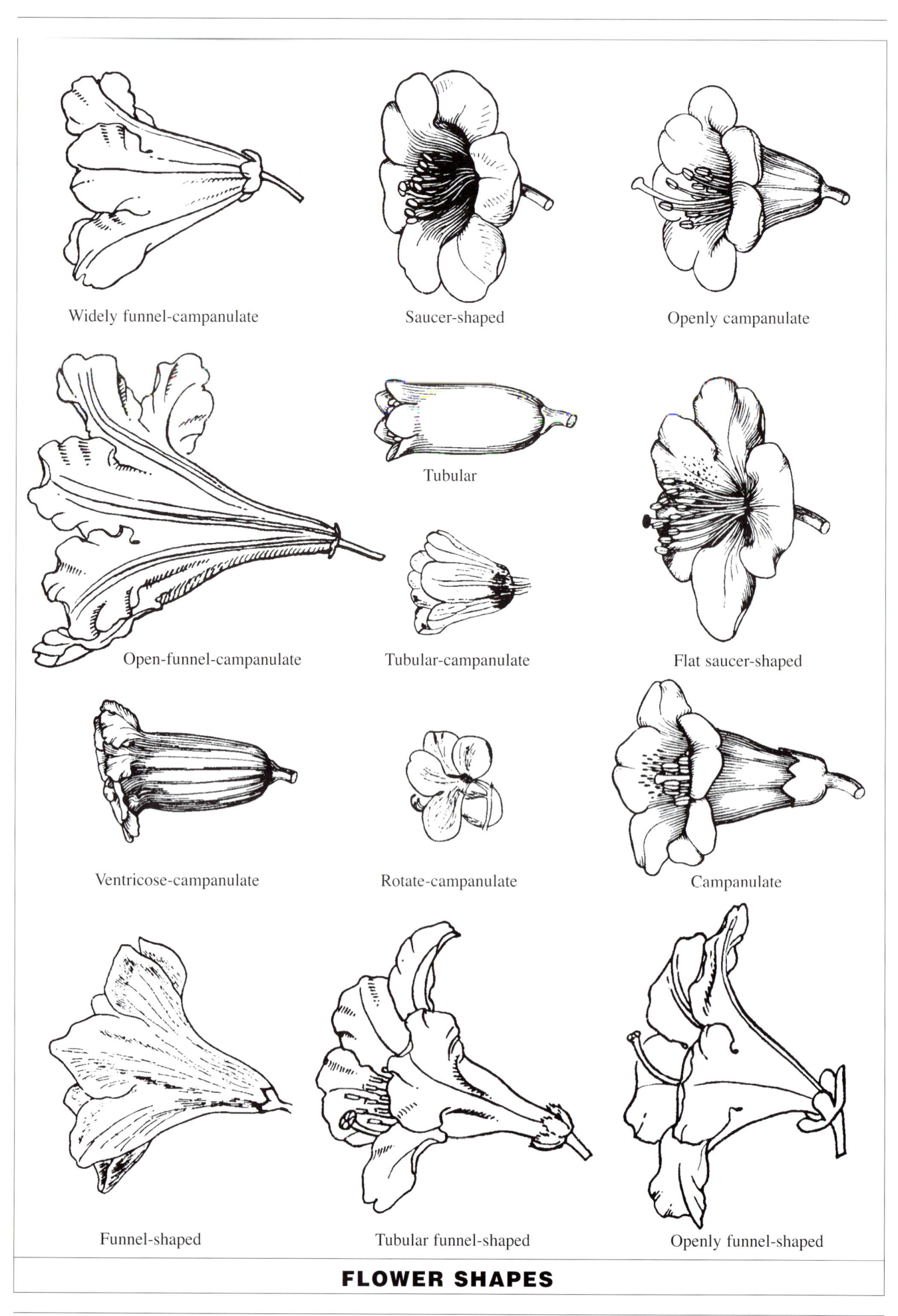

Saucer-shaped

Openly campanulate

Open-funnel-campanulate

Tubular

Tubular-campanulate

Flat saucer-shaped

Ventricose-campanulate

Rotate-campanulate

Campanulate

Funnel-shaped

Tubular funnel-shaped

Openly funnel-shaped

FLOWER SHAPES

Terms are defined as they apply to Rhododendrons; they are often used more widely in other botanical contexts.

Cross-referenced words in CAPITAL LETTERS.

acuminate (leaf) with a long slender point.

adpressed (hairs) held close to a leaf or stem and lying parallel to its surface, but not adherent to it; e.g. the leaf hairs of some of the members of Section Tsutsusi.

affinity or *aff.* affiliated, closely related. This term is used when a taxon does not quite fit the published description of, but obviously shows a relationship with a species. Used for taxa which have puzzled botanists or collectors!

agglutinate (indumentum) compacted and looking as if painted or glued on.

alternate (leaves) arising singly, one at each node; not opposite or WHORLED.

anther the uppermost part of a STAMEN containing the pollen.

apex (leaf) the tip.

apiculate (leaf) a short, sharp but not stiff point to the leaf APEX. e.g. *R. traillianum*.

auriculate a LOBE or ear at the base of the leaf of such species as *R. auriculatum* and *R. orbiculare*.

axillary (bud) coming from an axil which is the angle between a leaf and a stem. Species such as *R. racemosum*, *R. virgatum*, *R. albiflorum* produce flowers from axillary buds.

bark split a physical injury; the sap in the stem is frozen while the plant is in growth, causing the bark to split. It can be fatal in severe cases.

base (leaf) the lower end of the leaf where it joins the PETIOLE.

bistrate (indumentum) in two layers. Usually the lower layer is thin and PLASTERED and the upper layer is thicker.

bullate (leaf) having a blistered or puckered appearance, as in the upper surface of the leaf of *R. edgeworthii*.

calyx, calyces the outermost of the floral envelopes, particularly prominent in species such as *R. thomsonii*, *R. balfourianum*, *R. catacosmum*, *R. megacalyx*. The calyx is often significant in species identification.

candelabroid an upright inflorescence formed by a long RACHIS or/and upright PEDICELS as in *R. hyperythrum*.

capitellate (hairs) mop-headed, the hairs which make up the INDUMENTUM of *R. campanulatum* and *R. fulvum* for instance.

capsule the dry seed vessel.

chlorosis a yellowing of the leaves caused by mineral imbalance, too much or too little water, or climatic or genetic problems.

ciliate (leaf, calyx) fringed with hairs.

cline a population of plants showing a gradual gradation from one species, subspecies or variety to another.

clone A genetically uniform assemblage of individuals derived from a single individual by asexual propagation (i.e. not grown from seed). The finest forms of species are often given clonal names e.g. *R. calostrotum* 'Gigha' F.C.C.

compacted (indumentum) having the appearance of being flattened (often the lower layer of two-layered indumentum) in *R. sphaeroblastum* and *R. beesianum* for example.

contiguous touching (as with scales).

cordate (leaf) heart-shaped with the PETIOLE at the broad end.

coriaceous (leaf) leathery; the leaves of *R. calophytum* or *R. davidii* for example.

corolla the tube and LOBES (petals) of the flower.

crenate toothed, as in the leaf margins of *R. fletcherianum*.

crenulate (scales) with small teeth or notches, as in the scales of *R. baileyi*.

crisped (hair) strongly curved so that the tip lies near the point of attachment. A feature of *R. lanatum*.

cuneate (leaf base) wedge-shaped.

cuspidate tipped with a sharp, rigid point, e.g. the leaf of *R. hidakanum*.

declinate bent or curved downwards as in the style of *R. rubiginosum*.

decumbent more or less horizontal or prostrate for most of its length but erect or semi-erect near the tip.

decurrent (leaf) continued down the stem below the point of attachment as a ridge.

deflexed pointing downwards; the style of species in Subsection Lepidota for example.

dendroid (hair) tree-like or branching.

dimorphic occurring in two forms, used to describe the leaves of many species from Section Tsutsusi.

diploid having two complete sets of chromosomes per cell (2N=26), as opposed to haploid (N=13) which has one set.

discoid having the form of a disc.

discontiguous (scales) not touching

eglandular without GLANDS.

elepidote (opposite of LEPIDOTE) without SCALES. This distinction is of fundamental importance in the classification of rhododendrons.

elliptic (leaf) c. twice as long as broad, tapering equally both to the tip and the base.

emarginate having a notch cut out, as in the APEX of some forms of *R. semnoides*.

entire (leaf) with no teeth or indentations at the margins.

epapillate without PAPILLAE.

epidermis the outer layer of cells of leaves, stems, roots etc.

epiphyte a plant which grows on another plant but which does not derive any nutrient from it. Often growing in moss.

erose as though bitten or gnawed; the CALYX of *R. anthopogonoides* for instance.

evanescent soon disappearing; applies to hairs, INDUMENTUM or TOMENTUM on new growth which persists only a short time, e.g. in *R. facetum* and *R. brachycarpum*.

exserted projecting beyond, usually referring to the STAMENS and/or STYLE projecting from the COROLLA.

fasiculate composed of or growing in bundles.

fastigiate with branches growing more or less erect giving a plant a narrow, tower-like outline. (Confusingly it does not apply to *R. fastigiatum*.)

felted (indumentum) matted with intertwined hairs such as in the INDUMENTUM of *R. phaeochrysum* var. *phaeochrysum* and *R. sphaeroblastum*.

ferrugineous rust coloured.

fimbriate (hairs) fringed with slender processes or appendages: the hairs which make up the indumentum of *R. rex* ssp. *fictolacteum* for instance.

floccose (indumentum or tomentum) in discontinuous loose tufts, on the branchlets of *R. floribundum* and the leaves of *R. floccigerum* for instance.

folioliferous bearing leaves.

fulvous tawny; reddish-yellow.

glabrescent becoming GLABROUS or almost glabrous.

glabrous without hairs or glands of any kind.

gland a hair with a small spherical knob of sticky secreting tissue, hence glandular. Glands can be found on branchlets, leaves and parts of the flower and are an important diagnostic tool in identifying species.

glaucous (leaf) green, strongly tinged with bluish grey; with a greyish waxy bloom, for example in *R. campanulatum* ssp. *aeruginosum*, *R. lepidostylum*, *R. oreotrephes*.

hose-in-hose (flower) a second COROLLA within the first.

hypercrateriform goblet-shaped with spreading lobes; the corolla shape of *R. intricatum*.

impressed describes the way the style joins the ovary: the ovary widens abruptly at the base of the narrow style as opposed to tapering gradually into it.

inflorescence a number of flowers closely grouped together on a stem or axis to form a structural unit. In rhododendrons, this is usually a bud which opens to form a TRUSS of two or more flowers.

indumentum dense woolly or hairy covering found on leaves. The different types are very important for species identification.

lacerate torn or mangled as in the leaf scales of section Pogonanthum.

lamina the blade of a leaf.

lanate (leaf) clothed with woolly and intergrown hairs.

lanceolate (leaf) shaped like a lance head, 3-4 times as long as wide, tapering gradually towards the tip.

lax (inflorescence) one in which the flowers hang downwards, often between the foliage.

lepidote 1/ Subgenus Rhododendron whose species have scales on the leaves. 2/ with SCALES on the shoots, leaves and flower parts.
The opposite of ELEPIDOTE (without scales). This distinction is of fundamental importance in the classification of rhododendrons.

linear (leaf) narrow with parallel opposite sides, several times longer than wide. e.g. *R. trichostomum*.

lobe (part of flower or leaf) any division of an organ, especially rounded. Usually applies to outer parts of the COROLLA and CALYX.

locular having cavities, e.g. tri-locular having 3 cavities or compartments. This usually refers to seed capsules.

long-rayed (hairs) the ends of the hairs divide into several long strands. Several Subsection Taliensia species have such hairs.

midrib the central rib or ridge of a leaf.

monomorphic (leaf) in one form or shape only, as opposed to DIMORPHIC where two distinct shapes of leaf are produced.

mucronate (leaf) abruptly terminating in a sharp point.

mucronulate minutely MUCRONATE.

nectar pouches or ***nectaries*** sac-like vessels in the base of the corolla of some species containing a sweet substance, in most species of Subsections Irrorata and Thomsonia for example.

oblanceolate (leaf) widest near tip, tapering gradually to base, several times longer than wide, *R. fulvum*, *R. uvarifolium* for example.

oblique with unequal sides; the flowers of subsection Grandia for instance.

oblong (leaf) sides of leaf more or less parallel, ends obtuse or somewhat rounded, length of leaf at least twice its breadth, e.g. *R. cerasinum*.

obovate (leaf) as in OVATE but the PETIOLE is attached to the narrower end, in *R. mallotum*, *R. pocophorum* var. *pocophorum*.

obsolete scarsely apparent; often used to describe the CALYX.

orbicular (leaf) circular, e.g. *R. orbiculare*, *R. campylocarpum* ssp. *caloxanthum*.

oval (leaf) egg-shaped but more or less rounded at both ends, e.g. *R. callimorphum*.

ovary the seed-bearing part of the flower; the STYLE and STAMENS are attached to it. When swelling and starting to ripen it becomes the CAPSULE.

ovate (leaf) the approximate shape of a hen's egg, the PETIOLE attached to the wider end with the tip coming to a point.

pannose (leaf) having the appearance or texture of woollen cloth.

papillae nipple like projections of the outer walls of leaf surface cells such as on the leaves of *R. parmulatum*. Hence papillose: covered in papillae and papillate: bearing papillae.

pedicel the stalk supporting a single flower in an inflorescence.

perulae leaf bud scales which cover the leaf bud in winter. These usually drop off as the new leaves unfurl, but in some species such as *R. longesquamatum*, *R. pudorosum* and *R. cephalanthum* they persist and are a useful means of identification.

petiole leaf stalk.

phytophthora a genus of fungi containing many serious pathogenic diseases, especially root-rot, *Phytophthora cinnamoni*.

pilose (leaf) covered in fine soft hair.

plastered (indumentum) indumentum which appears to have been painted on, giving a smooth or polished finish, e.g. *R. insigne*. The lower layer of many species with BISTRATE indumentum is plastered.

powdery mildew a disease which attacks rhododendrons, causing leaf discoloration or leaf drop and which can be fatal in severe infections. Can be controlled with fungicides.

pruinose having a bloom or waxy powdery secretion on the surface.

pseudowhorl a circle of leaves which looks like a whorl. Common in Section Sciadorhodion.

puberulent covered with very fine soft hair.

pubescent covered in short, soft hairs.

punctate (leaf) having dots, depressions or translucent GLANDS, which normally require a magnifying glass to see. Usually on the lower leaf surface. e.g. in subsection Campylocarpa.

punctulate minutely punctate.

punctulation minute gland.

pustule a blister or pimple slightly larger than a PAPILLA. The lower leaf surface of *R. nakaharai* for instance.

raceme inflorescence with flowers evenly spaced on RACHIS, with PEDICELS of equal length with the terminal flowers opening last. Some forms of *R. racemosum* and its relatives have racemes. Hence racemose as in inflorescence of *R. micranthum*.

rachis the axis or main stalk of the INFLORESCENCE or TRUSS.

radiate spreading from a common centre.

ramiform (hairs) branched hairs, forming the INDUMENTUM of *R. lanatum* for instance.

recurved curved backwards or downwards.

reflexed bent sharply backwards from the base.

revolute (leaf) rolled back from the margin or APEX.

rosulate (hairs) collected into a rosette; the hairs in the indumentum of *R. dichroanthum* for instance.

rotate (corolla) wheel-shaped, circular and flat, with a short TUBE.

scabrid warty, as in the stems of the species of subsection Pseudovireya.

scales found on the leaves of LEPIDOTE rhododendrons; these can sometimes be seen with the naked eye but are better magnified so their form can be used to aid identification.

sericious clothed with small, soft straight hairs giving a silky effect.

sessile attached at the base without a stalk.

seta (plural **setae**) a bristle.

setose bristly.

setulose covered with small bristles.

short-rayed (hairs) the ends of the hairs divide into several short strands. Several subsection Taliensia species have such hairs.

shrub a woody plant with several stems or branches from near the base; of smaller stature than a tree.

splitting (indumentum) dividing into patches such as in *R. aganniphum* var. *flavorufum*.

stamen the male organ of the flower which bears the pollen.

stellate (hair) star-shaped.

stigma the part of the STYLE to which pollen adheres.

stoloniferous bearing stolons which are stems which creep along the surface of the ground, taking root at intervals. A few rhododendron species such as *R. canadense* and *R. lowndesii* can spread or increase this way.

style the tubular appendage to the OVARY which bears the STIGMA.

stipitate (glands) elevated on a stalk.

strigose rough with short, sharp, stiff hairs or bristles.

strigillose see STRIGOSE.

subagglutinate somewhat AGGLUTINATE.

suckering producing shoots from near or below ground level, often from rootstocks below the grafted union.

swarm a large group of natural hybrids between two species as in *R. x erythrocalyx*.

taxon (plural **taxa**) a general term for a taxonomic group of any rank e.g. genus, species, subspecies, variety. An entity with a distinct set of shared characters.

taxonomy the science of the identification, nomenclature and classification of plants.

terminal borne at the end of the shoot or branch.

testa the (hard) outer seed coating.

tetraploid having four times the haploid set of chromosomes 2N=52. This usually creates a genetic barrier with non-tetraploid plants.

tomentum PUBESCENCE composed of matted woolly hairs on stems, PETIOLE or OVARY.

tomentose covered with TOMENTUM.

truss a cluster of flowers on a single stalk or RACHIS.

tube the part of the COROLLA, usually narrow and tubular, which adjoins the PEDICEL.

unistrate (indumentum) single-layered.

velutinous velvety; a coating of fine hairs.

ventricose (corolla) inflated or swollen on one side; the flowers of subsection Grandia for instance.

vesicular (scales) bladder-shaped; the shape of the scales of *R. leucaspis* and *R. megeratum* for instance.

villose bearing long weak hairs, the leaves of some members of section Tsutsusi for instance.

viscous sticky.

whorl three or more leaves or branches at a node. Many rhododendrons produce flushes of growth terminating in a circle of leaves and buds known as a whorl.

winged (petiole) broad and flattened, often tapering into leaf base. Species such as *R. praestans* are characterised by this feature.

zygomorphic a COROLLA that can be divided into equal halves on one plane only, usually vertically. A feature of subsection Triflora.

Addendum April 1997

The following are taxonomic develpments included in the *Rhododendron Handbook, Species in Cultivation* R.H.S. 1997 which differ from coverage in this book.

R. arboreum ssp. *delavayi* var. *albomentosum* becomes *R. arboreum* ssp. *albomentosum*.
R. rothschildii may be a hybrid of *R. fulvum* (ssp. *fulvoides*).
R. serotinum Hutch. As originally described from a cultivated plant, this is a late-flowering, straggly plant in subsection Fortunea. It now appears that *R. serotinum* is a 'good' taxon and is a close relative of *R. decorum* from S. Yunnan, N. Vietnam.
R. uvarifolium becomes *R. uvariifolium*.
R. papillatum Balf.f. & Cooper subsection Irrorata is said to be in cultivation.
R. wasonii var. *wenchuanense* tentatively includes all *R. wasonii* with small white to pink flowers. We cannot accept this, as var. *wenchuanense* is described as having subagglutinate indumentum.
R. fragariflorum becomes *R. fragariiflorum*.
R. goreri Davidian (*R. nuttallii*) and *R. grothausii* Davidian (*R. lindleyi*) are both tentatively accepted.
R. parryae reinstated so includes *R. parryae* A.M. clone.
R. mucronulatum var. *chejuense* becomes var. *taquetii* (Lév.) Nakai.
R. keiskei Cordifolia Group becomes *R. keiskei* var. *ozawae* T. Yamaz.
R. mucronatum var. *ripense* becomes *R. ripense*.
R. mucronatum is considered a hybrid, perhaps with *R. stenopetalum*.
R. macrosepalum becomes *R. stenopetalum* (Hogg) Mabb.
R. nakaharai becomes *R. nakaharae*.

Bibliography

BEAN, W.J., *Trees and Shrubs Hardy in the British Isles*, Vol. III, 8th Revised Edition, John Murray, 1976.

CHAMBERLAIN, D.F. et al. *The Genus Rhododendron, its classification & synonymy*. Royal Botanic Gardens Edinburgh, 1996.

CHAMBERLAIN, D.F., *Notes from the Royal Botanic Garden, Edinburgh, Vol. 39, No. 2, Revision of Rhododendron II, Subgenus Hymenanthes*. H.M.S.O. 1982.

CHAMBERLAIN, D.F. & RAE, S.J., *Edinburgh Journal of Botany, Vol. 47, No. 2, Revision of Rhododendron IV, Subgenus Tsutsusi*. Royal Botanic Garden, Edinburgh.

COWAN, J. M. (ed.) *The Journeys & Plant Introductions of George Forrest V.M.H.* R.H.S. London, 1952.

COX, E.H.M. *Farrer's Last Journey*, Dulau, London 1926.

COX, P.A., *Dwarf Rhododendrons*, Batsford, 1973.

COX, P.A., *The Larger Species of Rhododendron*, Batsford, 1979, 1981.

COX, P.A., *The Smaller Rhododendrons*, Batsford, 1985.

COX, P.A., *The Larger Rhododendron Species*, Batsford, 1990.

CULLEN, J., *Notes from the Royal Botanic Garden, Edinburgh, Vol. 39, No. 1, Revision of Rhododendron I Subgenus Rhododendron, sections Rhododendron & Pogonanthum*. H.M.S.O., 1980.

DAVIDIAN, H.H., *The Rhododendron Species, Vol. I, Lepidotes*, Batsford, 1982.

DAVIDIAN, H.H., *The Rhododendron Species, Vol. II, Elepidotes, Series Arboreum to Lacteum*, Batsford, 1989.

DAVIDIAN, H.H., *The Rhododendron Species, Vol. III, Elepidotes continued*, Timber Press, 1992.

DAVIDIAN H.H., *The Rhododendron Species, Vol IV, Azaleas*, Timber Press, 1995.

FANG, WENPEI (edited) *Sichuan Rhododendron of China*, Science Press, Beijing, 1986.

FENG, GUOMEI (edited), *Rhododendrons of China, Vol. I*, Science Press, Beijing, 1989.

FENG, GUOMEI (edited), *Rhododendrons of China, Vol. II*, Science Press, Beijing, 1992.

FLETCHER H.R. *A Quest of Flowers*, (Ludlow & Sherriff journeys), Edinburgh University Press, 1975.

GALLE, FRED C., *Azaleas*, Timber Press, 1985.

HARJMA H. New names and nomenclatural combinations in Rhododendron (Ericaceae) *Ann. Bot. Fennici* 27, 1990

HARJMA, H. Taxonomic notes on *Rhododendron* subsection *Ledum* (*Ledum*, Ericaceae), with a key to its species. *Ann. Bot. Fennici* 28 pps 171-173, 1991

JUDD, W.S. & KRON, K.A., *Edinburgh Journal of Botany, Vol. 52(1) A Revision of Rhododendron, Subgenus Pentanthera (sections Sciadorhodion, Rhodora and Viscidula*, Royal Botanic Garden, Edinburgh.

KINGDON-WARD, F., *The Romance of Plant Hunting*, Arnold, 1924. + many other works.

KRON, K.A., *Edinburgh Journal of Botany, Vol. 50, No. 3, A Revision of Rhododendron section Pentanthera*, Royal Botanic Garden, Edinburgh.

LANCASTER R. *Travels in China, A Plantsman's Paradise*, Antique Collectors Club, 1989.

LEACH, D.G., *Rhododendrons of the World*, Allen & Unwin, 1962.

LESLIE, A. (compiler), *The Rhododendron Handbook 1980, Rhododendron Species in Cultivation*, The Royal Horticultural Society, 1980.

MILLAIS, J.G., *Rhododendrons and the Various Hybrids*. Longmans, 1917. Ditto Second Series 1922.

PUICHEUNG T. *A Survey of Genus Rhododendron in South China*. Worldwide Publications, 1983.

PHILIPSON, M.N. & PHILIPSON, W.R., *Notes from the Royal Botanic Garden, Edinburgh, Vol. 34, No. 1, A Revision of Rhododendron section Lapponicum*, H.M.S.O., 1975.

PHILIPSON, M.N. & PHILIPSON, W.R., *Notes from the Royal Botanic Garden, Edinburgh, Vol. 44, No. 1, A Revision of Rhododendron, Subgenera Azaleastrum, Mumeazalea, Candidastrum and Therorhodion*, H.M.S.O., 1986.

POSTAN, C. (ed.) *The Rhododendron Story*, Royal Horticultural Society, 1996.

SLEUMER, H. *Flora Malesiana ser.* 1 6: 474-668 1968.

STEVENSON, J.B., (edited) *The Species of Rhododendron*, The Rhododendron Society, 1930, 1947.

WILSON, E.H., *A Naturalist in Western China*, Methuen, 1913.

YOUNG, J., & LU-SHENG CHONG., (translated), *Rhododendrons of China*, Binford & Mort, 1980.

Index

Names in roman type refer to taxa at specific status. Those in italics are subspecies, varieties or sunk taxa.

aberconwayi	87
aberrans	181
achroanthum	274
acraium	231
adansonii	147
adenogynum	156
adenophorum	156
adenopodum	16
adenostemonum	91
adenosum	67
admirabile	92
adoxum	61
adroserum	92
aechmophyllum	351
aemulorum	113
aeruginosum	29
Afghanica subsection	234
afghanicum	234
aganniphum	157
var. aganniphum	157
var. flavorufum	157
agapetum	135
agetum	130
agglutinatum	169
aiolopeplum	169
aiolosalpinx	196
aishropeplum	176
alabamense	205
albertsenianum	128
albiflorum	8
albrechtii	220
algarvense	147
alpicola	270
alutaceum	159
var. alutaceum	159
var. iodes	160
var. russotinctum	159
amagianum	369
amamiense	5
amaurophyllum	320
ambiguum	336
amesiae	340
annae	87
ssp. annae	88
ssp. laxiflorum	88
annamense	381
anthopogon	225
ssp. anthopogon	225
ssp. hypenanthum	225
anthopogonoides	226
anthosphaerum	88
argyi	377
anwheiense	99
aperantum	114
apiculatum	339
apodectum	117
araiophyllum	89
araliiforme	61
Arborea subsection	9
arborescens	205
arboreum	9
ssp. arboreum	10
ssp. cinnamomeum	10
ssp. cinnamomeum var. cinnamomeum	10

ssp. cinnamomeum var. roseum	11
ssp. delavayi	12
ssp. delavayi var. delavayi	12
ssp. delavayi var. peramoenum	11
ssp. delavayi var. albomentosum	12, 390
ssp. delavayi aff.	12
ssp. nilagiricum	13
ssp. zeylanicum	13
argenteum	75
argipeplum	25
argyi	377
Argyrophylla subsection	15
argyrophyllum	16
ssp. argyrophyllum	16
ssp. hypoglaucum	16
ssp. omeiense	17
ssp. nankingense	17
var. cupulare	16
var. leiandrum	17
artosquameum	353
astrocalyx	35
arizelum	37
ashleyi	146
asmenistum	122
assamicum	299
asteium	119
asterochnoum	49
atentsiense	301
atlanticum	206
atrovirens	371
atroviride	339
aucklandii	55
aucubaefolium	7
Augustinii/Triflorum alliance	336
augustinii	337
ssp. augustinii	337
ssp. chasmanthum	338
ssp. hardyi	339
ssp. rubrum	339
aureum Franch.	330
aureum Georgi	136
var. aureum	136
var. hypopytis	136
auriculata subsection	24
auriculatum	24
auritum	327
australe	2
austrinum	207
axium	153
Azalea albiflora	8
aurantiaca	208
californica	212
coccinea	208
crocea	208
danielsiana	372
ferruginosa	268
flammea	208
fragrans	205
glabrius	217
japonica	217
kaempferi	373
ledifolia	377
liliiflora	377
macrantha	372
mollis	217
myrtifolia	1
parvifolia	268
pontica	211
semibarbatum	203
squamata	365
Azaleastrum albiflorum	8
Azaleastrum subgenus	1

Azaleastrum section	1
bachii	3
baeticum	147
Baileya subsection	234
baileyi	234
bainbridgeanum	150
bakeri	210
balangense	74
balfourianum	160
var. *aganniphoides*	160
balsaminiflorum	372
Barbata subsection	25
barbatum	26
basilicum	38
basfordii	292
batangense	270
bathyphyllum	161
beanianum	108
var. *compactum*	111
beesianum	185
beimaense	151
benthamianum	339
bergii	339
bhotanicum	292
bhutanense	162
bicolor Pursh.	209
bicolor P.C. Tam	381
blandfordiiflorum	244
blandulum	154
blepharocalyx	267
blinii	342
bodinieri	351
Boothia subsection	235
boothii	236
brachyanthum	251
ssp. brachyanthum	251
ssp. hypolepidotum	252
Brachycalyx section	364
brachycarpum	137
ssp. brachycarpum	137
ssp. fauriei	137
var. *roseum*	137
var. *rufescens*	137
var. *tigerstedtii*	137
brachysiphon	287
bracteatum	257
breviperulatum	371
brevistylum	257
brevitubum	287
breynii	372
brunneifolium	119
bullatum	247
bulu	260
bureavii	163
bureavioides	164
burmanicum	296
burmannii	377
butyricum	236
caeruleo-glaucum	240
caeruleum	349
caesium	331
calciphilum	317
calendulaceum	208
californicum	144
callimorphum	32
var. callimorphum	32
var. *myiagrum*	32
calostrotum	317
ssp. calostrotum	317
ssp. keleticum	319
ssp. riparioides	317
ssp. riparium	317
calophyllum	287

calophytum	49
var. calophytum	50
var. openshawianum	50
caloxanthum	33
Camelliiflora subsection	239
camelliiflorum	239
Campanulata subsection	29
campanulatum	29
ssp. aeruginosum	29
ssp. campanulatum	29
campbelliae	10
Campylocarpa subsection	32
campylocarpum	33
ssp. caloxanthum	33
ssp. campylocarpum	33
Campylogyna subsection	240
campylogynum	240
campylogynum var. *celsum*	240
camtschaticum	362
ssp. camtschaticum	362
ssp. glandulosum	363
canadense	219
Candidastrum subgenus	8
candidum	209
canescens	209
cantabile	275
capitatum Maxim.	261
capitatum Franch.	263
cardiobasis	57
cardoeoides	353
carneum	304
Caroliniana subsection	242
carolinianum	242
caryophyllum	379
catacosmum	108
cataplastum	258
catawbiense	138
caucasicum	139
cavaleriei var. *chaffanjonii*	7
cephalanthoides	231
cephalanthum	226
ssp. cephalanthum	226
ssp. platyphyllum	227
ceraceum	92
cerasiflorum	240
cerasinum	189
cerinum	238
cerochitum	94
chaetomallum	110
chamaethomsonii	125
var. chamaedoron	125
var. chamaethauma	126
var. chamaethomsonii	125
chamaetortum	226
chameunum	320
championae	4
chapaense	287
chapmanii	243
charalocladum	153
charianthum	348
charidotes	320
charitostreptum	252
charitopes	252
ssp. charitopes	252
ssp. tsangpoense	252
charopoeum	240
chartophyllum	351
chasmanthoides	338
chasmanthum	338
chawchiense	88
cheilanthum	262
chengianum	55
chienianum	21

chiengshienianum	336	cumberlandense	210	*dumulosum*	169	ssp. *forrestii*	127
Chioniastrum section	4	*cuneatum*	262	eclecteum	191	ssp. *papillatum*	127
chionophyllum	16	*cupressens*	169	var. *bellatulum*	191	Fortunea subsection	49
chlanidotum	115	*cuthbertii*	242	var. *brachyandrum*	191	*fortunei*	53
chloranthum	333	cyanocarpum	190	Edgeworthia subsection	247	ssp. *discolor*	53
chrysanthemum	124	var. *eglandulosum*	190	edgeworthii	247	ssp. *fortunei*	53
chrysanthum	136	var. *eriphyllum*	190	*elaeagnoides*	284	*foveolatum*	39
chryseum	275	*cyclium*	32	elegantulum	165	Fragariflora subsection	249
chrysodoron	236	*cymbomorphum*	151	elliottii	133	*fragariflorum*	249, 390
ciliatum	296	*dabanshanense*	175	*ellipticum*	6	*fragrans* (Adams) Maxim.	vii, 232
Ciliicalyx/Johnstoneanum alliance		Dalhousiae alliance	289	*emaculatum*	185	*fragrans* Franch.	233
	295	dalhousiae	289	*epipastum*	119	*franchettianum*	52
Ciliicalyx aggregate	303	var. *dalhousiae*	287	*eriandrum*	349	*fuchsiiflorum*	326
ciliicalyx	307	var. *rhabdotum*	287	*erileucum*	347	fulgens	63
cinereoserratum	365	*damascenum*	240	eriocarpum	372	Fulgensia subsection	63
cinereum	262	dasycladoides	154	eriogynum	134	*fulvastrum* var. *mesopolium*	119
Cinnabarina subsection	243	*dasycladum*	153	*eritimum*	88	Fulva subsection	63
cinnabarinum	244	dasypetalum	263	erosum	26	*fulvoides*	64
ssp. cinnabarinum	244	dauricum	314	x erythrocalyx	151	fulvum	64
ssp. tamaense	245	*dauricum* var. *mucronulatum*	315	esetulosum	151	ssp. fulvoides	64
ssp. xanthocodon	245	davidii	51	euanthum	61	ssp. fulvum	64
var. *pallidum*	245	davidsonianum	348	eucallum	151	*fumidum*	257
var. *pupurelllum*	245	decandrum	366	euchaites	130	galactinum	42
cinnamomeum	10	x decipiens	43	eudoxum	118	Genestieriana subsection	250
citriniflorum	115	decorum	51	var. brunneifolium	119	genestierianum	250
var. *aureolum*	115	ssp. decorum	52	var. eudoxum	119	gibsonii	299
var. *citriniflorum*	115	ssp. diaprepes	52	var. mesopolium	119	giganteum	81
var. *horaeum*	115	*decumbens*	372	euonymifolium	360	*giganteum* var. *seminudum*	81
clementinae	164	*deflexum*	345	eurysiphon	196	x giraldii	60
clivocolum	231	degronianum	140	exasperatum	27	*giraudiasii*	52
coccinopeplum	176	ssp. degronianum	141	excellens	290	*glabrius*	217
codonanthum	173	ssp. heptamerum	141	*eximium*	40	glanduliferum	54
coelicum	109	ssp. heptamerum var. heptamerum		*exquisetum* W.W. Sm.	353	*glandulistylum*	368
coeloneuron	165		141	*exquisitum* T.L. Ming	106	*glandulosum*	363
collettianum	228	ssp. heptamerum var. hondoense		faberi	166	*glaphyrum*	124
colletum	185		141	*faberioides*	166	Glauca subsection	251
colobodes	320	ssp. heptamerum		*facetum*	134	*glauco aureum*	240
comisteum	174	var. kyomaruense	141	Falconera subsection	37	glaucophyllum	253
commodum	238	ssp. yakushimanum	142	falconeri	40	var. album	253
compactum	273	var. *pentamerum*	141	ssp. eximium	40	var. glaucophyllum	253
complexum	261	dekatanum	vii	ssp. falconeri	40	var. tubiforme	254
concatenans	245	*delavayi*	11	*fargesii*	58	*glaucopeplum*	157
concinnum	339	*deleiense*	329	Farrerae alliance	365	*glaucum*	253
concinnoides	viii	dendricola	301	farrerae	365	Glischra subsection	67
confertissimum	268	*dendritrichum*	66	fastigiatum	263	*glischroides*	69
cookeanum	105	dendrocharis	309	faucium	192	glischrum	70
coombense	339	denudatum	18	*fauriei*	137	ssp. glischrum	70
cooperi	239	*depile*	353	var. *roseiflorum*	137	ssp. rude	70
cordatum	34	desquamatum	258	ferrugineum	311	var. adenosum	67
coreanum	384	x detonsum	157	*fictolacteum*	45	globigerum	159
coriaceum	39	*diacritum*	278	*fimbriatum*	266	*globigerum* as cultivated	176
coruscum	94	*diaprepes*	52	*fissotectum*	157	*gloeblastum*	35
coryanum	18	dichroanthum	116	flammeum	210	*glomerulatum*	280
coryi	216	ssp. apodectum	117	flavantherum	viii	*gnaphalocarpum*	365
coryphaeum	80	ssp. dichroanthum	117	*flavidum*	264	gongshanense	vi
cosmetum	320	ssp. scyphocalyx	117	*flavorufum*	157	*goreri*	293, 390
costulatum	342	ssp. septentrionale	117	*flavum*	211	grande	75
cowanianum	284	*dichropeplum*	169	fletcherianum	298	Grandia subsection	74
coxianum	297	*dictyotum*	181	fleuryi	308	gratum	45
crassum	287	dielsianum	326	flinckii	95	Griersoniana subsection	85
crebreflorum	226	dignabile	186	*floccigerum*	129	griersonianum	85
cremastum	240	dilatatum	366	var. *appropinquans*	130	*griffithianum*	55
cremnastes	284	*dilatatum* var. *boreale*	367	floribundum	18	*groenlandicum*	281
cremnophilum	231	diphrocalyx	72	fokienense	23	*grothausii*	292, 390
crinigerum	68	*discolor*	53	*fordii*	23	*gymnanthum*	92
var. *crinigerum*	68	*docimum*	151	formosum	298	*gymnocarpum*	119
var. *euadenium*	68	*dolerum*	153	var. formosum	299	*gymnogynum*	88
crispiflorum	372	*doshongense*	157	var. inaequale	299	*gymnomiscum*	231
croceum	35	*drumonium*	278	formosanum	19	habrotrichum	72
cruentum	163	*dryophyllum* part	169	Forrestii alliance	125	Haematodes alliance	108
cubittii	307	*dryophyllum* part	169	forrestii	126	haematodes	110
cucullatum	176	*dubium*	94	var. *repens*	127	ssp. chaetomallum	110
cuffeanum	299	*(x) duclouxii*	326	var. *tumescens*	127	ssp. haematodes	110

393

haemonium	225	inopinum	183	*laxiflorum*	88	*macrosmithii*	25
hallaisanense	384	insigne	20	*leclerei*	258	Maculifera subsection	98
hanceanum	328	x intermedium	312	*leucandrum*	349	*maculiferum*	100
hancockii	5	*intortum*	169	ledebourii	314	ssp. *anwheiense*	99
hannoense var. *lateritium*	372	intricatum	267	*ledoides*	233	Maddenia subsection	286
haofui	19	*ioanthum*	349	Ledum subsection	281	maddenii	287
hardingii	88	*iochanense*	324	Ledum subarcticum	283	ssp. *crassum*	287
hardyi	339	iodes	159	*californicum*	282	ssp. *maddenii*	287
harrovianum	343	Irrorata subsection	86	*glandulosum*	282	var. *longiflora*	287
headfortianum	294	irroratum	90	*groenlandicum*	281	magnificum	78
hedythamnum	32	ssp. *irroratum*	91	*hypoleucum*	282	*mairei*	186
var. *eglandulosum*	190	ssp. *pogonostylum*	91	*latifolium*	281	makinoi	145
heftii	31	ssp. *ningyuenense*	91	*macrophyllum*	283	Mallotum alliance	113
heishuense	350	*iteophyllum*	299	*minus*	283	mallotum	113
Heliolepida subsection	256	*ixeunticum*	68	*nipponicum*	282	*mandarinorum*	53
heliolepis	257	*jahandiezii*	349	*pacificum*	181	*manipurense*	287
var. *brevistylum*	257	*jangtzowense*	117	*palustre*	283	*mannophorum*	122
var. *fumidum*	257	*japonicum* A. Gray	217	*palustre* var. *decumbens*	283	*manopeplum*	151
var. *heliolepis*	257	*japonicum* (Blume) Schneider	141	*palustre* var *dilatatum*	283	mariesii	365
helvolum	169	jenkinsii	287	*palustre* var. *diversilobum*	282	*martinianum*	152
hemidartum	112	*Johnstoneanum aggregate*	301	*leilungense*	350	maximum	146
x hemigynum	113	johnstoneanum	302	*lelopodum*	6	*maximum* var. *purpureum*	146
hemitrichotum	322	*jucundum*	154	*lemeei*	342	mayebarae	367
hemsleyanum	55	kaempferi	373	*lepidanthum*	231	meddianum	194
heptamerum	88	var. *japonicum*	374	lepidostylum	331	var. *atrokermesinum*	195
herpesticum	117	kanehirai	374	Lepidota subsection	284	var. *meddianum*	195
hesperium part	349	kasoense	viii	lepidotum	284	megacalyx	295
hesperium part	351	kawakamii	360	*leprosum*	258	*megaphyllum*	38
hevolum	169	*kawakamii* var. *flaviflorum*	360	leptocarpum	236	megeratum	238
hexamerum	61	keiskei	341, 390	*leptocladon*	vii, 305	mekongense	332
hillieri	110	*keleticum*	319	*leptosanthum*	6	var. *longipilosum*	334
hidakanum	367	kendrickii	91	leptothrium	2	var. *mekongense*	333
hippophaeoides	265	kesangiae	76	*leucandrum*	349	var. *melinanthum*	333
var. *hippophaeoides*	265	var. *album*	77	*leucanthum*	377	var. *rubrolineatum*	333
var. *occidentale*	266	var. *kesangiae*	77	leucaspis	237	*melinanthum*	333
hirsuticostatum	338	keysii	246	*leucobotrys*	6	*mesopolium*	119
hirsutum	312	*kialense*	174	*leucolasium*	20	*metrium*	153
hirtipes	151	*kingdonii*	317	levinei	291	*metternichii*	141
hispidum	216	kiusianum	374	*levistratum*	169	ssp. *pentamerum*	141
hodgsonii	43	var. *kiusianum*	374	*liliiflorum*	291	var. *pentamerum* forma	
hodgsonii affinity	43	var. *sataense*	375	*linearifolium* var. *macrosepalum*	375	*angustifolium*	145
hongkongense	1	kiyosumense	367	lindleyi	292	microgynum	119
hookeri	193	*klossii*	6	liratum	117	microphyton	376
horaeum	115	komiyamae	382	lithophilum	334	Micrantha subsection	308
var. *aureolum*	115	kongboense	229	*litangense*	266	micranthum	308
horlickianum	304	kotschyi	313	*litiense*	35	*microleucum*	272
hormophorum	351	*kuluense*	67	*liukiuense*	380	*micromeres*	236
hortense	375	*kwangfuense*	53	longesquamatum	99	*microterum*	185
houlstonii	53	kyawii	135	*longicalyx*	34	mimetes	167
huianum	57	Lacteum alliance	184	*longifolium*	75	var. *mimetes*	167
hunnewellianum	20	lacteum	186	longipes	21	var. *simulans*	167
hutchinsonianum	339	*lacteum* var. *macrophyllum*	45	var. *chienianum*	21	miniatum	vii
hylaeum	193	laetevirens	339	var. *longipes*	21	minus	242
Hymenanthes subgenus	9	*lampropeplum*	173	longistylum	329	var. *chapmanii*	243
hypenanthum	225	*lamprophyllum*	3	*lophogynum*	334	var. *minus*	242
hyperythrum	143	Lanata subsection	94	*lophophorum*	169	*mishmiense*	236
hypoglaucum	16	lanatoides	95	*lopsangianum*	199	*mirabile*	250
hypoleucum	282	lanatum	96	lowndesii	285	missionarum	307
hypophaeum	350	*lancifolium* Hook.f.	25	*lucidum* Franch.	61	*miyazawae*	382
hypopytis	136	*lancifolium* Moench	147	*lucidum* Nutt.	239	modestum	296
hypotrichum	353	langbianense	91	luciferum	97, vi	molle	217
idoneum	278	lanigerum	13	ludlowii	355	ssp. *japonicum*	217
igneum	246	*laoticum*	6	ludwigianum	305	ssp. *molle*	217
imberbe	25	Lapponica subsection	259	lukiangense	92	*mollicomum*	322
impeditum	266	lapponicum	268	luteiflorum	254	var. *rockii*	322
imperator	354	*lasiopodum*	307	lutescens	342	*mollyanum*	79
inaequale	299	*laticostum*	341	luteum	211	*mombeigii*	66
indicum	372	latifolium	146	lyi	305	Monantha subsection	viii
var. *amoenum* f. *japonicum*	374	latoucheae	5	macabeanum	77	monanthum	viii
var. *formosanum*	381	var. *ionanthum*	5	*mackenzianum*	6	montroseanum	79
var. *ignescens*	381	laudandum	229	*macrogemmum*	373	morii	101
var. *japonicum*	374	var. *laudandum*	230	*macrophyllum*	144	*motsouense*	324
inobeanum	366	var. *temoense*	230	macrosepalum	375	moulmainense	6

Moupinensia subsection	309	ssp. cardiobasis	57	*pogonostylum*	91	*regale*	38
moupinense	310	ssp. orbiculare	57	*poilanei*	360	*repens*	126
mucronatum	377	*oblongifolium*	216	*polifolium*	279	Reticulatum alliance	365
var. mucronatum	377	*oreinum*	270	*poliopeplum*	122	*reticulatum*	366
var. ripense	377	oreodoxa	58	*poluninii*	97	rex	45
mucronulatum	315	var. fargesii	58	*polyandrum*	287	ssp. fictolacteum	45
var. chejuense	315	var. oreodoxa	58	polycladum	273	ssp. gratum	45
var. mucronulatum	315	var. shensiense	58	polylepis	343	ssp. rex	45
muliense	275	oreotrephes	353	Pontica subsection	135	*rhabdotum*	289
Mumeazalea subgenus	203	*oreotrephoides*	353	Ponticum section	9	*rhaibocarpum*	153
myiagrum	32	*oresbium*	270	ponticum	147	*rhantum*	61
myrtifolium	313	*oresterum*	35	var. baeticum	147	Rhododendron section	234
myrtilloides	240	orthocladum	272	var. brachycarpum	147	Rhododendron subgenus	224
nagasakianum	367	var. microleucum	272	*porphyroblastum*	176	Rhododendron subsection	311
nakaharai	377, 390	var. orthocladum	272	*porrosquameum*	257	*rhodora*	219
nakotaisanense	101	*osmerum*	275	*pothinum*	124	Rhodora canadensis	219
nanothamnum	153	*oulotrichum*	334	*praeclarum?*	231	Rhodora section	218
nanum	263	*ovatosepalum*	378	praestans	80	Rhodorastra subsection	314
nebrites	122	ovatum	3	praeteritum	59	*Rhodothamnus camtschaticus*	362
neglectum	206	*oxyphyllum*	6	praevernum	60	rhombicum	366
nematocalyx	6	pachypodum	306	*prasinocalyx*	35	rigidum	349
neoglandulosum	282	pachysanthum	102	prattii	170	*riparium*	317
Neriiflora subsection	107	pachytrichum	103	preptum	44	ririei	22
Neriiflorum alliance	128	*pagophilum*	153	primuliflorum	231	*rivulare*	317
neriiflorum	130	*paludosum*	270	*primulinum*	264	roseatum	307
ssp. agetum	130	palustre	268	principis	171	*roseotinctum*	122
ssp. neriiflorum	130	*pamprotum*	320	prinophyllum	214	roseum	214
ssp. phaedropum	130	*pankimense*	91	pritzelianum	308	*rosmarinifolium*	377
nigroglandulosum	168	*panteumorphum?*	151	*probum*	153	*rosthornii*	308
nigropunctatum	270	paradoxum	184	procerum	146	rothschildii	47
nikoense	221	Parishia subsection	133	*pronum*	172	*rotundifolium*	57
x nikomontanum	136	parmulatum	120	*prophantum*	135	roxieanum	176
nilagiricum	12	*parryae type only*	302	*proquinquum*	274	var. cucullatum	176
ningyuenense	91	*parryae A.M.*	307	prostratum	320	*var. globigerum*	159
niphargum	66	*parviflorum*	268	proteoides	173	var. oreonastes	177
nipholobum	196	*parvifolium*	268	protistum	81	*var. parvum*	177
nipponicum	224	*patulum*	354	var. giganteum	81	var. roxieanum	176
nishiokae	28	*pectinatum*	6	protistum affinity	vi	*roylei*	244
nitens	317	pemakoense	356	przewalskii	174	*rubiginosum*	258
nitidulum	269	pendulum	248	var. dabanshanense	175	*rubriflorum*	240
var. nitidulum	269	*pennivenium*	94	var. przewalskii	175	*rubrolineatum*	333
var. nubigerum	269	Pentanthera subgenus	204	pruniflorum	255	*rubroluteum*	335
var. omeiense	270	section	204	prunifolium	215	*rubropilosum*	379
nitidum	216	subsection	204	pseudochrysanthum	104	*rubropunctatum* Hayata	143
nivale	270	pentaphyllum	221	pseudociliipes	viii	rubro-punctatum Lev. & Vaniot	349
ssp. australe	271	*peramabile*	267	*pseudociliicalyx*	307	*rude*	70
ssp. boreale	270	*peramoenum*	11	Pseudovireya subsection	359	rufum	178
ssp. nivale	270	periclymenoides	213	*pseudoyanthinum*	339	rupicola	274
niveum	14	*persicinum*	88	pubescens	323	var. chryseum	275
nmaiense	226	*perulatum*	119	x pubicostatum	163	var. muliense	275
noriakianum	376	petrocharis	viii, 309	*pubigerum*	353	var. rupicola	274
notatum	301	*phaedropum*	130	pudorosum	82	rushforthii	361
nudiflorum	213	phaeochrysum	168	pumilum	356	russatum	275
nudipes	367	var. agglutinatum	169	punctatum	242	*russotinctum*	159
ssp. niphophilum	367	var. levistratum	169	puniceum	10	*sakawanum*	366
ssp. nudipes	367	var. phaeochrysum	169	*puralbum*	36	*salignum*	284
ssp. nudipes var. lagopus	367	affinity yellow	vii	purdomii	175	Saluenensia subsection	316
nuttallii	293	*phaeochlorum*	353	*purpureum*	146	saluenense	319
oblongifolium	216	*platyphyllum*	227	purshii	146	ssp. chameunum	320
oblongum	55	*pholidotum*	257	pycnocladum	278	ssp. saluenense	320
obtusum var. japonicum	374	piercei	111	quinquefolium	222	sanctum	369
obovatum	284	*pilicalyx*	306	racemosum	324	Sanguineum alliance	114
obscurum	349	pingianum	22	radicans	319	sanguineum	121
occidentale	212	*pittosporifolium*	7	radinum	233	*ssp. aizoides*	122
ochraceum	102	planetum	60	ramosissimum	270	*ssp. consanguineum*	122
odoriferum	287	platypodum	vi	ramsdenianum	93	ssp. didymum	122
officinale	136	plebeium	257	randaense	379	*ssp. mesaeum*	122
oldhamii	378	pleistanthum	352	rarosquameum	349	ssp. sanguineum	121
oligocarpum	vii, 101	pocophorum	112	*rasile*	52	ssp. sanguineum var. cloiophorum	
ombrochares	94	var. hemidartum	112	*ravum*	262		122
openshawianum	50	var. pocophorum	112	recurvoides	72	ssp. sanguineum var. didymoides	
oporinum	257	*poecilodermum*	176	*recurvum*	177		122
orbiculare	57	Pogonanthum section	225	redowskianum	364		

ssp. sanguineum var. haemalcum	122	sinolepidotum	284	var. albipetalum	119	venator	201
ssp. sanguineum var. himertum	122	sinonuttallii	293	var. rhodanthum	119	Venatora subsection	201
		sinovirgatum	358	Tephropepla subsection	327	vernicosum	61
ssp. sanguineum var. sanguineum	122	smilesii	307	tephropeplum	329	vesiculiferum	74
		smirnowii	148	thayerianum	23	vialii	3
santapauii	361	smithii	25	theiochroum	238	viburnifolium	381
saravanense	305	sonomense	212	theiophyllum	169	vicarium	270
sargentianum	232	sordidum	255	Therorhodion camtschaticum	362	vicinum	169
saxicolum	379	souliei	34	Therorhodion glandulosum	363	villosum	344
Scabrifolia subsection	321	sparsiflorum	239	Therorhodion subgenus	362	vilmorinianum type only	337
scabrifolium	325	speciosum Salis.	147	Thomsonia subsection	189	violaceum	270
var. pauciflorum	326	speciosum (Willd.) Sweet	210	thomsonii	198	Vireya section	359
var. scabrifolium	325	sperabile	131	ssp. lopsangianum	199	Virgata subsection	358
var. spiciferum	325	var. sperabile	131	ssp. thomsonii	198	virgatum	358
scabrum	380	var. weihsiense	131	var. candelabrum	199	ssp. oleifolium	358
schizopeplum	157	sperabiloides	132	var. cyanocarpum	190	ssp. virgatum	358
schlippenbachii	223	sphaeranthum	233	var. pallidum	200	viridescens	335
Sciadorhodion section	220	sphaeroblastum	179	thymifolium	279	viscidifolium	200
sciaphilum	247	var. wumengense	179	thyodocum	234	Viscidula section	224
scintillans	273	spilanthum	279	tliteteum	353	viscistylum	368
sclerocladum	262	spilotum	73	tolmachevii	283	viscosum	216
scopulorum	300	spinuliferum	326	tomentosum	283	wadanum	368
scottianum	306	spodopeplum	329	ssp. subarcticum	283	Waldemarea argentea	75
scyphocalyx	117	spooneri	52	ssp. tomentosum	283	wallichii	31
searsiae	343	squarrosum	258	torquatum	117	walongense	303
seguini part	351	stamineum	7	tosaense	382	wardii	35
seguini part	349	stenaulum	6	traillianum	181	var. puralbum	36
seinghkuense	248	stenophyllum Makino	145	var. dictyotum	181	var. wardii	35
Selensia subsection	150	stenoplastum	258	var. traillianum	181	warrenii	8
selense	153	stereophyllum	350	trichanthum	344	wasonii	182
ssp. dasycladum	153	stewartianum	196	trichocalyx	341	affinity	182
ssp. jucundum	154	var. tantulum	196	Trichoclada subsection	331	var. wenchuanense	182, 390
ssp. selense	153	stictophyllum	270	trichocladum	334	watsonii	84
ssp. setiferum	154	strictum	351	trichomiscum	119	websterianum	280
semanteum	266	strigillosum	106	trichophlebium	119	weldianum	178
semibarbatum	203	subansiriense	197	trichopodum	353	westlandii	6
semilunatum	333	subcoombense	339	trichostomum	233	Weyrichii alliance	368
semnoides	48	suberosum	351	Triflora subsection	335	weyrichii	370
sericocalyx	320	sublanceolatum	380	Triflora Species Nova	viii	var. amagianum	369
semnum	80	sublateritium	380	triflorum	345	var. sanctum	369
serotinum	vi, 390	subsessile	381	var. bauhiniiflorum	345	wightii	187
serpyllifolium	380	succothii	28	var. mahogani	345	hybrid	188
serrulatum	216	sulfureum	238	var. triflorum	345	Williamsiana subsection	202
setiferum	154	supranubium	306	trilectorum	121	williamsianum	202
setosum	276	surugaense	382	triplonaevium	159	wilsonae	5
sheltonii	61	sutchuenense	60	tritifolium	159	wiltonii	183
shensiense	58	sycnanthum	349	truncatulum	151	windsorii	10
shepherdii (type only)	91	syncollum	169	tsaii	279	wuense	166
sherriffii	195	taggianum	294	tsangpoense	252	wrayi	94
shikokianum	370	taiwanianum	360	var. pruniflorum	255	xanthinum	334
shimidzuanum	367	taliense	180	tsariense	97	xanthocodon	245
shojoense	365	Taliensia subsection	155	var. magnum	97	xanthostephanum	330
shweliense	256	tamurae	371	var. trimoense	97	yakuinsulare	380
siamensis	6	tanakae Hayata	6	tsarongense	231	yakushimanum	142
sichotense	viii, 314	tanakae Ohwi	384	tschonoskii	383	ssp. makinoi	145
sidereum	83	tanastylum	93	tsusiophyllum	384	yanthinum	339
siderophylloides	353	var. pennivenium	94	Tsusiophyllum tanakae	384	yaragongense	270
siderophyllum	349	var. tanastylum	94	Tsutsusi section	370	yedoense var. poukhanense	384
sieboldii	373	tapeinum	238	Tsutsusi subgenus	364	youngae	16
var. serrulatum	372	tapelouense	350	umbelliferum	365	yungningense	280
sigillatum	169	tapetiforme	277	ungernii	149	Yunnanense alliance	348
sikangense	105	taquetii	315	Uniflora subsection	354	yunnanense	351
var. exquisitum	106	taronense	301	uniflorum	357	zaleucum	346
var. sikangense	105	tashiroi	382	uvarifolium	65	var. flaviflorum	347
silvaticum	13	tatsienense	350	var. griseum	66	var. zaleucum	347
simiarum	23	tawadae	371	var. uvarifolium	66	zeylanicum	13
simsii	381	tawangense	130	vaccinioides	361		
simulans	167	telmateium	278	valentinianum	300		
Sinensia subsection	217	telopeum	33	valentinianum var. oblongilobatum	viii, 301		
sinensis	217	temenium	124				
sinofalconeri	48	ssp. dealbatum	124	vaseyi	219		
sinogrande	84	ssp. gilvum	124	veitchianum	307		
		ssp. temenium	124	vellereum	171